The Design of Prestressed Concrete Bridges

The Design of Prestressed Concrete Bridges

Concepts and principles

Robert Benaim

First published 2008
By Taylor & Francis
2 Park Square, Milton Park, Abingdon, Oxon OX14 4RN

Simultaneously published in the USA and Canada
By Taylor & Francis
270 Madison Avenue, New York, NY 10016

Taylor & Francis is an imprint of the Taylor & Francis Group, an informa business

Typeset in Sabon by
HWA Text and Data Management, Tunbridge Wells
Printed and bound in Great Britain by
The Cromwell Press Ltd, Trowbridge, Wiltshire

British Library Cataloguing in Publication Data
A catalogue record for this book is available from the British Library

Library of Congress Cataloging-in-Publication Data
Benaim, Robert.
The design of prestressed concrete bridges : concepts and principles / Robert Benaim.
p. cm.
Includes bibliographical references and index.
1. Bridges, Concrete–Design and construction. 2. Reinforced concrete construction. I. Title.
TG340.B3975 2007
624.2–dc22 2007004615

ISBN10: 0–415–23599–5 (hbk)
ISBN10: 0–203–96205–2 (ebk)

ISBN13: 978–0–415–23599–0 (hbk)
ISBN13: 978–0–203–96205–3 (ebk)

I would like to dedicate this book to my wife Simone who has supported me in all the phases of my professional life, from the initial decision to take the risk of starting my own practice, through the tensions and crises that are an integral part of the major international projects in which we were involved, to the long drawn out preparation of this book.

Contents

Figures

Acknowledgements

I would like to express my gratitude to all those who have helped me in the production of this book.

Professional help has been offered above all by Simon Bourne and Mark Raiss, my former colleagues and now directors of the Benaim Group, who in the midst of very busy schedules, found time to read attentively many of the chapters, make thoughtful, constructive comments, and check my mathematics in detail.

In my quest for attributable images I have been dependent on consultants' and contractors' staff searching back in their records for jobs, some of which are thirty years old, and I would like to thank all those who helped me in this way. In particular I would like to thank Pauline Shirley of Arup's library for her patience and helpfulness in identifying and making available photographs of the Runnymede Bridge and the Byker Viaduct, among other projects. I would also like to acknowledge the kind assistance of the librarians of the Institution of Civil Engineers.

As it is hoped that this book will be read by the technically minded who are not trained as engineers, I am grateful to my son David for his 'lay' reading of the early chapters, and for his constructive and helpful comments.

Disclaimer

One of the benefits of no longer being responsible for running a practice is that I am free to say what I think about any subject, without having to consider the commercial consequences. It should be clear that all views expressed in this book are my own, and do not engage my former practice, the Benaim Group.

Introduction

Concrete has been in use as a primary building material since Roman times. As it is strong in compression but weak in tension, it was used in arches, vaults and walls where it is stressed principally in compression.

In the mid-nineteenth century, it was discovered that iron and later steel bars could be embedded in the concrete, effectively giving it tensile strength. This allowed it to be used in beams and slabs, where it worked in bending. Buildings, bridges, retaining walls and many other structures were made in this reinforced concrete. However, although it is one of the principal building materials in the world, it has shortcomings. Reinforced concrete beams and slabs deflect significantly under load, requiring stocky sections to provide adequate stiffness; as it deflects it cracks which spoils its appearance and leaves the reinforcing bars vulnerable to corrosion; the large number of bars required to give the necessary strength to long span beams in bridges and buildings make it difficult to cast the concrete; it is labour intensive and slow to build.

In the 1930s, Eugène Freyssinet invented prestressed concrete. High tensile steel cables were substituted for the bars. These cables were tensioned by jacks and were then locked to the concrete. Thus they compressed the concrete, ridding it of its cracks, improving both its appearance and its resistance to deterioration. The cables could be designed to counter the deflections of beams and slabs, allowing much more slender structures to be built. As the cables were some four times stronger than the bars, many fewer were necessary, reducing the congestion within the beams, making them quicker to build and less labour intensive.

Most concrete bridges, except for small or isolated structures, now use prestressing. It is also being used ever more widely in buildings where the very thin flat slabs it allows afford minimum interference to services and in some circumstances make it possible to increase the number of floors within a defined envelope.

Despite its manifest advantages and widespread use in bridges, outside a minority of expert engineers, prestressing is not well understood by the profession, and is not well taught in most universities. Engineers have to learn as best they can as they practice.

The book has are three principal aims:

- The first is to help improve the quality of the design of prestressed concrete bridges.

  Throughout my career I have been amazed by the number of grossly uneconomical and sometimes virtually unbuildable concrete bridge designs produced by consultants. I was fortunate in this lack of competence, as it allowed me to launch my practice by preparing alternative designs for contractors bidding for work. In some cases, these alternative designs halved the materials in the bridge decks, produced very substantial savings in the cost of labour, reduced the construction programme and improved the appearance of the finished article.

A bridge must be suitable for its site and it must be of appropriate scale, it must be designed to be built efficiently and without unnecessary risk of failure, it must be economical and its appearance must be given a high priority. These attributes depend on the quality of the conceptual design. Design and analysis are often confused. Design requires engineering knowledge, skill and experience combined with imagination and intuition, while analysis is a more mechanical process.

I do not know of any other books that deal principally with the design of bridges as opposed to their analysis.

- The second aim is to explain clearly the basic concepts of prestressed concrete.

  Practising engineers are being pressurised to take responsibility for structures when they do not fully understand how they work. They can do this by using software packages that may be well written, but are dangerous in the hands of those who are not familiar with the underlying concepts.
- Finally, by concentrating on the concepts and principles underlying the design of bridges, it is hoped that this book will reinforce practising engineers' intuitive understanding of the subject.

  Most textbooks on the subjects of reinforced and prestressed concrete lose the essential simplicity of the concepts in a maze of mathematics. I hope this book will be accessible not only to experienced engineers, but also to students, to architects wishing to participate more in the design of bridges and to lay people interested in how bridges work.

When running my practice, I was frequently approached by younger engineers asking for guidance on some technical matter. I did not believe that my role was to tell them what to do, or how to solve a problem. To do so would have limited the outcome to my own experience and their creativity would have been sidelined. Furthermore, too often the quick answers to such questions are reduced to explaining the mathematical procedure to be followed to carry out the analysis, or which software package to use. Instead, I attempted to explain the underlying structural principles, and left them to find out for themselves precisely how to complete the design or to carry out the analysis. This book proceeds on the same principle. Its intention is not to tell the reader what to do, or how to do it, but to explain the structural principles underlying any action that needs to be taken.

I have put forward my best understanding of the many complex issues involved in design. My views are not always conventional, nor do they always comply with accepted wisdom. Although this understanding has been used for the design of many structures over a long career, it is necessary to exercise critical judgement when using this book. Specific guidance, for instance on the spans suitable for a certain type of bridge deck or the slenderness of slabs or cantilevers, should be considered as the starting point of design, not the conclusion.

The book is intended to be independent of any code of practice. Although the British code has been used for some examples, this was only to give them a basis of reality; they could just as well have been based on some other code of practice. Also, the text is intended to be jargon free; one should not need jargon to explain principles. If some has slipped in due to its familiarity making it difficult for me to distinguish it from real English, it is unintentional.

The illustrations have been produced to scale, except where distortion was necessary for polemic reasons. It is vital for engineers of all degrees of experience to draw and sketch to scale, particularly in the design phase of a project. A distorted scale changes one's appreciation of a problem and frequently leads to erroneous conclusions that are discovered later in the design process, wasting time, effort and credibility.

As the book is based principally on my own experience, the structures used as examples are those for which I was responsible when working for Europe Etudes or Arup, or were designed by the practice that I founded in 1980 and ran for 20 years. This practice was initially called Robert Benaim and Associates, or derivations of that name appropriate to the countries in which we had offices. It started as a one man band, and gradually expanded to over a hundred staff with offices in six countries. Since my withdrawal from the practice and its purchase by the senior managers, it is currently known as 'Benaim Group'. All the jobs referred to in the text that were carried out by the practice are credited to 'Benaim'.

The book is organised as follows:

- The meaning and nature of design as opposed to analysis is discussed in Chapter 1.
- Chapter 2 is an introduction to some basic structural engineering concepts and to the specialised vocabulary used in the book. It is for the convenience of non-engineers.
- Chapter 3 is an introduction to reinforced concrete as this is necessary to understand the later chapters.
- Chapters 4, 5 and 6 explain the principles of prestressing.
- Chapter 7 is concerned with the articulation of bridges and the design of substructure.
- Chapter 8 describes the logic that underpins the design of decks for girder bridges, and gives benchmarks for the material quantities that should be achieved.
- Chapter 9 analyses the function of each of the components of a bridge deck.
- Chapters 10, 11, 12 and 13 describe the different types of bridge deck.
- Chapters 14 and 15 are devoted to the methods of construction of bridge decks.
- Chapter 16 is a synthesis of the preceding chapters, describing how the scale of a bridge project influences the choice of the type of deck and its method of construction.
- Finally, Chapters 17 and 18 deal with arches and suspended decks which follow a different logic from girder decks.

Cross-referencing to sections elsewhere in the text is by section numbers shown in italics.

1

The nature of design

1.1 Design and analysis

The origin of the word design is the Latin 'designare', to draw. In classical times, the stability of a structure depended on its shape, which could be drawn by those with the special skills. Design now has a much-widened meaning embracing the concept of anything from bridges to floats for a carnival.
In the context of bridge engineering, design means the conceptual phase, where harmony is created out of the tumult of data which includes:

- the physical characteristics of the site;
- the technical aspects concerned with the strength of materials and the theory of structures;
- the specified design life of the bridge and the maintenance regime;
- the various regulations that must be complied with;
- the economic and time constraints that have to be met;
- the form of contract under which the bridge is to be built;
- the effect the new bridge will have on the community, either by its scale, its appearance, or by the changes it will make to the local environment;
- the wishes of the bridge owner.

An inspired designer may attain a state of grace, where original ideas combine with technical expertise and past experience to create the perfect solution that best fits all the data, and which in hindsight appears obvious.

Design must be followed by the detailed justification of a project, the analysis, to demonstrate that it is safe and complies with the relevant regulations. This analysis is followed by the preparation of drawings which are needed to communicate to the contractor the information required to build the structure, and the preparation of the contractual documentation. Although requiring skill and care, these latter phases of the process are different in nature to the initial conceptual design; they are more mechanical, and do not require the combination of technical expertise, aesthetic sense and imagination that are characteristic of conceptual design.

However, in many cases, design is the name given to the mechanical analysis of the structure, and even to the whole process. This is more than a semantic quibble. Analysing structures is principally a mathematical, mechanical procedure, whereas design is largely a matter of judgement in weighing up the importance of the many relevant criteria. If the two processes are given the same name, one tends not to notice the relative weight given to each. The mediocrity of many projects that are built is the result of the shortening, or virtual absence of the conceptual, the design phase.

Design and analysis are not strictly sequential. Although clearly design must start first, design development continues in parallel with analysis as ideas evolve or as the analysis gives rise to further insights into the behaviour of the structure, and opens new design possibilities. The analysis is an important part of understanding the structure. It needs to start with simple models which are easy to check, and only gradually build up to the final verification of the structure as a whole. As the analysis is carried out on a model of the structure, not the structure itself, the designer must always question whether the results represent reality or whether they have been distorted by the assumptions made in preparing the model.

The designer cannot delegate the analysis, he must remain in charge and needs the required knowledge and skills, in addition to his abilities to imagine, innovate and communicate.

1.2 A personal view of the design process

The nature of design is uniquely personal. For this reason the following description of how the author understands design may not be recognised by all readers.

Design is an interactive process, with signals shuttling between three notional centres in the brain, those responsible for appreciating beauty, for accumulating experience, and the brain's calculator. It most definitely is not just a sketching exercise, nor is it a logical, linear process.

When a client proposes a commission for the design of a new bridge, across say a river, with an outline description of the purpose of the bridge and the characteristics of the river, the first engineering response is to imagine a solution that appears to fit the facts as they have been described (often incorrectly), and is usually based on some idea that has been tried before, has been imagined or read about, or is the extrapolation of a previous idea, pushing it further towards some logical conclusion. As more information on the project becomes available, the suitability of this first idea is tested and then reinforced, modified or dropped in favour of another 'guide' idea. This process of imagining a solution and then subsequently confronting it with the facts is in the author's view the essence of creative design.

At the earliest stages of this design process, calculations are carried out. They are in general very simple, to compare the cost of alternatives, and to check on the sizes of members. For most bridges and other civil engineering projects, the structure can be notionally simplified to the point where the bending moments and shearing forces may be estimated by simple manual means, generally to within 15–20 per cent of the correct value.

Similarly, the loading on the bridge can be reduced from the pages of the code of practice definition to its simplest basics. From these simple beginnings, the size of members, the density of reinforcement, the intensity of prestress and the basic deflections of the structure can be calculated. Of course one makes use of books with charts of bending moments and deflections for beams and portals and safe load tables for columns and of codes of practice, not to check for detailed compliance, but to remind oneself of limiting stresses, load combinations, load factors etc. If one is very computer literate, simple computer models may be invaluable, as long as one can produce them almost automatically, without struggling to understand manuals, sign conventions etc. One must at all costs not engage one's brain into an 'analysis mode', or one's creativity will be swamped by one's intellect.

It is very important that, at this early stage, all calculations are kept conservative, so that one is not deluding oneself about the feasibility of a favoured option. These initial calculations allow the designer to develop his understanding of how the structure works, of how the forces flow. They also enable him to put sizes to members, and so to gain a first insight into the appearance of a structure. The aesthetics of structures are critically dependent on member sizes, and how size varies along a member.

The diagrams of normal forces, bending moments, shear forces and torques may be drawn along the members, and stresses calculated. As the structure is better understood, the logic of how it works becomes apparent. Member sizes may be refined to improve economy, to provide reserves of strength and to affect the appearance. However, one must not defy the basic logic of the structure; one cannot make a member excessively slender in defiance of the structural logic, just because it looks better; there must be a concordance between function and appearance. This does not mean that the size of members is dictated by their stress levels, but that one must not act contrary to the structural logic. This usually leaves a considerable margin for discretion in sizing members. For instance, two members that are equally stressed may be given different sizes for the sake of appearance.

At this stage, it may become clear that the structure is evolving in a way that is not satisfactory, either technically or aesthetically. When one embarks on this design process, it is frequently not clear what the nature of the destination will be. An essential part of design is the readiness to tear up what one has done and start again. An engineer who does not have the courage, or the time, to recognise that he is engaged in a dead end and to start again cannot pretend to be a creative designer.

1.3 Teamwork in design

Design is inevitably a team exercise. At its simplest, the bridge designer will have another engineer and one or two draftsmen working directly for him, while on large bridge projects the core team may include ten or more people. Generally, other specialist disciplines will also be involved for part of the design period, such as geotechnical engineers, quantity surveyors and the suppliers of proprietary products such as the bearings, expansion joints etc. An architect may also be involved, either as a partner in the concept or as a specialist involved in the design of finishes, handrails, lighting and other decorative aspects. Depending on the form of contract, some decisions are likely to require input from the client or the contractor.

If the design is to be anything other than banal, the team needs a leader who makes the project his own. There is no aspect of a bridge design that is not capable of more than one solution, whether it is the overall concept or the type of bridge bearing. There is thus great potential for diverging views and for indecision. This multitude of design decisions must be welded into a coherent project, and this can only be done successfully through one mind.

This need for a 'chief designer' is sometimes challenged by professionals, who claim that design is the result of collective decision making, with no one dominating the process. However, it is usually only necessary to consider what the effect on the design would be if each member of the team were to be substituted in turn. For most of the members, the result would be only minor changes to the finished design, while generally there is one team member whose substitution would change the project fundamentally.

In order to carry out his synthesising role, the designer must know enough about all the various specialities involved, so that he can understand the implications on the design as a whole of making one choice or another. Clearly, the designer is not expected to be skilled in all these various disciplines, but he must be able to question the specialists, understand the reasons for their choices, challenge their decisions, take second opinions and ultimately accept responsibility for them. In particular, the bridge designer should have a reasonable knowledge of soil mechanics, as decisions on foundations often determine the type of structure to be built.

This concept of the chief designer and the skills he requires is not new. In Chapter 1 of his first book, Vitruvius discusses the education of architects (which were not differentiated from engineers) in republican Rome, and puts forward the view that very few people can be expert in all the disciplines involved in construction, but that architects must deal with them all, with only imperfect knowledge. He goes on to ask the reader's forgiveness for his imperfect grammar, as he is an architect, not a gifted writer. Perhaps as a civil engineer, I may ask for the same indulgence!

1.4 The specialisation of designers

It is quite clear that society requires a large number of civil engineers to design, build, administer and maintain its roads, railways, water supply, sewerage system, power stations, ports and telecommunications infrastructure among other tasks. The great majority of those tasks do not require a deep and intuitive understanding of the behaviour of structures or the exercise of aesthetic judgement. It is important that these engineers be well trained, as they are in positions where they can make a significant contribution to society, and the more able among them are likely to attain positions of influence in the private sector or in government. Other engineers will become specialists in a wide variety of technical disciplines, such as geotechnical engineering, dynamics, information technology, wind engineering etc.

A minority of those who opt to train as civil engineers will become the designers of structures which, in addition to their utilitarian function become part of the built environment. This minority requires different training from the majority. They need to develop an intuitive understanding of the behaviour of structures, a thorough understanding of the nature of the various building materials, and an appreciation of the appearance of their structures.

This distinction is recognised to some extent by the profession in the United Kingdom. The Institution of Structural Engineers, with a membership of chartered engineers (MIStructE and FIStructE) in the UK of approximately 9,000, caters for the minority of designers, and the Institution of Civil Engineers, with a chartered UK membership (MICE and FICE) of approximately 36,000, represents the majority of more general civil engineers. In order to become a member of the Institution of Structural Engineers, suitably qualified graduates with about three years experience in industry have to pass an examination that tests their knowledge as designers of structures, while similarly experienced graduates applying for membership of the Institution of Civil Engineers are subjected to a written assignment that tests their more general suitability to take professional responsibility. However, the distinction is blurred, with many engineers being members of both institutions, and some designers being members of the Institution of Civil Engineers only.

The specialisation of designers is also recognised in many other countries, where engineering graduates wishing to take responsibility for the design of structures have to pass an additional examination some years after graduation, giving them a title such as Professional Engineer.

The distinction between general civil engineers and designers remains inadequately recognised by the profession, by the universities and by society. Being a designer is almost a separate profession from that of general civil engineer. If one were to imagine the spectrum of skills required in the building industry, extending at one end from an architect and at the other to a director of a civil engineering contracting firm, the bridge designer would cover a wide range, but his centre of gravity should be closer to the architect than to the contractor.

1.5 Qualities required by a bridge designer

Before the age of enlightenment engineers/architects built many splendid structures, soaring cathedrals, slender stone towers and daring arch bridges without knowledge of modern theory of structures or of analytical soil mechanics. Then in the eighteenth and nineteenth centuries, despite the primitive state of mathematics and structural theory, engineers built huge numbers of structures associated with the development of the canals, roads and railways, some of which were daring and dramatic, many of which have survived to the present day.

It is astonishing how little curiosity is shown by engineers and teachers of engineering about this huge body of successful structures which were built without the benefits of most of what is considered essential engineering training. It should make us question what are the basic skills required for a bridge designer.

The designer of engineering structures requires an understanding of how structures work and how they are to be built, an appreciation of how they look, and the communication skills required to describe his ideas to others. This understanding develops gradually, starting with his technical education and continuing with the feedback from completed projects, snippets of information read or overheard, back-of-the-envelope doodles or bath-time mental calculations. Sometimes, some item of information acts as the missing piece of a puzzle, suddenly illuminating an issue that was previously only partly understood. This process goes on throughout a career, and a creative engineer becomes progressively more creative until his faculties begin to decline. Clearly, some people are more gifted than others in this domain, and have an intuitive understanding of structures. Such natural engineers learn more quickly than others less talented, and make better use of their experience.

A designer's appreciation of beauty depends in part on his talent and in part on his training and experience. In the UK, prospective engineers concentrate on mathematics and science from the age of 16, and the appreciation and creation of beauty are absent from the majority of engineering courses. Mathematics and the other technical disciplines such as theory of structures and the properties of materials are the most tangible of the skills required by engineers, and thus are the ones that are given priority in their education. However they have become virtually the only skills that are taught, whereas the critical criterion that determines whether a structure will rise above the mediocre is the quality of the conceptual design. This requires in addition to technical knowledge and skill, imagination, aesthetic judgement and an appreciation of the context of the structure. These skills are much more difficult to teach.

Thus engineering designers have to rely on any innate talent for the vital aesthetic component of their practice, or alternatively seek input from architectural specialists. Enlisting the help of architects in the design of bridges is far better than simply ignoring the aesthetic component of design, but is much inferior to both aesthetic and technical components being in the mind of one person. An engineering designer should have an education that is reasonably balanced between the technical and the aesthetic.

1.6 Economy and beauty in design

An engineer designing a bridge has twin obligations, to his client to use his money wisely, and to society to produce a structure that will enhance the built environment. In fact, beauty in engineering design has its roots in the tension that exists between designing for economy and designing for appearance.

Economy in this context is not simply saving money; it is a concept of rationality and frugality. It is fundamental to engineering design that the designer is constantly planning how he can save materials, and how he can make the construction process simpler, even if many of these design decisions in isolation would not register on the overall balance sheet of a project.

An example of this tension between appearance and economy is given by the design of an access ramp to a high level bridge, Figure 1.1. The main bridge consists of a trapezoidal box section, 2.4 m deep, allowing it to span 60 m or more. The access ramp must climb from ground level to merge with the main structure. At the point of merger, the ramp has the same depth and shape as the main bridge. However, the 2.4 m depth would be out of scale for a deck close to ground level. Consequently, the ramp is given a depth that gradually reduces to 0.7 m as it approaches the ground, with the spans shortening correspondingly. This is clearly not the most economical choice, as the formwork for the downstand webs of the ramp will be continually changing. In order to mitigate this additional cost of formwork, the geometry of the ramp deck may be defined by keeping the length of the web shutters constant and equal to those of the main bridge, but changing their angle. Thus if the ramp is built span-by-span, the side shutters of the webs may be re-used for each span. This is an intellectual concept based on an attempt to rationalise the construction method and save cost, which gives rise to a distinctive appearance. Finally this appearance must be judged on its own merits.

When an engineer designs, whether it is the overall concept of a bridge or an individual member, he first must understand the structural behaviour, and then seek rationality and economy. The search will usually leave him many options, which allows him to make choices concerning the appearance of the structure.

A very simple example is the design of the bridge pier carrying a single bearing, Figure 1.2. The pier is subjected to a vertical load and to a horizontal load at the top which produces a bending moment that increases linearly to a maximum at the base of the pier, Figure 1.2 (a). The size of the pier at the top will be limited by the size of the bridge bearing, while at the bottom it will be governed by the combined effect of the compression force and the bending moment. The engineer has a choice between, for instance:

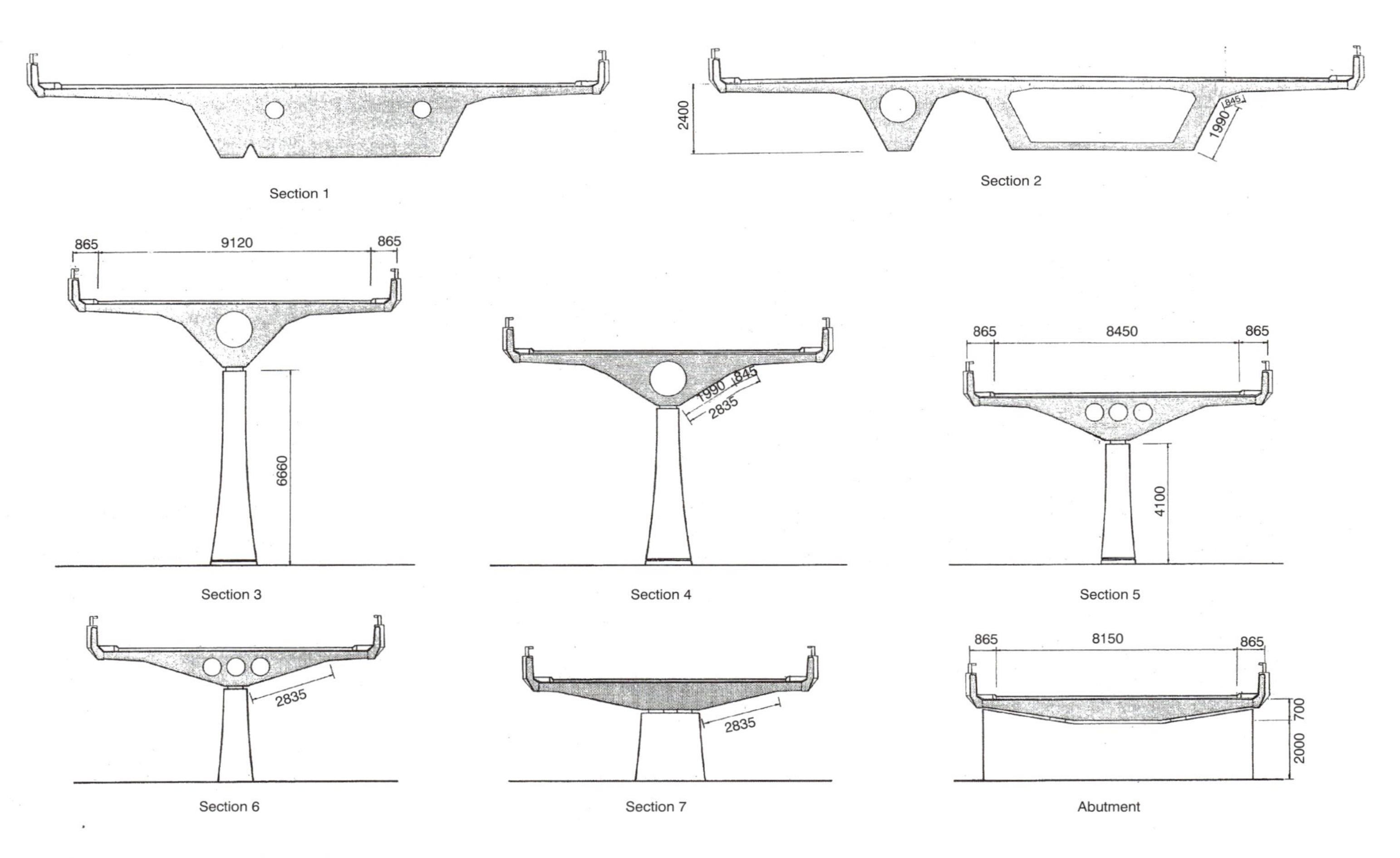

Figure 1.1 Cross sections of slip road merging with main carriageway

- a prismatic column of a generous size that allows minimum reinforcement to be used throughout, Figure 1.2 (b);
- a smaller prismatic column that needs minimum reinforcement at the top, but heavy reinforcement at the base, Figure 1.2 (c);
- a column that is as small as possible at the top and tapers uniformly to the bottom, Figure 1.2 (d);
- a column which is as small as possible at the top and whose width then varies such that the minimum reinforcement may be used throughout, Figure 1.2 (e);
- some combination of any of these.

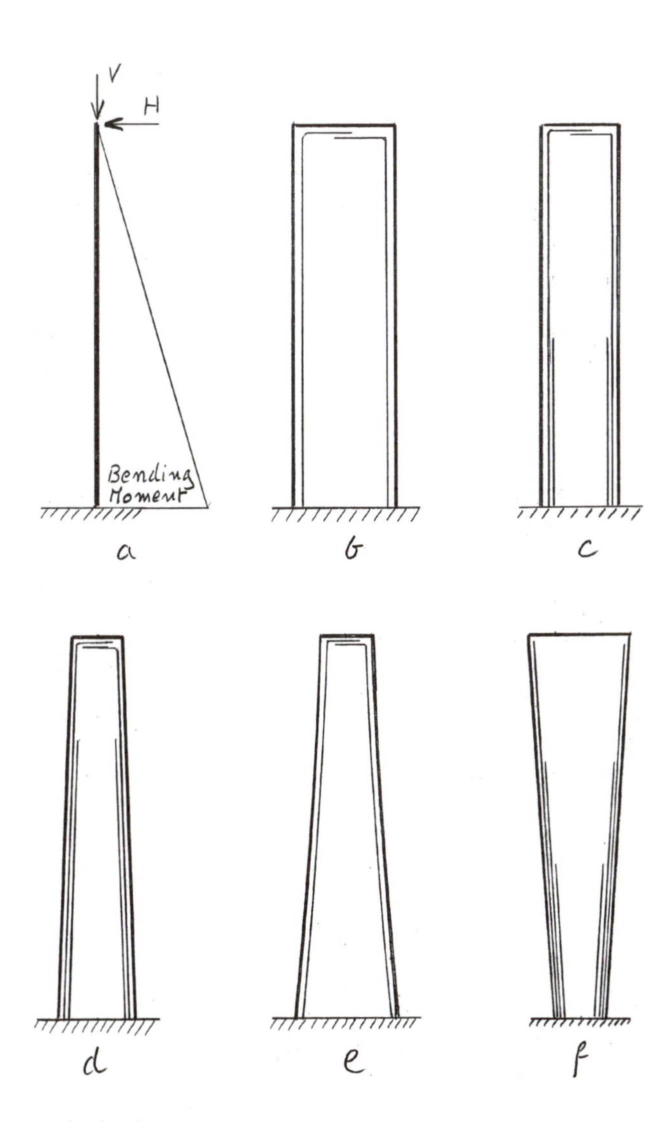

Figure 1.2 Options for bridge pier

His choice will be informed by other aspects of the project, for instance:

- the number of similar columns in the project;
- the range of heights of such columns;
- the need for variations on the basic column size to cater, for instance, for bridge expansion joints, anchor piers or different length spans;
- the need for a family of columns to cater for other bridges forming part of the same project;
- the architectural context of the bridge.

As the engineer considers the economy of the various choices to be made, he will most probably find that several options have costs that are within the margin of estimating error. Consequently, although the search for economy is at the heart of his design, it cannot be used as an alternative to aesthetic judgement; the engineer must choose the shape he considers is best in all the circumstances.

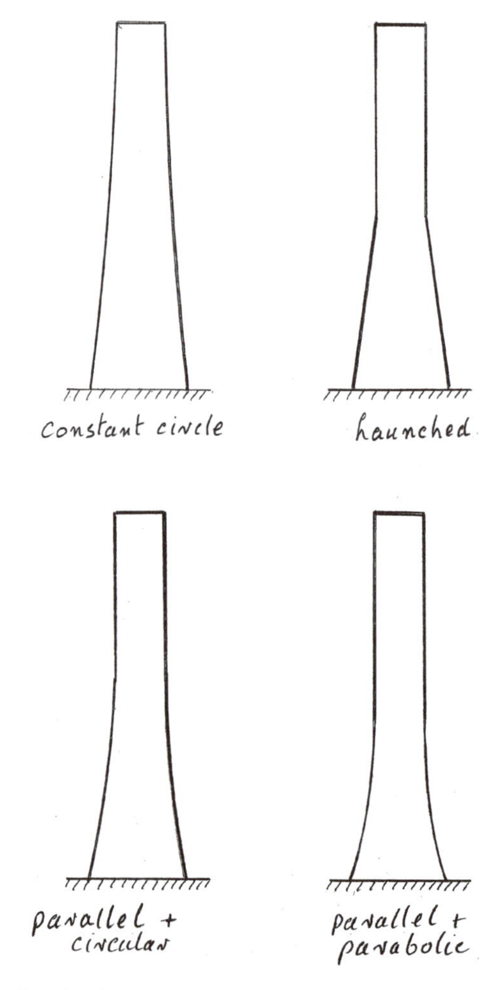

Figure 1.3 Options for flared column

Figure 1.4 STAR Viaduct: typical pier (Photo: Benaim)

Figure 1.5 Byker Viaduct: pier finishes (Photo: Harry Sowden/Arup)

Once he has made his basic choices, he then has to refine his design, both for economy and appearance; small changes of shape can greatly affect the appearance, as may be seen in comparing the options for a column of varying width shown in Figure 1.3. Reinforced concrete detailing considerations may also suggest minor dimensional changes, to give a rational arrangement of bars, or to make best use of standard bar lengths and minimise waste.

What an engineering designer cannot do and retain the integrity of his design is to fly in the face of rationality and economy, and design a heavily loaded column that, for instance, tapers towards the bottom, Figure 1.2 (f), creating an artificial problem that then needs to be solved by misdirected engineering ingenuity. This is true even if the additional cost as compared with a rational design is negligible.

There is no reason that the column should not be decorated, with corners cut off, the sides faceted, Figure 1.4, or with ribs or other decorative finish, Figure 1.5 (*7.15.4*), as long as the cost of this decoration is reasonable in the context of the project. Some aspects of such decoration may be functional, for instance to reduce the apparent bulk of the column by changing the way light reflects off it or to control water runs to improve its weathering, while some may be just to make it more attractive.

Engineering design is thus driven by the simultaneous consideration of rationality, economy and appearance. Designing economically alone is not enough. There is no automatic linkage between economy and beauty; aesthetic judgement is required at every step of a design.

Engineers have been known to put their faith in the idea that if they design honestly, and reflect in their structure the flow of forces, the result will inevitably be aesthetically satisfactory, or even beautiful: the idea that 'form follows function'. Unfortunately, this is not sufficient. Within the confines of honesty and economy, the engineer is left with a wide choice, which requires aesthetic judgement. A useful analogy is to consider the design of the human face, which is well defined by its function, but which gives rise to an infinite number of outcomes.

If bridge designers are not confident of their aesthetic ability, they should request the assistance of an architect, who should be involved from the earliest stages of the design. If they are lucky, they will find one who understands the special quality of engineering design, and who does not take over the project with his own, non-engineering taste. Such collaboration can be very creative, but success depends firstly on the engineer being skilled and confident in the technical domain, and secondly in the architect having a genuine interest and feeling for engineering structures. Even engineers who have confidence in their aesthetic judgement can find collaboration with a talented architect very creative, with the architect questioning the engineer's choices, and proposing different ways of seeing the design.

1.7 Expressive design

An important part of engineering design is that it should be expressive of the forces in play. In the example of the bridge pier given above, all other things being equal, it is better for the greater moment at the base to be resisted by increasing the size of the pier, and thus acknowledging the greater strength required, rather than by keeping the size constant and increasing the reinforcement that is hidden within the concrete envelope.

Some types of structure are more expressive than others. A beam of constant depth is singularly inexpressive of the fact that the greatest bending moment is at mid-span. Various designers, including Morandi, have adopted 'fish belly' beams, which do express the fact that they need greater strength at mid-span, and allow the removal of redundant web material near the ends, Figure 1.6, despite the fact that they most probably cost slightly more than a beam of constant depth; the additional complication of the formwork and the reinforcement is likely to outweigh any savings in the volume of concrete. The designer was justified in attempting to make the structure more expressive, and hence more interesting, although the final judgement must be whether it works aesthetically, as well as expressing engineering values.

An engineer may be justified in choosing a type of bridge that is not the most economical, but that is more expressive. Clearly, this needs the approbation of the client and cannot be done as part of a competitive tender where no credit is given for appearance. For instance, it may be established that the most economical bridge for a site is a girder bridge on vertical supports. However, the site may be just right for an arch, Figure 1.7, with the strong foundations required, although the arch would be more expensive. Such an arch bridge design can only achieve distinction if it is a rational choice and if the designer then strives for and achieves economy in the design as described above. It could not be considered a good design if the designer had imposed an arch on the site, despite its unsuitability; for instance, if the arch required massive and expensive foundations to resist the thrust, even if these foundations could not be seen by the public.

This demonstrates that good engineering design has an esoteric component only to be appreciated by professionals. A good example is the series of valley piers of the Byker Viaduct in Newcastle, which were flared to resist the wind and centrifugal forces of the twin-track railway it carried. The central part of each pier was cut away at ground level to save materials and to reduce the impact on walkers in the park. The size of these cut-outs was made just large enough to allow the precast units of the bridge deck to fit through, in an economical and innovative construction method, Figure 1.8. This fact is only known to those who remember the bridge being built, but forms part of the intellectual justification for the shape and size of the cut-out. The final judgement on a design must rest with the combined response of public and profession.

The choice of a solution that is expressive of the flow of forces applies to the design of members and details of a bridge as well as to the structure as a whole.

The detailed shape of the towers for the two-level Ah Kai Sha cable-stayed bridge designed by Benaim to cross the Pearl River close to Guangzhou, further illustrates the relationship between technical and aesthetic decisions, Figure 1.9. The bridge has

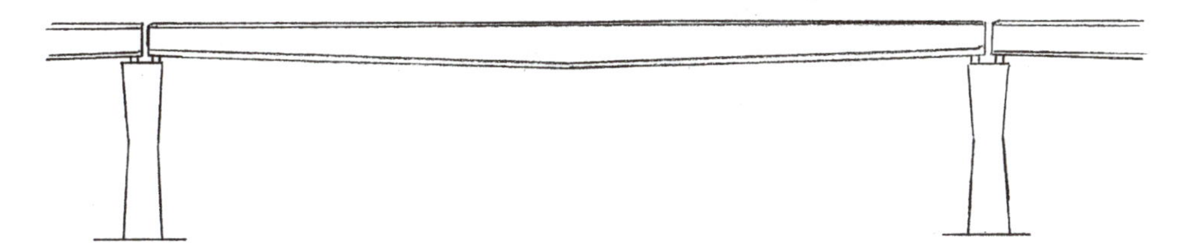

Figure 1.6 Fish belly beams: simply supported beams of Maracaibo Bridge

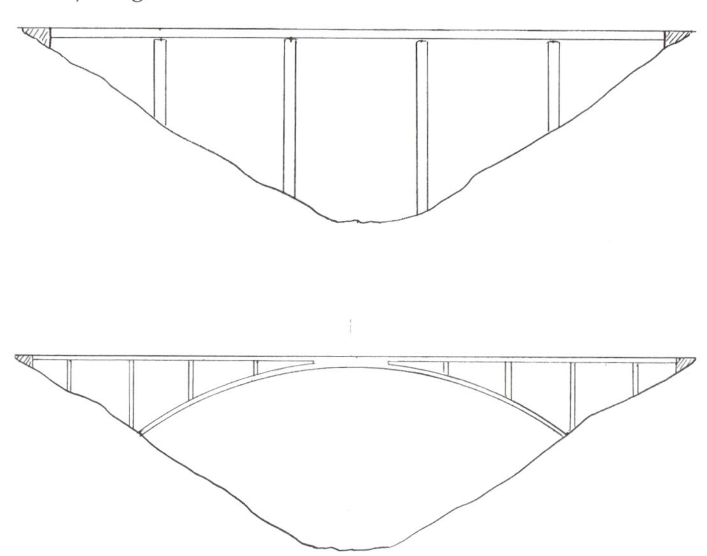

Figure 1.7 Beam or arch?

Figure 1.8 Byker Viaduct under construction (Photo: Arup)

Figure 1.9 Ah Kai Sha Bridge towers (Image: David Benaim)

a main span of 360 m, and is exceptionally wide, at 42 m. The bridge is described in more detail in Chapter 18.

The towers rise 100 m above the ground, some 70 m above the deck. Their functions are to provide the height necessary to attach the cable stays that support the deck, to carry the loads down to the foundations and to give stability to the deck under the effects of typhoon winds and earthquake.

The tapering shape of the towers has been determined by the progressive increase in weight applied to them by the stay cables and due to their own considerable self weight, by the longitudinal forces imposed by earthquakes, wind and the expansion and contraction of the deck, and by the lateral forces due to earthquakes and wind. As the towers cantilever from the foundations, all these forces increase their effect towards their base, particularly below the level of the deck. Consequently, the rate of taper increases below deck level. This tapered shape of the towers expresses the forces acting on them as well as providing the necessary strength.

Most cable-stayed bridge towers have at least two cross-beams, one of which supports the deck, which make the towers into portals. These cross-beams are always heavily reinforced and slow down the construction of the towers, particularly when it is intended to build the towers by slip-forming, as in this case. Furthermore, the great width of the deck would have required the beams to be very substantial. As a consequence it was decided to omit the cross-beams, and to provide the necessary stability by designing the towers as vertical cantilevers, built into the pile caps.

Support for the deck, in the absence of cross-beams, is provided by powerful concrete brackets which are attached by prestress after slip-forming the columns.

All bridges expand and contract with changing temperature and either the deck has to be separated from the piers by movable bearings, or the piers have to be made flexible enough to permit this movement. Here, the deck is fixed to the piers, which have been made flexible by splitting them into two leaves, Figure 18.18.

Each leaf of the tower has a dumbbell shape, which represents the most efficient use of material. The dumbbells become solid at the junction with the deck to resist the concentration of forces that occur in that zone, Figure 18.19.

The towers are situated outside the deck and, as a result, the stay cables all pull slightly inwards. The combined effect of these pulls is very significant, and the towers would require larger, more expensive columns if they were not propped apart at the top. The prop is designed to be made on the deck and winched up into place, and so has to be as light as possible. For this reason it is made in an 'I' section. It must be stiff enough to resist buckling under compression and bending under its own weight as it spans between the columns, but the connections of the prop with the towers must be sufficiently small so that they do not attract bending moments under the effect of lateral loads. The prop is designed with the depth increasing towards mid-span, its shape expressing these constraints and actions.

Every significant dimension of these towers had a technical, rational justification, and none were chosen for appearance alone. However, the appearance of the towers was present in the mind of the designer at all times, and was chosen to express the forces and actions acting on them.

The splendid Alex Fraser Bridge in Canada, designed by Buckland and Taylor, Figure 1.10, solved some of the same problems differently. The bridge is narrower

Figure 1.10 Alex Fraser Bridge towers (Photo: Henning J. Woolf/Buckland & Taylor)

than Ah Kai Sha and only has one level of traffic. The transverse stability of the towers is provided by the portal action of two cross-beams, and consequently, the legs of the towers do not need to increase in size substantially below the deck. The deck is carried by bearings on the lower cross-beam. Although the columns are outside the deck as at Ah Kai Sha, the designers have cranked them inwards above the deck, so that the stay cables pull concentrically, and a top strut is not necessary. The horizontal forces created by the crank are carried by the two cross-beams.

This comparison is not intended to imply that one solution is better than the other but to emphasise how the engineering and aesthetic decisions interact to produce two quite different solutions to a problem that has similar components.

1.8 Bridges as sculpture

Bridges are in general the purest expression of the art of the structural engineer. The core disciplines in their design are the theory of structures, the behaviour and strength of materials and the search for economy in materials and in methods of construction. In order to be able to resolve the tension between appearance and economy that is the source of creativity in the design of bridges, it is essential that the designer should have a deep and intuitive understanding of these disciplines, which leads one to the conclusion that he should be an engineer.

However, bridge owners may well have reasons for building a bridge that are other than purely functional. For instance, it is not uncommon that a bridge is required to be a 'landmark', or a symbol of the regeneration of an area. In these circumstances, it ceases to be principally an engineering artefact, and becomes a cultural artefact, a form of functional sculpture. The structural disciplines cease to be dominant, and in particular, the search for economy and rationality that is at the heart of engineering design, will be downgraded as a constraint. It may well be better for such a project to be led by an architect, who can best make the synthesis that is necessary to meet the client's needs.

For instance, the Hungerford Footbridge across the Thames in London, Figure 1.11, is a cable-stayed structure where the fan of stays only supports about two-thirds of the length of each span. This quite defeats the logic of providing expensive pylons and stays, as it leaves the deck with substantial bending moments, and also gives rise to very uneven loads in the stays, some of which must be virtually unloaded and hence redundant. Although this is a parody of a bridge, using bridge vocabulary out of context, it may be very successful with the public.

A good example of the difference between the architectural and the engineering design of a bridge is given by the Runnymede Bridge that carries the M25 and the A30 across the River Thames near Staines.

The original Runnymede Bridge was designed by Sir Edwin Lutyens, and built posthumously. It appears to be a stone and concrete arch with brick spandrels, Figure 1.12 (a). In fact the spandrels conceal a steel portal structure, and the brick and Portland stone towpath arches conceal massive reinforced concrete abutments, transferring the thrust of the portal to the clay foundation, Figure 1.12 (b) [1]. Consequently, the bridge is in no way expressive of the structural actions, and is not at all economical or rational, principally because the clay foundations are not well suited to carry the thrust of the 55 m span portal subjected to highway loading. However, it is well loved and admired, and is one of only two major bridges designed by Lutyens.

Figure 1.11 Hungerford Footbridge (Photo: Robert Benaim)

When the M25 came to be built in the 1970s, the author, then working for Arup, was responsible for the design of a parallel bridge, Figure 1.12 (c). This bridge has precisely the same function and span as Lutyens' original, although it works quite differently. For instance, the arch thrust is balanced by the thrust of the rear, raking strut, so only vertical loads are applied to the foundation, Figure 1.12 (d). Although it was more expensive than a simple girder bridge, this was justified by the precedent set by the adjacent structure. The design has been guided by the principles of reconciling economy, rationality and appearance, as described above, and expresses its structural action. This is no guarantee that it will be as well loved as the original.

The other major bridge designed by Lutyens is the Hampton Court Bridge, also across the Thames, Figure 1.13. This is a series of Portland stone arches with brick spandrels, or is it? Does it matter?

Figure 1.12a Runnymede Bridge: original Lutyens design (Photo: Arup)

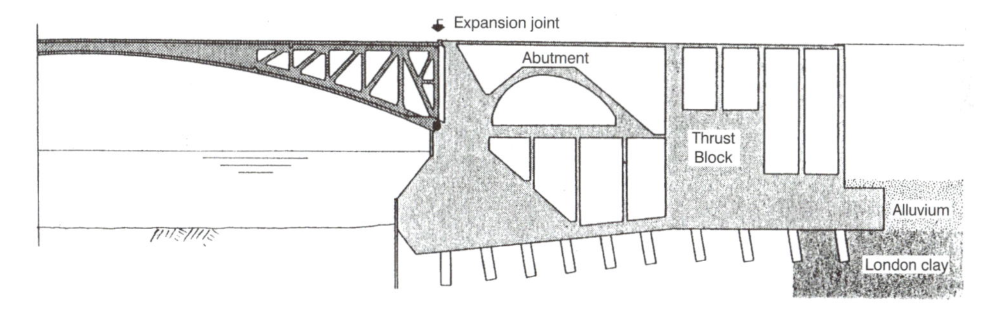

Figure 1.12b Long section of Lutyens bridge (Source: Thomas Telford Ltd, adapted from D.W. Cracknell, *Proceedings of the Institution of Civil Engineers*, Vol 25, July 1963, pp. 325–44)

Figure 1.12c New Runnymede Bridge (Photo: Harry Sowden/Arup)

Figure 1.12d New Runnymede Bridge: load testing the bridge model (Photo: Robert Benaim)

Figure 1.13 Hampton Court Bridge by Lutyens (Photo: Robert Benaim)

1.9 Engineering as an art form

The effective start of modern building may be defined as the construction of the Pantheon in Rome, in 120 AD, Figure 1.14. This building exhibits engineering design in the modern sense. The roof is a 44 m diameter spherical concrete dome, resting on concrete walls that are some 30 m high. In order to reduce the thrust of the roof, it is made of lightweight concrete, of which the density reduces towards the crown. Also, the underside of the roof is relieved by caissons, which fulfil the three roles of saving materials, reducing weight and hence the thrust on the walls, while maintaining stiffness, and decorating the interior.

The walls are 6 m thick, in plain dense concrete. Near their base they are relieved by chapels on their inside face, made within their thickness. Again, these chapels have multiple functions, economising on materials, displacing the centroid of the wall outwards to improve its resistance to the thrust of the roof, providing a useful facility and decorating the temple. The upper lifts of the wall are more massive, the weight helping to resist the roof thrust.

Thus the appearance of this building is critically integrated with the engineering concept, and in the author's opinion constitutes an art form which David Billington [2] has called 'structural art'. This form of art depends on an engineering understanding of the forces acting on the structure and the expression of these forces.

The design of the great mediaeval cathedrals shows a similar fusing of form and function, where the roles of architect and engineer had not yet been separated, Figure 1.15. Viollet-le-Duc, commenting in his *Dictionnaire de l'architecture française* on the design of these cathedrals, coined the phrase 'architecture raisonée', which might be translated as 'analytical architecture'. This concept is appropriate to the design of bridges, as described in the previous sections of this chapter.

As in any form of art, there is a continuum between the multitude of artisans, and the small minority who reach the summits of skill and inspiration, who are known as artists. There is no clear dividing line between the superb artisan and the artist, but great artists are acknowledged as such by most people.

The young engineer who is designing the reinforcement for 100 pile caps for a long viaduct may not see himself as a future artist. However, depending on how he goes about this design, he may well be preparing himself for much more creative engineering later in his career. He can do a boring, routine job, or alternatively there are many ways in which he can carry out these mundane tasks creatively. For instance, he can think about the mechanism whereby the vertical loads and bending moments are transferred from the pier stem to the piles, he can compare strut-and-tie solutions to classical bending theory, reconcile his understanding of good engineering with the stipulations of the code of practice, question the purpose of each bar, find reinforcement arrangements that minimise off-cuts and waste and he can create a modular reinforcing cage that expands easily with the span and depth of the pile cap.

There must of course be a chief designer who oversees his tasks, and who sets the tone of the type of design he expects. If this chief designer is interested in producing a design of quality, he will be prompting his assistant to think creatively about his task, and giving him ideas to refine and develop. Although the majority of such chief designers will never aspire to being recognised as artists, the best will.

It is common that artists are not recognised as such in their own lifetimes; it is often in hindsight that their qualities are put in perspective and given the accolades

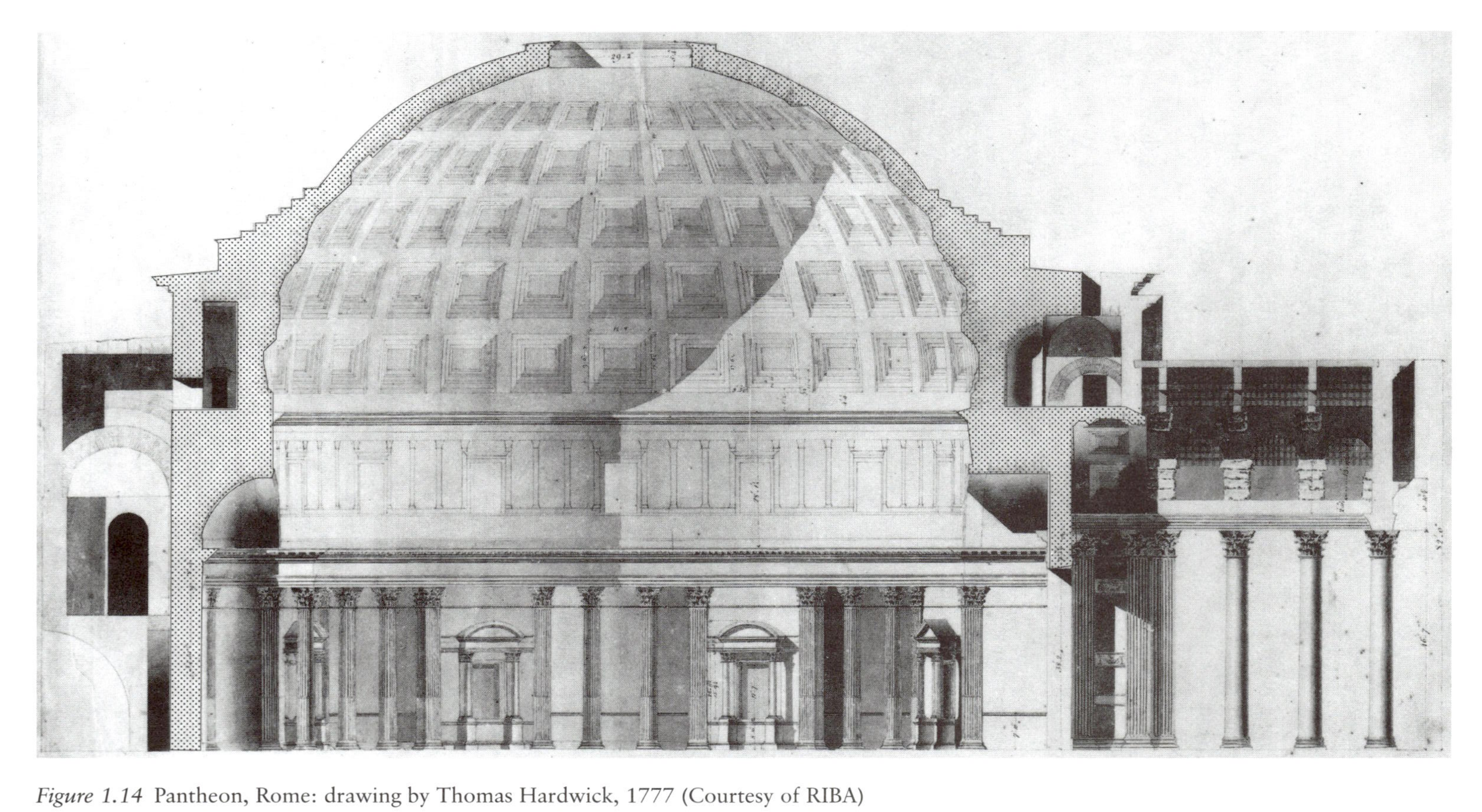

Figure 1.14 Pantheon, Rome: drawing by Thomas Hardwick, 1777 (Courtesy of RIBA)

Figure 1.15 Beauvais Cathedral (Courtesy of the ICE)

Figure 1.16 Saltash Bridge by Brunel (Courtesy of the ICE)

Figure 1.17 Palazzetto dello Sport by Nervi (Source: Enrico Hoepli, Tav XCVI of Costruire Correttamente by Pier Luigi Nervi)

Figure 1.18 Salginatobel Bridge by Maillart (Courtesy: Madame Marie-Claire Blumer-Maillart)

they deserve. Who would now deny the exceptional quality of some of the structures designed by Brunel, Figure 1.16, Nervi, Figure 1.17, or Maillart, Figure 1.18. These are works of art, but they are the work of engineers; they could not have been carried out by architects without engineering training.

If the artistic content of engineering were to be more generally understood, the language of the appreciation of engineering would change; it would become obvious that the training of engineering designers must be more than just technical; that this training should include components that develop creativity and imagination. A wider range of skills would be attracted into engineering, less mathematical, more visionary.

2
Basic concepts

2.1 Introduction

It is hoped that this book will be read both by engineers who wish to deepen their understanding of prestressed concrete and bridge design, and by those without engineering training but who are interested in bridges. This chapter is for the benefit of the latter group, and is intended to offer a very brief introduction to some of the basic principles that govern the design of concrete structures, and to introduce much of the specialist vocabulary used in the book.

2.2 Units

Length is measured in millimetres and metres.

Load and force are measured in Newtons (N), kilonewtons (1 kN = 10^3 N) and Meganewtons (1 MN = 10^6 N). 1 kN is the weight of a heavy person, while 10 kN is approximately 1 ton.

Stress is measured in Megapascals, MPa (1 MPa = 1 N/mm^2 = 1 MN/m^2).

Moment is measured in kilonewton metres, kNm, or meganewton metres, MNm.

2.3 Loads on bridge decks

The loads on a bridge deck are made up of:

a) Self weight; the weight of the bare concrete structure.
b) Superimposed dead loads; the weight of permanent loads applied to the bare concrete structure, such as parapets, footpaths, road surfacing etc. These loads do not contribute to the strength of the deck.
c) Live loads; transient vehicular, rail or pedestrian loads applied to the deck. Live loads may be uniformly distributed along the deck (referred to in the text as udl), corresponding to a busy traffic lane or to a long train, or concentrated, corresponding to a single heavy axle, lorry or locomotive.
d) Environmental loads; principally wind and earthquake.

Loads (a) and (b) together are referred to in the text as ‘dead loads’.

2.4 Bending moments, shear force and torque

Consider the cantilever shown in Figure 2.1 (a), of length L metres and carrying a load of W kN at its extremity. The cantilever consists of thin top and bottom flanges, and of a web joining them together.

The load W creates a bending moment in the cantilever, which at a distance l from the end is $W \times l$ and which reaches a maximum of $W \times L$ at the root of the cantilever, Figure 2.1 (b). The bottom flange of the cantilever is stressed in compression, and the top flange is stressed in tension, Figure 2.1 (c). The top and bottom surfaces of the cantilever are called the top and bottom extreme fibres, or the extrados and the intrados, respectively. The bending moment is designated as hogging, as it causes tension on the top fibre. Bending moments that cause tension on the bottom fibre, for instance due to an upwards load on the end of the cantilever, are designated as sagging. The distance between the centres of the compression and tension flanges is h. The compression and tension forces in the flanges, $\pm F$, create an internal couple which must balance the external applied moment (ignoring very small longitudinal forces in the thin web). Consequently at the root of the cantilever, $Fh = WL$. As stress = force/area, the stress in each flange $\sigma = F/A$, where A is the cross-section area of each flange.

The cantilever tip deflects downwards as the top flange extends and the bottom flange compresses. The amount of deflection depends on the height and thickness of the web, on the width and thickness of the flanges and on the stiffness of the material of which the cantilever is made.

The deflection $\delta = WL^3/3EI$

where

W is the load on the cantilever end;

L is the length of the cantilever;

I is the 'moment of inertia' of the cross section of the cantilever, and is a measure of the strength given by its geometry;

E is the Young's modulus (also called the 'modulus of elasticity' or just the 'modulus') of the constituent material of the cantilever, and is a measure of the stiffness of the material. For instance, concrete has a modulus of, typically, 30,000 MPa, while steel, which is much stiffer, has a modulus of 200,000 MPa.

If the load W was applied suddenly, the cantilever would vibrate as well as deflecting.

There are two ways in which the cantilever may collapse. The first is in bending; either the top or bottom flanges may not be strong enough, when they would fail by extension or crushing, respectively, Figure 2.1 (d). The other is if the web joining them is not strong enough to hold the two flanges together, when the failure would be in shear, Figure 2.1 (e).

Shear force is always proportional to the slope of the bending moment diagram, shown in Figure 2.1 (b). As here this slope is constant, the shear force is also constant, Figure 2.1 (f), and equal to the load W. The action of shear is best represented by an analogy, in which the web is considered to be an 'N' truss, whose diagonal members

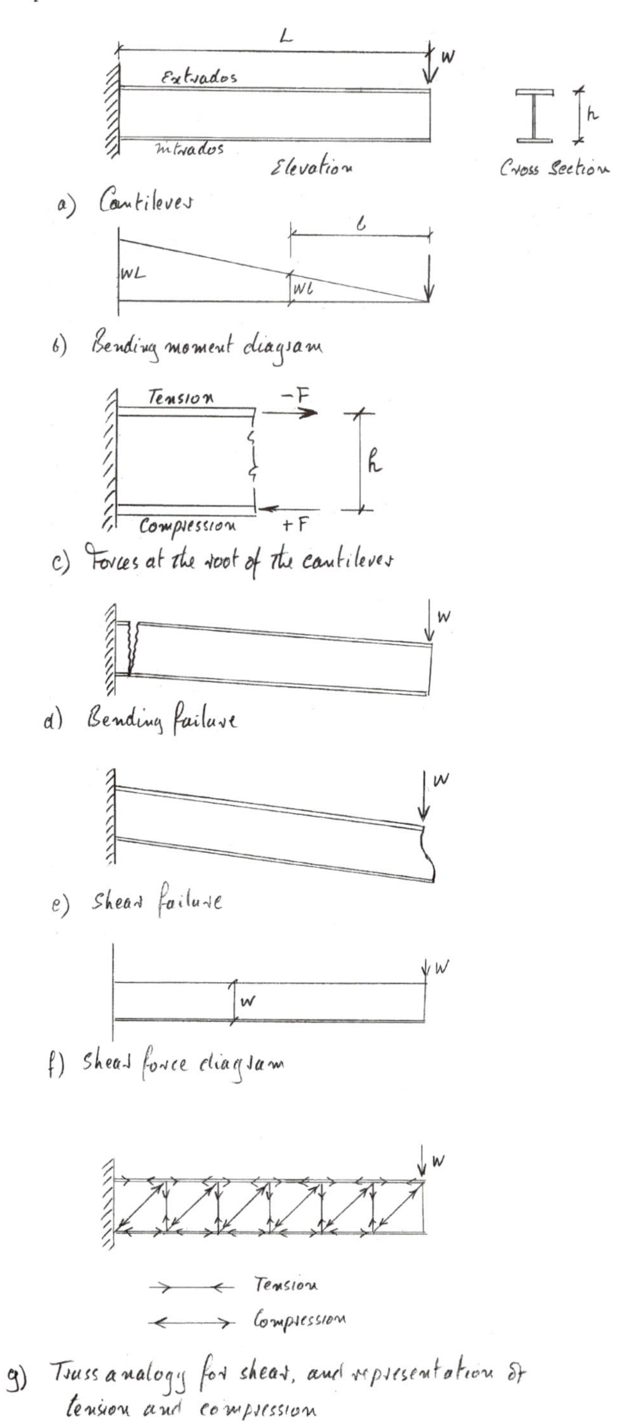

Figure 2.1 Bending moment and shear force on cantilever

are stressed in compression and the vertical members in tension, Figure 2.1 (g). This diagram also shows the conventional representation of tension and compression forces in members which is used in this book. Truss analogy is explained in more detail in *3.10*.

The cantilever could have consisted of a beam of rectangular cross section, Figure 2.2 of width b and height h. The tensile and compressive stresses caused by the bending moment are zero at what is termed the neutral axis, which is at the centre of the beam for a symmetrical cross section, and they are proportional to their distance from this neutral axis. Consequently they are a maximum at the top and bottom extreme fibres where they are represented by the symbol $\pm\sigma$. As force = area × stress, the forces of tension and compression forming the internal couple are equal to the average stress in the top or bottom half of the beam multiplied by the cross-section area of half the beam, $F = \pm(\sigma/2)\times(bh/2) = \pm\sigma bh/4$. The lever arm of the internal couple is the distance between the centroids of the tension and compression forces. As the force diagrams are triangular, this lever arm is $2h/3$. The stress on the top and bottom extreme fibres of the cross section created by the applied moment WL can be found by equating the external moment with the internal couple, $WL = F\times 2h/3$. Substituting for F, $WL = (\sigma bh/4)\times 2h/3$, or $WL = \sigma bh^2/6$, or $\sigma = WL/(bh^2/6)$. The term $bh^2/6$ is known as the elastic modulus of the rectangular cross section (not to be confused with E, the modulus of elasticity).

For bridge decks with cross sections that are unsymmetrical about a horizontal axis, Figure 2.3, the elastic moduli corresponding to the top and bottom extreme fibres are not equal. They are conventionally designated by z_t and z_b for the top and bottom extreme fibres respectively. Thus the stresses on the top and bottom extreme fibres of the bridge deck, subjected to an external bending moment M are M/z_t and M/z_b.

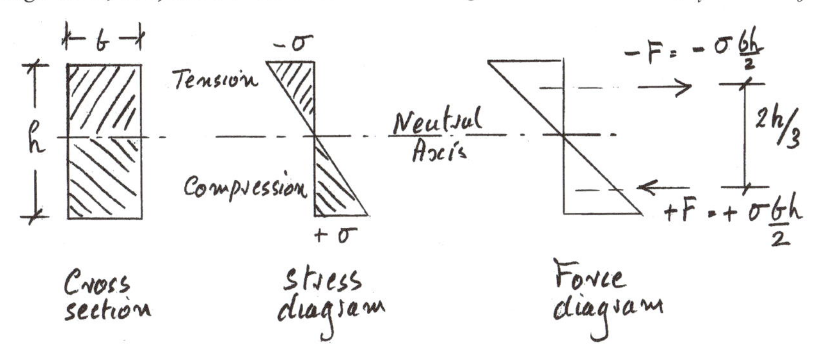

Figure 2.2 Rectangular cross section cantilever

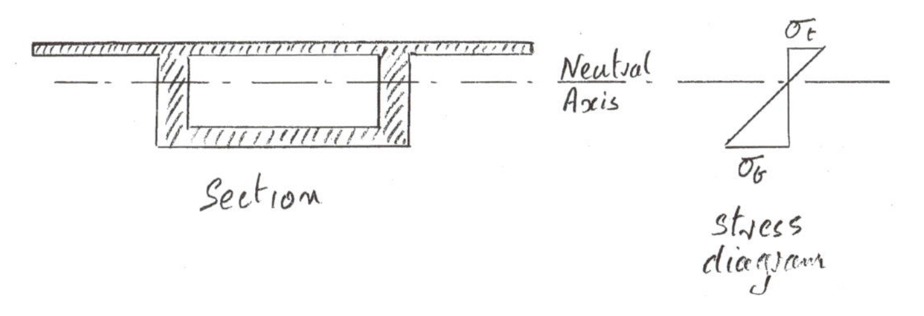

Figure 2.3 Section unsymmetrical about a horizontal axis

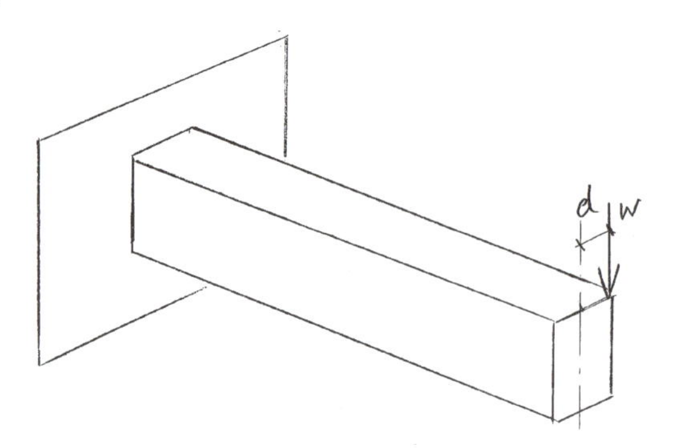

Figure 2.4 Eccentric load creating torsion

If the load *W* had been applied to the cantilever off centre, Figure 2.4, in addition to the bending moment and the shear force that remain unchanged, the beam would be twisted. If the eccentricity of the load is *d*, the twisting moment, or torque, is W×d.

2.5 Limit states

In the past, structures were designed to respect limiting stresses. In the cantilever described above, the tensile and compressive stresses at the top and bottom of the cantilever would have to remain below values that were considered safe for the material used in its construction. This would govern the maximum load *W* that the cantilever may safely carry.

The more enlightened modern tendency is to design to respect 'limit states'. Structures must perform satisfactorily in normal service, and must also have a sufficient margin of safety against collapse. The normal service condition is called the Serviceability Limit State (SLS). The SLS includes criteria on the deflection of structures when this may affect performance, such as damaging floor finishes or interfering with the drainage of rainwater, on their susceptibility to vibrate excessively and, critically, on their durability. Reinforced concrete deteriorates principally by the corrosion of the steel reinforcement within it. The useful life of a reinforced concrete member is controlled by the thickness and the quality of the protective concrete layer outside the reinforcement, called the cover, and by the width of the cracks in the concrete that occur in service (*3.5*, *3.6*).

The collapse condition is called the Ultimate Limit State (ULS). The structure must not collapse when the loads applied to it are factored up by Load Factors, and the strength of its constituent materials is factored down by Material Factors. The magnitude of the Load Factors depends principally on the uncertainty that exists over the likely intensity of any load. For instance, the self weight of a bridge deck may vary only within narrow limits, due to variations in the density of materials and to tolerance on the size of the members, and as a result the Load Factor applied to self weight is low, typically 1.15. On the other hand, the traffic loading on a bridge deck is inherently variable and there is the possibility that under exceptional circumstances it may be heavier than expected, leading to a higher Load Factor, typically 1.3 – 1.5.

The Material Factor on steel reinforcement is low, typically 1.05 – 1.15, as steel is produced by an industrial process and there is little variation in its strength. On the other hand concrete is mixed on site and is inherently a variable product, leading to the need for a higher Material Factor, typically 1.5.

2.6 Statical determinacy and indeterminacy

The concepts of statical determinacy and indeterminacy re-occur in the text. A short explanation, or reminder, is given below.

Structures are determinate if the forces applied on their supports, called the support reactions, can be calculated using the two basic equations of equilibrium:

- the moments about any point sum to zero; Equation 1
- the forces in any direction sum to zero. Equation 2

Consider the statically determinate beam shown in Figure 2.5 (a), with a span L carrying two loads W_1 and W_2 which are situated respectively at distances l_1 and l_2 from point A. The beam is assumed to rest on bearings that allow rotation, the reactions at the bearings being R_1 and R_2.

Take moments about point A:

Equation 1 $W_1 \times l_1 + W_2 \times l_2 - R_2 \times L = 0$, which gives $R_2 = (W_1 \times l_1 + W_2 \times l_2)/L$.

Resolve the forces and reactions in the vertical plane:

Equation 2 $R_1 + R_2 - W_1 - W_2 = 0$, which gives $R_1 = W_1 + W_2 - R_2$.

As the loads and their positions on the span are known, the equations may be simply solved to yield the value of the reactions. Once the reactions are known, the bending moments and shear forces at any point of the beam can be calculated.

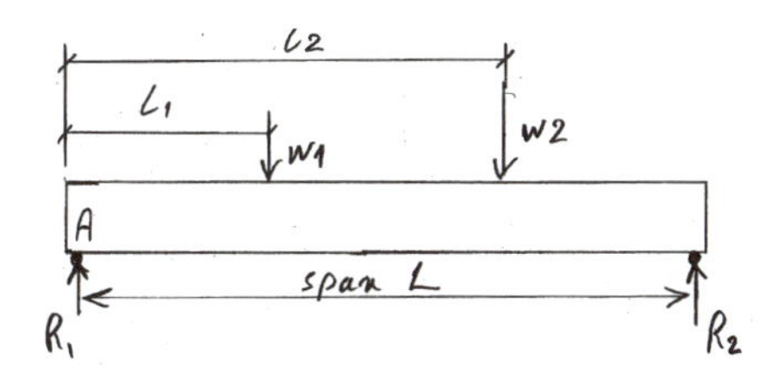

a) statically determinate beam

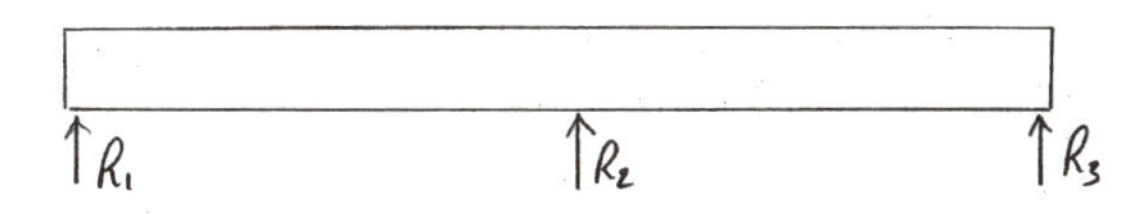

b) statically indeterminate beam

Figure 2.5 Statically determinate and indeterminate beams

Statically determinate structures may also be called 'isostatic', while determinate single span beams may be called 'simply supported' or 'simple beams'.

Structures are classed as indeterminate when their support reactions cannot be calculated by considering only the two equations of equilibrium. For instance, the two-span beam shown in Figure 2.5 (b) has three support reactions, and this requires three equations to solve for the value of the reactions. The third equation may be generated by a variety of means that are the scope of specialist books on structural analysis [1]. Indeterminate structures are also called 'hyperstatic' or 'redundant', while monolithic beams with more than one span are called 'continuous beams'.

In statically determinate structures, the reactions are known absolutely; if one of the supports of the beam shown in Figure 2.5 (a) was to settle, the support reactions would not be affected, and in consequence the bending moments and shear forces in the beam would also not be changed. In indeterminate structures, the support reactions and the bending moments and shear forces in the beam depend on the rigidity of the supports. For instance, if the central support of the two-span beam shown in Figure 2.5 (b) was to settle, some of its load would be shed onto the end supports, and additional bending moments and shear forces would be set up in the beam.

In real life, no support is completely unyielding, and consequently the exact reactions and the bending moments in any indeterminate structure, and in particular in a continuous beam, are subject to a degree of uncertainty. The best one can do is to make assumptions as to the likely settlement of the supports and then calculate the support reactions and the bending moments in the beam that result from those assumptions.

A three-legged stool is determinate, as the load on each leg is independent of the unevenness of the ground. A four-legged table is indeterminate, because the load will always be unequally distributed between its legs, unless they are exactly the same length and the ground on which it rests is perfectly flat, impossible conditions in reality.

Most real structures are indeterminate, unless specific measures are taken to create determinacy, such as introducing hinges into the structure, or using simply supported beams. As a result of this indeterminacy, the external reactions of structures and their internal stresses cannot be known precisely. This is only a problem for those who believe that for a structure to be safe it must comply strictly with the limiting stresses or conditions set by the codes of practice. In reality, structures are safe or unsafe depending on the quality of their designers, and code compliance is a box that has to be ticked to provide some minimum standards of public safety.

In many structures, indeterminacy is a desirable attribute despite the uncertainty it creates, as it reduces the vulnerability of a structure to accidental damage. A discussion on the relative merits of determinacy and indeterminacy specifically for bridge structures may be found in *7.14*.

3
Reinforced concrete

3.1 General

There are many excellent books on all aspects of the design and use of reinforced concrete. It is not the intention of this chapter to attempt to summarise such a wide subject in a few pages. Its purpose is, after a brief historical perspective, to act as an introduction to the following chapters on prestressed concrete, and to highlight specific aspects of the subject particularly relevant to the design of bridges, and which the author in his practice has concluded are inadequately understood by many engineers.

3.2 The historical development of reinforced concrete

The first known use of concrete is for a floor in Israel, dated to approximately 7000 BC. The Egyptians used concrete as an infill for stone-faced walls from about the second millennium BC, and the Greeks used it as a mortar or render from about 500 BC. The earliest Roman use of concrete dates from about 300 BC, initially as a core material, between masonry facings. These early uses of concrete probably used cement made from burnt lime (quicklime).

In the second century BC, the Romans discovered that adding pozzolana to the lime produced a much stronger concrete, which could be used as a building material in its own right. This discovery allowed them to revolutionise construction by designing large span concrete domes. The two most well known and well preserved Roman concrete buildings with a domed roof are the Pantheon in Rome, Figure 1.14, built in 120 AD, and Hagia Sophia in Istanbul built in the sixth century, Figure 3.1.

Both of these buildings are in good condition, having resisted the ravages of the Mediterranean weather, and the seismicity of both sites. The most recent earthquake in Turkey destroyed many modern buildings, but left Hagia Sophia intact. Their survival is not an accident; both buildings are extremely well designed, and demonstrate great virtuosity in the knowledge and use of concrete. For instance, the domed roof of the Pantheon is made of lightweight concrete to reduce the thrust on the dense concrete walls. When they were erected, their designers had centuries of accumulated experience in the material; and it shows!

After the collapse of the Roman Empire, the knowledge of how to make and use concrete as a primary structural material was lost for many centuries. Although lime concrete was used for foundations and as a filling material for walls throughout the middle ages and the renaissance, it was not until the end of the eighteenth century

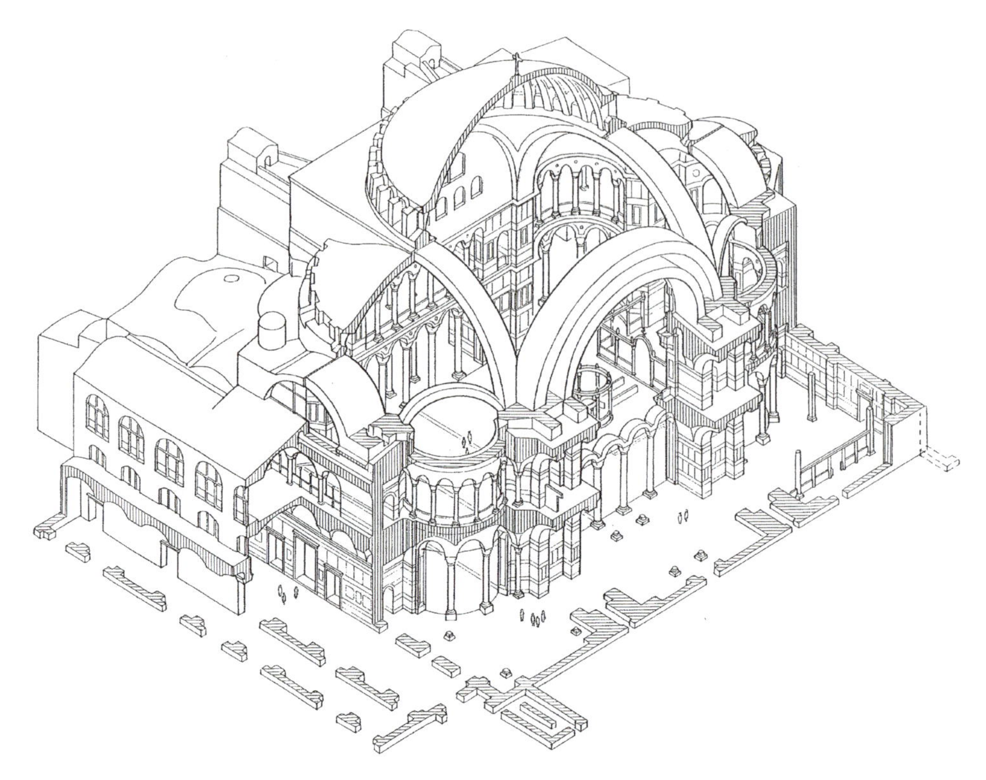

Figure 3.1 Partly cut-away isometric drawing of the sixth-century church of Hagia Sophia in Constantinople (Source: R.J. Mainstone, *Hagia Sophia*, Thames and Hudson 1988)

that organised experiments were conducted to improve the strength and speed of hardening of cement. John Smeaton experimented with mortars to find a material suitable for the construction of the Eddystone lighthouse, the construction of which commenced in 1756. Then in 1824, Joseph Aspdin took out a patent for Portland cement, to begin the modern development of concrete [1].

Concrete is brittle in tension, and cannot sustain significant tensile stresses. The Romans attempted to reinforce their concrete using bronze bars, but this was not successful, due to the different coefficients of thermal expansion of the two materials. Consequently, its use was limited to structures that were primarily in compression: walls, arches and domes. The great discovery of the nineteenth century was the reinforcement of concrete using iron or steel, which have virtually the same coefficient of expansion as concrete, and which effectively gave it tensile strength. The first patent in the United Kingdom for reinforced concrete was taken out in 1854 by William Wilkinson, although the technique was only used widely towards the end of the century. So, even now, we have only something over a hundred years of experience in the use of modern reinforced concrete, and despite the undoubted greater speed of learning and change characteristic of our age, we are still finding out which aggregates are safe to use, and how to control the chemical make up of cement.

Although the inclusion of reinforcement has given to concrete a vastly greater versatility, it has also brought with it the majority of the problems that the modern designer must understand if he is to make durable structures, safely and economically. The principal problem introduced by reinforcement is the corrosion of the steel leading to the deterioration of concrete structures, and to their need for maintenance.

Modern designers appear to have forgotten that concrete in compression does not need to be reinforced, and that leaving out the steel not only saves money but improves durability. The concrete side walls of the Byker tunnel, for instance, are not reinforced, Figures 17.6 and 17.7. However, there is no doubt a great future for reinforcement that does not corrode.

3.3 General principles of reinforced concrete

3.3.1 The composite material

Concrete is an artificial stone. Like stone, it is strong in compression, but weak and brittle in tension. Also like stone, its strength in tension may be further compromised by fissures or cracks. Thus, unreinforced concrete can only be used in circumstances when the material is in compression, or when the tensile stresses are very low.

Embedding steel bars in the concrete creates a composite material, where the concrete provides the strength in compression, while the bars carry any tensile forces.

3.3.2 Concrete

The strength of concrete in compression is measured by crushing standard samples. In the UK these samples are in the form of 150 mm cubes, while on the continent of Europe and in the USA they are in the form of cylinders of 150 mm diameter and 300 mm height. In general, the stress at which a cylinder fails is approximately 85 per cent of the strength of a cube made of the same material, although the difference is less the stronger the concrete [2]. As the strength of concrete increases with time, it is conventionally determined at an age of 28 days after casting, although the 90 day strength is also relevant. The degree to which the strength increases with time depends on the nature of the cement. For instance, concrete made with blended cement including a proportion of pulverised fuel ash (PFA), increases in strength more than that of unblended Portland cement. This enhanced strength is significant in establishing the true factor of safety of a structure.

The strength of a concrete test specimen depends on the rate at which it is loaded, the faster the application of the load the higher the strength. Thus tests for concrete need to specify a rate of loading [3].

Concrete used in bridges typically has a 28 day cube strength of between 40 MPa and 60 MPa. Strengths up to 100 MPa have been used exceptionally while even stronger concretes, up to 150 MPa and beyond, are still the subject of research involving the construction of some trial structures.

Concrete is not an elastic material. As the stress/strain curve does not have a straight portion, one cannot define a unique Young's modulus for concrete and different codes of practice use different definitions. The most common definitions are the tangent modulus, which is the slope of the curve at one point and the secant modulus which is the slope of the line that connects the origin to a point on the curve, Figure 3.2 (a) [4]. If the concrete is loaded by a pulse, it will exhibit a dynamic modulus that is similar to the initial tangent modulus [5]. The Young's modulus of concrete increases with its

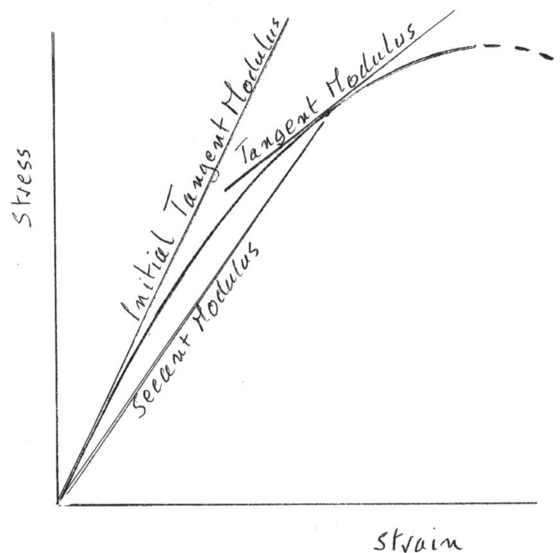

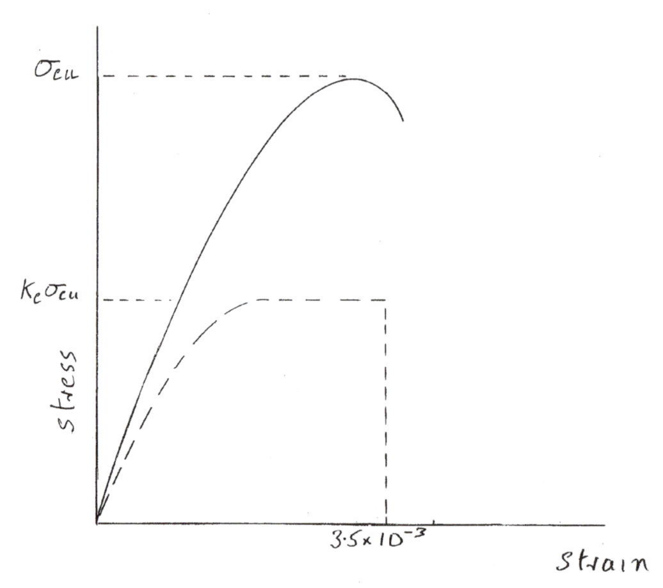

Figure 3.2 Stress–strain curve for concrete

strength, but is typically, as defined by the British code, of the order of 34,000 MPa for loads applied for a short time on bridge quality concrete.

When concrete fails in compression it exhibits a degree of ductility. The strain of plain concrete at failure in compression is conventionally assumed to be of the order of 2–4 $\times 10^{-3}$ (3.5×10^{-3} in the UK code of practice), Figure 3.2 (b) [6]. This ductility in compression may be greatly increased by the provision of reinforcement specifically designed for that purpose.

When concrete is loaded in compression and the load is maintained, the instantaneous deformation is followed by a deferred deformation known as creep. This deferred deformation is frequently up to twice the magnitude of the instantaneous value, and in a large structure such as a bridge deck, may take several years to complete.

The corollary of this behaviour is that if a fixed deformation is imposed on a concrete member, the force required to maintain that deformation would reduce with time. This is known as relaxation or load decay. Creep and relaxation of concrete are discussed in more detail in *3.9*. The effect of creep on the bending moments, deflections and stresses in bridge decks is discussed in *6.21* and *6.22*.

Concrete has a tensile strength that is of the order of 10 per cent of its compressive strength. This tensile capability is essential to its strength in shear and to create bond with reinforcement. However, as concrete fails in tension in a brittle manner, and may be cracked before loading due to internal stresses, this tensile strength is usually ignored in calculating the bending strength of beams and slabs.

3.3.3 Steel

The steel generally used as reinforcement has a yield strength which lies between 400 and 500 MPa and a Young's modulus that lies between 190,000 and 210,000 MPa. The typical stress/strain curve is shown in Figure 3.3, and demonstrates a high degree of ductility; usually the strain at rupture is in excess of 12 per cent. The reinforcing bars are deformed by rolling ribs onto the bar during manufacture. These deformations improve the bond of the bars with the concrete. Higher strength steel is available, but cannot in general be used effectively as it causes excessive cracking of the concrete. Undeformed mild steel, with a yield stress of 250 MPa is used occasionally, generally as starter bars where its greater ductility allows it to be bent and re-bent.

3.3.4 The modular ratio

The ratio of the Young's moduli of steel to concrete is called the modular ratio. This ratio varies between about 6 for short-term loads, to 15 or over for sustained loads, and is important, as it determines how the steel and the concrete in a structural member share the applied loads between them.

For instance, consider a concrete column reinforced with steel bars that represent 3 per cent of the total cross-section area. When subjected to an axial compressive load, in the short term the stress in the steel will be six times higher than the stress in the concrete, and the steel would carry some 16 per cent of the total load. If the load is maintained, the modular ratio increases as the concrete creeps, and load is shed from the concrete onto the steel, which eventually will carry nearly one-third of the load.

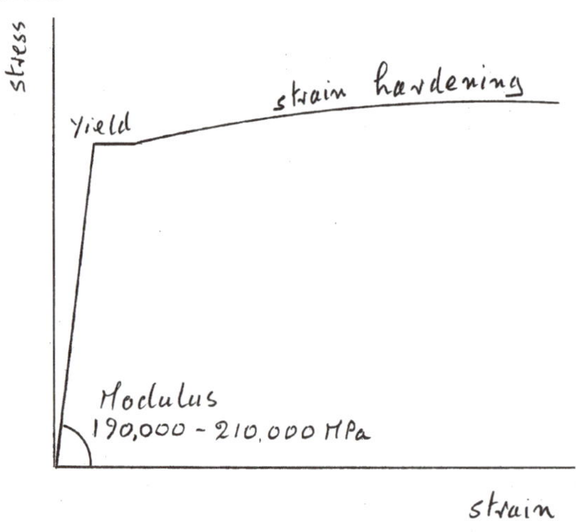

Figure 3.3 Stress–strain curve for high yield reinforcing steel

3.4 Reinforced concrete in bending

3.4.1 General

A reinforced concrete member has to satisfy both the Serviceability Limit State and the Ultimate Limit State. The SLS criteria include considerations of deflection, vibration and of control of the width of cracks (3.5), while the ULS is principally the consideration of the collapse of the member in bending, by yield of the steel or crushing of the concrete, or in shear. Although beams are usually designed at the ULS, the beam actually spends its life at working loads, and its performance at the SLS is just as important.

Most correctly designed beams are 'under-reinforced' in bending. This means that the steel reinforcement yields before the concrete crushes, and the mode of failure is ductile. The simple methods described below are aimed at demystifying the preliminary sizing of under-reinforced concrete members at the ULS. Clearly the detailed design and verification of a beam involve much more extensive and careful calculations, at both the SLS and the ULS.

3.4.2 Preliminary sizing of bending reinforcement in beams and slabs

When designing a reinforced concrete beam to resist specified superimposed loads, the first step is to guess a size. For a lightly loaded beam, the depth is likely to be of the order of the span/20, while for more heavily loaded beams it may be twice that figure. The width of the beam will be determined by the space required to accommodate the reinforcement, by the shear stresses or by the need to provide an adequate area of concrete in compression. This guessed size allows a dead load to be calculated, to give a total load and a total bending moment. A preliminary calculation is then necessary to make a first estimate of the area of reinforcing steel required, and to check that the depth and width of the beam are adequate.

The bending moment applied to a beam is resisted by an internal couple. In a reinforced concrete beam, the tensile component of this couple is provided by the reinforcement, while the concrete above the neutral axis, for a sagging moment, provides the compressive component, Figure 3.4.

The force F to be provided by both the concrete and the steel reinforcement is given by the equation $F = M/l_a$, where M = the bending moment applied at the section, and l_a is the lever arm of the internal couple. The lever arm of the internal couple is the distance between the centroid of the tension reinforcement and the centroid of the compression forces in the concrete.

Figure 3.4 (b) shows a typical stress diagram for a beam at working load, when the steel stress is still within the elastic range. It may be seen that there is an area just below the neutral axis where the concrete is in tension. Lower down the section, as the tensile stress in the concrete would exceed its limiting value the concrete cracks, and the stress in the concrete falls to zero. The contribution of the zone of uncracked concrete in tension to the overall bending strength is conventionally ignored in calculation.

The accurate calculation of the lever arm at working load is complicated, as the position of the neutral axis is not fixed but depends on the magnitude of the bending moment. For a beam that is rectangular in cross section, a rough approximation may be made by assuming that the depth of the neutral axis n is $0.5d_e$ below the top fibre of the beam, where d_e is the depth of the beam measured to the centroid of the tensile reinforcing steel, and is known as the 'effective depth'.

Then, as the centroid of the triangular compressive force diagram is located at $n/3$ below the top fibre,

$$n \approx 0.5d_e \text{ and } l_a \approx d_e - 0.5d_e/3 \approx 0.83d_e.$$

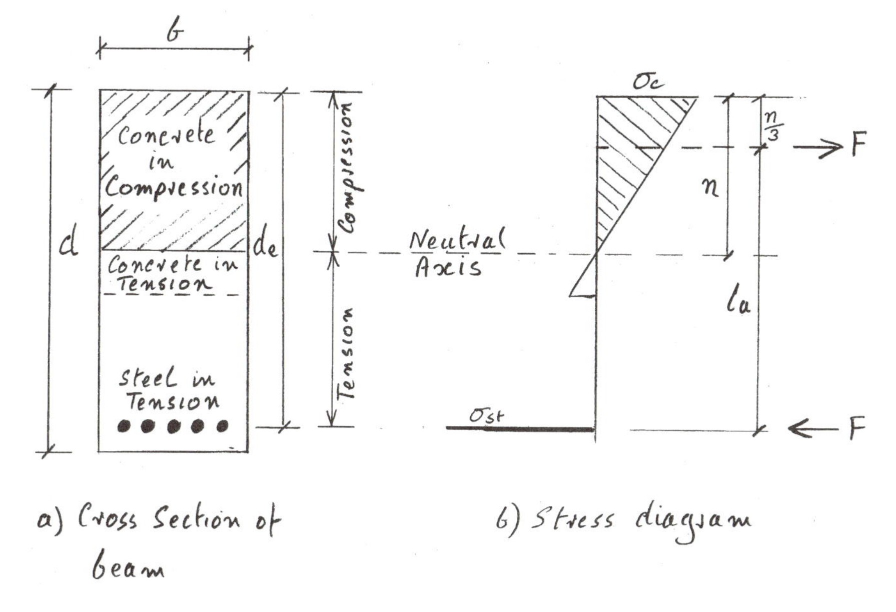

Figure 3.4 Beam in bending at working load

Once the force F is known, the maximum compressive stress in the concrete is $2F/bn$, where b is the breadth of the beam, and the required area of reinforcing steel $A_{st} = F/\sigma_{st}$, where σ_{st} is the permissible working stress in the steel.

To summarise:

$$F = M/l_a$$

$$l_a \approx 0.83d_e$$

$$\sigma_c \approx 4F/bd_e$$

$$A_{st} = F/\sigma_{st}$$

For a more precise calculation, graphs and computer programs are available which allow the position of the neutral axis, the working stress in the steel and the cross-section area of the reinforcement to be calculated accurately.

Fortunately, most codes of practice now require reinforced concrete beams to be designed in bending at the ULS, which is a much simpler calculation for the normal case when the beam is under-reinforced. The steel reinforcement may be assumed to be working at its yield stress, σ_y, reduced by a factor k_{st} that is defined by the code of practice. The concrete compressive stress is assumed to be in the plastic range, with a stress block typically of the shape shown in Figure 3.5 (a). The shape of this stress block is derived directly from the factored stress/strain curve shown in Figure 3.2 (b). The strength of concrete in compression for this purpose will be $k_{c1}\sigma_{cu}$, where σ_{cu} is the crushing strength of the concrete and k_{c1} is a reduction factor defined by the code of practice (*3.4.3*). For preliminary calculations, this stress block may be simplified to a rectangular shape, working at a stress of $k_{c2}\sigma_{cu}$ as shown in Figure 3.5 (b), where k_{c2} is a factor further reducing the strength of the concrete to take account of this simplification.

The depth of the rectangular stress block that corresponds to an applied bending moment must be calculated. Initially, this depth may be guessed, as may be the arrangement of the tensile reinforcement, yielding a first approximation to the lever arm. The tensile force in the reinforcement and the compressive force in the concrete corresponding to the guessed lever arm may then be calculated from $F = M/l_a$, where M is the applied bending moment at the ULS and l_a is the distance between the centre of the tensile reinforcement and the centre of the guessed rectangular concrete stress block. The depth of the compressive stress block required to provide this force, and the number and size of the reinforcing bars may be then calculated, and a second approximation made to their arrangement. A corrected lever arm is calculated, and a revised F derived. Usually, convergence is achieved after very few repetitions of this cycle, yielding a check on the adequacy of the width and depth of the beam, and a good approximation to the steel area required to resist a known bending moment.

If the cross section is not rectangular, the centroid of the compressive force is simply calculated by comparing the rectangular compressive stress block with the cross section, Figure 3.5 (c).

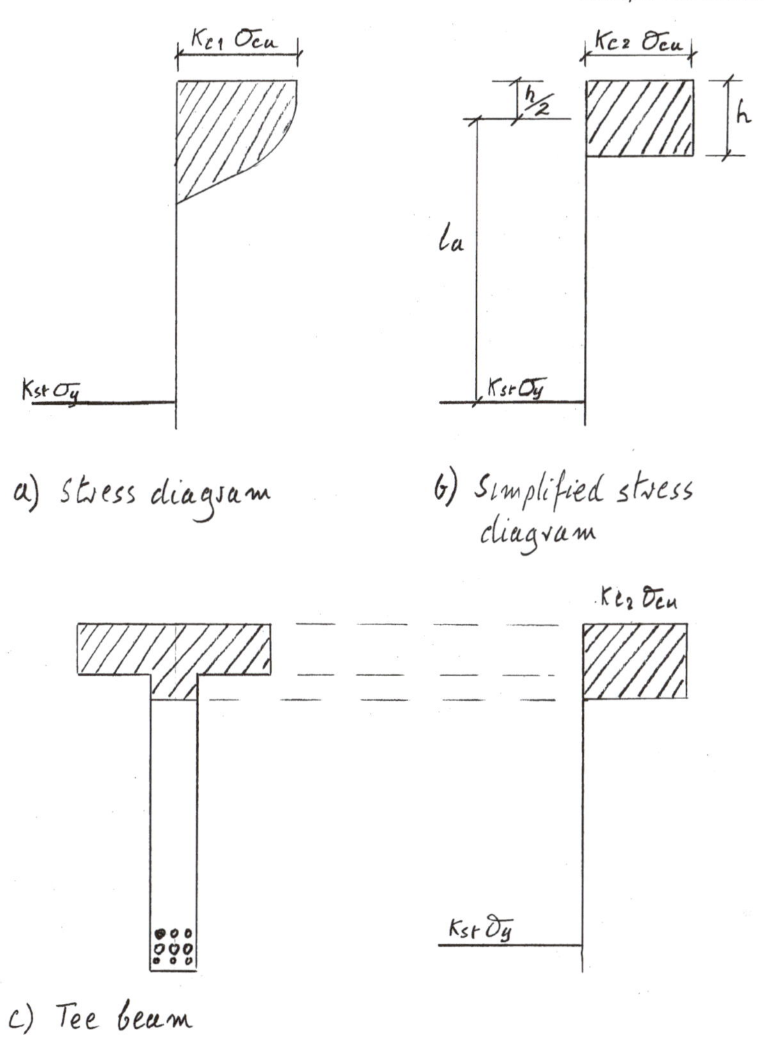

Figure 3.5 Beam in bending at the ULS

3.4.3 Examples of the preliminary assessment of bending reinforcement

Example 1

Consider a 15 m span Tee beam as shown in Figure 3.6 (a), made of concrete with a 28 day cube strength of 50 MPa (σ_{cu} = 50 MPa), and reinforced with steel with a yield stress of 460 MPa. The depth of the beam is 1,000 mm, the width of the 150 mm thick top flange is 1,100 mm and the width of the web 350 mm. The beam is loaded with a uniformly distributed load of 50 kN/m, producing a working bending moment at mid-span of 1.4 MNm and an ultimate bending moment of 2 MNm.

The calculations will be carried out using the material factors and general methods that correspond to BS5400: Part 4: 1990. However, this example is equally valid

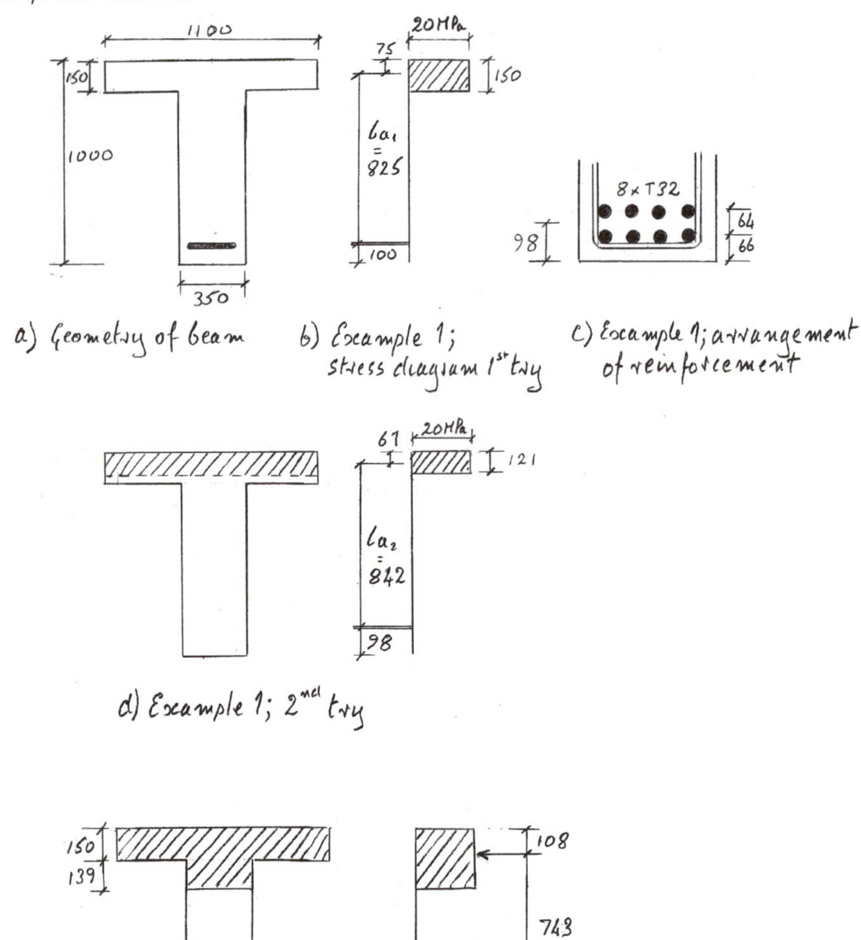

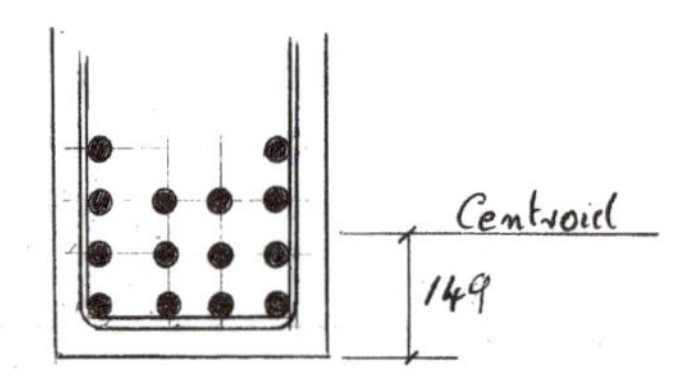

Figure 3.6 Examples 1 and 2

for other codes of practice, substituting suitable material factors for the steel and concrete.

The ultimate stress in the steel is assumed to be $460/\gamma_m$, where γ_m is the material factor (2.5). With $\gamma_m = 1.15$, the ultimate stress to be used in the calculation is 400 MPa (k_{st} in Figure 3.5 $= 0.87$).

The strength of concrete in a bending member is known to be less than its strength as described by the compression test. The British code BS5400 defines this reduction as a factor of 0.67 with respect to σ_{cu}. The concrete strength must be further factored down by the material factor γ_m, which covers the risk that it may not be as strong as specified, which for concrete is equal to 1.5. Thus, k_{c1} is $0.67 / 1.5 = 0.45$.

When a simplified rectangular stress block is considered, as in this example, the ultimate strength of the concrete should be further reduced to $0.4\ \sigma_{cu}$, or 20 MPa for the quality of concrete considered ($k_{c2} = 0.4$).

A first guess at the lever arm l_{a1} may be made by assuming that the centroid of the reinforcement is 100 mm from the bottom fibre of the beam and that the depth of the compressive stress block equals the depth of the top flange, Figure 3.6 (b).

Hence

$$l_{a1} = 1 - 0.1 - 0.075 = 0.825 \text{ m.}$$

The force required in both the steel and concrete to provide the internal couple would be

$$F = M/l_{a1} = 2 / 0.825 = 2.42 \text{ MN}$$

The area of reinforcing steel A_{st} required to produce an ultimate force of 2.42 MN is:

$$A_{st} = 2.42 / 400 = 6.05 \times 10^{-3} \text{ m}^2\text{, or } 6{,}050 \text{ mm}^2.$$

This area may be provided by approximately 7.5 bars of 32 mm. As a first try adopt 8 bars, yielding 6,434 mm^2, arranged as shown in Figure 3.6 (c). As recommended in a former French code, it is good practice to limit the size of bars to 10 per cent of the minimum dimension of the member containing the reinforcement, 350 mm in this case; this is respected by adopting 32 mm bars. It should also be noted that the true size of the deformed bars normally used for reinforcement is about 10 per cent larger than their nominal size, due to the protrusion of the ribs. The bars are spaced vertically at twice their nominal diameter and it has been assumed that the cover to the 12 mm stirrups is 35 mm. The centroid of the reinforcement, adopting these rules, is 98 mm from the beam soffit.

The depth of the rectangular stress block in the 1.1 m wide flange working at 20 MPa required to produce this force is $2.42/(1.1 \times 20) = 0.110$ m, less than the thickness of the slab.

The lever arm may now be recalculated as $1.0 - 0.098 - 0.110 / 2 = 0.847$ m. The force required of the internal couple becomes $F = 2 / 0.847 = 2.36$ MN, which is close enough to the first guess. Thus the choice of reinforcement consisting of 8 bars of 32 mm is adequate as a preliminary design.

It should be noted that as the compressive zone is entirely contained within the top slab, this calculation would have been identical for a beam of rectangular cross section.

Example 2

The ultimate bending moment is increased to 3.2 MNm. As a first guess, assume as before that the whole of the top slab is compressed, and that the centroid of the reinforcement is at 100 mm from the bottom fibre, yielding the same lever arm $l_{a1} = 0.825$ m.

Following the same calculation yields a force

$$F = 3.2 / 0.825 = 3.88 \text{ MN}$$

A_{st} would be

$$3.88 / 400 = 9.7 \times 10^{-3} \text{ m}^2 = 9{,}700 \text{ mm}^2$$

corresponding to just over 12 bars of 32 mm. Using the same rules as before, these would be arranged in three rows of four bars, and their centroid would be 130 mm from the bottom fibre.

The area of concrete required would be 3.88 / 20 = 0.194 m^2, and the depth of the compressed area, 1.1 m wide would be 0.194 / 1.1 = 0.176 m. As this is greater than the thickness of the slab, the compression block will have to extend into the web.

The slab can supply a force of 1.1 × 0.15 × 20 = 3.3 MN, leaving 3.88 – 3.3 = 0.58 MN to be found in the web. The area of compressed concrete required is 0.58 / 20 = 0.029 m^2, extending down the web a distance of 0.029 / 0.35 = 0.083 m. The centroid of the compressed concrete is calculated as 0.092 m from the top of the section. Consequently the lever arm becomes 1.00 – 0.130 – 0.092 = 0.778 m, considerably less than the first guess of 0.825 m.

If this new lever arm is used in a new round of calculation, it will clearly increase the force required in the steel and concrete, increasing the number of reinforcing bars required, and extending the compressed concrete down the web. Both these developments would reduce the lever arm still further, requiring yet another round of calculation. Experience shows that when the lever arm from the first round of calculation is smaller than that guessed, there is a danger of chasing one's tail in a calculation that converges only slowly. Consequently, it is necessary to make a second guess that attempts to bracket the correct lever arm.

Guess a new lever arm $l_{a2} = 0.7$ m, when

$$F = 3.2 / 0.7 = 4.57 \text{ MN}$$

A_{st} becomes

$$4.57 / 400 = 0.0114 \text{ m}^2 = 11{,}400 \text{ mm}^2$$

which corresponds to about 14 bars of 32 mm. These will be arranged as shown in Figure 3.6 (f), and have a centroid which is 149 mm from the bottom fibre.

The force to be supplied by the web is now 4.57 – 3.3 = 1.27 MN, requiring an area of 1.27 / 20 = 0.0635 m^2, and a depth of compressed concrete in the web of 0.0635 / 0.35 = 0.181 m. The centroid of the compressive force is now 0.121 m from the top surface, giving a calculated lever arm of 1 – 0.149 – 0.121 = 0.73 m. This is larger than the guessed figure. Consequently, the correct lever arm must be between 0.787 m and 0.73 m.

For the third guess try l_{a3} = 0.75 m. Then

$F = 4.27$ MN

$A_{st} = 10{,}675$ mm^2

corresponding to 13.3 bars; maintain 14 bars for preliminary design; the force to be supplied by compressed concrete in the web is 4.27 – 3.3 = 0.97 MN, requiring a depth of 0.139 m; the centroid of compressed concrete is now at 0.108 m from the top surface, giving a calculated lever arm of 1 – 0.149 – 0.108 = 0.743 m, Figure 3.6 (e) and (f). This is close enough to the guess of 0.75 m.

Any further increase in moment could only be carried by an additional compressed area further down the web with an ever-diminishing lever arm. It is clear that the bending moment is close to the maximum that the section can carry.

The total depth of compressed concrete is 0.289 m. As the depth of compressed concrete increases, the ductility of the beam decreases. It is generally accepted that the depth of compressed concrete should not exceed 40 per cent of the depth of the beam. The moment capacity and ductility of the beam could be increased by using stronger concrete, or by adding reinforcing bars working in compression in the top slab.

One of the functions of a quick preliminary calculation is to make clear when a more careful calculation is required. For this case, which is clearly approaching the limit of the capabilities of the beam, the detailed design will have to be particularly careful.

For the detailed design of both of these examples, it would be necessary to check that the SLS conditions, in particular the control of the width of cracks, are satisfied by these reinforcement arrangements. The crack control criterion may well govern the amount of reinforcement required and its arrangement, particularly for heavily loaded beams, although in general the size of the beam is controlled by the ULS calculation described. The minimum depth of beams and slabs may also be limited by the need to control long-term deflection under permanent loads. For the preliminary design, it would also be necessary to check that the width of the web is adequate to carry the shear force.

3.5 The cracking of reinforced concrete

3.5.1 General

When concrete is stressed in tension, it exhibits very limited ductility before cracking. Tensile stresses in concrete are caused by structural actions such as bending and shear in beams, and by internal stresses generated as it sets. This cracking affects the appearance of the member, may reduce its durability and causes water-retaining structures to leak. The cracking of concrete is its greatest drawback as a construction material, and is consequently one of the principal concerns of the designer. It is essential that designers understand the causes and the effects of such cracking and take the measures necessary to restrict it to acceptable proportions. Most codes of practice on the use of reinforced concrete set out acceptable limits on the width of cracks and give formulae for calculating their width and spacing for bending members. However the designer should, while respecting the codes, not assume that they are providing him with a reliable and adequate basis for guiding him in his quest to build a good structure.

The cracking of concrete is only relevant to the SLS, that is at working load. Consequently, although the proportions of reinforced concrete structures are defined principally at the ULS, the designer must understand their behaviour at working load.

3.5.2 Cracking due to bending of normally proportioned beams

As concrete cannot follow the extension of the steel reinforcement stressed in tension, a bending member will inevitably crack on its tension face. Most codes of practice are based on the interpretation of tests. These tests give reasonably reliable results for the spacing and width of cracks in areas where the bending moments are changing slowly, such as the bottom fibre of beams subjected to sagging bending moments. However, they are unlikely to give accurate predictions where bending moments are changing rapidly, such as over intermediate supports.

There are instances where the designer would be advised to be more cautious than permitted by the codes. One particular area is the cracking of structures that have a very high proportion of dead load and where the steel stresses under these loads are above 250 MPa. Examples of such structures are the roofs of cut-and-cover tunnels, and footbridges. Cracks that are permanently open tend to become highlighted as the concrete weathers, severely compromising its appearance.

3.5.3 Effect of cracking on durability

Codes of practice recommend limiting the width of cracks in concrete in order to protect the steel reinforcement from corrosion, with the acceptable width of cracks depending on the aggressiveness of the environment. There has been conflicting experimental evidence on the deleterious effect of cracks on the durability of reinforced and prestressed concrete. However, the designer should err on the side of caution, and of common sense.

It seems most probable that cracks that are permanently open will reduce the protection of the steel reinforcement. The deleterious effect of cracks that only open under the effect of live loads will depend on the frequency of application of the live loads. For highway bridges for instance, which are designed to resist exceptional loads that may only occur at rare intervals in the life of the bridge, cracks that open under these loads are most unlikely to be damaging. At the other extreme, bridges carrying mass transit railways, where the design live load may occur every few minutes round the clock should be treated more conservatively.

Bending cracks are associated with a compressed layer of concrete, and hence do not allow water to penetrate through the member. On the other hand, cracks due to the restrained heat of hydration cooling (3.6) penetrate through the member and will allow water leakage, Figure 3.7. Such through cracks, even when very fine, are likely to be more deleterious than bending cracks for the durability of a concrete structure that retains water such as a reservoir, or that excludes water such as a tunnel.

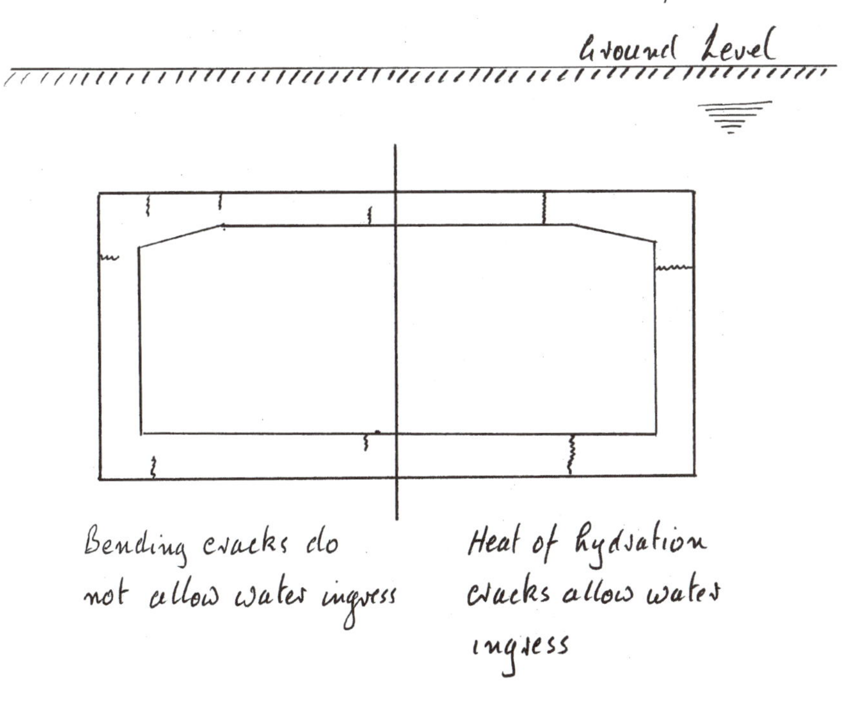

Figure 3.7 Typical highway tunnel below the water table

3.5.4 Alternative types of reinforcement

Corrosion of carbon steel reinforcement is the principal cause of deterioration of reinforced and prestressed concrete. Consequently much research is carried out into alternative materials.

The starting point should be to examine whether reinforcement may be omitted, either from one face of a member, or altogether. The walls of the Byker Tunnel, Figure 17.6, were unreinforced, while some columns and walls carrying only compression and small bending moments may also be designed as plain concrete. Designers may draw inspiration from the many classical and mediaeval structures where columns, slender towers, arches and vaults were built of unreinforced concrete or masonry. It is also debatable whether the compressed faces of slabs need reinforcement. However, as always when stepping beyond conventional wisdom, the designer must make sure that he has considered every eventuality, such as exceptional load conditions that may induce tension into otherwise compressed members.

Stainless steel reinforcement is becoming more frequently specified for concrete at particular risk of corrosion, such as in the splash zone of marine bridge piers. Stainless steel reinforcement is several times more expensive than conventional carbon steel, but may well be justified when considering the whole life cost of a structure. It may also be used sparingly, to reduce the risk of a brittle failure due to hidden corrosion. For instance, the precast parapets of the East Moors Viaduct, designed by the author in 1982, are attached to the deck by a series of links. 20 per cent of the links in each precast panel are stainless steel, so that if the conventional links were to corrode due to a failure of the waterproof membrane, the stainless steel links would guarantee a ductile failure mode.

Conventional carbon steel may be coated with epoxy resin to protect it from corrosion. This was popular some years ago, but is falling out of favour. There are reports that the adhesion of concrete to such coated steel may be reduced, with considerably increased bond lengths. Also, the coating is prone to damage during delivery and installation.

Much research has been carried out on reinforcement consisting of fibreglass rods. Conventional glass is attacked by the alkali in the cement, but special alkali-resistant glass is being devised. Although such products are not yet in the mainstream, this research may well bear fruit.

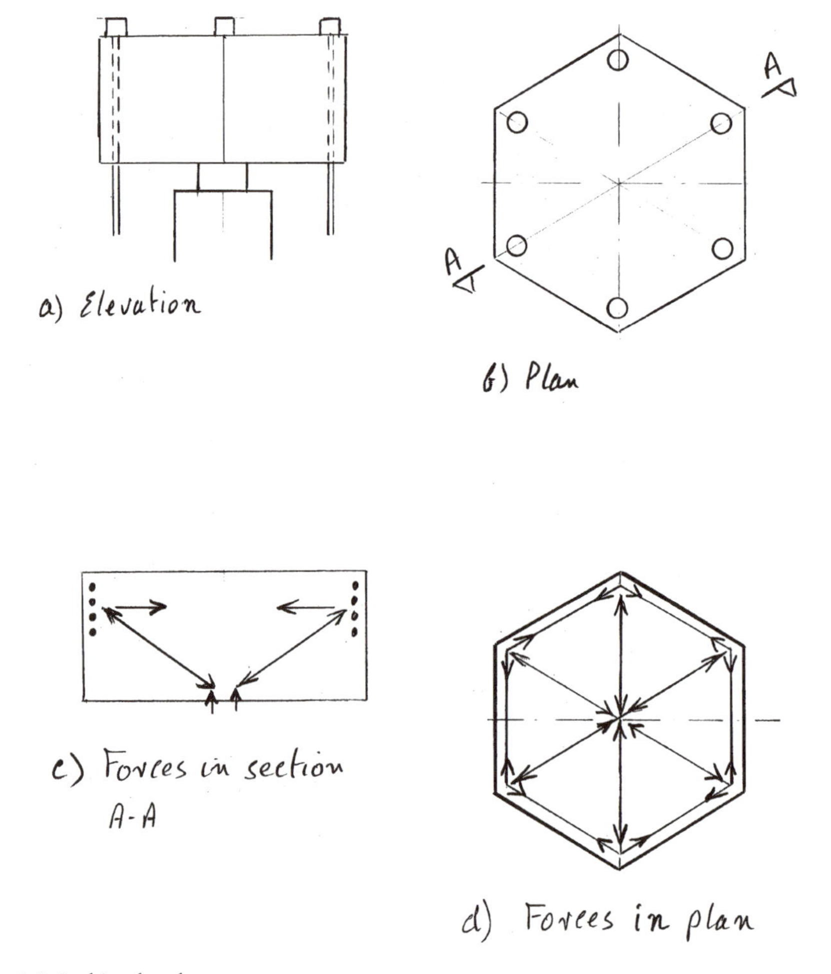

Figure 3.8 Jacking head

3.5.5 Cracking of sections where bending theory does not apply

The code formulae are not applicable to areas of reinforced concrete structures that do not work in bending. The designer is left with little guidance. Some elements of the author's personal experience may be of use.

The cracking of concrete is not a linear function of the stress in the steel. In the author's experience, if this stress remains below some critical figure, the concrete does not appear to crack, even though its theoretical tensile stress is several times higher than its tensile strength.

One example of this is a jacking system designed by the author. Hexagonal reinforced concrete jacking heads were used to lift a heavy load. The jacking heads rested on a central hydraulic jack, and six prestressing grade bars suspended the load from the head, Figure 3.8. The reinforcement for the jacking head was designed as hexagonal hoops, resisting six radial concrete struts. As the stress in the steel would be constant all round the head, it was feared that the concrete may crack unacceptably, and consequently the working stress in the steel was limited to 150 MPa. The theoretical concrete stress adjacent to the steel may be found by dividing the steel stress by the modular ratio, and would be some 25 MPa under the effect of short-term loads, approximately six times its tensile strength. The strain in the steel, neglecting its bond to the concrete, would be $150 / 200{,}000 = 7.5 \times 10^{-4}$, or 0.75 mm/m. The bond to the concrete would reduce this strain slightly. If the strain in the steel is assumed to be reduced by 20 per cent, it would become 0.6 mm/m. Hence one would expect to see cracks in the concrete totalling 0.6 mm/m. However, no cracks in the concrete were discernible to the naked eye. Neville [7] states that cracks finer than 0.13 mm are not visible without magnification. Either the concrete had cracked but the cracks were too small to see, and hence to worry about, or the concrete had yielded in tension without cracking.

Similar experiences elsewhere have convinced the author that limiting the working stress of the reinforcing steel to about 150 MPa generally results in an apparently crack-free concrete, while if the steel stress does not exceed 250 MPa, cracking will remain within tolerable limits.

3.6 The exothermic reaction

3.6.1 General

As cement sets, it heats up. The temperature rise of the setting concrete depends on the cement content, on the fineness of grinding of the cement which governs the speed of the chemical reaction, on the type of cement and on the thickness of the member. Thick sections of concrete with a high content of finely ground cement can attain setting temperatures of up to 90°C. If cement is delivered to site still hot from the furnace of the cement works, temperatures of the setting concrete even in quite thin sections may be unexpectedly high.

The time taken for the temperature to rise to its peak and then to cool down depends principally on the thickness of the structural element, Figure 3.9. In general terms, as the concrete is heating up and expanding it is still plastic. By the time it starts to cool and shrink, it has hardened, with rapidly increasing strength and modulus of elasticity. Consequently the heating phase does not cause the concrete to be stressed significantly in compression. On the other hand, any restraint to the shrinkage of the

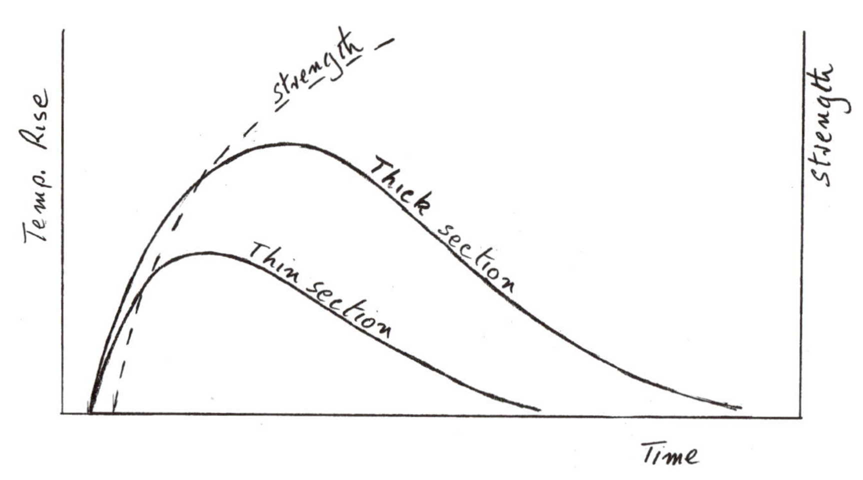

Figure 3.9 Diagrammatic representation of temperature rise and strength gain of setting concrete

concrete as it cools will stress it in tension, and may cause it to crack. It is ironic that this exothermic reaction is by far the most significant cause of prejudicial cracking of concrete, while the greatest design time, and code activity, is devoted to limiting the less-damaging cracking in bending. It is most important that the designer understands this characteristic of concrete. (It should be noted that this cooling-induced shrinkage of concrete is quite distinct from long-term drying shrinkage.)

3.6.2 The cracking of thick sections of concrete due to temperature gradients within the mass

When a concrete member is cast, the heat of the reaction is dissipated by surfaces in contact with the air or the shuttering. In thick sections (anything over about 800 mm) there will be a significant temperature gradient through the thickness of the member, with the skin being cooler than the core. Once the concrete has set, the hotter core has to cool down from a higher temperature than the skin, and thus is put into tension and tends to crack. It has been reported that the strength of concrete in the core of thick sections is weaker than small test specimens of the same mix, and this is probably due to this effect. The author does not believe that designers need to worry overmuch about such a loss of strength; the material factors of the codes of practice may be assumed to include for this effect. This internal cracking is likely, however, to create leakage paths within the concrete that may cause problems for underground structures subject to water pressure.

The temperature gradients that cause this cracking may be reduced by insulating the free surfaces of the pour, although this can have the effect of increasing the maximum temperature attained by the setting concrete, and may exacerbate some of the other causes of cracking described hereafter.

3.6.3 The cracking of concrete caused by the presence of reinforcement

If the steel reinforcement in a concrete mass is entirely enclosed, there is no reason why it should not follow closely the rise and fall of the concrete's temperature. If it does so, as the coefficient of thermal expansion of steel is similar to that of concrete, it cannot place any stress on the concrete. However, if the reinforcement is in contact with the air, as is the case for starter bars, the high thermal conductivity of steel will cause the reinforcement to remain cooler than the surrounding concrete, at least in the vicinity of construction joints. The concrete will then cool from a higher temperature than the steel. This can give rise to prejudicial cracking parallel to the reinforcing bars, particularly when large-diameter bars are used in thin concrete sections. Such cracking is also made worse by the natural drying shrinkage of the concrete, and it is wise to limit the diameter of reinforcing bars to one tenth of the local concrete thickness. This limit is also important for development of sound bond between the concrete and the steel.

3.6.4 The cracking of a concrete structure built in stages due to the restraint by previously hardened concrete

This is probably the most widely experienced form of cracking due to the exothermic reaction. It is most commonly encountered in walls cast onto mature concrete footings. Most engineers will have seen the characteristic vertical cracks, usually at about 2–3 m centres, that disfigure these walls. As before, the concrete heats up while it is still plastic, becomes progressively harder and cools, shortening as it does so. This shortening is restrained by the footing, and consequently the wall cracks. The theoretical maximum amount of cracking is given by the simple expression:

$d = \alpha \times t$ where d = total shortening; α = coefficient of expansion of concrete; t = temperature drop.

Thus if the temperature drop were to be 40°C and $\alpha = 12 \times 10^{-6}$ the total potential shortening is 0.48 mm/m.

However, the immature concrete of a reinforced wall is capable of straining plastically in tension, and it only cracks once this tensile strain capacity has been exhausted. Thus the actual cracking is likely to be less than half the maximum figure calculated above.

The restraint to shortening reduces with height above the footing, due to the shear flexibility of the wall. Consequently at some height above the footing the cracks will disappear. Just above the footing, the bond of the wall to the footing tends to spread the tensile strain evenly along the wall, totally inhibiting cracking or creating fine closely spaced cracks. Thus the characteristic crack pattern due to restrained heat of hydration cooling is as shown in Figure 3.10.

Cracking due to this cause is particularly prejudicial to concrete that must resist water pressure, such as cut-and-cover tunnels, because the cracks penetrate through the full thickness of the wall and thus create leakage paths. Problems also arise for bridge decks cast in-situ, in short segments, such as those built by the free cantilever method (*9.3.8*). The wide top slabs that are cast against each other are in effect horizontal walls, and tend to crack due to the restraint of the preceding segment.

The provision of horizontal reinforcement in the wall cannot stop this tendency to crack, but the bond between the immature concrete and the steel can control the width of the cracks. If the reinforcement consists of large-diameter bars, the bond lengths are too great to limit effectively the crack widths. If, on the other hand, there is too little reinforcement, the concrete will ignore it and behave as if it were unreinforced (*3.7.2*). The optimum reinforcement consists of small bars, generally T12 or T16 at

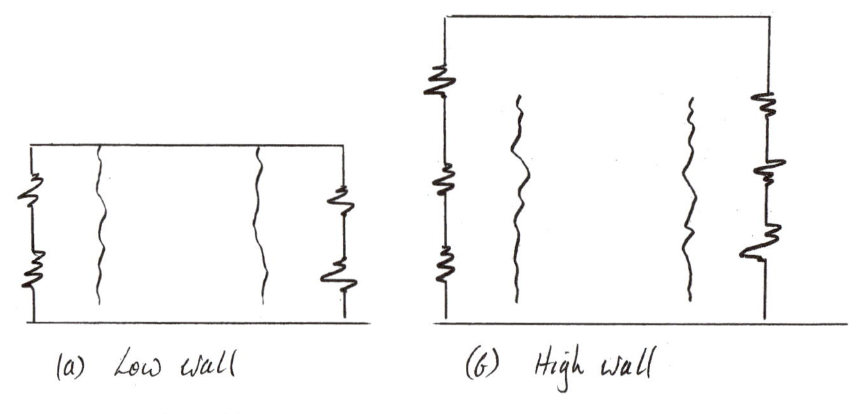

Figure 3.10 Typical heat of hydration cracking of walls

relatively close spacing, typically 100–150 mm, on each face. The various codes of practice give empirical formulae for justifying the reinforcement required to limit the cracks to some nominal width. The usual aim of design is to limit the width of the cracks to 0.2 mm. Conventional wisdom holds that this width of crack will be sealed by autogenous healing of the concrete or clogging of the cracks.

3.6.5 *The cracking of concrete due to different thickness in the same pour*

If a concrete cross section includes areas of significantly different thickness, they will have quite different heating and cooling curves, Figure 3.9. The result is that the thicker areas will still be cooling when the thinner areas have stabilised, causing cracking of the former. The example is quoted in *12.6* of a bridge deck consisting of a thin top slab and thick downstand ribs that suffered severe cracking from this effect.

This mode of cracking is similar to that which occurs in cast iron. Iron founders know that the detailed design of a section should avoid sudden changes in thickness, and stress-raising sharp re-entrant corners. Similar care is required in the design of concrete sections.

Another example of the practical significance of this effect is in the preliminary project designed by Benaim for 164 m concrete spans for the Storabaelt crossing, Figure 9.12 (e). The variable-depth bridge deck was to be built by the counter-cast segmental method, using 8 m long segments. As there was a significant thickening of the base of the web of the deepest segments, it was feared that the delayed cooling and shortening of this thick area may destroy the accuracy of the counter casting. Consequently water pipes were to be cast into the thicker areas to keep them at the same temperature as the thinner slabs during the critical cooling period.

3.6.6 *Cracking caused by the concrete jamming onto an inflexible mould*

If concrete is cast in a rigid mould that is trapped between concrete members, the mould must be released before the concrete starts to cool down, otherwise it will cause the concrete to crack. A typical example is a segment of a box section deck poured in a steel mould for the counter-cast method of construction, Figure 14.4. The steel mould

Figure 3.11 Singapore Central Expressway (Photo: Robert Benaim)

for the soffit of the top slab is trapped between the webs and must include a flexible component or be released to allow the slab to shorten as it cools.

3.6.7 Measures to limit heat-of-hydration cracking

The problem for the designer is that cracking due to this cause is not reliably predictable. The immature concrete can absorb significant tensile strains without cracking. Relatively trivial factors may push the concrete over its limit, and trigger cracking. Benaim designed the twin-cell cut-and-cover tunnel of the Singapore Central Expressway, Figure 3.11, which was many hundred metres long, with walls and slabs generally 1 m thick or more. The roof slabs of both cells were cast simultaneously, but whereas one cell was virtually uncracked, the other exhibited significant restraint cracking. The only difference between the two cells was the degree of shading from the sun offered by the sheet piling lining the excavation!

The designer has the choice between:

- specifying considerable tonnages of reinforcement as a precaution to control cracking that may not occur;
- modifying the concrete mix to reduce the peak temperature reached;
- cooling the concrete during mixing;
- cooling the concrete after casting;
- adding admixtures which seal the cracks;
- in a repetitive structure, adopting an observational technique, and adding reinforcement as necessary to control the observed crack pattern.

a) Precautionary reinforcement

The technical literature or the appropriate codes of practice define the area of reinforcement required as a function of ambient temperature, concrete mix design and thickness of the concrete. It should be noted that this reinforcement will control the width of any cracking, but will not eliminate it. This option may involve remedial measures to seal or hide the cracks.

b) Modifying the mix design

- The cement content of the mix should be the lowest required that provides adequate strength and durability.
- The cement must not arrive hot on site.
- It may be possible to specify the fineness of grinding of the cement, with a coarser particle size slowing down the rate of gain of strength, and hence reducing the peak temperature of the concrete.
- Some of the Portland cement may be replaced by pulverised fuel ash (PFA) or ground, granulated blast furnace slag (GGBFS). (Although this lowers the heat of hydration, there is some evidence that it does not reduce cracking as expected.)

c) Cooling the concrete during mixing

- Iced water may be added to the concrete mix to reduce the starting point of the temperature rise. However, as the weight of water in the mix is relatively small, this measure has a limited effect.
- Wet the aggregate and blow air through it. This cools the aggregate by stimulating evaporation. This is of limited effectiveness.
- Cool the aggregate with liquid nitrogen. This is very effective in cooling the concrete, but approximately doubles its cost.

d) Cooling the concrete after casting

Pipes may be cast into the concrete so that water may be circulated to remove the heat of the setting concrete. Such a measure, sometimes associated with other options as described above, is used for structures that must be prevented from cracking, such as immersed tube tunnels.

e) Adding admixtures

Admixtures can be very effective but very significantly increase the cost of the concrete.

f) The observational technique

In a repetitive structure, this entails casting the early prototypes, observing where unacceptable cracking occurs, and then adding reinforcement to successive pours until the cracking is satisfactorily controlled. For instance, in the typical wall problem described above, it may be seen that cracks start at a certain distance above the base of the wall, widen to a maximum, and fade out below the top. Reinforcement may then be added only in the areas where the cracking is wider than is considered acceptable.

3.7 The ductility of reinforced concrete

3.7.1 General

Not only would a plain concrete beam have very little strength in bending, but as concrete fails in tension in a brittle mode, there would be no warning of impending collapse. These problems are overcome by reinforcing the concrete beam with ductile steel bars. When over-stressed the steel yields, and before failing will undergo a degree of strain hardening. Consequently, if a suitably reinforced concrete beam is subjected to overloading, as the steel yields the deflection of the beam will increase notably, and it will crack grossly on the tension face. Furthermore, due to the strain hardening of the steel reinforcement, the beam could sustain a small further increase in load before collapsing. This ductile behaviour of reinforced concrete is a very important characteristic, not only giving visible warning of overloading, but also making it very tolerant of simplified methods of analysis, and allowing practical arrangements of reinforcement.

3.7.2 Minimum reinforcement

For the reinforced concrete to exhibit ductile behaviour, it must contain a minimum percentage of reinforcement. This may be understood by considering an unreinforced concrete beam of rectangular cross section. When subjected to a bending moment, this beam will fail in a brittle mode when the tensile stress in the concrete reaches its limiting value. We may call this limiting bending moment $M_{ult,c}$.

If the beam were to be reinforced with less than a critical percentage of steel, the moment of resistance of the reinforced section, assuming that the concrete is cracked in tension and all the bending strength is provided by the steel reinforcement, $M_{ult,st}$, would be less than that of the unreinforced beam ($M_{ult,st} < M_{ult,c}$). The beam would still fail at $M_{ult,c}$ in a brittle mode, the steel instantly yielding and rupturing. If the amount of reinforcement is greater than this critical value, when the moment reached $M_{ult,c}$ the concrete would rupture, but the moment would be held by the reinforcement. The critical percentage of high yield reinforcement for a bending member is generally assumed to be about 0.15 per cent of the concrete area, placed close to the extreme fibre in tension. If mild steel is used, the percentage is increased in proportion to the yield strength of the bars.

A similar consideration arises for members stressed in direct tension, where the minimum percentage of high yield reinforcement required to give ductile behaviour is widely accepted to be 0.6 per cent of the concrete area, distributed symmetrically in the section.

3.7.3 Redistribution of bending moments

In continuous beams made of reinforced concrete, the bending moments over the supports are generally greater than the moments in the span. When the loads are increased from working values towards the ULS, it is convenient to allow hinges to

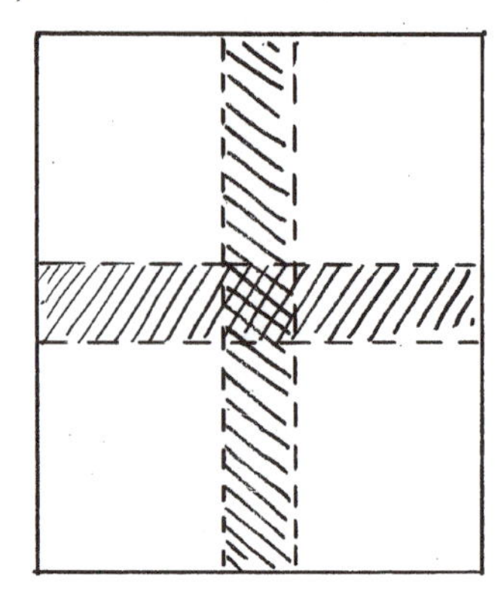

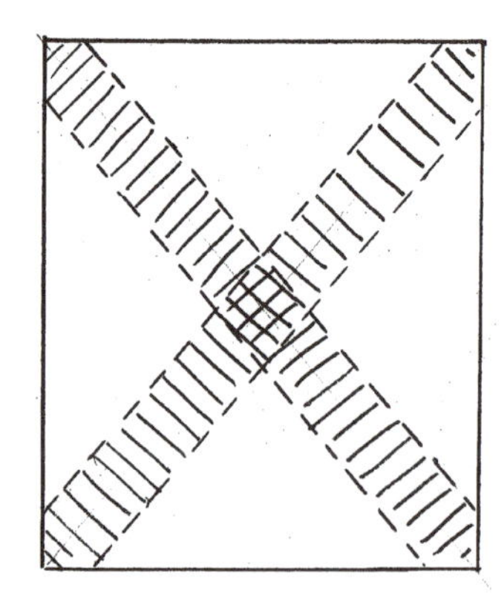

Figure 3.12 Alternative reinforcement of two-way slab with beam strips

form at the supports, and then to carry the further increase in moments in the spans. This is only possible due to the ductile behaviour of reinforced concrete.

However, this implies that the concrete beam will crack over the supports earlier and more severely than in the spans. The designer must decide if such cracking is acceptable.

3.7.4 *Flexibility in the arrangement of reinforcement*

An example of the flexibility in design given by the ductility of reinforced concrete is the two-way spanning slab. This slab may be analysed elastically, and the reinforcement arranged according to the results of that analysis. It is also possible to arrange the reinforcement in ways that take no account of the elastic analysis, for instance in orthogonal cruciform beam strips, or in two diagonal bands, Figure 3.12. As long as the analysis takes into account the pattern of the reinforcement, this will be quite safe. As the slab is loaded up, initially the stresses will follow the elastic pattern. However, once the concrete starts to yield in tension or to crack, the loads will be transferred to the strong bands created by the reinforcement. The slab has effectively been transformed into a beam and slab system.

Another example is the deep beam described in *3.11.2*. If it had been made of a brittle material, the elastic analysis would have described the only possible mode of behaviour. Once the stresses calculated by this analysis reached the breaking stress of the material, the beam would have failed. The ductility of the steel used to reinforce concrete makes it possible to organise the reinforcement in a variety of ways.

If the pattern of reinforcement for either the deep beam or the slab is poorly chosen, although the strength may be adequate, there is a possibility that the structure will crack excessively. It requires experience to choose sound reinforcement arrangements.

3.8 Imposed loads and imposed deflections

There is a fundamental difference between imposed loads and imposed deflections that frequently is understood inadequately by practising engineers. For instance, when a load P is applied to the end of a reinforced concrete cantilever it will deflect, Figure 3.13. If it is inadequately reinforced it will yield and collapse. Assume on the other hand

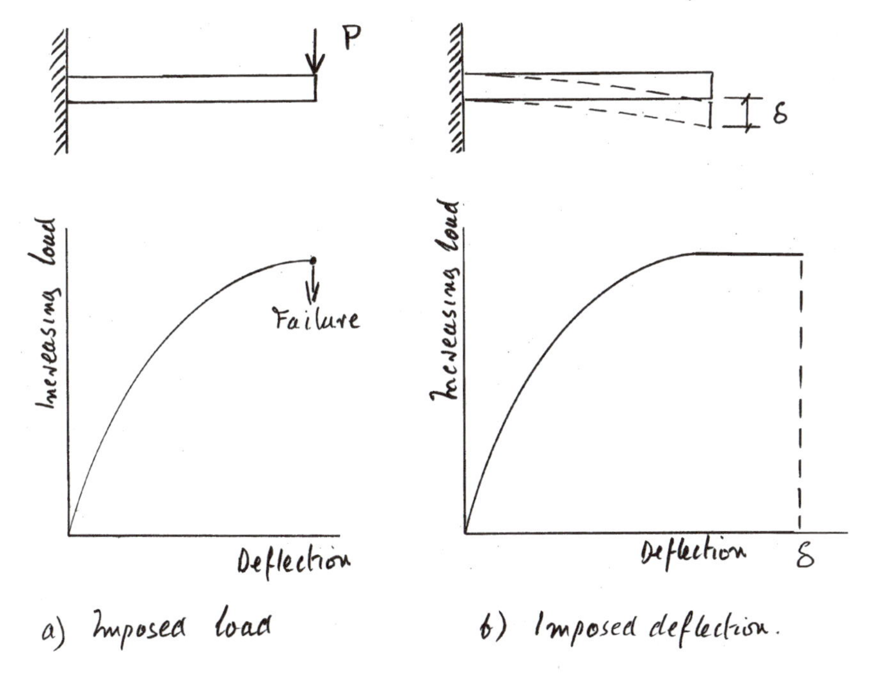

Figure 3.13 Imposed loads and imposed deflections on a cantilever

that the end of the cantilever is displaced downwards by a load applied at the end to a predetermined deflection δ. If the deflection is large enough the reinforcement will yield, but the cantilever will not fail, as it is not allowed to deflect any further. In fact, the end load will be sustained at that required just to maintain the deflection. This is of course a result of the ductility of reinforced concrete; if the cantilever had been made of brittle material, imposing a large deflection on it would also have caused it to fail.

There are many instances in the design and analysis of structures where such fixed displacements are the relevant form of loading. The structures may be damaged by them, but as long as they have sufficient ductility they cannot fail. However, it is possible that the damage caused may be such as to render the structure unserviceable. Consequently, they need to be considered differently; the ULS is not the relevant criterion governing design.

For instance, the settlement of one pier of a bridge deck may well cause it to crack, but if the deck respects normal design criteria and is ductile, it will not fail. It may however be damaged to the point where it becomes unserviceable or requires repair. Another very common occurrence of this situation is where a bridge deck is pinned to a series of piers (7.9), and the piers are subject to bending due to the changes in the length of the deck. The change in length, whether it is caused by temperature variations or by creep and shrinkage of the concrete, is a defined displacement and cannot pull the piers over, although it can damage them.

An instance where confusion may well arise is where the deck rests on the pier through a sliding bearing. The bearing exerts a longitudinal force on the pier equal to the vertical reaction multiplied by the friction coefficient. Although the action of the bearing on the pier is often assumed to be an applied force, and the pier designed to the ULS, in fact the movement of the bearing is limited by the length change of the deck, and the pier could not be pulled over by this force. The critical design criterion becomes one of limiting the cracking of the pier stem.

3.9 Creep and relaxation of concrete

3.9.1 Creep

Creep is the deferred deformation of concrete. When a concrete structure is loaded, it undergoes an instantaneous elastic deformation. This will then gradually increase over a number of years, usually reaching a value that is two or three times the elastic value, Figure 3.14 (a).

The ratio of deferred deformation to elastic deformation is known as the creep coefficient, or ϕ. The total deformation of a concrete specimen is thus

$$\delta_t = \delta_e + \phi \times \delta_e, \text{ or } \delta_e(1+\phi),$$

where δ_e is the elastic deformation. The value of ϕ depends on the thickness of the concrete, the relative humidity of the air, the concrete mix design and on the age of the concrete at first loading. Appendix B of BS5400: Part 4: 1990 gives guidance on the calculation of ϕ. Practical values of ϕ vary between 0.8 – 1, for mature precast concrete, to 1.5 – 2.5 for cast-in-situ concrete.

This formulation of creep is known as 'linear', in that it is proportional to the elastic deformation. As a result, the deferred deflection of a continuous beam does not change the distribution of bending moments in the beam (*6.21*).

3.9.2 Relaxation

A closely related phenomenon is concrete relaxation. When concrete is subjected to a fixed displacement, the force or moment generated initially is related to the short-

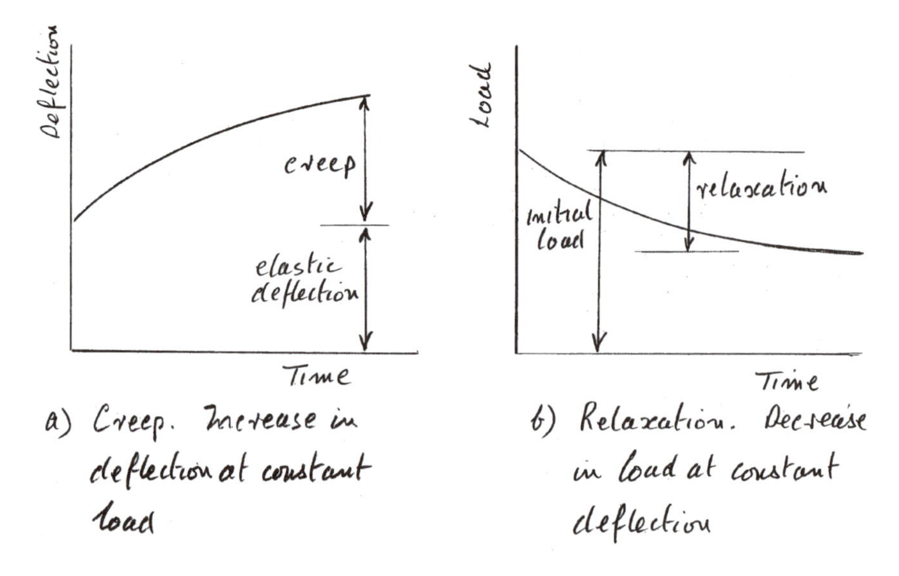

Figure 3.14 Creep and relaxation of concrete

term Young's modulus. If the displacement is maintained the load then gradually drops off, with the final load being related to the creep characteristics of the concrete, Figure 3.14 (b).

The degree to which the load relaxes depends on the rate at which the displacement is applied. For instance, if the displacement is applied quickly, in one phase, the long-term final load will typically be only 20 per cent of the initial load. It is for this reason that it is not worth trying to alter the bending moments in a concrete beam by jacking the supports, as is routinely done for steel bridges; the jacked-in bending moments creep away.

If, at the other extreme, the displacement is applied very slowly, over a period of months or years, the long-term final load will typically be of the order of 40 per cent of the load calculated on the assumption of elastic behaviour.

In most instances that affect prestressed concrete bridges, the rate of application of the displacement is between these two extremes. For instance, piers that are pinned to a bridge deck, as described in 7.9, are subjected to imposed displacements by the shortening of the deck. This shortening has an initial rapid component as the prestress is applied to the deck, followed by a slow component as it continues to shorten under the influence of creep and shrinkage.

All such calculations are approximate, and for preliminary calculations at least, one could do worse than to assume that the final load will be, for a rapidly applied displacement, between upper and lower bounds of 20 per cent and 30 per cent of the initial load, and for slow application of the displacement, between 40 per cent and 50 per cent of the load calculated on the assumption of elastic behaviour. The piers will also be subjected to cyclical movements of the deck, some of which are short term, and some seasonal. The effect of the short-term loading should be calculated using a Young's modulus that is slightly below the value used for live loading, while the effect of seasonal movements should be calculated with a lower E. A reasonable compromise is to use an E of 75 per cent of the short-term value for all temperature movements.

For a more mathematical definition of these two extremes, see reference [8]. Alternatively, the data given in Appendix B of BS5400: 1990: Part 4 can be used as the basis for repetitive calculations that appear to give an accurate representation of the evolution of the forces. However, it should be borne in mind that the reliability of the result of this calculation depends on the validity of the initial data, which is at best one view of a very extensive body of experimental data.

3.10 Truss analogy

3.10.1 Truss analogy applied to beams

Truss analogy is very useful in providing a tool to design the shear reinforcement for reinforced and prestressed concrete beams. The concrete beam is likened to a truss, in which the compression boom and the inclined compression web members are in concrete, while the tension boom and the tensile web members consist of reinforcement.

In a reinforced concrete beam, the compressive web members are usually assumed to be inclined at 45°, as this is the inclination of the principal compressive stress. In a prestressed concrete member, the longitudinal compression flattens the angle of the principal compressive stress, and the web struts may be assumed to lie at a flatter

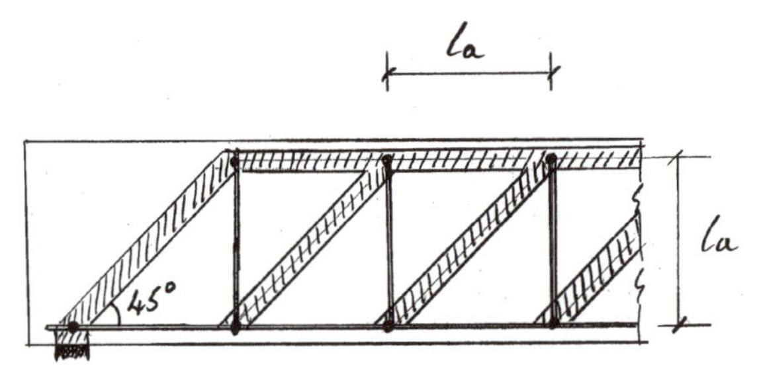

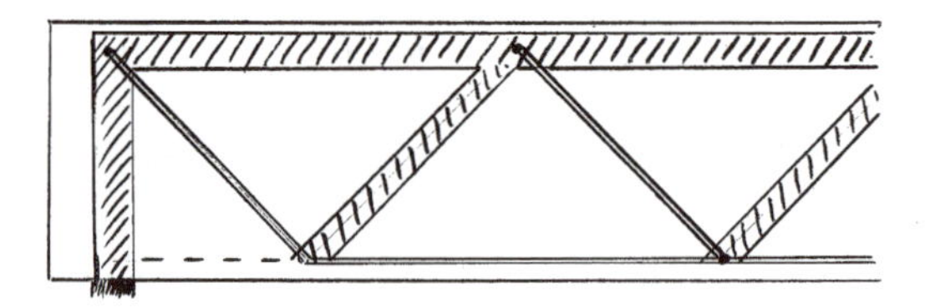

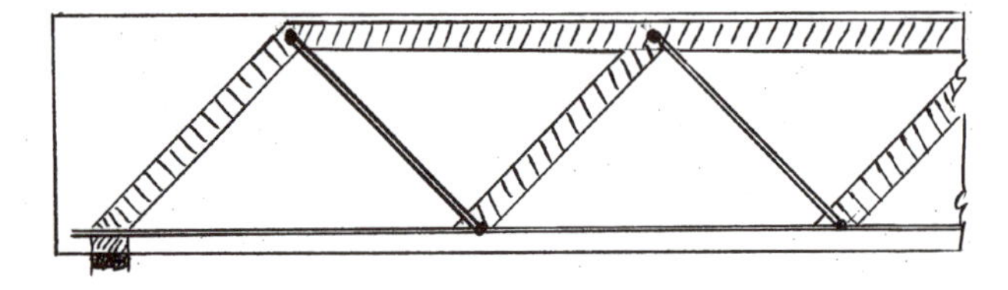

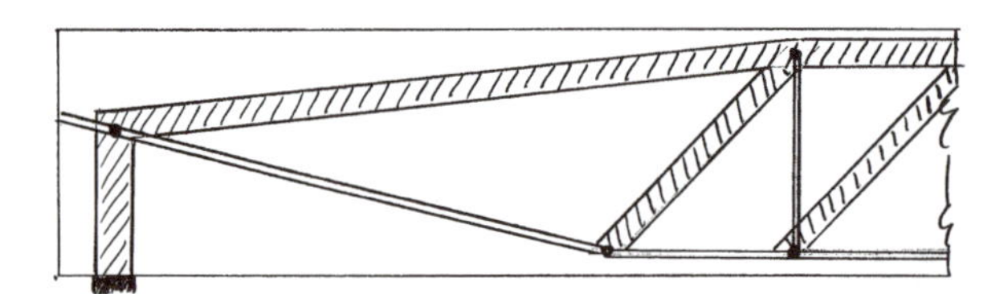

Figure 3.15 Truss analogy for beams

angle, although in practice 45° is generally adopted for ULS design as a conservative simplification. In short deep beams, such as pile caps, the angle of the struts may be steeper than 45°. The web tension members may be vertical, creating an 'N' truss, or inclined creating a Warren truss. The pitch of the truss is related to the bending lever arm; that is the distance between the centroid of the compression boom and the centroid of the tensile reinforcement. For an 'N' truss with the compressive members inclined at 45°, the most common assumption, the pitch of the truss will be equal to this lever arm, Figure 3.15 (a).

As is inherent in the action of the truss, each vertical tensile web member carries the full shear force acting at that section. The reinforcement required to carry this shear force is not usually concentrated at the web nodes, but spread out over the pitch of the truss. Consequently, the force per metre run of beam to be provided by the shear reinforcement is basically defined by Q/l_a, and the area of this reinforcement per metre run of beam is $Q/l_a\sigma_{st}$, where Q is the applied shear force, l_a the lever arm (= the pitch of the truss), and σ_{st} the stress at which the steel is allowed to work. This basic simplicity may be modified by the codes of practice which introduce empirical factors based on experience derived from testing, and which also introduce minimum quantities of shear reinforcement.

If bent-up bars are to be used to carry some of the shear force, together with the inclined compression struts they form a Warren truss, Figure 3.15 (b) and (c). The 'N' truss described above and the Warren truss may be superimposed, each carrying their proportion of the shear. Bent-up bars went out of fashion, because if they were arranged in a dense array, they made it difficult to cast and compact the concrete. However, if the designer knows what he is doing, there is no fundamental reason why they should not be used. A variation of the bent-up bar analogy is when the beam is prestressed by tendons that are swept up towards the beam supports to carry a proportion of the shear, a reinforced concrete 'N' or Warren truss carrying the remainder, Figure 3.15 (d).

3.10.2 Beam end problems

In a simply supported beam, the bending moment falls to zero at the bearing, which could lead to the erroneous conclusion that the requirement for reinforcement also falls to zero. However, the 'N' truss analogy as shown in Figure 3.15 (a) makes clear that the load is carried to the bearing by the last inclined compression strut. From this analogy, it is clear that the reinforcement bars that constitute the last bottom boom tension member are stressed at the beam end, and must be fully anchored beyond the centre line of the bearing. This requirement is not well represented by some codes of practice.

If the bars lie directly above the bearing, they will be compressed by the vertical reaction of the bearing, and will anchor in a short length. However, in many cases, the beam is wider than the bearing, and some of the bars will not be so confined, and must be fully anchored, preferably by a hook.

There are alternative methods of terminating a beam. For instance if the Warren truss model shown in Figure 3.15 (b) was to be shifted half a pitch leftwards, the load to the bearing would be carried by the last inclined tension member, Figure 3.15 (c). No design bottom boom tension reinforcement would then be required in the last bay; in fact the concrete below the last bent-up bar could be removed. Clearly the last inclined tension member would have to be fully anchored beyond the truss node at the beam end.

Truss analogy allows many of the problems that arise at beam ends to be given a logical solution.

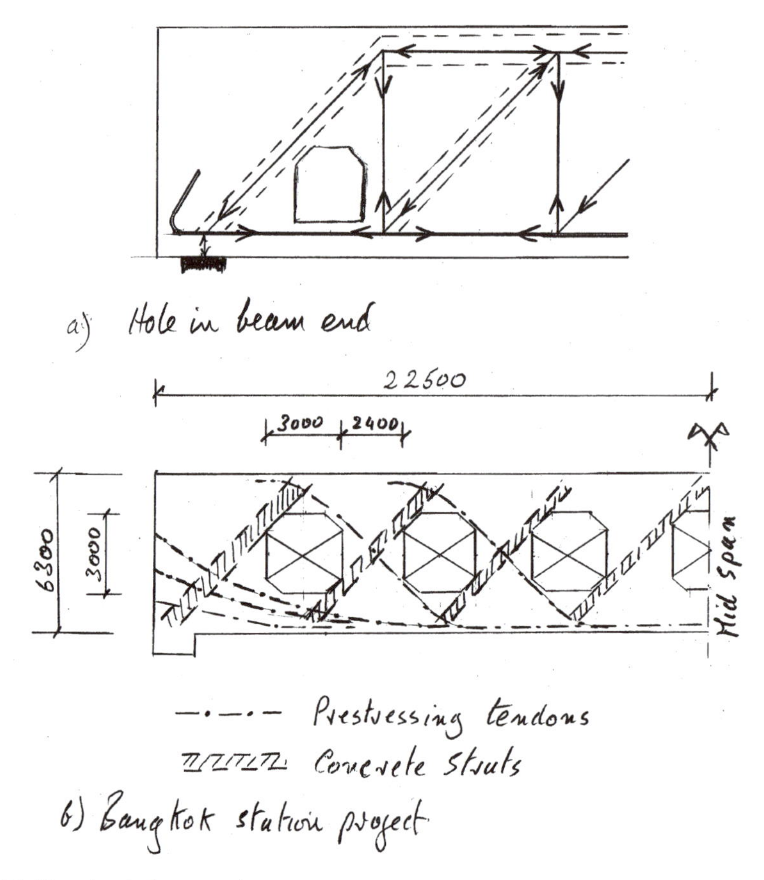

Figure 3.16 Opening in beam web

3.10.3 Openings in beams

It may be necessary to locate an opening in the web of a beam near the support. If it is possible to choose the shape and position of this opening, the strut-and-tie analogy described shows where it will have the minimum effect, Figure 3.16 (a). However, if the opening has to go in an inconvenient place, the loads may have to be carried round the hole in Vierendaal action.

A further example of this was a project for a bridge deck carrying an elevated urban railway shown in Figure 3.16 (b). The deck was in the form of a large prestressed concrete box girder, with the railway inside and traffic on the top flange. At stations it was necessary to create regular openings for passengers to exit from the train. The detailed shape of these openings, as well as their exact spacing, could be tuned so that they left a viable Warren truss system intact, with prestressed concrete tension members and compressed concrete struts. If this had not been possible, it would have been necessary to adopt a far more heavily reinforced Vierendaal girder system.

3.10.4 Halving joints

Halving joints are used frequently at the ends of statically determinate bridge beams, in order to reduce the depth of the support system. Truss analogy is the best tool for designing the reinforcement in these complicated zones, and gives direction on its arrangement and curtailment. A typical truss analogy for a halving joint is shown in Figure 3.17 (a). The analogy makes clear the following points which may not be obvious:

- At point A, the bottom boom tension steel must be fully anchored, by turning up the bars for instance.
- Member B, the bottom boom tension steel for the reduced height section must be carried back so that it can transfer its force into the truss system. Note that anchorage lengths have not been shown, and this steel would need to be fully anchored beyond point C. This force also increases the compression force in member D, the lower part of the last compression strut of the deeper section.
- Member E, the last web tensile member, has its tension significantly increased due to the increased compression in D.

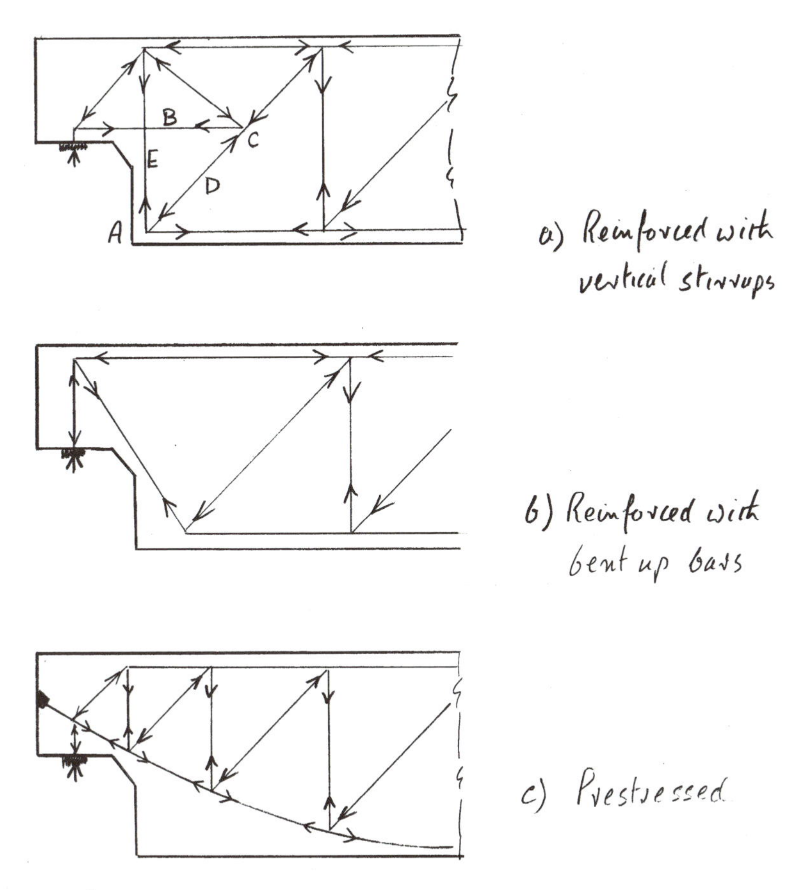

Figure 3.17 Halving joint

This example makes clear that it is essential to follow through all the forces that arise from the analogy. The force in the lower tension boom of the reduced-height section must not be vaguely anchored into the mass of the deeper section; it is essential that the force in it be correctly transferred to the truss system.

Frequently, there is not just one truss solution. Figure 3.17 (b) shows an alternative truss analogy, in which the shear forces are carried up by bent-up bars, while Figure 3.17 (c) shows a similar halving joint which is prestressed. It is possible to share the shear force between several different trusses, which may help in reducing congestion of reinforcement.

In dealing with complex areas where there are large stress concentrations, it is wise to be conservative. For the design of halving joints in prestressed concrete bridges, the author tended to superimpose a prestressed and a reinforced concrete truss, and to ensure that each system carried say 65–70 per cent of the total load, giving an additional 30–40 per cent safety factor. For instance, to limit the risk of cracking at the re-entrant angle in service, the prestressed truss may be designed to carry the whole working load shear force, with the reinforced concrete truss providing the safety factor. In being conservative, the designer must make sure he is not causing such congestion of reinforcement that it is too difficult to cast and compact the concrete. Sound, well-compacted concrete is the essential precondition to durability and strength.

For such complicated details it is advisable to check that the reinforcement derived from the truss analogy bears a reasonable correlation with that derived from conventional bending and shear theory.

3.10.5 Loads applied close to a support

The effects of loads applied close to a support are best understood with truss analogy. When loads are applied within approximately a lever arm, l_a, from the support, the load travels directly to the support without needing to be suspended by stirrups, Figure 3.18. Theoretically, no additional shear reinforcement is required to carry this load. However, load tests have shown that beams with large loads close to the support may fail in shear with a failure plane that is steeper than 45°, and that shear steel may well be necessary. The inclined compression strut must be balanced by additional compression in the top boom and tension in the bottom boom.

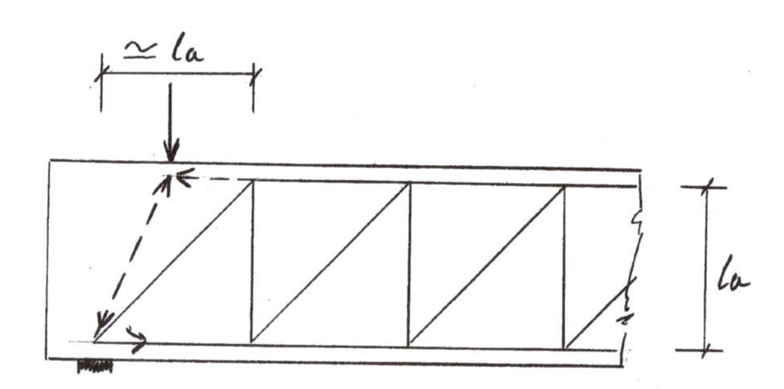

Figure 3.18 Load applied close to support

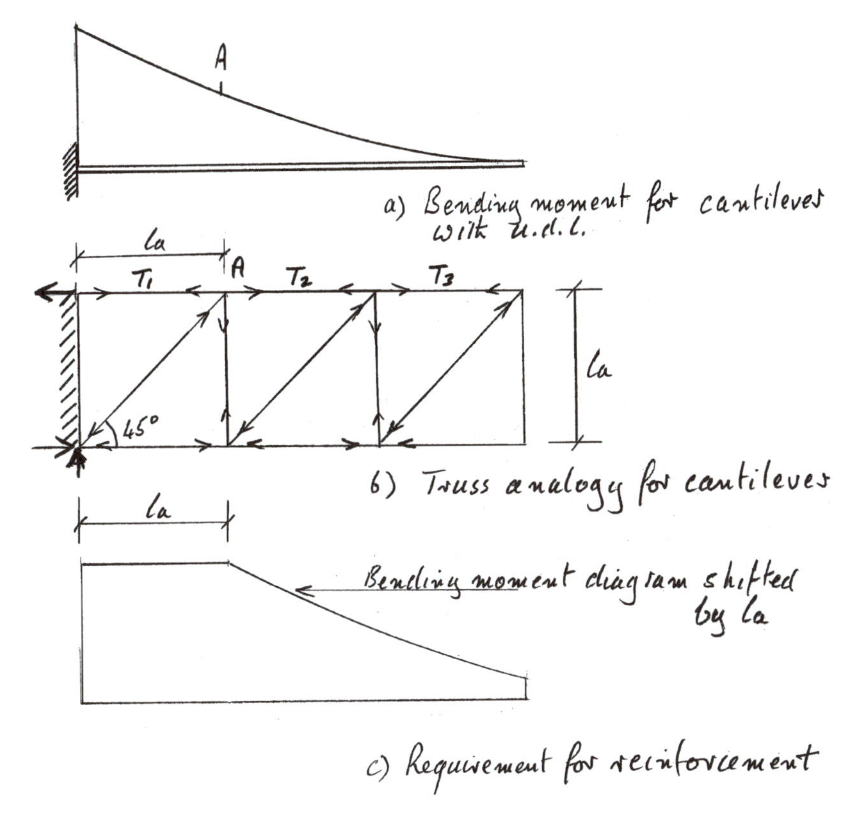

Figure 3.19 Curtailment of reinforcement

3.10.6 Curtailment of reinforcement

The strut-and-tie analogy assists in understanding the mechanics of the curtailment of reinforcement. Consider the bending moment diagram for a uniformly loaded reinforced concrete cantilever, Figure 3.19 (a). By point A, the bending moment will have fallen considerably, and hence the theoretical requirement in reinforcement will also have fallen, with several bars being curtailed. However, according to the truss analogy, Figure 3.19 (b), the maximum tension T_1, and consequently the maximum reinforcement, are both constant up to point A. Hence the reinforcement requirement needs to be shifted horizontally by a lever arm with respect to the bending moment diagram, Figure 3.19 (c). The same reasoning holds true for simply supported and continuous beams, and is the basis of the requirements of some codes of practice.

It should be noted that shifting the bending moment by a lever arm is only rational if the beam has been reinforced with vertical stirrups, and the 'N' truss analogy is applicable. If bent-up bars are being used with a Warren truss analogy, or if a shear field analogy is being used, the mechanics of curtailment should be reviewed.

3.10.7 Hanging steel

When loads are applied at the bottom of a reinforced concrete beam, reinforcement must be provided, over and above the shear reinforcement, to carry the forces up

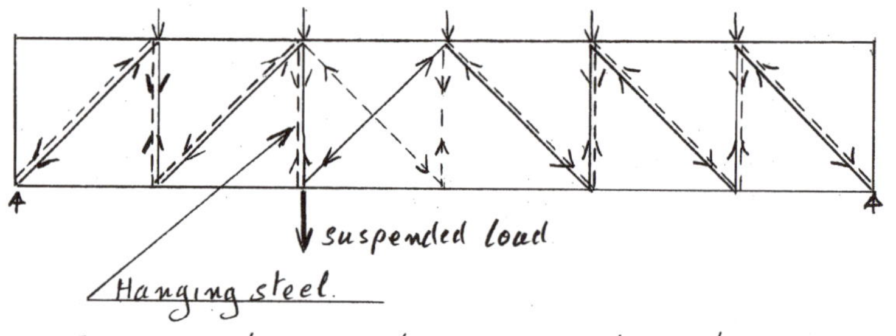

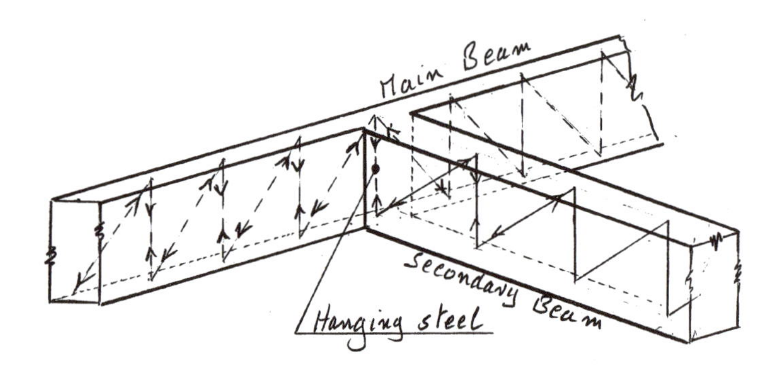

Figure 3.20 Hanging steel

into the beam. Truss analogy provides the best tool for understanding the problem, and detailing the necessary steel. Figure 3.20 (a) shows that the hanging steel must be taken right up to the top of the beam for the forces to be correctly introduced into the truss.

A common case of the need for hanging steel is when a secondary beam frames into a main beam. Truss analogy indicates that the shear from the secondary beam is applied by the last compression strut to the bottom of the secondary beam, at its junction with the main beam. Consequently, reinforcement must be provided at the junction of the two beams to hang this shear force, and carry it up to the top of the main beam, Figure 3.20 (b). A typical example of this in bridge design is where the webs of a prestressed concrete deck meet a pier diaphragm that is cantilevering out beyond the bridge bearings (9.6).

3.10.8 Shear fields

For beams conventionally reinforced with stirrups, an extension of the concept of truss analogy is to consider a large number of superimposed trusses, where the vertical tension members each consist of a single row of stirrups associated with their inclined compression member, Figure 3.21 (a). In the simple truss analogy, each web member

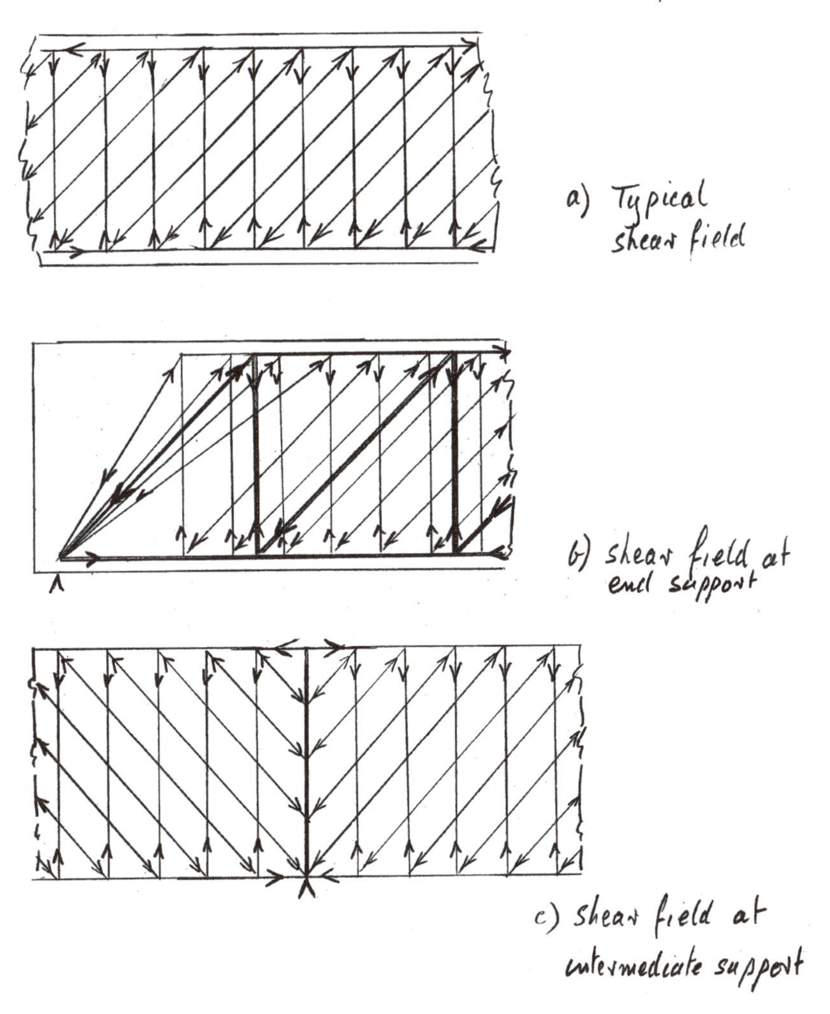

Figure 3.21 Shear fields

then represents the resultant of a number of sub-trusses. At the end of simply supported beams, this version of the analogy leads to a possible arrangement of the truss as shown in Figure 3.21 (b).

At intermediate supports, where the shear force either side of the support is equal, the arrangement shown in Figure 3.21 (c) is appropriate. If the shear force is unequal either side of the support, the equal portion may be considered as in Figure 3.21 (c), while the unbalanced portion would be as in Figure 3.21 (b). In particular, this concept is used in the design of bridge deck diaphragms (9.6) and in other cases when one beam frames into another and it is helpful to consider that the end reaction is distributed up the end face of the beam.

Clearly, the simple logic leading to the rules on curtailment of reinforcement over the piers of continuous beams, described in *3.10.6*, is upset by this concept of a shear field. It would be interesting to see the results of tests of beams where the reinforcement is curtailed in accordance with the logic of shear fields.

3.11 Strut-and-tie analogy

3.11.1 General

The strut-and-tie analogy is closely related to the truss analogy described in *3.10*. While truss analogy is applicable to regular beams, strut-and-tie models are used in complex areas where bending theory, which assumes that plane sections remain plane when stressed, is not applicable. The designer must imagine a viable strut-and-tie system, where the struts are in concrete, and the ties are materialised by reinforcing bars or prestressing tendons. It is essential that the strut-and-tie system is set up rigorously, with all forces resolved and all nodes in equilibrium. It is also essential that concrete struts have a cross-sectional area adequate to carry the loads, that the tensile members are large enough to fit in all the required reinforcing bars and that all bars are fully anchored beyond the truss node.

In general, if a viable strut-and-tie system can be drawn with members of adequate size, and all the reinforcement can be anchored, the structure will be safe, even if the flow of forces is quite different to that defined by elastic analysis. An example is the deep bracket designed by the author, where the loads of a column in a tall building were

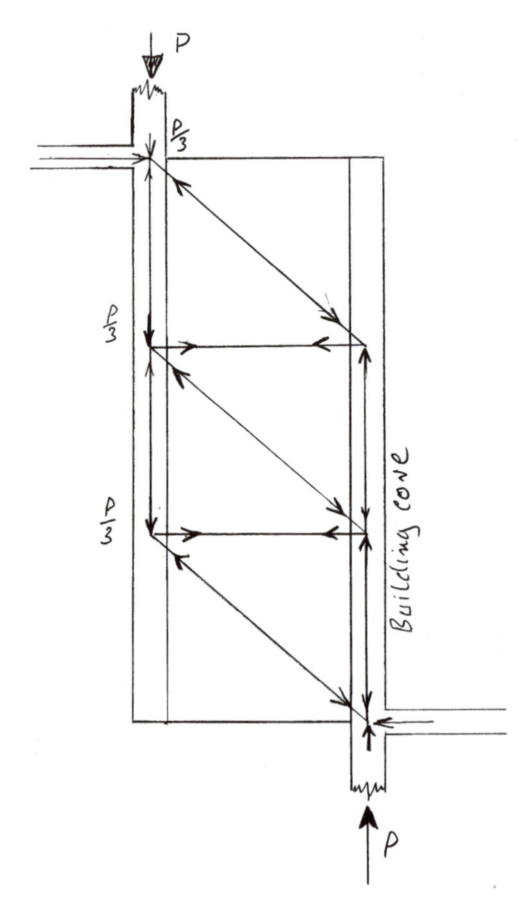

Figure 3.22 Tiered strut-and-tie brackets

transferred to the building core by a tiered series of strut-and-tie brackets, each sharing the total shear force equally, Figure 3.22. This adaptability of reinforced concrete is due to the ductility described in 3.7. However, concrete is only ductile in this sense if it cracks due to extension of the reinforcement before it fails in concrete compression, emphasising the importance of ensuring that the compressive members of the truss are adequately dimensioned.

Although concrete is adaptable and a variety of strut-and-tie models may be used to solve any particular problem, the model that mirrors most closely the elastic flow of forces is least likely to suffer prejudicial cracking, although it may not be the most economical or buildable option. Choosing the most appropriate model requires some experience.

The designer of reinforced and prestressed concrete structures will be faced with many examples of these special areas, such as bridge deck diaphragms, brackets and halving joints, sudden changes in section or the zones immediately behind concentrated loads such as the attachment of stay cables to a bridge deck.

3.11.2 Limits of elastic analysis

The stresses in an elastic material may be analysed using elastic methods, such as finite element analysis. However reinforced concrete is not an elastic material, such as steel for instance, but a composite and such methods are not strictly applicable. An example is the analysis of a deep beam loaded with two concentrated loads, Figure 3.23 (a). An elastic analysis of the beam would show a stress diagram that had a very deep compression zone, and a shallow tensile zone, yielding a lever arm that did not exploit the full depth of the beam. If the beam was to be reinforced in accordance with this elastic stress diagram, it would require an area of tensile reinforcement spread over the bottom quarter of the beam.

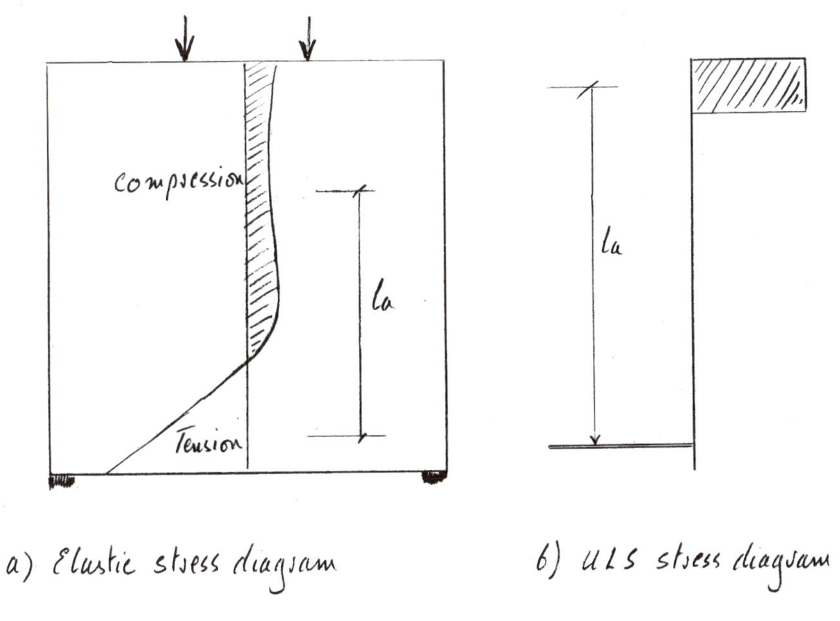

Figure 3.23 Stress diagrams for deep beam

However, if one were to design this bending reinforcement at the ULS, as described in *3.4*, using the full height of the beam to define the lever arm, Figure 3.23 (b), the beam would be safe even though the reinforcement required would be concentrated at the bottom fibre of the beam, and would, due to the greater lever arm, be significantly less than that defined by the elastic analysis. The elastic analysis described the flow of forces in the deep beam up to the moment when the concrete started to yield in tension. From this time on, the behaviour of the beam was no longer elastic, and the forces found new routes. When the beam is not reinforced in accordance with the elastic stresses, the yielding of the concrete in tension may cause it to crack, although in many instances the tensile strains are so small that cracks do not actually appear.

3.11.3 Deep beams

The principal bending reinforcement required for the deep beam may be calculated as for a beam of normal proportions as described in *3.4*, although intuitively this does not seem right. Another way of describing the behaviour of the deep beam shown in Figure 3.23 is to imagine the loads being transferred from their positions of application directly to the beam supports by inclined concrete struts. In order to provide equilibrium to these struts, they need to be held apart at their summit by a horizontal compression member, and tied together at their feet as shown in Figure 3.24 (a).

If the two loads were the only ones to be applied to the beam, for instance if the deep beam was in fact a spandrel in a building, carrying two columns onto more widely spaced supports, the majority of the concrete for the deep beam could be omitted, leaving only a frame consisting of the struts and ties, Figure 3.24 (b). It would be necessary to allocate a suitable size to the structural elements, which would determine the inclination of the two principal struts, and which would govern the forces required in the struts and ties.

Similarly, for the deep beam the geometry of the strut-and-tie model needs to be established by sizing the struts so that they work at an acceptable concrete stress and sizing the ties to accommodate the required amount of reinforcement. It should never be forgotten that these struts and ties are an analogy, not a description of the reality. Just as in the frame of Figure 3.24 (b) the struts would shorten and the ties lengthen, in the deep beam the compressed zones shorten and the tensile areas extend. This creates secondary stresses in the adjacent concrete that may cause cracking. Consequently, the designer should be conservative when dimensioning the struts, generally using compressive stresses that are below the limiting values that would apply to the frame, and he should not omit secondary anti-cracking reinforcement.

It is advisable, when using a strut-and-tie analogy, to carry out a second analysis using conventional bending theory, to confirm a reasonable agreement between the area of steel calculated for the tie of the former with the tensile reinforcement of the latter.

The concept of the strut-and-tie analogy is extremely useful in solving many otherwise intractable problems in the design of reinforced concrete. It also gives the designer a clear concept of how the structure may be designed to work and how to arrange the reinforcement.

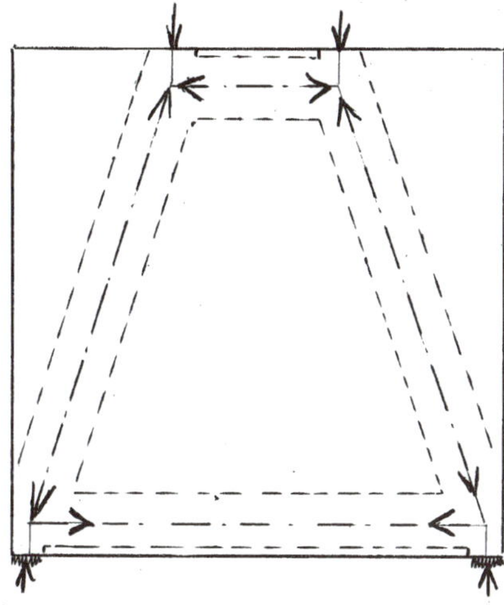

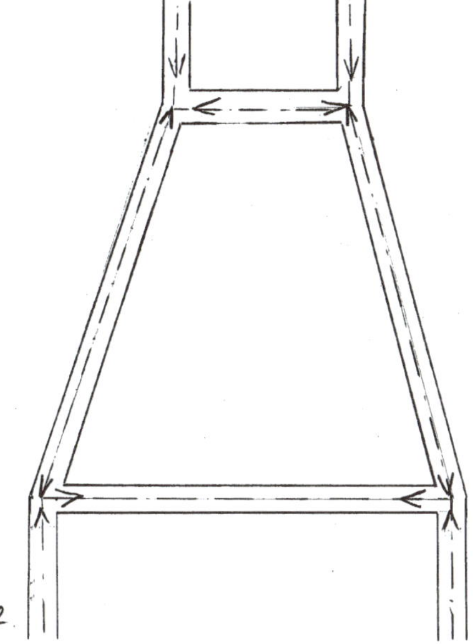

Figure 3.24 Deep beams

3.11.4 Shear friction

The strut-and-tie analogy for the behaviour of reinforced concrete may be applied to the common problem of carrying shear force through a construction joint. A very crude way of carrying the shear force across this joint is to use large-diameter dowels that work in shear and bending, and that apply highly concentrated local compressive forces to the concrete, Figure 3.25 (a). However, this is not a good solution, as dowels deflect significantly when under load, cracking the joint and losing the homogeneity of the concrete.

The concept of shear friction is a far more elegant, economical and successful method of making the joint. The basic concept is that the shear force is applied across the joint by angled concrete struts, Figure 3.25 (b). The force in these struts must be balanced by reinforcement acting in tension across the joint, effectively clamping it tight.

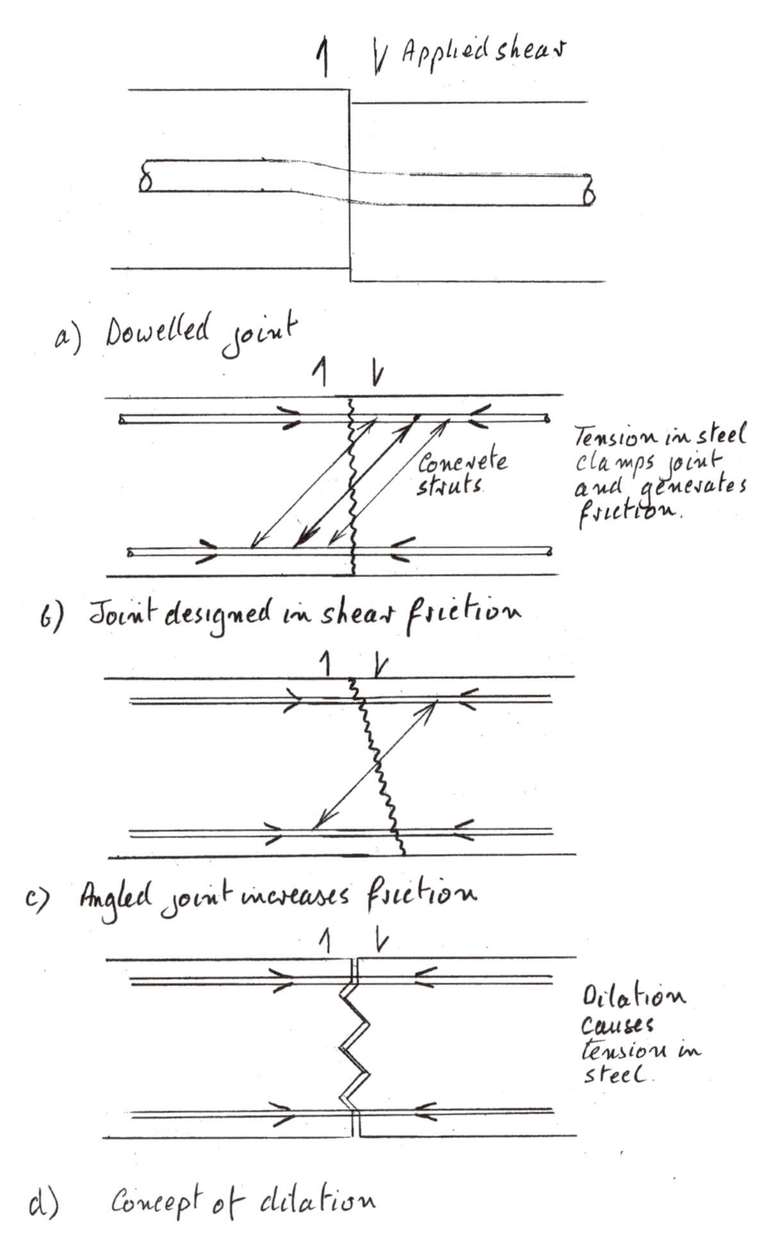

Figure 3.25 Shear friction

The angle for the struts may be chosen as 45°, which corresponds to a well-roughened interface, and to a coefficient of friction $\mu = 1$. In this case the reinforcement across the joint must be designed to apply a clamping force equal to the applied shear. Clearly the reinforcement must be sized using adequate safety factors. The various codes of practice treat this action in a variety of ways, including factoring the effective coefficient of friction up or down. However, a basic understanding of the phenomenon makes clear the importance of roughening the surface of the old concrete to achieve a high coefficient of friction. It also makes clear that a significant reduction in reinforcement, or increase in the factor of safety, may be achieved where it is possible to angle the joint, even slightly, so that the shear force applies a component of direct compression across it, Figure 3.25 (c). The reinforcing steel may also be angled so that a component of the tensile force in the steel directly resists the sliding shear.

If the joint is angled, it must be ascertained that there are no load cases that reverse the shear, when the steel would be put into compression, relieving the clamping force on the concrete and reducing the shear friction.

Another way of understanding this action is to imagine the surface of the joint as serrated. If the concrete to the right of the joint in Figure 3.25 (d) is to slide downwards, it must move outwards as it slides over the rugosities. This outwards movement, or dilation, will stress the reinforcement and clamp the joint.

3.11.5 Bond

A reinforcing bar buried in concrete clearly has no force at its end, and must pick up force from the concrete over a bond length. The best model for understanding the bond of reinforcing bars to concrete is to consider that the forces are transferred between the bar and the concrete by an array of 45° struts arranged around the bar, and distributed along the bond length, Figure 3.26 (a). These compression struts tend to split the concrete along the axis of the bar.

If the bar is anchored into a thick section of concrete that is reinforced in both perpendicular directions to resist splitting, it is clear that the bar will have the shortest possible bond length. An old French code considered that full bond would be achieved in just 12 bar diameters under these conditions. If, at the other extreme, the bar is placed close to the surface of the concrete, it is not surprising that a bond length of 50 to 60 diameters may be necessary to allow the bar to work at full stress. This model of bond also makes clear how important it is that bars should be enclosed along their anchorage length by links that can carry the splitting forces.

This view of the mechanics of bond also illustrates why it is important to limit the size of bars within thin sections of concrete. As stated in *3.6.3*, the diameter of a reinforcing bar should in general not exceed a tenth of the minimum concrete dimension.

This concept further explains that plates bonded to the surface of the concrete transfer their load by 45° struts, which tend to push the plates off the concrete surface. Such plates should be tied into the concrete along their bond lengths to avoid failure by peeling, Figure 3.26 (b).

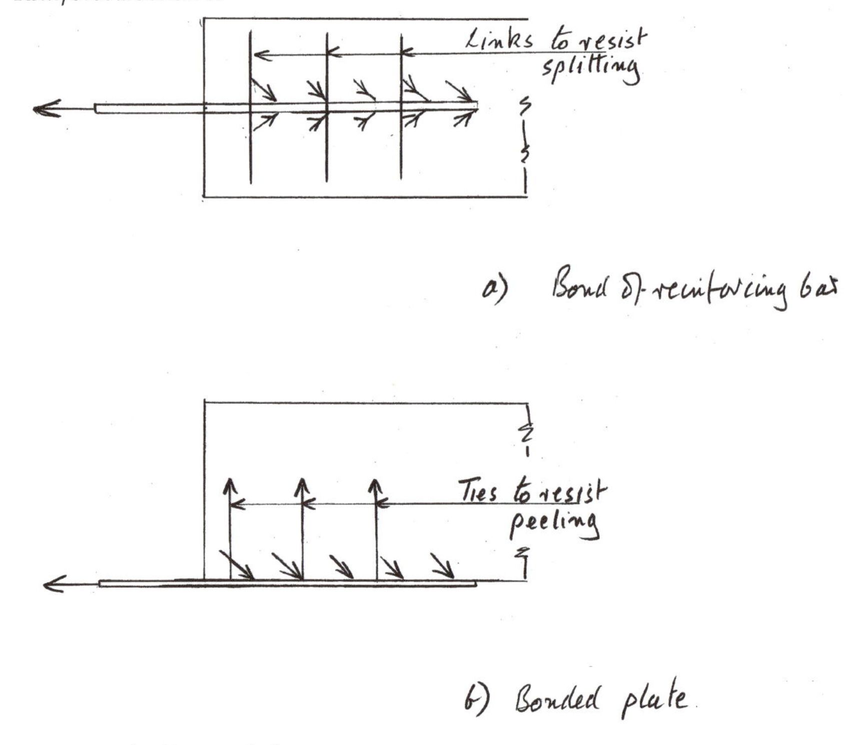

Figure 3.26 Bond of bars and plates

3.11.6 Conclusions

Strut-and-tie analogy is more a philosophy than a method of calculation. It gives to the designer a tool with which he can make sense of complicated actions, and find a rational organisation of reinforcement.

In complex areas where it may be difficult to decide on an arrangement of struts and ties, it is often a good idea to run a finite element analysis first. This will give a picture of the direction of the compressive and tensile forces, and give clues to the strut-and-tie pattern. It may also highlight areas where the concrete may be overstressed in compression, and where strut dimensions need to be increased, perhaps by thickening the concrete slabs or webs.

Using this analogy requires considerable expertise. The worst possible way of carrying out the design of a complex structure would be to pass the task to an inexperienced engineer armed with a finite element program. One criterion for deciding whether an engineer is sufficiently experienced to tackle a particular difficult piece of design is that he should be capable of drawing up a strut-and-tie analogy for it.

For further information on the design of strut-and-tie analogies, see [9] and [10].

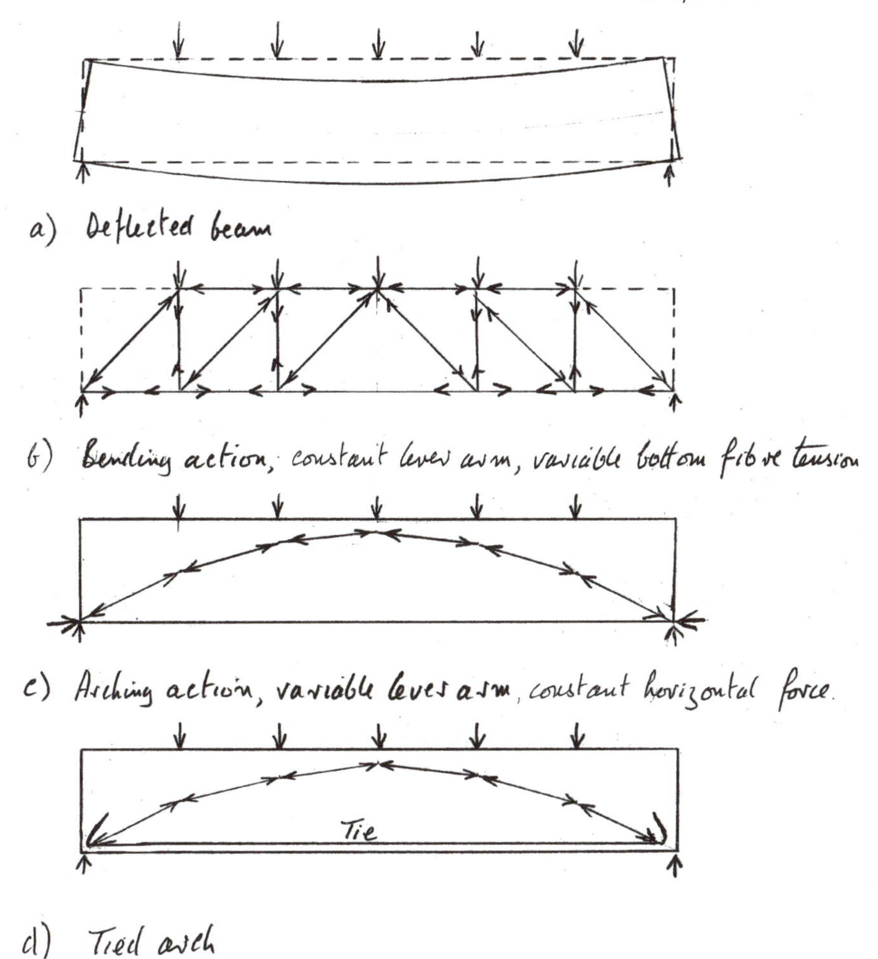

a) Deflected beam
b) Bending action, constant lever arm, variable bottom fibre tension
c) Arching action, variable lever arm, constant horizontal force.
Tie
d) Tied arch

Figure 3.27 Arching action in beams (Note: compression members are shaded)

3.12 Continuity between the concepts of bending and arching action

3.12.1 General

When a simply supported beam carries a series of loads it is subjected to bending moments; the bottom fibre extends as it is stressed in tension, and the compressed top fibre shortens, causing the beam to deflect and the beam ends to rotate, Figure 3.27 (a). The loads are carried to the supports by shear forces and bending moments, which may be described by the truss analogy, Figure 3.27 (b).

If the extension of the bottom fibre is impeded by external restraints, the loads are carried to the supports in a different manner. Figure 3.27 (c) shows how a segmental arch may be drawn within the depth of a stocky beam. As long as the restraints to the thrust of the arch are sufficiently rigid, it will carry all of the loads applied to the beam and no bottom fibre tensile reinforcement will be required, other than nominal anti-crack steel. For more slender beams and for less than perfect restraint, the load will be shared between arching and bending actions. In general, if it is possible to draw within the beam, an arch with suitable thickness, and with a rise of at least 1/10, a significant proportion of the load is likely to be carried in arching action if adequate restraint is present.

The restraint to the spreading of the arch may also be provided by reinforcement, creating a tied arch system, Figure 3.27 (d) (*17.4* and *17.10*).

Loads are carried in arching quite differently from bending. In bending, the forces comprising the internal couple are proportional to the bending moment and peter out at supports, while the lever arm of the internal couple remains constant. Consequently, the tension reinforcement at the bottom fibre of the beam may be curtailed as the moment falls. Furthermore, the beam relies on its shear strength in order to carry its loads to the supports.

In an arch, there is also an internal couple that consist of the horizontal component of the compression in the arch, and the horizontal restraint at the level of the springings. The horizontal component of the force in the arch remains constant along its length, the lever arm petering out towards the springings. Consequently, if the restraint is provided by reinforcement, this will be constant along the length of the arch, and will need to be fully anchored beyond its connection with the arch ring. In an arch, there is no primary shear force; the load is carried to the supports by the inclination of the compression in the arch. (There may well be secondary shear forces in the arch ring due to local bending moments.)

The arch ring that is in compression will shorten. Even in the presence of rigid abutments, this shortening of the arch will cause the crown to drop, and any concrete present below the arch ring will crack if it is not adequately reinforced. If the reinforcement provided to resist this cracking is greater than a nominal amount, as it is stressed in tension it will create an internal lever arm, giving bending strength to the beam. Thus the arch system converges towards a beam, although for the beam to carry a significant proportion of the load it must also be provided with shear reinforcement.

3.12.2 Examples of arching action

Arching action is an important component of many structures, and it is necessary for the designer to understand it. An example of how this was applied in reality is loosely based on the design of the concourse level of the underground Central Station on the Hong Kong Mass Transit Railway, carried out by the author when with Arup, and shown diagrammatically in Figure 3.28 (a).

The escalators to the trains drop down through long slots in the slab. Consequently, the very substantial thrust due to earth and water on the side walls has to span around the slots, Figure 3.28 (b). If the slab had been designed using bending action, the load from the walls would have been carried by a pair of fixed ended beams consisting of the slab strip either side of the slot, framing into the full-width slab at either end of the slot. This would have required substantial quantities of reinforcing steel both for the hogging and sagging areas.

However, it is possible to draw an arch to carry the thrust of the walls. The reaction at the arch springings may be resolved into a 'lateral' component or thrust which is resisted by the slab at either end of the slot, and a 'normal' component which represents the earth pressure which is transferred to the symmetrical arch on the opposite side of the slot. There must be, beyond the last of the slots in each direction, a structure that

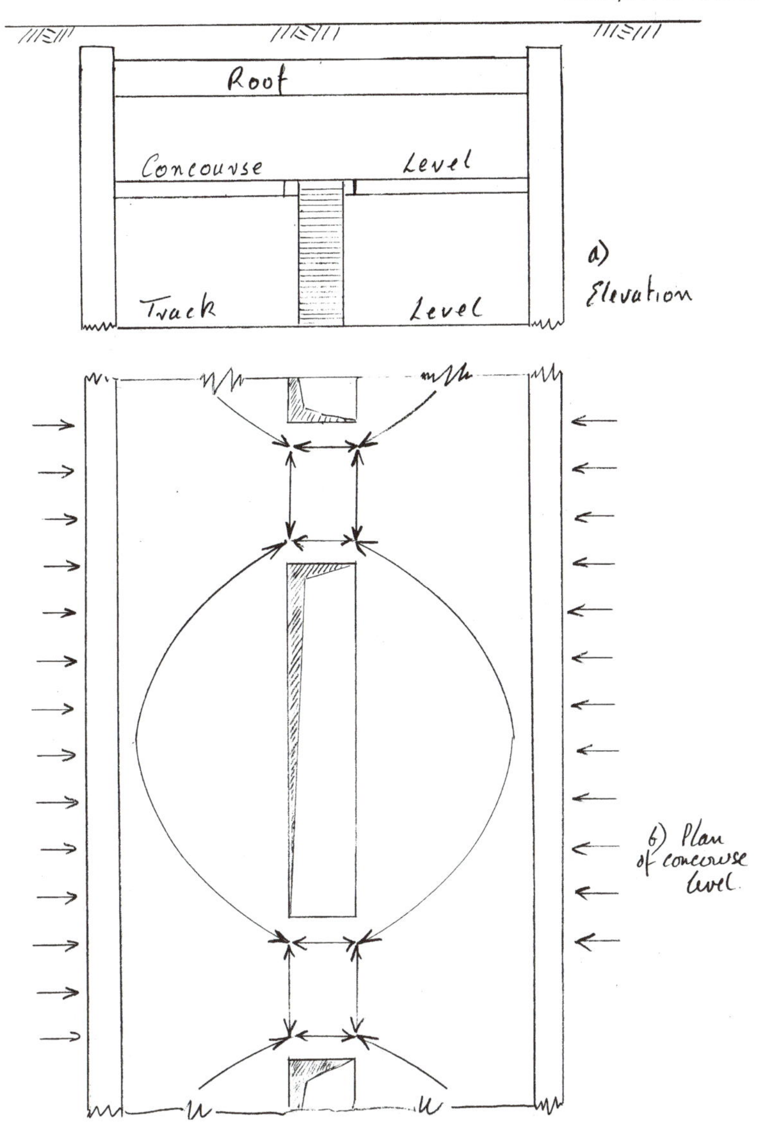

Figure 3.28 Arching action in reinforced concrete floor

can carry the final unbalanced arch thrust. In this case, this abutment consisted of a length of concourse slab which could transfer the thrust to the perimeter walls at the ends of the excavation.

3.12.3 Bridge deck slabs

A further example of the practical importance of arching action is the design of the deck slab of concrete bridges. Although such slabs, which have to carry the wheel loads of traffic, are usually designed as bending members, significant savings in reinforcement may be realised by considering their arching action. This is discussed in more detail in 9.3.5.

4

Prestressed concrete

4.1 Introduction

4.1.1 General

Prestressing is more a philosophy than a specific technique. It means preparing a structure to receive a load by applying a pre-emptive countervailing load. For instance, if it is known that a column will be deflected 100 mm to the left by applied loads, the designer can arrange to bend it 50 mm to the right; the column then only has to be designed to resist a deflection of ±50 mm, rather than the full 100 mm. A more substantial example is a pin-ended concrete arch that the designer knows will be stressed by sagging moments at mid-span caused by the shortening due to creep, shrinkage and temperature drop. He can pre-empt this load case by jacking the springings together causing moments of opposite sign, Figure 4.1. Clearly, this has to be done under careful control, as he has to take into account all the other load cases, some of which might combine unfavourably with the loads he has jacked into the arch. For instance, the designer must consider the possibility that the arch may be subjected to a temperature rise immediately after his jacking operation, before it has experienced shortening due to shrinkage and creep.

Road and airport runways may be compressed to improve their resistance to localised heavy loads by installing jacks in regular joints; the concrete box sections forming the roadway of the Mont Blanc tunnel were prestressed in this manner, by regular live joints housing banks of flat jacks. Dams have been stabilised by anchoring them to the ground using prestressing cables, and retaining walls are routinely supported by prestressed ground anchors. Beams made of steel and timber, as well as of concrete, may be prestressed to improve their performance. The definition of prestressing is thus very wide. The principal concern of this book is the prestressing of concrete members with steel cables in tension.

4.1.2 Definition of prestressing

One may consider the prestressing of concrete members in two different ways. First, consider the following explanation:

> Concrete is similar to stone, in that it is strong in compression, but weak and brittle in tension. A simple beam of rectangular cross section is subject to compressive

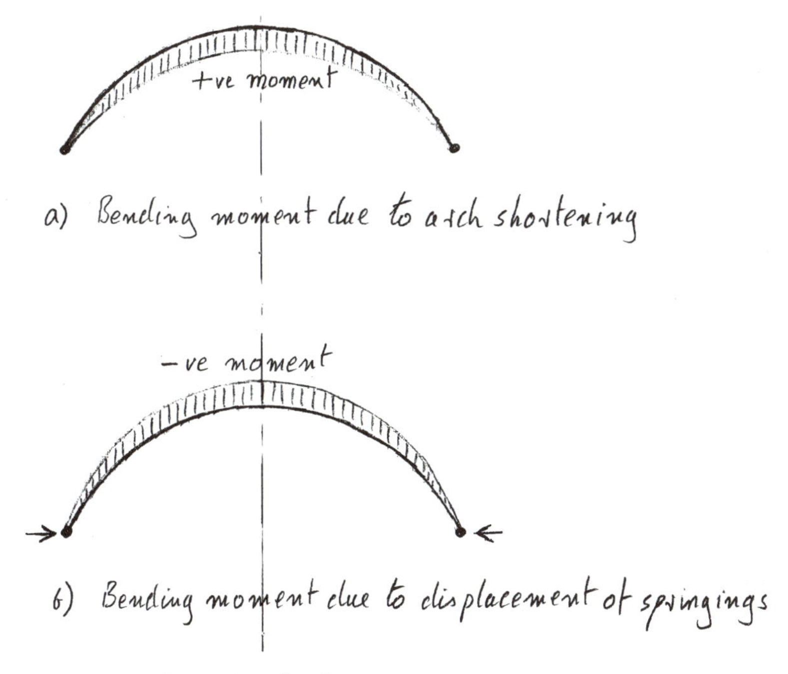

+ve moment
a) Bending moment due to arch shortening
–ve moment
b) Bending moment due to displacement of springings

Figure 4.1 Prestressing of two pinned arch

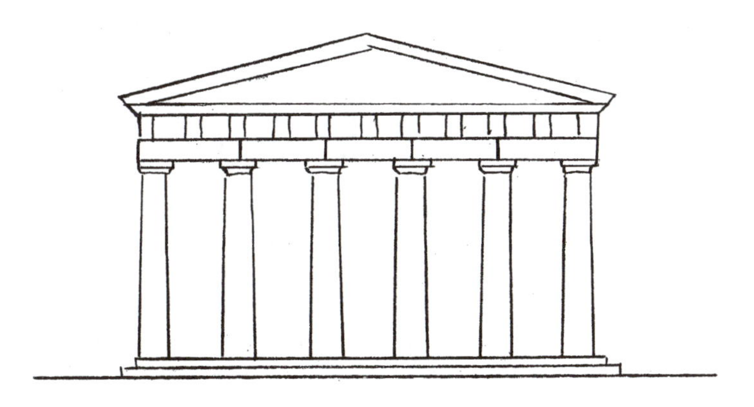

Figure 4.2 Greek temple

stresses on its top fibre and tensile stresses on its bottom fibre that are equal. Thus the span of a beam made of stone or of plain concrete is limited by the material's low tensile strength. The rules on the safe span for stone beams have been at the origin of classical architecture, requiring closely spaced columns and suitable proportions for lintels typical of Greek temples, Figure 4.2.

Figure 4.3 shows the bending stresses in such a rectangular beam when it carries a total load (self weight plus applied load) of w kN/m. One may assume that the compressive stress at the top fibre and the tension stress at the bottom fibre of the beam have values of +5 and –5 units of stress respectively. One may further assume that this tensile stress is the maximum safe value; if any more load is added

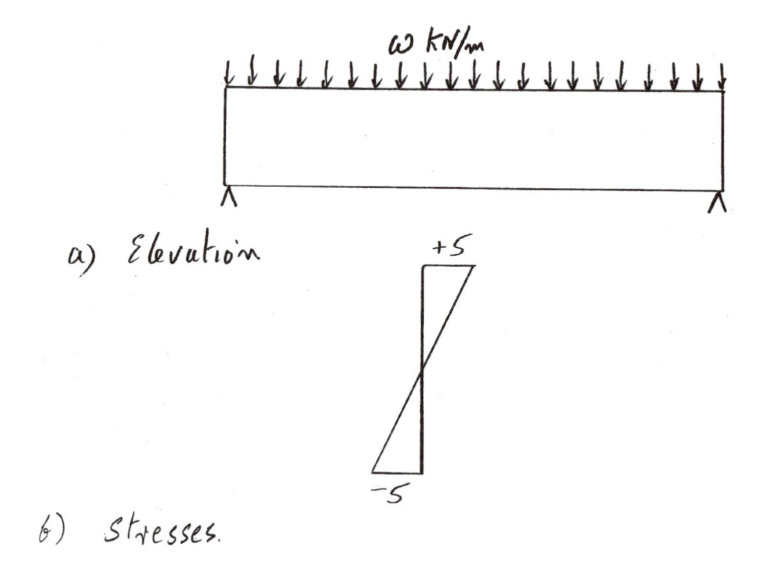

Figure 4.3 Plain concrete beam

to the beam it would break in tension and collapse. However, the compressive stress of +5 units is well below the crushing strength of the material.

Now assume that a steel bar is passed through the centre of the beam and is stressed by jacks reacting against the beam. Thus the bar is put into tension and the beam is compressed by the same amount. Assume that the uniform compressive stress in the beam due to this prestress is +5 units, Figure 4.4. When this compressive stress is combined with the bending stresses caused by the load of w kN/m, they produce a compressive stress of 5 + 5 = 10 units on the top fibre and 5 – 5 = 0 units on the bottom fibre.

A further load of w kN/m may now be added to the beam, giving a total of $2w$ kN/m, creating additional bending stresses of ±5 units at the top and bottom fibres, taking the combined stresses to +15 units on the top fibre and –5 units on the bottom fibre. Thus the load that can be carried by the beam without exceeding the maximum allowable tensile stress has been doubled, but at the cost of tripling the compressive stress on the top fibre. Alternatively, the load may be kept at w kN/m, but the span increased by a factor of $\sqrt{2} = 1.41$.

In a further rationalisation, the prestressing bar may be displaced downwards, placing it closer to the bottom fibre to increase its effectiveness. For instance, if it were displaced by just one-sixth of the depth of the section to the lower edge of the 'middle third' (*5.4*), it would create stresses that were zero on the top fibre, and +10 units of compressive stress on the bottom fibre. This would allow the load carried by the beam to be increased to a total of $3w$ kN/m before the bottom fibre stress attained the safe limit of –5 units, or the span to be increased by a factor of $\sqrt{3} = 1.73$. The top fibre stress would be +15 units once again, Figure 4.5.

Thus in this explanation, prestressing is a technique that enhances the capacity of a beam made of a material that is weak in tension but strong in compression to carry

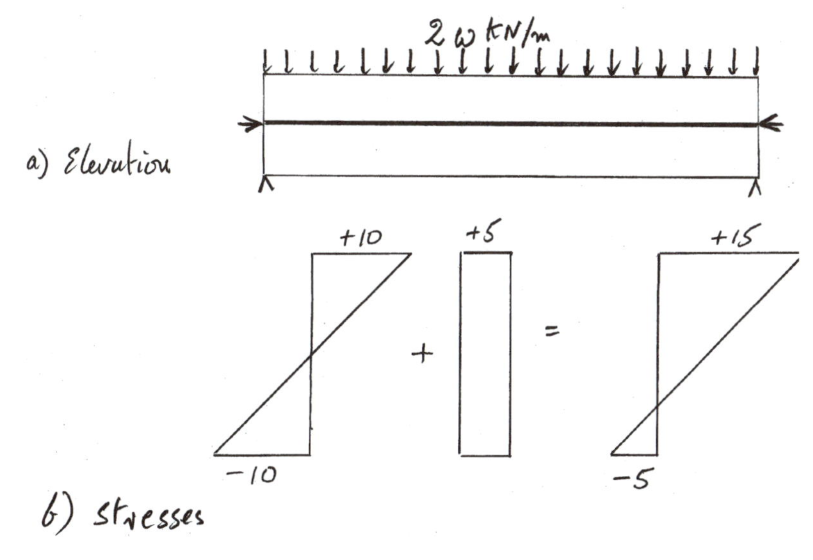

Figure 4.4 Centrally prestressed beam

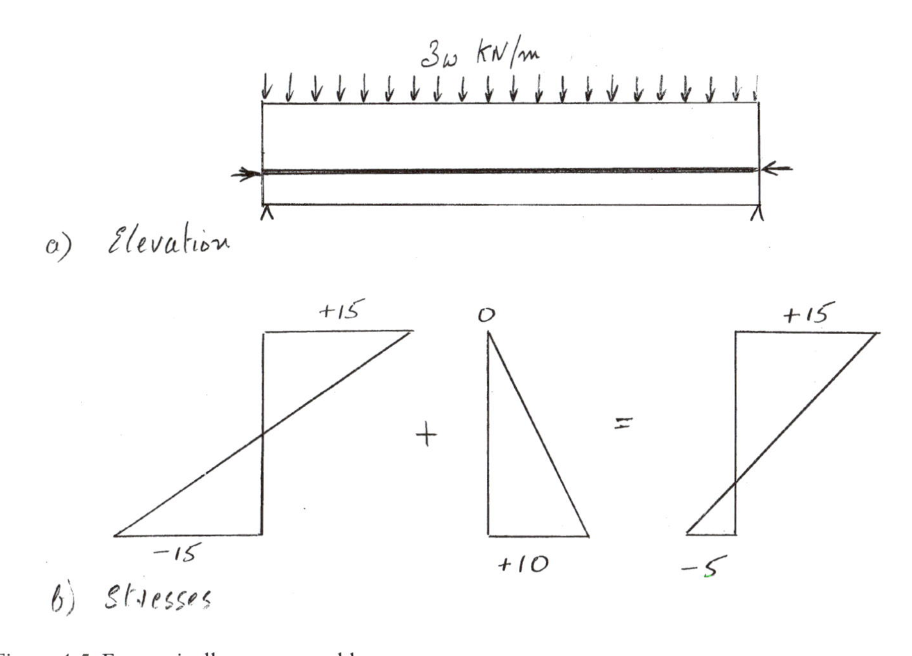

Figure 4.5 Eccentrically prestressed beam

loads or increase its span. It effectively creates a new construction material which is strong in tension.

It is this view of prestressing that is the basis for most codes of practice, where tensile bending stresses in concrete are either proscribed completely, or limited to very low values which are below the cracking threshold of concrete.

Alternatively, consider the following explanation;

> The weakness of concrete in tension may be remedied by reinforcing it with embedded steel bars. In a simple beam, these bars are placed near the bottom fibre that is in tension. The beam thus becomes a composite, with the reinforcement carrying the bottom fibre tension and the concrete near the top fibre in compression. The steel bars are protected against corrosion by the alkaline environment of the surrounding concrete. However, as the bars are stressed in tension, they elongate and crack the adjacent concrete, weakening the corrosion protection and spoiling the appearance of the member. The deflection of the beam also increases as the concrete cracks, particularly under permanent loads, and this can cause problems for floor finishes and facades in buildings, or for the alignment of bridges. The tensile stress in the reinforcement under working loads must be limited to around 250 MPa in order to keep these effects under tolerable control, leading to the use of steel with a yield strength that may not usefully exceed 460–500 MPa. Consequently, reinforced concrete cannot take advantage of the more economical higher strength steel available, with the result that large numbers of bars are required, causing congestion and difficult construction, and beams need to be stocky and slabs thick to limit deflections.
>
> By prestraining the reinforcement, steel with an ultimate strength of 1,860 MPa may be used, working at a stress of around 1,000 MPa, greatly reducing congestion. By compressing the concrete, cracking at working loads may be completely eliminated, or reduced to any desirable degree and by giving eccentricity to the prestressing tendons, self-weight deflections may be completely eliminated or even reversed.

Thus in this explanation, prestressing is an improvement to the technique of reinforced concrete.

4.2 A comparison between reinforced concrete and prestressed concrete

4.2.1 Concrete tie for a tied arch

A graphic illustration of the difference between reinforced and prestressed concrete may be given using the example of a 20 m long concrete tension member that is the tie of a tied arch. Assume it has a cross section of 0.35 m × 0.35 m = 0.1225 m^2, made of concrete with a compressive strength of 50 MPa and a tensile strength of –4 MPa, Figure 4.6. The Young's moduli of steel and concrete are assumed to be 200,000 MPa and 34,000 MPa respectively. It is assumed that this member is to be designed to carry a working load tension of 1.2 MN, and an ultimate load of 1.9 MN.

The concrete on its own would have a tensile strength of only 4 × 0.1225 = 0.49 MN. However, as concrete is unreliable in tension, this would have to be drastically factored down to provide a safe load.

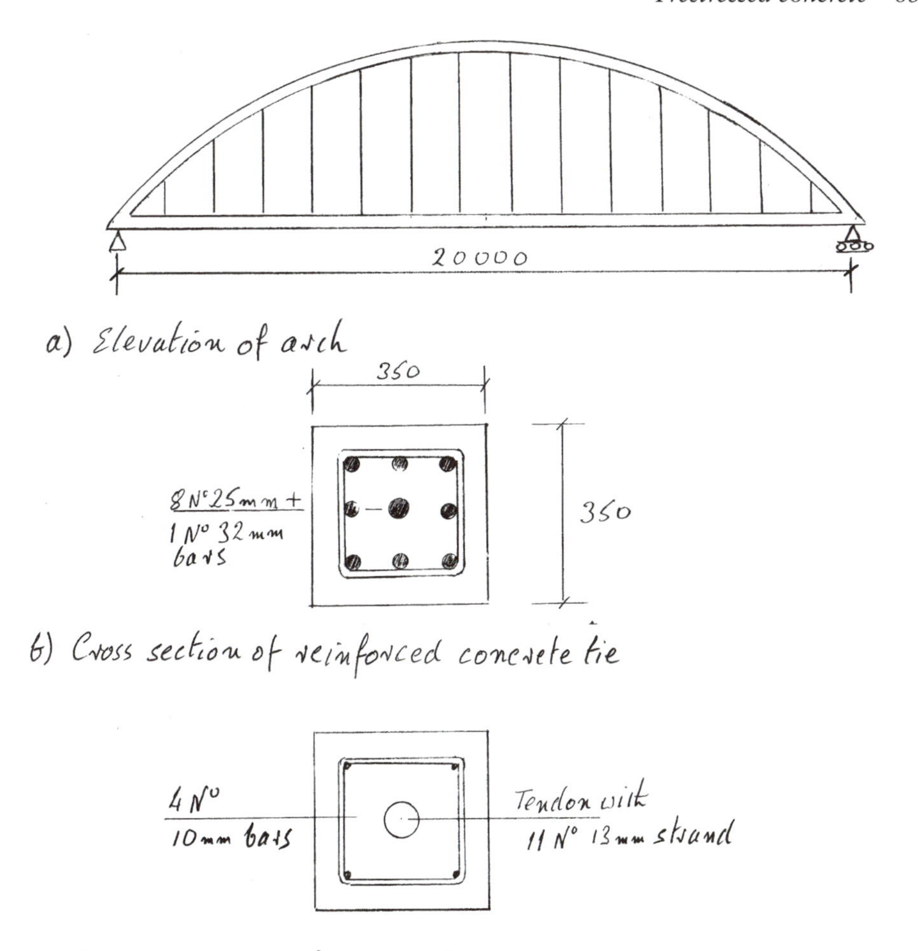

Figure 4.6 Tied arch

4.2.2 *Reinforced concrete option*

As the concrete cannot be relied upon to provide a safe tension member, all the tension in the tie must be carried by steel reinforcement. The regulations in force in most countries require that reinforced concrete members are designed for the ultimate load (that is at the ULS), and are then checked for their performance at working load (at the SLS). Thus using reinforcing bars with an ultimate strength of 460 MPa, and a material factor of 1.15, the area of steel required is $1.9 \times 10^6 \times 1.15 / 460 = 4{,}750$ mm^2. This may be provided by 8 No 25 mm bars and 1 No 32 mm bar giving a steel area of 4,731 mm^2, marginally short. The strength of the reinforced concrete tension member is $4{,}731 \times 460 \times 10^{-6} / 1.15 = 1.89$ MN. This area of reinforcing steel represents 3.9 per cent of the concrete cross-section area of the tie, Figure 4.6 (b), which is close to the maximum practicable, above all as 20 m long bars are not available, and the bars have to be lapped or coupled.

Under the working load of 1.2 MN, the stress in the steel would be 254 MPa and the extension of the steel reinforcement on its own would be $254 \times 20 \times 10^3 / 200{,}000 = 25.4$ mm. In fact, the bond of the concrete to the steel would reduce this extension to approximately 80 per cent of that of the bare steel, that is to approximately 20 mm. The tension member would be heavily cracked, with clearly visible transverse cracks at relatively close centres. For instance, if cracks were assumed to be spaced at 350 mm along the length of the member, their average width would be of the order of $254 \times 350 \times 0.8 / 200{,}000 = 0.36$ mm.

Such a frequency and width of cracks would be considered unacceptable, both from considerations of durability and of appearance. In addition, such a large extension of the tie would introduce significant secondary bending moments into the arch. Consequently, the SLS check would certainly involve providing additional steel to reduce its stress and control the cracking. This would most probably require a bigger tie to contain this extra steel. However, in the interests of maintaining the clarity of this comparison between the techniques of reinforced and prestressed concrete, the SLS check has been set aside.

The load/deflection graph characteristic of the member in tension will be approximately linear until it starts to crack, when its stiffness reduces substantially. When the member is unloaded, it would not recover its original length. This is principally due to the fact that some slippage of the bond between the steel bars and the concrete will have occurred.

The weight of reinforcing steel required over the 20 m length would be 740 kg.

4.2.3 *Full prestressing*

The most conservative design for a prestressed concrete structure is to assume that the concrete must have no tensile stresses due to bending or to normal force at working load. This is normally called Class 1 prestressing, and is used when it is considered essential to avoid all cracking of the concrete, either due to the aggressiveness of the environment or the sensitivity of the structure. Class 1 and Class 2 (see *4.2.4*) prestressed concrete structures are designed at the SLS to control working load stresses in the concrete, and are then checked at the ULS.

Assume the section to be prestressed with a tendon consisting of a number of 13 mm steel strands. Each strand has an ultimate tensile stress of 1,860 MPa, a working stress of 1,120 MPa, a crosssection area of 100 mm^2 and a working load of $1{,}120 \times 100 \times 10^{-6} = 0.112$ MN per strand. As the aim of Class 1 prestressing is to eliminate tensile stresses in the concrete tie at working load, the prestress must supply a compressive force equal to the applied tension, that is 1.2 MN. This requires a steel area of $1.20 \times 10^6 / 1{,}120 = 1{,}071$ mm^2 which may be satisfied by providing 11 strands, giving an area of 1,100 mm^2, and a prestressing force at working load of $0.112 \times 11 = 1.23$ MN, slightly more than required.

The compression on the tie due to the prestress force alone is $1.23 / 0.1225 = 10$ MPa. Once the arch is loaded and applies its working tensile force of 1.2 MN to the tie, the concrete will become virtually unloaded; all the tension force is carried by the prestressing strand.

The unloading of the 20 m long concrete tie from a compression of 10 MPa to zero causes an extension of $10 \times 20{,}000 / 34{,}000 = 5.9$ mm. The load/deflection graph will be entirely elastic, and on unloading the member will recover its original length. (For a discussion on the effects of concrete creep, see *3.9*.)

The ultimate strength of the prestressing strand, using a material factor of 1.15 as before, would be 11 × 100 × 1,860 × 10^{-6} / 1.15 = 1.78 MN, which is 0.12 MN short of the required figure. In fact, a light reinforcing cage will be necessary to hold the prestress tendon in place, and to give some strength during construction, before the tendon is stressed. This may be assumed to consist of 4 No 10 mm bars, providing an additional ultimate force of 0.12 MN, and weighing about 49 kg, Figure 4.6 (c). The weight of prestressing strand is 173 kg, giving a total steel weight of 222 kg.

This prestressed version of the tension member will be much easier to build as the section is uncluttered, needing only one tendon and a light cage to install as compared with 9 heavy bars. It is much less deformable, extending by only about a quarter of the reinforced concrete version. It will also carry the working load of 1.2 MN without cracking, and thus conforms to the first description of prestressing given in *4.1.2* above. However, there are two other possible designs for this tension member.

4.2.4 Reduced prestressing

For structures where the risk of a small amount of cracking of the concrete is acceptable, the prestress may be reduced, leading to some tensile stresses in the concrete at working load. For the 50 MPa concrete used for this example, a tensile stress of approximately 2.8 MPa is generally considered to be the threshold at which cracking is likely, and is accepted as the design limit. Prestressing designed in this way is called Class 2.

If the number of prestressing strands is reduced to 8, the prestressing force applied to the concrete would become 0.89 MN, and the compressive stress on the concrete would be 7.3 MPa. When the 1.2 MN tensile force is applied, there will be a residual tensile stress on the member of (0.89 – 1.2) / 0.1225 = –2.5 MPa. As cracking will be non-existent or minimal, the extension of the member would be very similar to the fully prestressed version.

The ultimate strength of the prestressing steel is now only 8 × 1.78 / 11 = 1.29 MN, a shortfall of 1.9 – 1.29 = 0.61 MN as compared with the specified requirement. This may be made up from so-called 'passive' or unstressed reinforcement. The 4 No 10 mm bars of the previous example need to be increased to 8 No 16 mm reinforcing bars that have an area of 1,608 mm^2 and provide an additional ultimate force of 0.64 MN, slightly more than enough. The weight of prestressing strand has been reduced to 126 kg, and the 8 reinforcing bars now weigh 252 kg, giving a total steel weight of 378 kg.

An important additional advantage of this design is that the initial compressive stress on the concrete has been reduced from 10 MPa to 7.3 MPa. Such a reduction in stress is beneficial, as it should always be one of the aims of a designer to ensure that the structure at rest, that is under permanent loads only, is stressed as lightly as possible. This ensures that it has the greatest reserves of strength to cope with unexpected loading or damage.

4.2.5 Partially prestressed concrete

Finally, the structure may be designed as partially prestressed. Here cracking is accepted, although the presence of even a reduced prestressing force will substantially reduce its severity. The tension member may be designed to use any convenient mixture of prestressed and passive reinforcement. For instance, it may be convenient to use a standard size of prestressing cable consisting of 7 strands with a working force of 0.78 MN, generating a compressive stress on the concrete of 6.4 MPa. The tensile stress on the tie under working load becomes 3.4 MPa. The ultimate strength provided by the prestress becomes 1.13 MN, and additional passive reinforcement of 4 No 25 mm bars is necessary, giving a total ultimate strength of 1.91 MN. The prestress weighs 110 kg and the reinforcement 308 kg, with a total steel weight of 418 kg.

A tensile stress of 3.4 MPa under working load is likely to lead to some cracking of the tie, but much less than for the reinforced concrete member. At this low level of working tensile stress, the behaviour of the member will remain close to elastic, and the extension will be only slightly greater than the 5.9 mm of the fully prestressed member due to this cracking. The tension member is now transitional between reinforced and prestressed concrete. However, if the proportion of prestressed steel were to be further reduced, the member would behave more like the reinforced concrete example.

It should be noted that prestressing strand is more susceptible to metal fatigue than reinforcing bars. Consequently, if the live loading regime is such that cracks may open and close a very large number of times over the life of the structure, a check on the fatigue resistance of the tendons would be necessary.

4.2.6 *Summary of advantages offered by prestressing*

The practical advantages offered by prestressing in this example are:

- the smaller weight of steel to be handled and fixed;
- the reduced congestion of the section leading to easier and less error prone casting of the concrete;
- the greater stiffness and the elastic behaviour of the member;
- the greater durability due to the absence, or greatly diminished incidence, of cracking at working load;
- the improved appearance due to the absence of cracking.

4.2.7 *Cost comparisons*

Typical installed costs in the UK for reinforcement and prestressing steel are £700 per ton and £2,000 per ton respectively, a ratio of approximately three, whereas the ratio of the ultimate strengths is approximately four. When the weight of steel is governed by the ULS, as in this example, the prestressed version would show a saving of some 30 per cent.

However, whereas the cost of reinforcing steel varies generally by not more than about ±10 per cent between applications of a similar size, more than 50 per cent of the cost of prestressing consists of site labour, of which a large part is due to the operations of stressing and then grouting the tendons. Consequently the cost of prestressing varies considerably, depending principally on the length of the tendons, with the cost rising steeply as the length falls below about 25 m. A 10 m long tendon costs approximately twice as much per ton of steel as one 40 m long. One of the aims of a designer should be to make tendons as long as is practicably possible, although the lesser cost of the tendons needs to be balanced against their lower force due to friction losses (see *5.17.2*). In countries with labour costs that are lower than those of the UK, prestressing becomes relatively cheaper in comparison with reinforcing.

Furthermore, in many cases, the total weight of steel does not depend on the ULS but on the SLS. For instance, if the arch tie described above had been built in reinforced concrete, as described in *4.2.2*, it would have been necessary to increase the weight and cost of the reinforcement, and probably also to increase the size of the tie. On the other hand, it would have been possible to reduce the size of the prestressed tie, in particular for the partially prestressed option, reducing its cost and giving the designer further aesthetic options.

In addition to reduced cost, prestressing offers other advantages which will be described in later chapters, including considerations of buildability and speed and security of construction, as well as enhanced quality due to cancelled dead load deflections, elastic behaviour and the absence of cracking.

4.2.8 The buckling of prestressed members

This example highlights a very important characteristic of prestressed members. It would be preferable to prestress the tie before the arch itself is erected. Alternatively, it would be necessary to build the arch and apply its thrust in increments, adding a balancing increment of prestress each time, a slow and more costly procedure. If the prestress were applied to the tie in the absence of arch thrust, the tie would be compressed with the prestress force of 1.23 MN. The tie has a slenderness of 0.35 m / 20 m = 1/57; a column of these proportions subjected to an external force of 1.23 MN would be unstable and would buckle.

However, if the tendon applying the compressive force is in a duct inside the tie, or is otherwise attached to it so that any deflection of the tie takes the tendon with it, the tie cannot buckle under the prestress force. As it attempts to buckle sideways under the effect of the end compressions, the tendon in tension is also deflected by the same amount and thus creates a correcting force that is equal and opposite to the disturbing force.

The designer must be careful when using this concept. If the prestressing tendons had been external to the tie, and not connected to it along its length, the tie would indeed buckle. There is an intermediate situation where the duct carrying the tendon is unusually large for whatever reason, and the compressed tie can develop a significant lateral deflection before the tendon is deflected. Under this circumstance, the countervailing force applied by the tendon may be significantly smaller than the disturbing force, and the member may become unstable.

4.3 Pre-tensioning and post-tensioning

There are two principal families of prestressed bridge decks: pre-tensioned and post-tensioned. Pre-tensioning involves tensioning the cables before the concrete is cast. The cables are anchored to a strengthened mould. Once the concrete has hardened, the cables are released, and maintain their tension by their adherence to the concrete. This technique is used principally for the construction of relatively short span bridge decks using standard bridge beams.

Post-tensioning involves first casting the concrete deck, and then installing cables, which are stressed by proprietary jacks reacting against the concrete. The cables are then locked to the concrete by anchors. This book is principally concerned with the design of post-tensioned structures, although mention is made of the option of pre-tensioning when appropriate.

4.4 Conclusion

Designing a prestressed concrete structure requires a higher degree of skill from the designer than if he was designing a conventional, 'passive' structure. He has to understand more fully the range of actions on his structure, the effects of creep and shrinkage of concrete, the difference between internal and external forces and between loads and imposed deformations. Whereas an unskilled designer of a reinforced concrete structure can hide his lack of ability by over-reinforcing it, over-prestressing a structure may reduce its safety. For instance, increasing the prestressing in the tie of the above example 'for safety' would in fact increase the compressive stress in the unloaded tie, and increase the risk of failure rather than providing any additional margin of safety.

The example of a tension member was chosen for its simplicity and clarity. However, the benefits of prestressing are even greater in beams and slabs, as in addition to those summarised in *4.2.6*, the self-weight deflections of the member may be cancelled, and, in continuous beams, manipulation of the prestress secondary moments allows a degree of optimisation of the ratio of support and span moments. Prestressing also frequently allows a reduction in the depth and in the cross-section area of beams and a reduction in the thickness of slabs as compared with reinforced concrete.

In Chapter 5, the general principles of prestressing are illustrated by following step by step the preliminary design of a statically determinate beam. In Chapter 6, these principles are extended to the design of continuous beams.

5

Prestressing for statically determinate beams

5.1 General

In this chapter, the general principles of prestressing are explained by following step by step the preliminary design of a statically determinate beam. The beam used as an example is loosely based on the post-tensioned beams for the crossing of the Pearl River Delta carrying the Guangzhou–Shenzhen–Zuhai Superhighway in southern China. A description of this bridge may be found in *10.3.8*. We will consider one typical 2.3 m deep bridge beam of 32 m span, Figure 5.1. The example will simplify the calculations, principally in that it is assumed that the whole beam, including the associated top slab, is cast in one phase (Chapter 10), and the live loads have been simplified.

5.2 Materials employed for the example

The concrete is assumed to be of grade 50/20 (a crushing strength of 50 MPa and a maximum aggregate size of 20 mm), and to have a Young's modulus (E) of 34,000 MPa, a density of 25 kN/m^3 and a creep coefficient of 1.5 (*3.9*).

The prestressing strand is 13 mm superstrand, with a cross section of 100 mm^2, a breaking stress of 1,860 MPa and an E of 200,000 MPa. Passive reinforcement has a yield stress of 460 MPa. The calculations are done broadly in accordance with BS5400: Part 4: 1990.

5.3 Section properties

Section properties are normally calculated using computer programs. However, it is very useful to be able to carry out this function rapidly by hand, and a suitable method is given in the Appendix. For this section, shear lag is not an issue. However, see *6.10.2* for comments on the effect of shear lag.

The beam has the following section properties:

A = Area = 1.470 m^2
I = 2nd moment of area = 1.055 m^4
y_t = top fibre distance = 0.828 m
y_b = bottom fibre distance = 1.472 m
a_t = top kern = 0.488 m
a_b = bottom kern = 0.867 m

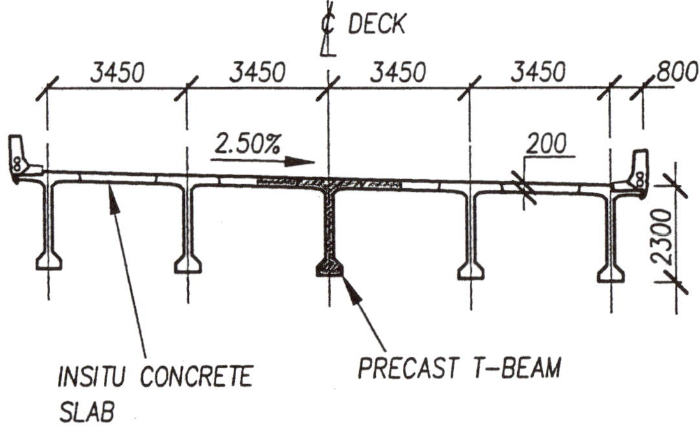

a) Cross Section of Deck

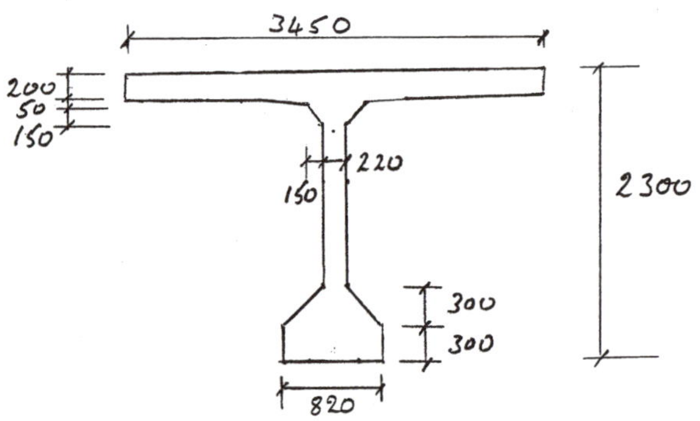

b) Cross section of beam

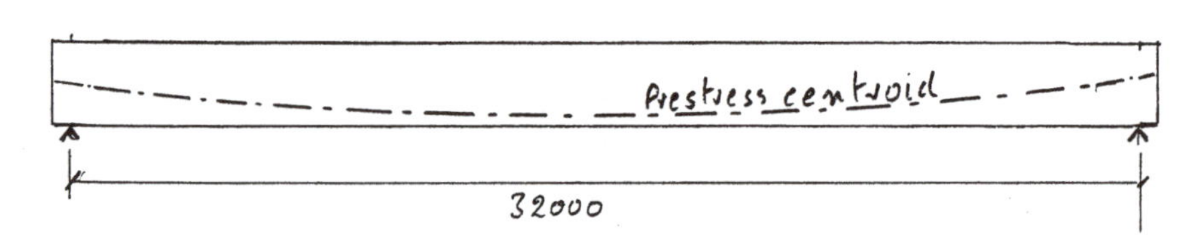

c) Elevation of beam

Figure 5.1 Statically determinate beam

z_t = top modulus = 1.274 m^3
z_b = bottom modulus = 0.717 m^3
η = efficiency of section = 0.589

where:

$z_t = I/y_t$ m^3
$z_b = I/y_b$ m^3
$a_t = \eta \times y_t$ m
$a_b = \eta \times y_b$ m
$\eta = I/Ay_t y_b$ = proportion of depth of section occupied by the central kern.

5.4 Central kern and section efficiency

The inner zone of any cross section is called the central kern, which is a very important concept in prestress design, Figure 5.2 (a). A compressive force applied within the kern will cause only compressive stresses on the cross section. When the force is applied at the limit of the kern, the stress on the remote fibre will be zero, Figure 5.2 (b). A compressive force applied outside the kern will cause tensile stresses as well as compressive stresses, Figure 5.2 (c). The upper and lower limits of the central kern are called a_t and a_b and are equal to $\eta \times y_t$ and $\eta \times y_b$ respectively.

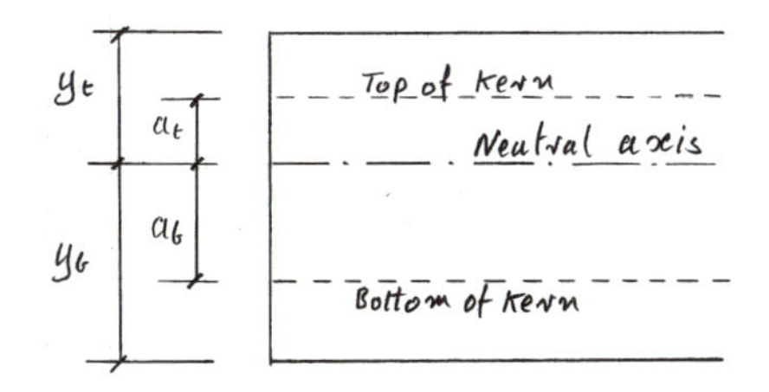

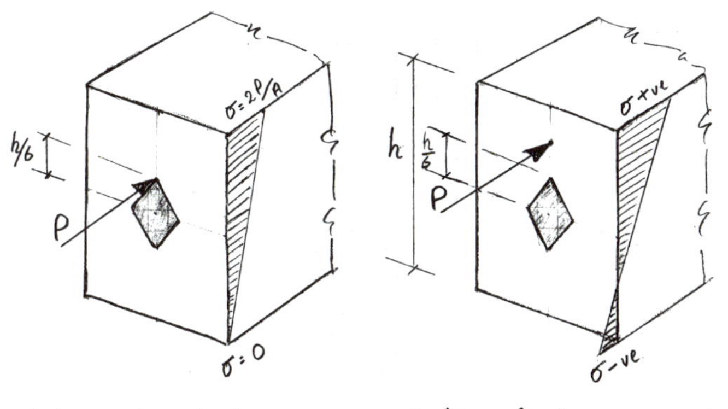

Figure 5.2 The central kern

The central kern is the middle third ($\eta = 0.33$) for a rectangle and the middle quarter ($\eta = 0.25$) for a solid circle. For concrete bridge decks, η varies from approximately 0.3 for a solid slab with a profile that follows the road crossfall, Figure 5.3 (a), to 0.45 – 0.5 for a typical twin rib deck, Figure 5.3 (b), to 0.65 for a well-proportioned box section with thin webs, Figure 5.3 (c).

With experience, it is possible to guess the approximate value of η for a proposed bridge deck. If one then calculates the cross-section area A and guesses the values of y_t and y_b as suitable proportions of the depth, using $I = A\eta y_t y_b$, it is possible to come to a quick first approximation of the second moment of area for a bridge deck, which is very helpful in the initial sizing calculations.

For instance, for the box section shown in Figure 5.3 (c), one may calculate the cross-section area as 5.88 m^2 and then guess that the neutral axis was at 0.9 m from the top fibre and that the efficiency was 0.6. Then $I = 5.88 \times 0.9 \times 1.3 \times 0.6 =$ 4.13 m^4. The correct, calculated values for y_t, y_b and η are 0.791 m, 1.409 m and 0.636 respectively, and $I = 4.171$ m^4. The approximate calculation was reasonably close.

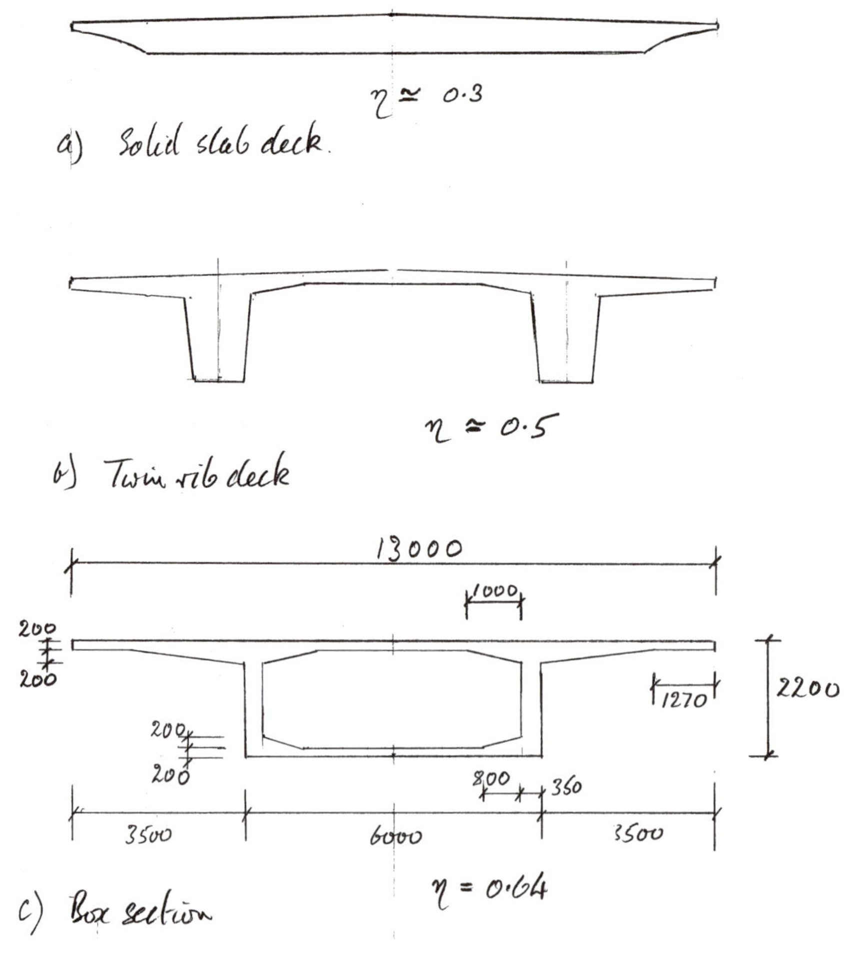

Figure 5.3 Efficiency of typical deck types

Furthermore, the calculation of section efficiency constitutes a very useful check on the calculation of section properties. An untypical value for η stands out, whereas it is difficult to have a feeling whether the calculated inertia is right or wrong.

5.5 Loads

The beam has to carry its own self weight of 1.47 $m^2 \times 25$ kN/m^3 =37 kN/m, a superimposed dead load consisting of 100 mm of deck surfacing weighing 0.1 m × 22 kN/m^3 × 3.45 m = 7.6 kN/m, and a live load, which, under UK rules would consist principally of 45 units of HB loading consisting of a single four-axle vehicle weighing 1,800 kN (including a dynamic factor of 1.2).

In a detailed design, and indeed in a careful preliminary design, a grillage analysis would be carried out to determine the proportion of the live load carried by any beam. For the purpose of this illustration, it will be assumed that the live load on the beam, for both bending and shear, consists of 100 per cent of the HB vehicle.

5.6 Bending moments, bending stresses and shear force

The critical sections for checking the suitability of the beam to carry the loads described above are at mid-span for bending and for sizing the prestress, and at the beam ends for shear force and for checking the width of the web.

The prestress is designed at the Serviceability Limit State. The SLS bending moments and stresses at mid-span are given in Table 5.1.

The width of the web and the shear reinforcement are designed at the Ultimate Limit State. The shear force at the end of the beam is given in Table 5.2.

Note: The load factors have been taken from BS5400: Part 4: 1990. However, these load factors are not of the essence of this illustration, and could be substituted by others corresponding to alternative codes of practice.

Table 5.1

	SLS moment MNm	*Top fibre stress* MPa	*Bottom fibre stress* MPa
Self weight	4.74	+3.72	–6.61
Finishes	0.97	+0.76	–1.35
Live load	10.80	+8.48	–15.06
Total	16.51	+12.96	–23.02

Table 5.2

	SLS shear force kN	*Load factors*	*ULS shear force kN*
Self weight	592	1.1 × 1.15	749
Finishes	122	1.1 × 1.2	161
Live load	1,520	1.1 × 1.3	2,174
Totals	2,234		3,084

5.7 Centre of pressure

In order to calculate the prestress force required to resist the bending moments tabulated above, it is necessary to understand the concept of centre of pressure. The centre of pressure is the point at which a normal force effectively acts on the section.

Consider the column shown in Figure 5.4 (a), where it is shown loaded only by an axial force P. In the absence of any external bending moment, the centre of pressure is coincident with the point of application of this force, and the stress on the column is P/A, where A is the cross-sectional area of the column. If a bending moment M is now applied to the column, Figure 5.4 (b), the centre of pressure is displaced by a distance $e = M/P$. The bending moment has added stresses to the column, which at the extreme fibres are $\pm M/z$, where z is the section modulus. The total stresses on the extreme fibres of the column are now $P/A \pm M/z$. The column is in exactly the same state of stress as if the load P had been applied at an eccentricity e in the first place, Figure 5.4 (c).

Thus, if a prestressing force is applied to a beam, in the absence of any external bending moments, the centre of pressure will be coincident with the centroid of the prestress force. If the prestress is applied at a distance e from the neutral axis, it applies a bending moment to the section equal to Pe. The stresses caused by the prestress are the sum of the axial compressive stress and the stresses due to this internal bending moment, or $P/A \pm Pe/z$. If the section is not symmetrical and consequently the moduli of the extreme fibres are different, the stress on the fibre remote from the prestress will be $P/A - Pe/z_{remote}$, and that closer to the prestress will be $P/A + Pe/z_{close}$.

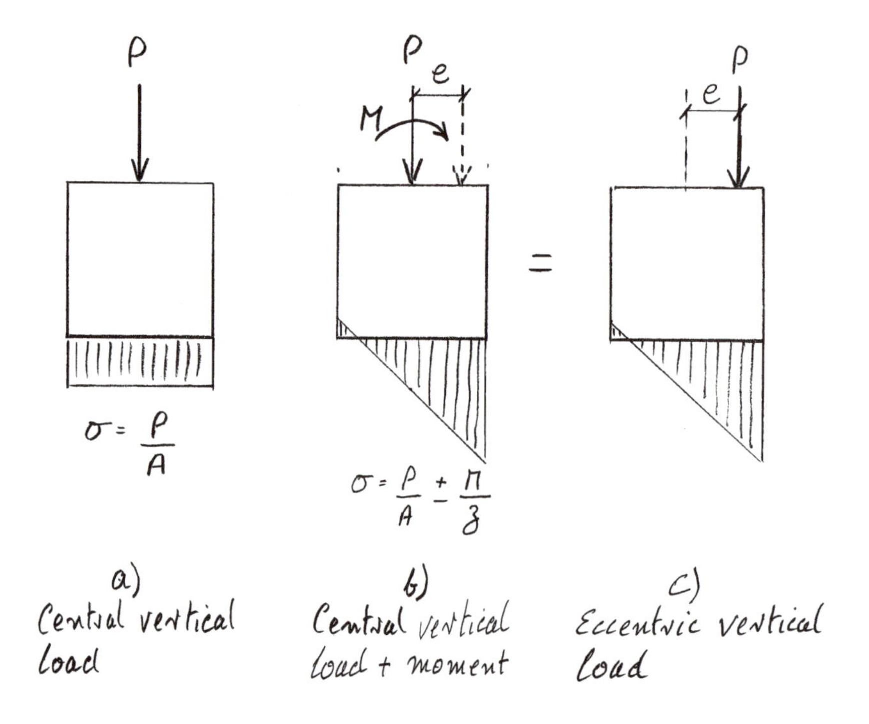

Figure 5.4 Centre of pressure

5.8 Calculation of the prestress force

5.8.1 General

It will be assumed initially that the design is carried out to Class 1, that is the prestress must be sized so that no tensile stresses arise under the effects of the applied bending moments.

5.8.2 Calculation using the concepts of central kern and centre of pressure

For greatest effectiveness, the tendons will be placed as low as possible in the section, but until the number of tendons has been calculated, their exact arrangement and proximity to the bottom fibre is not known. Assume that the centroid of the tendons is at 0.2 m from the bottom of the section. The essential geometry of the design section is shown in Figure 5.5 (a).

When the beam is considered to be weightless and subject to no bending moment, the centre of pressure will be coincident with the centroid of the prestressing tendons. When an external sagging bending moment M, that will tend to compress the top fibres is applied to the beam, the centre of pressure moves upwards by a distance of M/P, where P = prestress force.

For the stress on the bottom fibre of the beam to be zero when the maximum bending moment due to external loads is applied to the beam, the centre of pressure of the prestress must be located at the upper limit of the kern. The distance between the centroid of the tendons and the upper limit of the kern is:

$$(y_b - 0.2) + a_t \quad \text{or} \quad 1.472 - 0.2 + 0.488 = 1.76 \text{ m}$$

see Figure 5.5 (b).

The distance the centre of pressure may move under the maximum moment at a section is normally referred to as l_p, the prestress lever arm.

$$l_p = M_{max}/P \quad \text{or} \quad P = M_{max}/l_p.$$

Thus in this case, where the maximum moment = 16.51 MNm, and l_p = 1.76 m, the prestress force required is

$$P = 16.51 / 1.76 = 9.38 \text{ MN}.$$

The distance of the prestress tendons from the neutral axis is known as e, the eccentricity of the prestress. Here e = 1.472 – 0.2 = 1.272 m. The bending moment applied to the beam by the prestress force is Pe.

It is also necessary to check that when the beam is subjected to the minimum bending moment there are no tensile stresses on the top fibre. For this condition to be satisfied the centre of pressure must be at or above the bottom limit of the kern, under the actions of prestress force plus the minimum bending moment. For this example, the minimum bending moment exists when the beam is subjected only to its own self weight, in addition to the prestress.

The distance between the assumed prestress centroid and the bottom kern limit is:

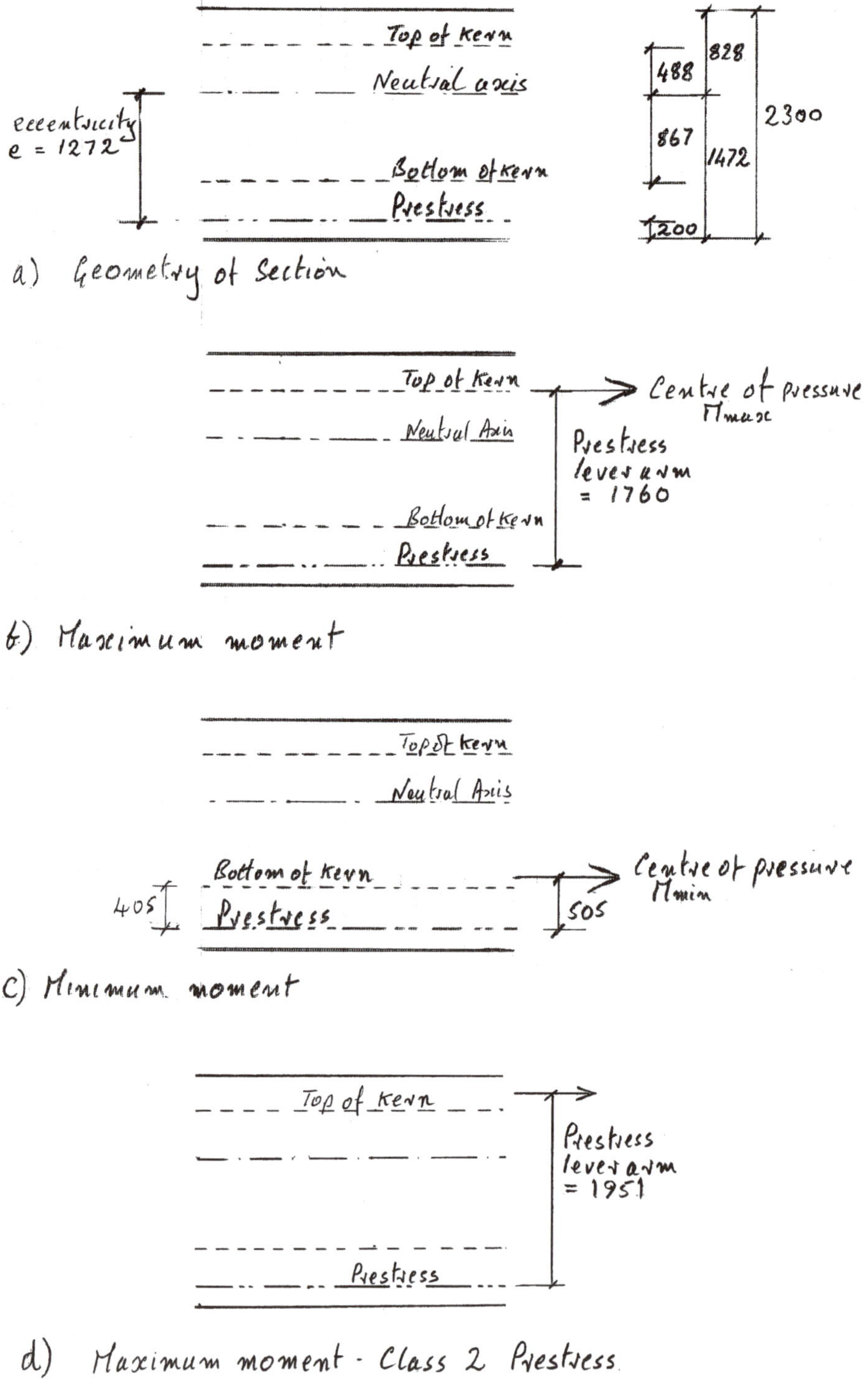

Top of Kern
Neutral axis
eccentricity
e = 1272
Bottom of Kern
Prestress
488
828
867
1472
2300
200

a) Geometry of Section

Top of Kern
Neutral Axis
Bottom of Kern
Prestress
Centre of pressure
Mmax
Prestress
lever arm
= 1760

b) Maximum moment

Top of Kern
Neutral Axis
Bottom of Kern
Prestress
405
505
Centre of pressure
Mmin

c) Minimum moment

Top of Kern
Prestress
Prestress
lever arm
= 1951

d) Maximum moment - Class 2 Prestress

Figure 5.5 Calculation of prestress force using kern and centre of pressure

$y_b - a_b - 0.2$, or $1.472 - 0.867 - 0.2 = 0.405$ m.

Under the effect of the self-weight moment, which is 4.74 MNm, the centre of pressure moves up from the centroid of the tendons by 4.74 / 9.38 = 0.505 m, Figure 5.5 (c). As this is above the bottom kern limit, no tensile stresses will occur under combined permanent prestress force and self weight alone. The addition of the deck finishes will further compress the top flange.

Figure 5.5 illustrates these calculations graphically. This graphical representation is extremely useful, and a diagram based on Figure 5.5 should be sketched *to scale* and referred to whenever this type of calculation is being done. Use of this figure is much more effective than a sign convention for clarifying the direction of movement of the centre of pressure, in particular when moments of different sign are applied to the section, or for complicated structures such as portals and frames.

5.8.3 Calculation by direct consideration of stresses

Finding the required prestress force using the direct calculation of stresses operates as follows. The function of the prestress is to just cancel out the tensile stresses created on the bottom fibre of the beam by the maximum applied moment. This tensile stress is shown in Table 5.1, and is –23.02 MPa.

Assume that a prestress force of 1 MN is applied 0.2 m above the soffit. The stresses caused by that force acting alone would be as for the column described above, $P/A \pm M/z$, where M = the moment due to prestress = Pe. As the section is not symmetrical, z is different for the top and bottom fibres, and is given in the list of section properties in *5.3*, together with A.

Eccentricity $e = 1.272$ m, and $Pe = 1 \times 1.272$ MNm.

The stresses on the extreme fibres due to unit prestress acting alone are:

Top fibre $1 / 1.47 - 1 \times 1.272 / 1.274 = 0.680 - 0.998 = -0.318$ MPa

Bottom fibre $1 / 1.47 + 1 \times 1.272 / 0.717 = 0.680 + 1.774 = +2.454$ MPa

Consequently, in order to counteract the bottom fibre tensile stress of –23.02 MPa the prestress force must be 23.02 / 2.454 = 9.38 MN. This is of course identical with the former calculation.

The stresses on the extreme fibres due to 9.38 MN of prestress alone are thus:

Top fibre $9.38 \times -0.318 = -2.98$ MPa

Bottom fibre $9.38 \times +2.454 = +23.02$ MPa

Table 5.3

	Top fibre		*Bottom fibre*	
	σ partial	*σ cumulated*	*σ partial*	*σ cumulated*
Self weight	+3.72		–6.61	
Prestress	–2.98	+0.74	+23.02	+16.41
Finishes	+0.76	+1.50	–1.35	+15.06
Live load	+8.48	+9.98	–15.06	0

5.9 Table of stresses

It is always necessary to check the stresses after any sizing exercise in order to confirm the suitability of the beam under the applied loads and to assist in checking for mistakes in the calculation. By far the best format for tabulating stresses is shown in Table 5.3.

All stresses are in MPa. 'σ cumulated' gives the running total of the 'σ partials'.

This presentation has the great advantage that it shows visually how the stresses build up stage by stage, and makes it very easy to spot mistakes in the calculation, or to assess the effects of any change, of load or of prestress.

Many of the calculations required for designing a prestressed concrete bridge can be readily carried out using spreadsheets. However, it is very important that the spreadsheets are set out in such a way that the tabulation of stresses maintains the clarity of the presentation shown above.

One principal aim of the preliminary design calculations described in this book is to illuminate the changes that may be made to the design to improve its performance. Methods of calculation or modes of presentation of calculations that simply publish the final result are of no use in this endeavour. It is essential for the engineer to understand the difference between calculations that are aimed at checking the compliance of an established design, and those essentially exploratory calculations carried out during the true 'design' phase.

Table 5.3 shows that when the prestress is combined with self weight alone, the stresses on the top and bottom fibres respectively are +0.74 MPa and +16.41 MPa. The maximum compressive stress on the top fibre under the effect of live load is +9.98 MPa, while on the bottom fibre the stress falls to zero, in accordance with the sizing calculations.

Table 5.4

	Top fibre stresses		*Bottom fibre stresses*	
	σ partial	*σ cumulated*	*σ partial*	*σ cumulated*
Self weight	+3.72		–6.61	
Initial prestress	–3.43	+0.29	+26.47	+19.86
Prestress loss	+0.45	+0.74	–3.45	+16.41
Finishes	+0.76	+1.50	–1.35	+15.06
Live load	+8.48	+9.98	–15.06	0

This is clearly a very simple example, and the benefits of this presentation of the stresses become more apparent as the reality becomes more complex. For instance, in a bridge beam such as this, when the tendons are first stressed, the force in them will be higher than its final value (losses of prestress are described in *5.17*). This may be included as shown in Table 5.4, which assumes that the initial prestress is 15 per cent higher than its final value (note that the stresses due to prestress are proportional to the prestress force).

This shows that when the tendons were first stressed the compression on the bottom fibre was +19.86 MPa, which is high enough to cause concern. This is typical of this form of beam.

The table flags up the areas of concern, and allows the designer to concentrate on them. For instance, with such a high stress on the bottom fibre, the effect of the empty cable ducts on the section properties of the beam should be investigated, as should the possibility that the density of the concrete of the beam may be less than assumed, or that the span of the beam when temporarily supported in the casting yard may be less than the design span. If the stresses had been well within limits, such refinements in the calculation would not have been necessary.

If it is concluded that this stress is, in the opinion of the designer, unsafe or is above the limit allowed by the relevant code of practice for concrete at the age when the last prestressing cable is stressed, it would be necessary to take one of the following steps:

- increase the size of the heel of the beam. This would increase I and z_b and thus reduce the bending stresses on the bottom fibre;
- increase the strength of the concrete;
- delay the stressing of some cables until the concrete had gained adequate strength;
- provide compression reinforcement in the heel, increasing the transformed section properties. This is not an economical option, and would only be adopted if it were impossible to increase the size of the heel.

Books on prestressed concrete which are more analytical than this one give equations that allow one to calculate the prestress force directly, and to check that limiting compressive stresses are not exceeded. The author of this book prefers to use the concepts described in this section, as they induce the designer to think about the physical reality of the prestressing and the concrete stresses involved rather than their becoming mathematical abstractions.

5.10 Non-zero stress limits

It is quite simple to adapt the method of determining the prestress force by direct consideration of stresses to situations where the limiting tensile stresses are not zero. For instance, if the design specification had called for Class 2 prestress, and the limiting stress on the bottom fibre had been –2.5 MPa, rather than zero, the prestress force may be calculated as follows.

We know that $P/A + Pe/z_b = 23.02 - 2.5 = 20.52$ MPa (*5.8.3*).

Table 5.5

	Top fibre stresses		*Bottom fibre stresses*	
	σ partial	*σ cumulated*	*σ partial*	*σ cumulated*
Self weight	+3.72		–6.61	
Initial prestress	–3.06	+0.66	+23.59	+16.98
Prestress loss	+0.40	+1.06	–3.07	+13.91
Finishes	+0.76	+1.82	–1.35	+12.56
Live load	+8.48	+10.3	–15.06	–2.50

With $A = 1.47$, $e = 1.272$ and $z_b = 0.717$, the new prestress force may be calculated as $P = 8.36$ MN.

The table of stresses, including the effect of the initial overstress of the tendons, is given in Table 5.5.

One of the most significant benefits of reducing the prestress in this way, in addition to the reduction in cost of the prestressing, is the reduction in compressive stress on the bottom fibre of the beam. Whereas before, the stress was close to 20 MPa, this falls to close to 17 MPa, which is much easier to accommodate.

One may also calculate the new prestress force by considering the movements of the centre of pressure as per *5.8.2*. The stress on the bottom fibre due to a prestress force of 9.38 MN acting alone at an eccentricity of 1.272 m has been calculated as 23.02 MPa. A movement of the centre of pressure of 1.272 m would bring it up to the neutral axis of the beam, and the stress on the bottom fibre would have fallen to $P/A = 9.38 / 1.47 = 6.38$ MPa. Hence a movement of 1.272 m of the centre of pressure results in a reduction of stress on the bottom fibre of 23.02 – 6.38 = 16.64 MPa. Consequently, assuming a linear relationship, in order to reduce the stress on the bottom fibre by a further 6.38 + 2.5 = 8.88 MPa, it is necessary for the centre of pressure to move a further $8.88 \times 1.272 / 16.64 = 0.679$ m. This gives a new prestress lever arm of 1.272 + 0.679 = 1.951 m, from which one may calculate a new prestress force of 16.51 MNm / 1.951 m = 8.46 MN, which is close to the correct figure of 8.36 MN. The centre of pressure is now outside the central kern by 0.679 m – a_t = 0.679 – 0.488= 0.191 m, see Figure 5.5 (d).

This method is approximate as the relationship of stress with the movement of the centre of pressure is not quite linear, but as can be seen, gives a result that is very close to the previous exact calculation, and further illustrates the concepts of the kern, the centre of pressure and the prestress lever arm.

5.11 Compressive stress limits

It should be noted that the calculations described above only govern the attainment of the minimum stress criteria. The prestress so defined may overstress the concrete in compression. There exist equations which allow one to adjust the prestress eccentricity to respect maximum stresses as well as minimum stresses. However, in the author's experience, they are not of much practical use for design. If the concrete is overstressed in compression, in most cases the remedy is not to adjust the cable profile but to increase the concrete dimensions, or its specified strength. Sizing the concrete section is far too complex an activity to be resolved by manipulating equations.

5.12 Sign convention

A simply supported beam is a very simple example of the calculation of prestress force and position, as the bending moments due to applied loads are all of the same sign. However, there are more complex examples, such as continuous beams where any section is subjected to moments of different sign, or portals and frames where the meaning of positive and negative moments must be defined carefully. For this reason, it is useful to set out the sign convention that is adopted for this book, and that is used widely in the industry:

- Compression is positive, tension negative.
- A positive moment causes tension on the bottom fibre of a beam, or the inside fibre for a portal.
- The bending moment (*Pe*) due to tendons situated below the neutral axis stresses the top fibre in tension. Consequently eccentricity *e* is negative below the neutral axis, and positive above. However, the sign for eccentricity is frequently omitted when there is no ambiguity about its location, such as for statically determinate beams when the cables are always below the neutral axis.

In the author's experience, there is no effective substitute to thinking about the physical reality, understanding in which direction the centre of pressure moves when a moment is applied and drawing diagrams similar to those shown in Figure 5.5 to visualise the limits of the cable zone.

5.13 Arrangement of tendons at mid-span

The prestress force of 9.38 MN now needs to be split up into individual tendons. A useful rule of thumb for preliminary design is to assume that the working stress for strand is 1,000 MPa, which means that modern 13 mm strand which has an area of 100 mm^2, may be assumed to have a force of 100 kN per strand, and 15 mm strand, with an area of 150 mm^2 per strand, 150 kN. (Clearly, for detailed design or for estimating quantities, it is necessary to carry out a calculation of the prestress losses and hence to determine more accurately the long-term prestress force in each strand.) Hence, 9.38 MN represents 63 of the larger strands or 94 of the smaller size. As standard anchorages are available for 12 or 19 strands, a reasonable choice would be to adopt 8 tendons each consisting of 12 of the 13 mm strand, providing 96 strands. These tendons will be housed in ducts with internal/external diameters of 65 mm/72 mm. (If plastic ducts are adopted they will be somewhat larger.) (See also Chapter 9 for the determination of web thickness and methods of compacting the concrete.)

The force provided will thus be 96×0.1 MN = 9.6 MN, slightly more than required. These 8 tendons need to be arranged at mid-span in such a way that filling the heel of the beam with concrete is facilitated, and then the eccentricity used for this sizing exercise should be checked. Figure 5.6 shows a possible arrangement that is in accordance with the rules of the British Standard, and which maximises the eccentricity of the tendon group. The centroid of the force is 0.136 m from the soffit. Thus the value of 0.2 m adopted for the calculation was conservative, and a small reduction in prestress force would be achieved by increasing the eccentricity adopted for the

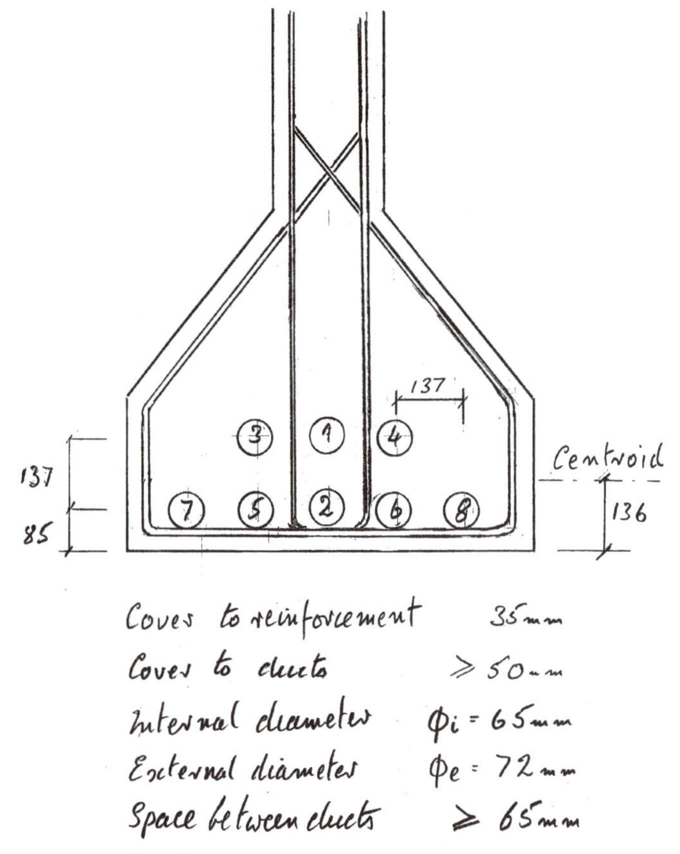

Figure 5.6 Arrangement of tendons

calculation. However, it is not worth carrying out such a recalculation until a more accurate figure for the force in the cables has been found.

5.14 Cable zone

Clearly the bending moment varies along the length of the beam and reduces to zero at its ends. Once the prestress force has been defined at mid-span, it is necessary to calculate the safe eccentricity for the prestress centroid at all points along the beam. This safe eccentricity is defined such that the specified minimum stresses on the extreme fibres are not exceeded under any combination of dead and live loads. The safe zone for the centroid of the cables is called the cable zone.

If the prestress centroid is at the upper limit of the cable zone, under maximum applied bending moments, the stress on the bottom fibre of the beam will be equal to the minimum allowable stress. When the prestress centroid is at the lower limit of the zone, the stress on the top fibre will be at the permitted minimum.

The cable zone at the mid-span section of the beam may be found by considering the moments given in Table 5.1. First, assume that the cable force is the theoretically defined value of 9.38 MN, and the limiting stresses are zero on both extreme fibres. Under the effect of the maximum moment of 16.51 MNm the centre of pressure must

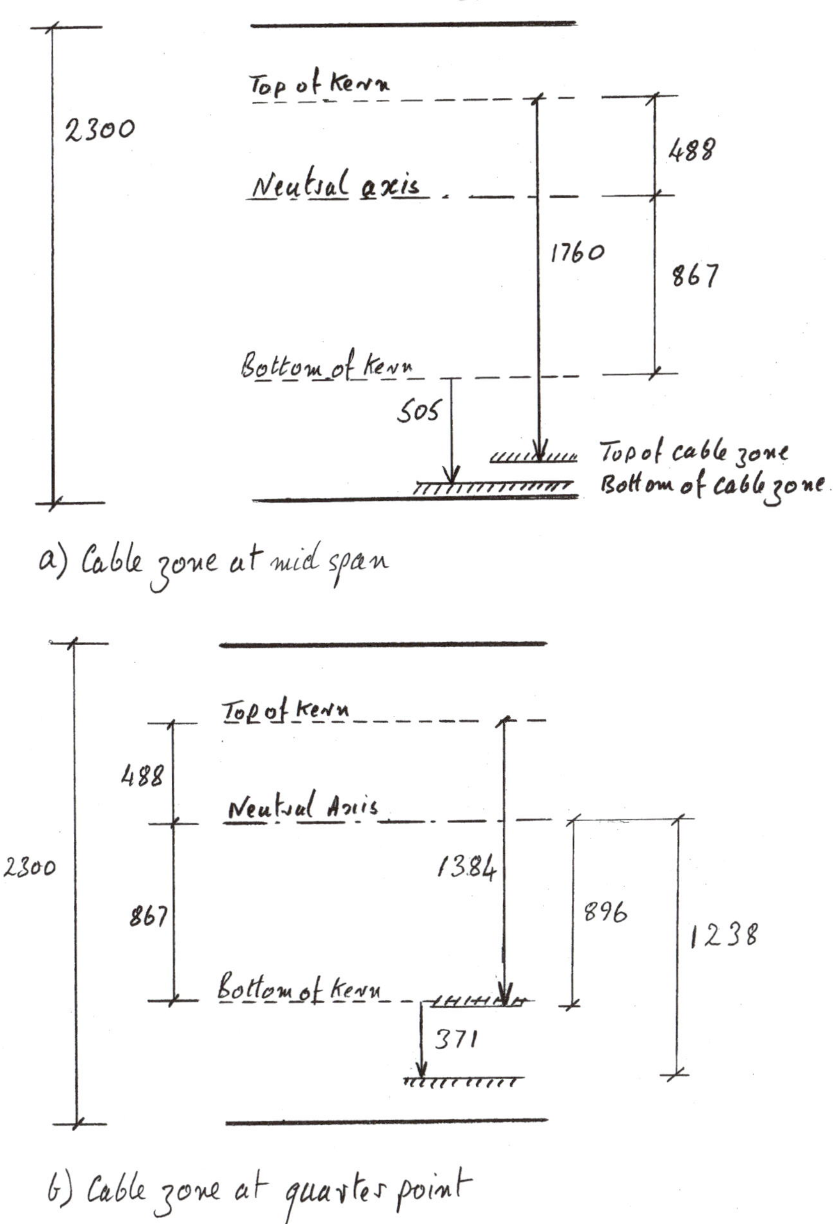

Figure 5.7 Plotting the cable zone

not rise above the top of the central kern if the stress on the bottom fibre is not to fall below zero. The top of the cable zone can thus be found by plotting a point that is 16.51 / 9.38 = 1.76 m down from the top of the kern. As a_t = 0.488, the top of the cable zone is at an eccentricity of 1.76 – 0.488 = 1.272 m. This is shown in Figure 5.7 (a).

Under the effect of the minimum moment, which is 4.74 MNm, the centre of pressure must not lie below the bottom of the kern, if the stress on the top fibre is not to be below zero. Thus one may find the bottom of the cable zone by plotting a point that is 4.74 / 9.38 = 0.505 m down from the bottom of the kern. As $a_b = 0.867$, the bottom of the cable zone is at an eccentricity of 0.867 + 0.505 = 1.372 m. The cable zone at mid-span may thus be defined as lying between eccentricities of 1.272 m and 1.372 m. The zone has a width of 0.1 m. The bottom of the cable zone may in some cases lie below the maximum achievable eccentricity of the prestress centroid, defined by the minimum cover that must be provided to the tendons.

If the adopted cable force of 9.6 MN is used for this calculation, the top of the cable zone lies at e = 1.232 m and the bottom at 1.361 m, a width of 0.129 m. When the prestress force is increased, the cable zone is always widened and its eccentricity reduced.

If we consider the quarter point of the beam described above, the bending moments and stresses are given in Table 5.6.

For the purpose of this demonstration, we will assume that the prestress force is identical with the adopted force at mid-span, that is 9.6 MN. In reality, this is a reasonable assumption for preliminary design, but for final design the variation of force along the length of the prestressing cables will need to be considered (see *5.17*).

Under the effect of the maximum moment of 13.29 MNm the centre of pressure must not rise above the top of the central kern if the stress on the bottom fibre is not to fall below zero. The top of the cable zone can thus be found by plotting a point that is 13.29 / 9.6 = 1.384 m down from the top of the kern. As $a_t = 0.488$, the top of the kern is at an eccentricity of 1.384 – 0.488 = 0.896 m.

Under the effect of the minimum moment, which is 3.56 MNm, the centre of pressure must not lie below the bottom of the kern if the stress on the top fibre is not

Table 5.6

	Moment MNm	*Top fibre stress* MPa	*Bottom fibre stress* MPa
Self weight	3.56	+2.79	–4.97
Finishes	0.73	+0.57	–1.02
Live load	9.00	+7.06	–12.55
Total	13.29	+10.42	–18.54

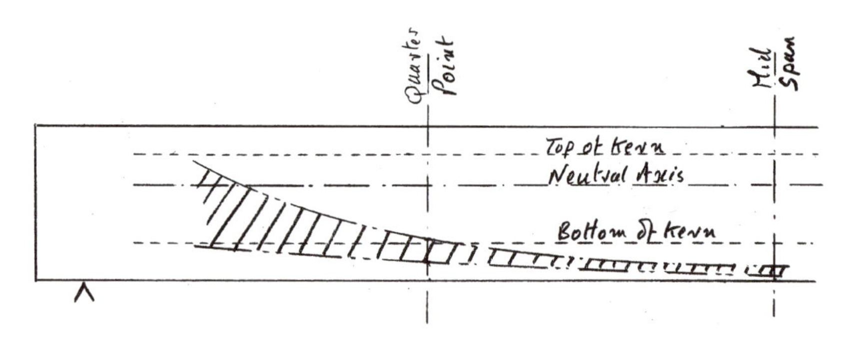

Figure 5.8 Elevation of cable zone

to be below zero. Thus one may find the bottom of the cable zone by plotting a point that is 3.56 / 9.6 = 0.371 m down from the bottom of the kern. As a_b = 0.867 the bottom of the kern is at an eccentricity of 0.867 + 0.371 = 1.238 m, Figure 5.7 (b).

The cable zone at the quarter point may thus be defined as lying between eccentricities of 0.896 m and 1.238 m.

The cable zone may be plotted in this way at any number of points along the beam, Figure 5.8. For a simply supported beam, few points are necessary. However, for a continuous beam, it is normally necessary to plot the cable zone at every tenth point of each span at least.

5.15 The technology of prestressing

In order to understand the terms used in the following sections, it is necessary to describe briefly the technology of prestressing. This subject is covered extensively by the brochures of the manufacturers of prestressing equipment; in the UK these are principally CCL, VSL and Freyssinet for cables made of strands, BBR for cables made of strands or wires, and Macalloy and Dywidag for prestressed bars. However, many other suppliers exist worldwide. Consequently, only a brief outline of the technology will be given here.

Strand consists of sub-cables made up of wires that are twisted together. Usually, each strand consists of seven wires, as may be seen in Figure 5.13 (a). Most commonly, strand comes in 13 mm nominal size (12.5 mm or 12.9 mm actual size) made up of wires that are approximately 4 mm in diameter, or 15 mm nominal size (15.2 mm or 15.7 mm actual size) made up of wires that are approximately 5 mm in diameter. Wire

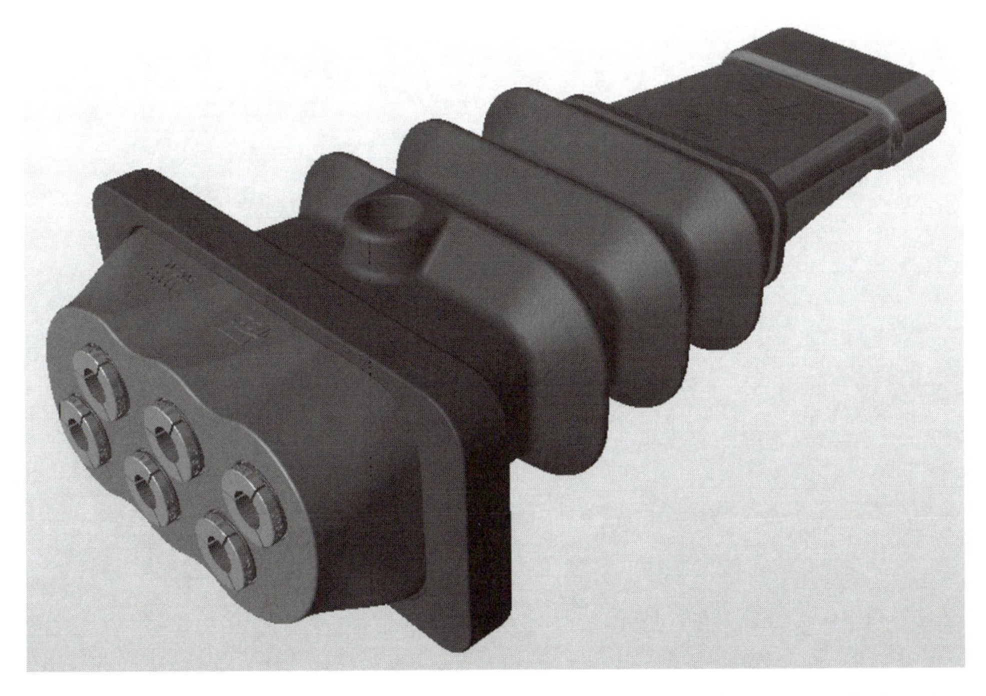

Figure 5.9a Typical prestress anchors: CCL slab anchor for 6 strands (Image: CCL Stressing Systems Ltd)

Figure 5.9b Typical prestress anchors: CCL anchor for 19 No 15.7 mm strands (Image: CCL Stressing Systems Ltd)

Figure 5.9c Typical prestress anchors: CCL anchor for 37 No 15.7 mm strands (Image: CCL Stressing Systems Ltd)

Figure 5.9d Typical prestress anchors: bar anchors (Photo: Benaim)

Figure 5.9e Typical prestress anchors: buried dead anchors (Photo: Benaim)

tendons are generally made up of 7 mm wires, while bars range from 16 mm up to 75 mm diameter.

The tendon anchors consist of steel fabrications or cast iron forgings which are sized to apply acceptable pressure to the concrete. Figure 5.9 (a) shows a CCL anchor for six strands designed to fit into a thin slab, Figure 5.9 (b) shows a CCL anchor for a typical bridge deck tendon with 19 No 15.7 mm strands and Figure 5.9 (c) shows one of the most powerful CCL anchors available, with 37 No 15.7 mm strands and an ultimate force of 10,323 kN.

A stressed anchor is known as 'live', while unstressed anchors are known as 'dead'. Each strand is locked to a live anchor by hardened steel wedges that fit into conical holes. Strand at dead anchors may also be locked by wedges that are pushed home before the tendon is stressed, or by metal sleeves that are swaged onto the individual strands. Wires are locked to the anchorage by button heads forged onto each wire, while the anchorage body is threaded to take up the extension of the tendon. Bars are

Figure 5.10 Prestressing jack (Photo: Benaim)

threaded, either over their full length or just at the ends, and are locked by nuts usually bearing on steel plates, Figure 5.9 (d).

Savings in anchorage hardware and in labour may be achieved by adopting buried anchors for the unstressed end of tendons. These anchors generally work by exploiting the bond of the concrete to the steel tendons, or by looping the individual wires or strands, Figure 5.9 (e). Buried anchors make it essential to prefabricate the tendon and to place it in the reinforcing cage before the concrete is cast, rather than threading the tendon into the concrete after it has been cast, which is the preferred method of construction.

The prestressing tendons must be tensioned by hydraulic jacks, Figure 5.10, and then their force transferred from the jack to the concrete by anchors. The tendons may be stressed from one end only, or from both ends. Clearly, single-end stressing is preferable, as it minimises the cost of labour and reduces the number of jacks required on site. The decision on whether to stress from one end or two depends principally on the friction losses in the cable (*5.17.2*). When tendons are less than about 50 m long, they are normally single-end stressed. However, much longer cables may be single-end stressed, for instance when their alignment is very flat and friction losses are low, or when it is not practical to fit a jack to one end. When the tendons are stressed they extend, typically by 6 mm/m for wire and strand, and 3.5 mm/m for bars.

Double-end stressing generally employs two jacks acting simultaneously. However, it is also possible to stress from one end and then to transfer the jack to the other end to complete the process. The main limitation on this latter procedure for cables that are anchored by wedges is that if the force at the unstressed end is too great, it may be impossible to release the wedges.

5.16 Cable profile

5.16.1 Arrangement of cables

Once the cable zone has been defined, it is necessary to arrange the individual cables, such that their centroid lies within the zone. The simplest cable profile would consist of keeping all the cables in single file in the web. Although this is both simple to design, and simple to build, it is clear that when compared with Figure 5.6 the eccentricity of the cable group would be significantly less than was assumed in the calculation of cable force, and consequently a substantially greater prestress force would be necessary.

In the cable arrangement shown in Figure 5.6 some of the cables are housed in the heel of the beam and have to pass across the plane of the links in the web. As the cables will cross this plane of reinforcement at a shallow angle, several links will be interrupted, and the designer has to devise a suitable reinforcement detail that maintains the strength of the web. It is most definitely not satisfactory to leave this detail to be defined by the contractor, or the site staff.

For a simply supported beam such as our example, a third fixed point, other than the arrangement of tendons at mid-span and at the quarter point, is the arrangement of the anchors at the beam end (*5.24*). The designer's first choice would be to spread the anchorages out evenly up the full height of the web, Figure 5.11 (a). This will stress the concrete evenly, and reduce the amount of reinforcement required. In general, the web will have been thickened to accommodate the size of anchor chosen.

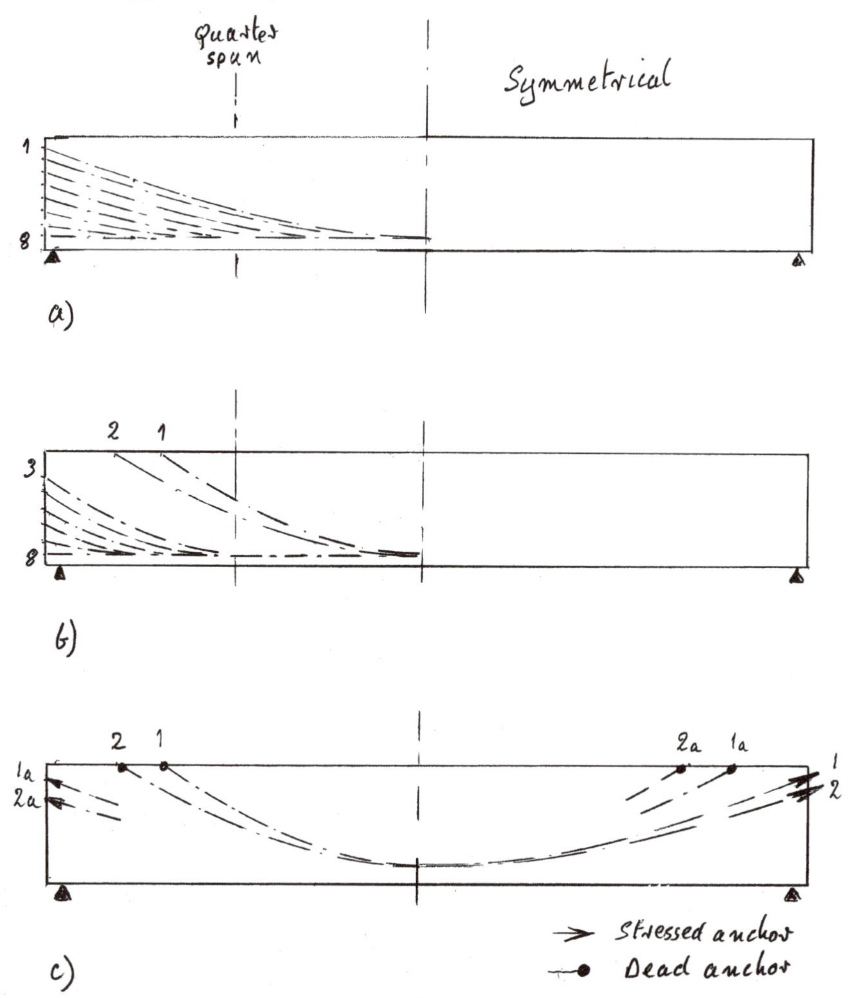

Figure 5.11 Cable arrangements

It would be possible to define the locus of the centroid of the prestressing cables as a parabola and then to arrange that each cable follows flatter or steeper versions of this curve. Although this is an option frequently adopted, it is not a very practical idea. The curve will inevitably be very flat near the centre of the beam, and every cable support will be at a slightly different level. It is much better practice to adopt a pragmatic cable profile, where each cable stays parallel to the beam soffit for a length near mid-span, and then peels off as required, as shown in Figure 5.11 (a and b). The overall centroid of the cable group may still be arranged to lie close to a parabola, if necessary. A simple cable profile, used on the Sungai Kelantan Bridge, is shown in Figure 5.12.

As the strands or wires making up the prestress tendon are usually pushed or pulled into the empty ducts, these have to be a relatively loose fit around the tendon. Generally, the cross section of the duct is approximately twice the area of steel of the

Figure 5.12 Cables in a typical beam (Photo: Benaim)

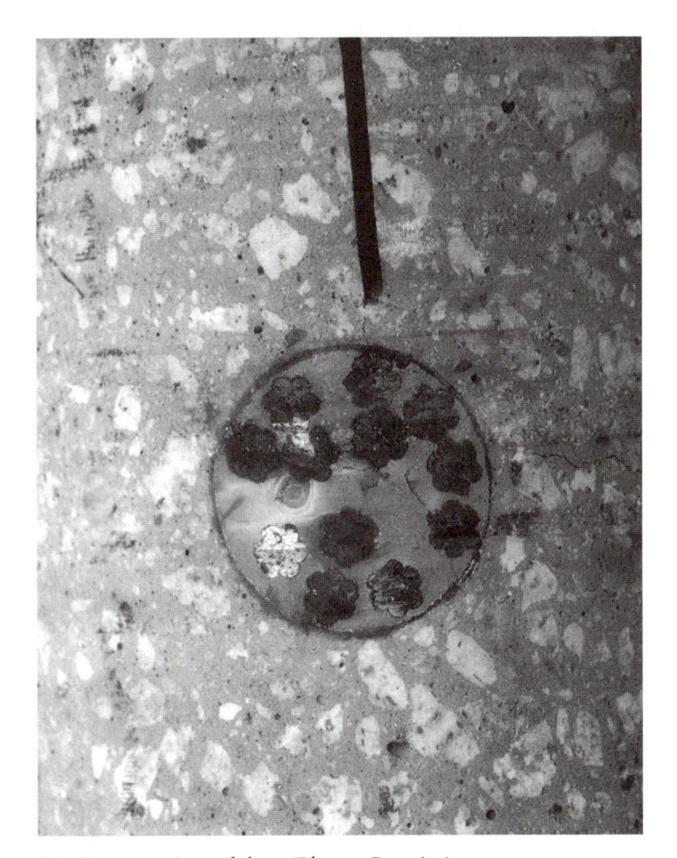

(a) Cross section of duct (Photo: Benaim)

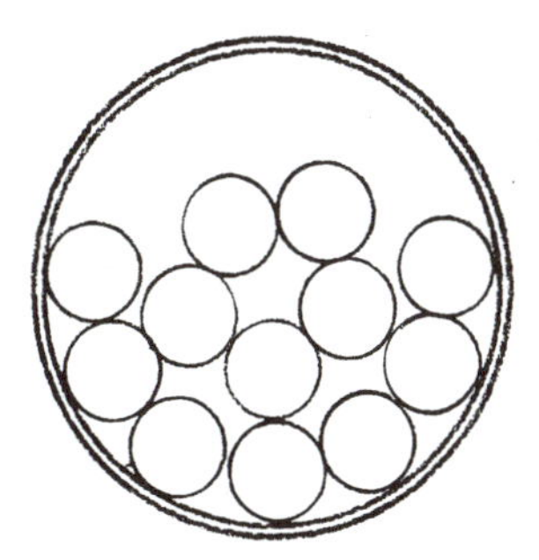

(b) Bunching of strands in a curved duct

Figure 5.13 Eccentricity of cables within their ducts

tendon, Figure 5.13 (a). Consequently, when the duct is curved, the tendon will bunch up to one side. The design of the prestress should take into account the position of the centroid of the bundle of strands or wires within the duct, Figure 5.13 (b), rather than the axis of the duct itself.

Reference [2] is useful for all aspects of prestress technology.

5.16.2 Stopping off cables

In order to save prestressing steel, it is possible to reduce the prestress force in zones of the beam where bending moments are lower. For instance, the maximum bending moment at the quarter point of the beam is 13.29 MNm, down from 16.51 MNm at mid-span. Rather than raising the prestressing cables, they could have been kept at their maximum eccentricity and the force reduced. For a stress limit of zero on the bottom fibre, the prestress lever arm would be the same as at mid-span, that is 1.76 m. The prestress force required at the quarter point would thus be 7.73 MN, corresponding to 77 strands.

Practically, with the cable arrangement shown in Figure 5.11 (b) it would be possible to reduce the number of tendons in the heel from 8 to 7. Further cables may be stopped off as the bending moment falls. Cables 1 and 2 are shown swept up and anchored in pockets in the top flange of the beam. Although reducing the total weight

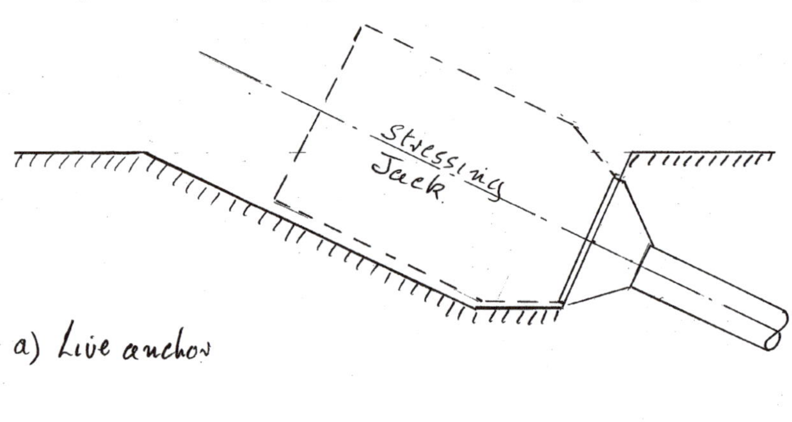

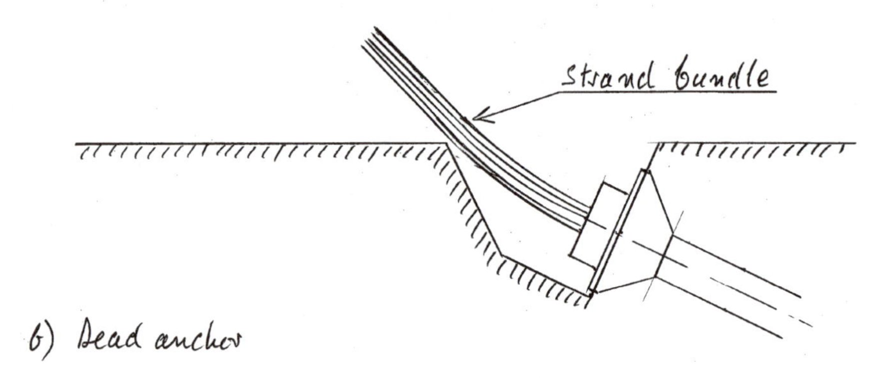

Figure 5.14 Arrangement of swept up anchors

of prestressing steel, this solution is not often adopted in modern precast beam decks. In order to fit a stressing jack onto the anchor, it is necessary to provide a wide and deep pocket, Figure 5.14 (a), which would require a local thickening of the web/slab junction, complicating the fabrication of the form and the design and placement of the reinforcement, and affecting the appearance of the beam. Also, the pocket would interrupt the transverse reinforcement of the top slab over a considerable length, requiring relatively heavy trimming steel.

In fact, as the tendons are short, they may be tensioned from one end only. At the unstressed end, where no stressing jack has to be fitted to the anchor, a dead anchor may be used, needing a much smaller pocket, Figure 5.14 (b).

An alternative tendon arrangement would have been to design cables 1 and 2 as an anti-symmetrical pair, with each being anchored at the end of the web at one end, and swept up to the top flange at the other, Figure 5.11 (c). This would allow the cables to be stressed from the beam end, with a dead anchor in the top flange. In fact, within a 220 mm web this arrangement is not possible, as cable 1 must pass cable 2. However, in bridge decks with thicker webs, such anti-symmetrical tendon arrangements are frequently used in designs for both isostatic and continuous prestressed structures. It is always necessary to check that there is room for the cables to pass each other.

Although the pocket for a dead anchor is much smaller than that for a live anchor, it is still necessary to provide an adequate thickness of concrete beneath the anchor plate. This is unlikely to be available in the node created by the junction of the 220 mm thick web and the 200 mm thick slab of this example, leading to the need for a local thickening of the web.

In order to avoid this complication, the designer may use, for the cables that are swept up, less powerful tendons whose anchorages are compatible with the thickness of concrete available. The disadvantage of this solution would be imposing on the contractor the use of two different sizes of stressing jack. If the project includes a large number of beams, which would in any case require several sets of prestressing equipment, this may not be a problem.

Tendon anchorages housed in top pockets as described have been out of favour as there is a risk that water may infiltrate into cracks between the parent concrete of the beam and the second-stage concrete used to fill the pocket after stressing the tendons. In particular, if this water is contaminated with de-icing salts, there would be a risk of corrosion of the tendons. During construction, before the pockets have been concreted, the ducts may fill with rainwater, corroding the prestressing tendons. Furthermore, in some climates the water in the ducts may freeze, splitting the beams.

All of these risks are real. However, it is quite possible to overcome them, in the temporary state by good specification and site procedures, and in the permanent works by covering the pocket with an additional layer of waterproofing, such as a film of epoxy resin.

To economise on the cost of a dead anchor and to avoid the need for a top pocket and a web thickening, the designer could opt for a buried anchor, Figure 5.9 (e), with the disadvantages described in *5.15*.

The tendons that have been swept up apply forces to the concrete right up to the anchorage itself. For the example shown in Figure 5.11 (b), the calculation of the effect of the prestress at the quarter point needs to include the P and the e of the swept up tendons. While they are within the central kern, they add compression to the concrete cross section, but no tension. However, as they rise towards their anchorage points, they move above the kern, and will reduce the compressive stress on the bottom flange of the beam. Clearly, when tendons are stopped off in this way, more design sections need to be checked. The calculation of the cable zone must take such stopping-off tendons into consideration. Furthermore, stopping-off cables reduces the relief of shear force provided by the prestress (*5.20*).

Cutting short some cables as described will save prestressing steel, although it will not economise on prestress anchorages. Adopting buried anchors would save on the cost of hardware. However, these savings have to be balanced against any additional costs to other aspects of the project, including those associated with the changes in working procedure. This balancing of pros and cons is the essence of the work of the prestressed concrete bridge designer.

5.16.3 Steeply inclined cables

It should be noted that the prestress force P used in the calculations of *5.8 et seq.* should strictly be the horizontal component of the force in the tendons, that is $P\cos\alpha$, where α is the angle between the tendon and the neutral axis of the beam.

In most prestressing schemes the tendons have a relatively flat profile, where α rarely exceeds 10°. Consequently, the approximations that $P\cos\alpha = P$ and $P\sin\alpha = P\tan\alpha$ are generally accepted.

However, in cases when $\alpha > 10°$, such as when cables are swept up as described above, a more exact calculation may be appropriate.

5.17 Losses of prestress

5.17.1 General

When prestress was first attempted, tentatively in the late nineteenth century and more seriously in the 1930s, high strength steel was not available. The strain ε in mild steel prestressed to 200 MPa would have been

$$\varepsilon = \text{stress} / E = 200 / 200{,}000 = 1{,}000 \times 10^{-6}.$$

However, it was not appreciated that much of the extension of the steel would be lost by shortening of the concrete due to shrinkage and creep. For instance, concrete shrinkage strain after hardening is of the order of 200×10^{-6}, and concrete creep, when subjected to a compressive stress of say 5 MPa, is also of the order of 200×10^{-6}. Further losses of prestress would occur due to the elastic shortening of the concrete as multiple tendons were stressed, to the loss of force as the tendon force was transferred from the jacks to the concrete, and to the relaxation of the stress in the steel. Consequently the long-term force in the prestress tendons would have been less than 50 per cent of the initial force, destroying both the performance of the prestressed beam and the viability of the system.

Modern strand, with a breaking stress up to 1,860 MPa may be stressed at between approximately 1,300 MPa and 1,600 MPa, depending on the national regulations, giving rise to extension strains typically of some 6500×10^{-6}. Consequently, the effects of concrete shortening due to elastic stresses, creep and shrinkage are proportionately much less, although they are still very significant. In every project, it is necessary to calculate the prestress force in the tendons that is actually present, both immediately after anchoring the tendons, and in the long term, at each design section along the beam. The forces in the tendons are less than the force at the stressing jacks due to the various sources of losses.

Prestress losses occur due to the following causes:

- friction
- anchor set
- shrinkage
- elastic shortening of the concrete
- creep
- relaxation of the steel.

5.17.2 Friction

Friction in ducts is due to two closely related effects, duct curvature and wobble. In a curved duct the tendon will bear on the inside of the curve, and friction will be generated as the tendon extends during stressing. The friction coefficient will depend on the nature of the duct and the condition of the tendon. It lies between about 0.1 for clean tendons bearing on smooth plastic ducts, to 0.2 for clean tendons in steel ducts and up to 0.4 for tendons bearing on concrete or for rusty tendons in steel ducts. The loss of force in the tendon will depend on the friction coefficient and on the angle turned through by the tendon. The force at any point in the cable

$$P_x = P_0 e^{-\mu\Sigma\alpha},$$

where:

P_x is the force at point x

P_0 is the force applied by the jack at the anchorage

μ is the friction coefficient

$\Sigma\alpha$ is the cumulated angle turned through at point x, in radians.

The angular deviation of a tendon is very much under the control of the designer, who should generally attempt to minimise friction losses in tendons by adopting the flattest profiles, and by avoiding unnecessary points of inflexion, both in elevation and in plan. Compare the two alternative alignments shown in Figure 5.15 (a). The two alignments have the same eccentricities at the centre and end of the beam, but the left-hand one has twice the angular deviation of the right-hand alignment, and consequently greater loss of force due to friction. The calculation of angular deviation must include the effects of plan curvature, as well as those in elevation.

Wobble is an additional component of frictional loss, and is related to the length of the cable. It is due to unintended variations in the duct alignment, caused by inaccuracies in placing the ducts and in their displacement during concreting. A high level of workmanship will reduce the wobble coefficient whereas forms of construction

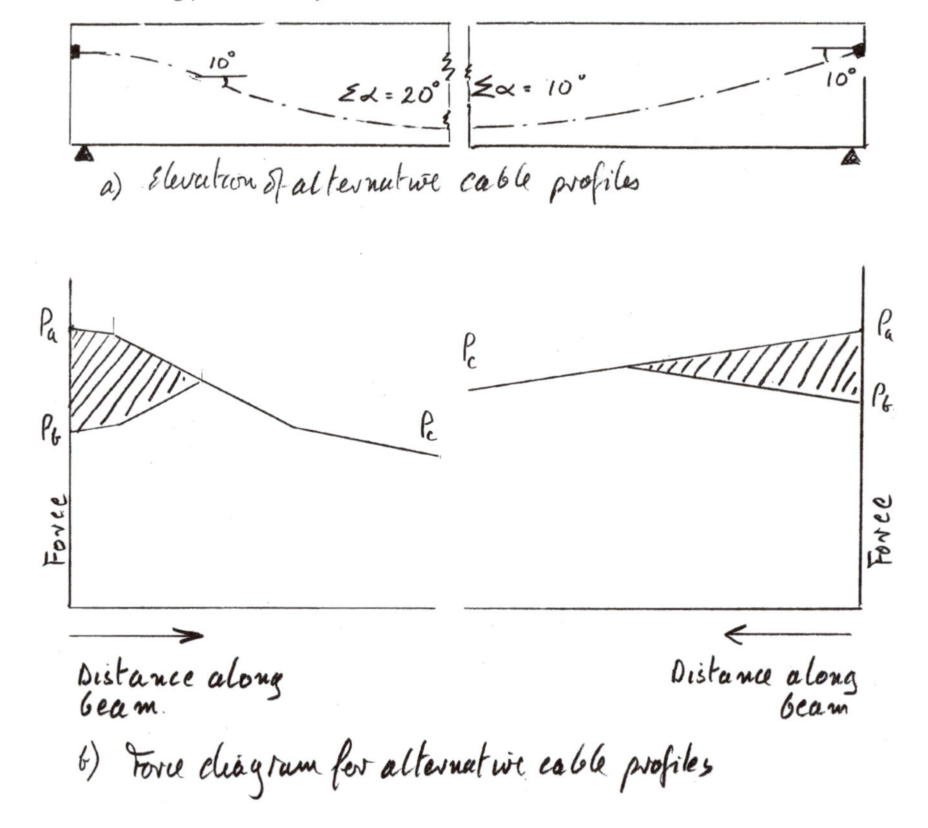

Figure 5.15 Loss of prestress due to friction and anchor set

where the beam is precast in short lengths will usually increase the wobble. The most rational way to calculate the loss of force due to wobble is to add a notional angular deviation to the intended angular deviation. The equation then becomes

$$P_x = P_0\, e^{-\mu(\Sigma\alpha + \omega x)},$$

where:

ω = a wobble coefficient

x = the distance along the tendon.

ω is expressed in radians/m, and has a value that normally lies between 0.005 and 0.01 for internal prestressing

5.17.3 Anchor set

Tendons are stressed by jacks that grip the tendon and react against the concrete. When the load is transferred from the jack to the anchorage unit, there is bound to be a loss of extension. The size of this loss depends principally on the type of tendon and anchor used.

When the tendon consists of threaded bars anchored by nuts it is possible to minimise the loss by tightening up the nut with a torque wrench before releasing the jack. However, one can only apply about 20 per cent of the full force of the bar using a torque wrench, and there will still be some set in the threads of the nut, and some bedding in of the anchor plate onto the underlying concrete. Under the best circumstances, the loss is likely to be of the order of 1–2 mm for bars with a fine thread, and 2–3 mm for a bar with a coarse rolled thread.

When the tendon consists of strands anchored by wedges, the anchor set is likely to be much greater, of the order of 6–10 mm, and more variable.

This anchor set is very significant for short tendons. A strand tendon extends on stressing by approximately 6.5 mm/m, while a bar tendon which has a lower steel strength extends by approximately 3.5 mm/m. Thus for a tendon that is only 2 m long for instance, it is not efficient to use wedge anchored strands, as at least half the extension is likely to be lost in anchor set. Even when using bars, for such short tendons fine threads should be used, and the actual force remaining in tendons is likely to vary significantly, depending on the cleanness of the threads, and the quality and consistency of the workmanship exercised in stressing and anchoring. It is advisable to re-jack short tendons to ensure consistency. (It is possible to re-stress and shim wedge anchors to eliminate part of the anchor set, but this is a time-consuming process.)

In a curved tendon, the length over which the anchor set will affect the force is limited by friction, giving a force diagram as shown in Figure 5.15 (b). In this diagram, the initial force at the stressing jack is P_a. Friction and wobble in the ducts reduces the force at the centre of the beam to P_c. It is clear that the more tortuous alignment leads to a greater loss of force due to friction. After transferring the jacking force to anchorages, anchor set reduces the force in the tendon at the anchor to P_b. The loss of force at the anchor is greater for the more tortuous alignment, but its effect extends less far along the cable. The shaded areas of the graph are proportional to the anchor set, and so are equal for the same set. The extension of the tendon is equal to G/AE, where G is the area beneath the graph, A is the cross-section area of the tendon and E is its Young's modulus.

5.17.4 Shrinkage

Whereas one would expect that the shortening of concrete due to shrinkage would be a reasonably well-documented characteristic, there remain a very wide variety of values recommended, or indeed imposed, by various national rules. The codified value of total concrete shrinkage strain lies between approximately 200×10^{-6} and 600×10^{-6}. The rate at which shrinkage occurs is very important for assessing the consequent loss of prestress. Most cast-in-situ concrete is stressed at between 2 and 7 days from casting, while precast concrete is likely to be several weeks old before it is stressed. It is necessary to make the best estimate of the amount of shrinkage remaining after stressing to calculate the loss of prestress.

The rate of development as well as the total amount of shrinkage depend on a variety of factors, which include the thickness of the concrete, the humidity of the air, the quantity of and type of cement, which are well documented in Appendix C to BS5400: Part 4: 1990, for instance.

5.17.5 Elastic shortening of the concrete

As the stressing force is applied to the concrete it shortens. If the prestress force is applied with just one tendon, this shortening would not cause any loss of prestress. However, if the prestress consists of more than one tendon, those that have already been anchored will shorten with the concrete under the effect of the stressing of subsequent tendons.

5.17.6 Creep

The delayed shortening of concrete due to the compression induced by prestress will affect all the tendons. It is the compression at the level of the tendons that creates the loss. Consequently, for a tendon that is not at the neutral axis, the stress under which the concrete creeps is affected by bending in the beam. The loss should be assessed under the long-term dead load condition of the deck. If the cables are bonded to the concrete, the loss due to creep will be local to a particular concrete section, and will not be averaged out along their length. If the tendons are unbonded, the creep loss will be averaged out over the length of the tendon.

The total amount of creep will be affected by the same factors that affect the total amount of shrinkage. However, creep is also strongly affected by the age at which the concrete is first loaded, being less for older concrete. Consequently, when stressing tendons early to allow a rapid turn-round of falsework and a short construction cycle, it is important to stress as few tendons as possible in the first phase, delaying the stressing of the remainder to as late as possible in the cycle.

5.17.7 Relaxation of the steel

Most modern strand has a relaxation that does not exceed 2.5 per cent of the initial stressing force at 1,000 hours, and that is taken as the total relaxation. However, some bars and strand may lose up to 7–8 per cent of their stress due to relaxation.

5.18 The concept of equivalent load

There is yet another method of explaining and understanding prestressing. Consider the cable in tension subjected to a concentrated force at its mid-point, as shown in Figure 5.16 (a). For equilibrium, the cable must be deflected into a ‘V’ shape. If the cable force is P, the central load W and the cable is deflected by an angle α, $W=2P\sin\alpha$. The cable is applying an upwards force that is equal and opposite to the downwards load.

Consider now the prestressed concrete beam of rectangular cross section shown in Figure 5.16 (b) of length $2l$. The prestressing cable with a force P is anchored at the neutral axis at the ends of the beam, and is deflected in a ‘V’ shape, with a deflected angle of α and an eccentricity at mid-span of e_c. The effect on the beam is of a compressive force at its neutral axis of $P\cos\alpha$, a downwards vertical force at its ends of $P\sin\alpha$, and an upwards vertical force at mid-span of $2P\sin\alpha$, Figure 5.16 (c).

As in most prestressed beams the angle of the prestressing cables with respect to the beam neutral axis is usually less than 10°, it is assumed that $\cos\alpha=1$, and that

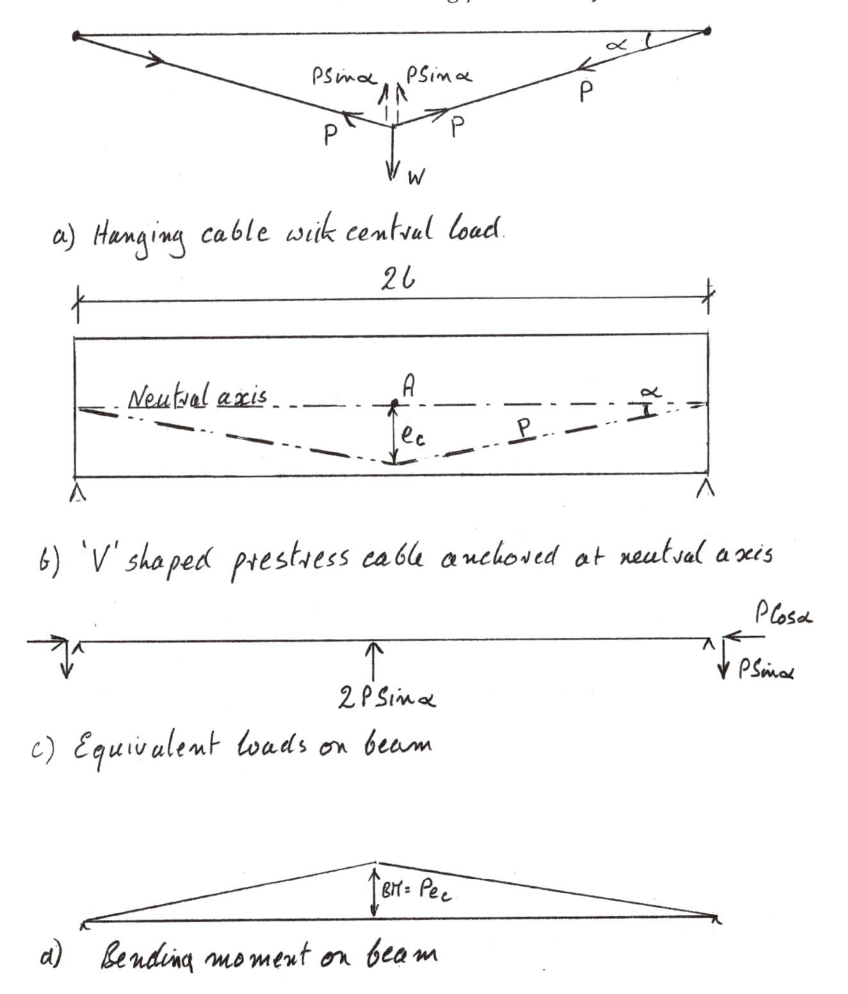

Figure 5.16 The concept of equivalent load for a 'V' profile cable

sinα=tanα. Thus the compressive force on the beam ≈ P, and the upwards force at the centre of the beam ≈2Ptanα.

The upwards force of the prestress at mid-span combined with the forces applied at the ends of the beam are known as the equivalent loads.

The bending moment at the centre of the beam, Figure 5.16 (d), caused by the prestress may be calculated in three ways:

- by considering the effect of the equivalent load at mid-span. $M_p = WL/4 = (2P\tan\alpha)\times 2l/4$. As $\tan\alpha = e_c/l$, the moment $M_p = Pe_c$;
- by taking moments about the point A, which is at the centre of the beam and on the neutral axis. The bending moment caused by the prestress alone is $l\times P\tan\alpha$, which when e_c/l is substituted for tanα also comes to Pe_c;
- by considering force × eccentricity = Pe_c.

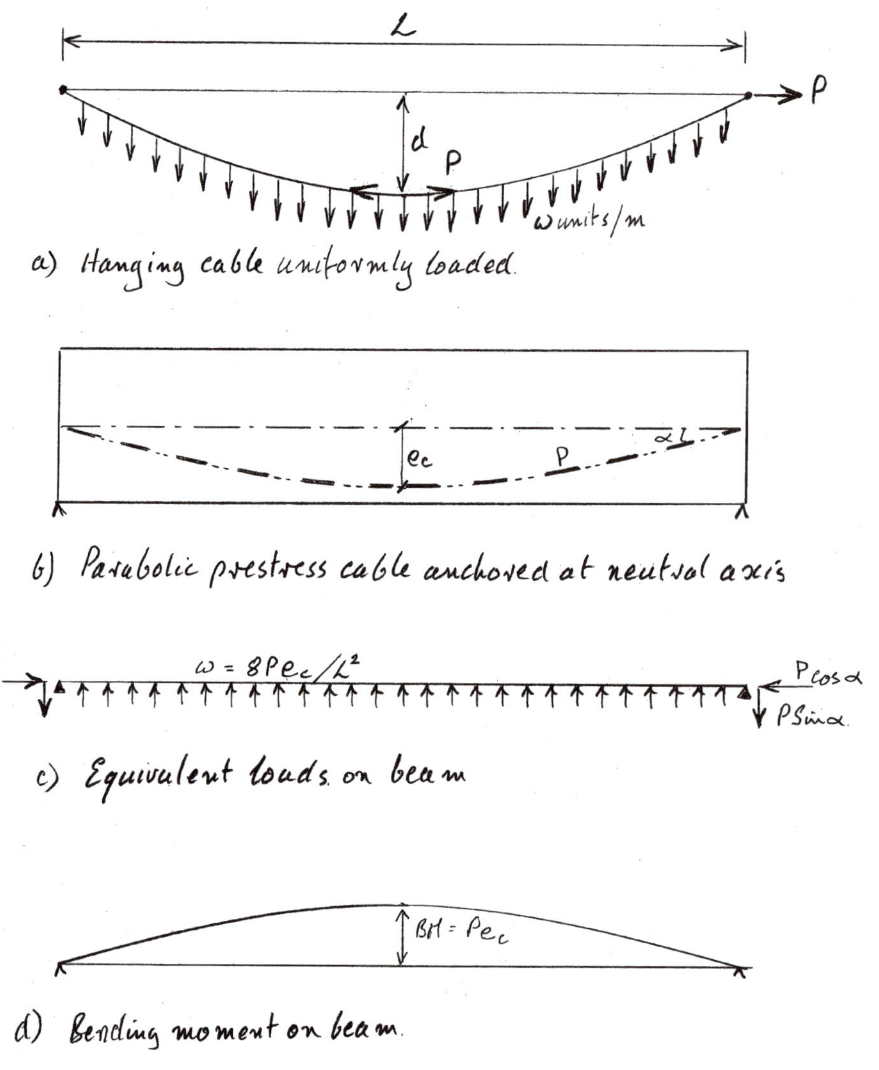

Figure 5.17 Equivalent load for a parabolic cable

The moment due to prestressing on the beam at points anywhere along the beam span may be calculated by any of these three methods.

If the prestressing cable had not been anchored at the ends of the beam on the neutral axis, but at an eccentricity e_0, in addition to the previous equivalent loads there would be an end moment of Pe_0.

A tensioned cable that is loaded by a uniformly distributed load w units per metre will adopt a parabolic deflected shape, Figure 5.17 (a). The horizontal component of the force in the cable $P = wL^2/8d$, where d is the amplitude of the cable and L is the distance between anchorage points. (This is the same equation that is used to find the thrust in a parabolic arch subjected to a uniform load.) Consequently, a parabolic prestressing cable will exert a uniformly distributed load on a beam, Figure 5.17 (c). The equivalent distributed load w kN/m is given by:

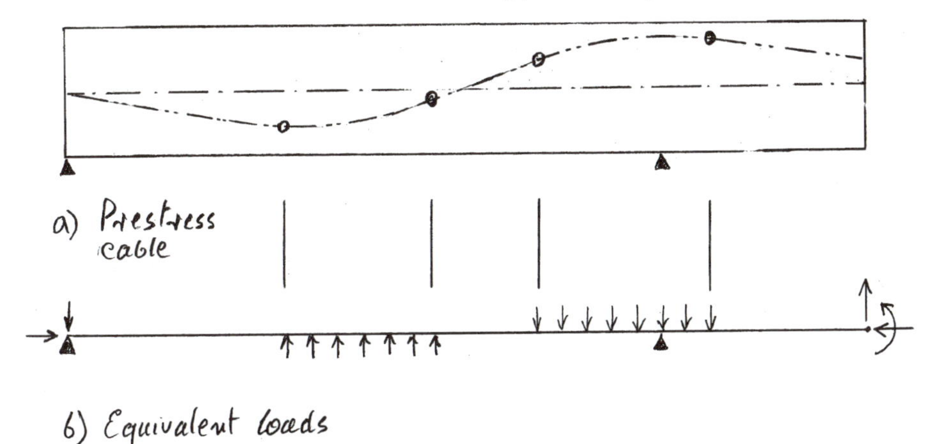

Figure 5.18 Equivalent loads for general case

$$w = 8Pd/L^2 \text{ kN/m},$$

where P kN is the force in the cable and d and L are the amplitude of the cable and the span of the beam in metres.

This is often a useful concept in preliminary calculations, for calculating the deflection of a beam due to prestress, for instance. (If the cable is attached at the ends of the beam at the level of the neutral axis, amplitude d = eccentricity e_c.)

Any shape of prestressing cable may be broken down into a series of angular deviations, straight lengths and parabolas, and thus, when combined with end moments and forces, one may convert the effect of the cable into a set of equivalent loads, Figure 5.18. This is particularly useful when analysing a structure as a grillage or space frame, as the prestress can then be treated as any other load.

However, care has to be taken about two aspects of this idealisation. First, the shape of a prestressing cable must be referred to the neutral axis of the beam. Thus even if the cable is straight and the neutral axis is deflected, the effect is the same as for a straight neutral axis and a deflected cable. For instance, consider the roof beam shown in Figure 5.19. The neutral axis is deflected in a 'V' shape, while the cable is straight. The equivalent load due to the prestress is identical with that of the straight neutral axis and the deflected tendon. Similarly, if the cross section of a beam changes along its length, with the result that the neutral axis is deviated, equivalent point loads are created by a straight cable, Figure 5.20.

Second, the force in a prestressing cable is not constant along its length, due to friction between the cable and its sheath. Thus for an accurate translation of a cable into equivalent loads, it is necessary to include the forces tangential to the cable that correspond to the friction forces imparted by the cable on the structure. Although this effect may be ignored in simple preliminary calculations, it must be considered for detailed design.

Consequently, equivalent forces are not easy to calculate by hand for a structure of any complexity. There exist computer programs that will transform any cable geometry and structural shape into equivalent loads, but, as always, the designer should be wary

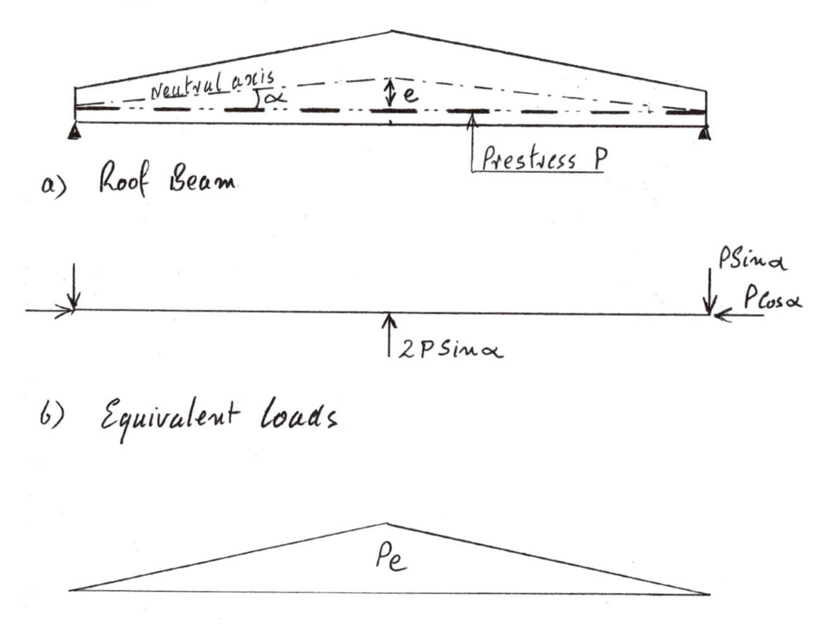

Figure 5.19 Equivalent loads; straight cable and deflected neutral axis

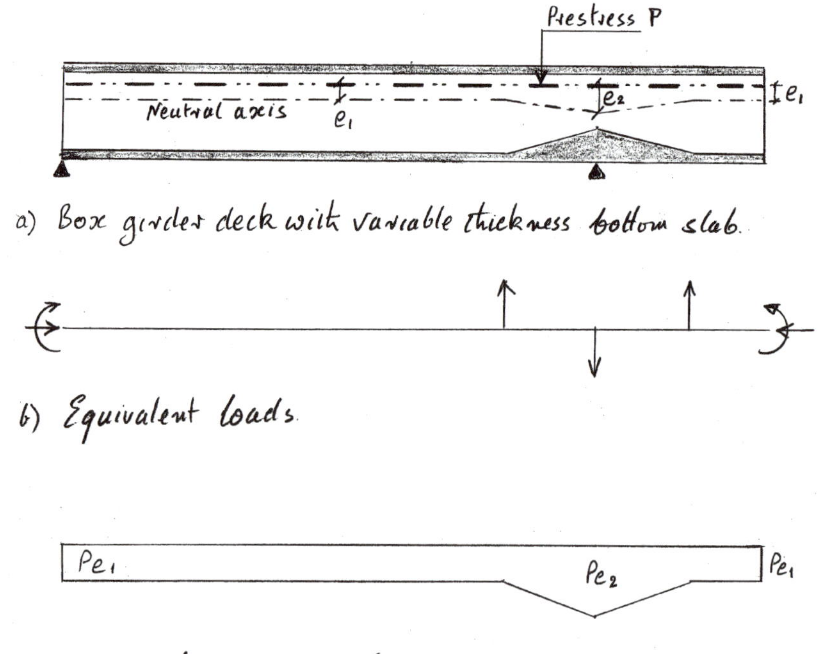

Figure 5.20 Equivalent loads; variable thickness flange

of any 'black box' unless he understands the basis on which it has been established, or has an alternative method of calculation to check the results.

The concept of equivalent loads is extremely valuable for understanding the effects of prestressing on a structure, and for carrying out preliminary designs or simple final designs. It is also very useful for the mechanised calculation of prestressed structures, but must be used with considerable caution by the designer.

Reference [5] is useful for design based on equivalent loads.

5.19 Internal and external loads

It is important that the designer of prestressed structures understands clearly the difference between external and internal loads. The self weight, other dead loads and traffic loads applied to a beam are external loads. They generate reactions that are equal and opposite to the loads. If the support for a beam is not strong enough, when a load is applied to the beam, the support may fail; wind load on a girder may overturn it.

The loads applied by prestressing cables to a beam are internal. They normally give rise to bending moments and shears in the beams, but they do not add to the external reactions. The equivalent loads shown in Figures 5.16 to 5.20 are in equilibrium; all forces must resolve to zero, as must all moments taken about any point.

In continuous beams, discussed in more detail in Chapter 6, the prestress is likely to shift reactions between supports, but these reactions still resolve to zero.

5.20 Prestress effect on shear force

The angle of the prestress cables with respect to the neutral axis provides a very significant shear force that may be designed to counteract the shear force due to dead and live loads. If the cable centroid of the 32 m long beam described in *5.1 et seq.* and in Figure 5.11 (a) had followed a parabolic shape, with an eccentricity of $y_b - 0.2 = 1.272$ m at mid-span, and zero at the beam ends, its slope at the beam end would have been $2 \times 1.272 / 16 = \tan^{-1}0.15 = 9°$. The shear force due to the prestress force of 9.38 MN acting at 9° to the horizontal would be 1.49 MN.

The ultimate shear force of 3.08 MN (*5.6*) is reduced to 3.08 – 1.49 = 1.59 MN. In some codes of practice, the prestress shear force to be used for this calculation is factored down, typically by a factor of 1.15. Hence the net shear force to be carried by the beam becomes $3.08 - 1.49 \times 0.87 = 1.78$ MN. This shear force is compatible with the web thickness of 0.22 m adopted for the beam. Consequently, it is not necessary to thicken the web to carry shear force. The web would in fact be thickened for a short distance from the beam end to accommodate the prestress anchors and to resist the concentration of force they create.

The shear force applied by the prestress is under the control of the designer. For instance, the average angle of the prestress cables at the end of the beam in the scheme shown in Figure 5.11 (b) is clearly steeper than the parabola mentioned above. The designer may adjust the slope of the cables at critical points along the beam to optimise the shear relief provided.

5.21 Anchoring the shear force

In the equivalent truss model for shear force, described in *3.10*, a proportion of the main tension reinforcement in the bottom boom of the beam must be anchored beyond the centre line of the end support to resist the horizontal component of the last inclined compression strut. This tie must provide a force that is equal to the shear force. This is equally valid in a prestressed beam, and requires that sufficient tendons be anchored close to the bottom fibre to anchor the net shear force, that is the applied ULS shear force less the ULS prestress shear force.

In our example, the ultimate net shear force is 1.78 MN. The ultimate strength of the 12/13 mm tendons adopted is $1{,}860 \times 12 \times 100 \times 10^{-6} / 1.15 = 1.94$ MN. (1.15 is the material factor corresponding to BS5400: Part 4: 1990.) Thus one tendon must be anchored near the bottom of the end face of the beam. Alternatively, the net shear force may be anchored using an appropriate section of passive reinforcing steel.

5.22 Deflections

5.22.1 General

In many structures, the control of deflections is a critical aspect of their performance. The cancellation of dead load deflections is one of the main benefits provided by prestressing. A typical example is the design of floors for standard office blocks. The thickness of the floor for a reinforced concrete scheme is governed by its long-term dead load deflections. Excessive deflections give problems to floor finishes, partitions and facades. Consequently floor slabs have a span/depth ratio that typically does not exceed 25. Prestressing the slab allows the dead load deflections to be completely cancelled, allowing span/depth ratios of 40 and above. In a multi-storey block, this may allow a lower building height, or more floors.

Reinforced concrete bridge decks must also be relatively stocky to limit their dead load deflections, while prestressed decks tend to deflect upwards under permanent loads and consequently may be much more slender. This will be demonstrated using the beam described in Figure 5.1. Generally, the slenderness of prestressed concrete decks is limited only by considerations of vibration under the effect of live loads, by the strength of the concrete, or by cost, as slenderness is expensive.

5.22.2 Approximate calculation of prestress deflections

The deflection of a simply supported beam subject to uniform loads may be calculated from the formula $\delta = 5Ml^2/48EI$ where M is the bending moment at mid-span. As described in *5.18*, a parabolic prestressing cable of force P with eccentricity zero at the beam ends and e at mid-span, applies an equivalent uniformly distributed load of $8Pe/L^2$, and a moment at the beam centre of Pe. Thus the deflection due to prestress may be calculated by substituting Pe for M in the above equation. The mid-span deflection is not sensitive to minor deviations of the prestressing cable away from the parabola. If the prestress has an eccentricity at the beam ends, an additional deflection due to end moments will need to be added.

When it is considered necessary to calculate the deflections or beam end rotations due to prestress with more precision, the calculation must be carried out from first principles, using the exact shape of the centroid of the prestress cables. However, it should always be remembered that too much precision is likely to be illusory, as the greatest variables in this calculation are the Young's modulus of concrete, and the effects of creep.

5.22.3 Example

For the beam described in Figure 5.1, the long-term permanent moment at mid-span due to the combined effect of self weight and finishes is +5.71 MNm, due to live loads is +10.8 MNm (*5.6*), and the moment due to prestress, *Pe* is –11.93 MNm (*5.8*). A Young's modulus for concrete of 34,000 MPa will be adopted for the deflection due to live load, and 34,000 / 2.5 =13,600 MPa, compatible with a creep coefficient of 1.5, for the long-term deflections due to self weight and prestress. As the section is fully prestressed, it will be uncracked, and the full value of second moment of area has been used in the calculation.

The deflections are calculated as follows:

Self weight + finishes	42 mm downwards
Prestress	89 mm upwards
Live load	32 mm downwards
Net deflection under dead load	47 mm upwards
Net deflection under live load	15 mm upwards

It is clear that the self-weight deflection has been more than compensated for by the prestress, and consequently, in order to produce a beam that is flat in the long term under dead loads, it is necessary to build it with a downwards pre-camber of 47 mm. This is typical of such prestressed bridge beams.

The value of *E* and of the creep coefficient may vary from beam to beam in a series, depending on variations in the composition of the concrete and on the delay between casting and stressing the beam. However, this uncertainty only acts on the difference between the dead load and the prestress deflections. This is a very important feature of prestressed concrete structures that has to be understood by the designer.

It would for instance, be possible to design the prestress such that it gave an upwards deflection exactly equal and opposite to the dead load deflections. Under these circumstances, the net dead load deflection would be zero, and this would be independent of the value of Young's modulus. For the beam used in this example, it would involve increasing the prestress force and reducing the eccentricity. The new values of *P* and *e* may be found by considering the following two equations.

First, for their deflections to be equal and opposite, the prestress moment must equal the dead load moment (on the assumption that the prestress has a parabolic profile and is anchored on the neutral axis at the beam ends).

Hence $Pe = 5.71$ MNm.

Second, for there to be zero stress on the bottom fibre under the effect of maximum bending moment, the centre of pressure must lie at the top of the kern.

Hence $P(e+a_t) = 16.51$ MNm, with $a_t = 0.488$ (refer to Figures 5.5 and 5.7).

Solving these two equations for P and e gives the answer that $P = 22.1$ MN and $e = 0.258$ m.

Of course, it would not be economical to more than double the prestress force in this beam. However, it is a good example of how familiarity with the concepts of centre of pressure and central kern allow one to control the effects of prestressing. There are in fact situations where designing the prestress to control dead load deflection is economical. One such is for precast car park beams, where any difference in the deflection of adjacent beams would cause difficulty in connecting them.

5.23 The shortening of prestressed members

Typically, prestressed members are compressed at 5 MPa, although this may be as low as 2 MPa or as high as 10 MPa. A member compressed at 5 MPa will shorten elastically by approximately 0.15 mm/m and will then continue shortening for, typically, a further 0.25 mm/m due to creep, giving a total shortening of 0.4 mm/m. This is in addition to shrinkage of the concrete and temperature movements.

It must never be forgotten that prestressed members must be free to shorten for the prestress to be active. Although this is generally the case for the longitudinal prestress of bridges that are carried by sliding bearings or flexible piers, it is not so obvious

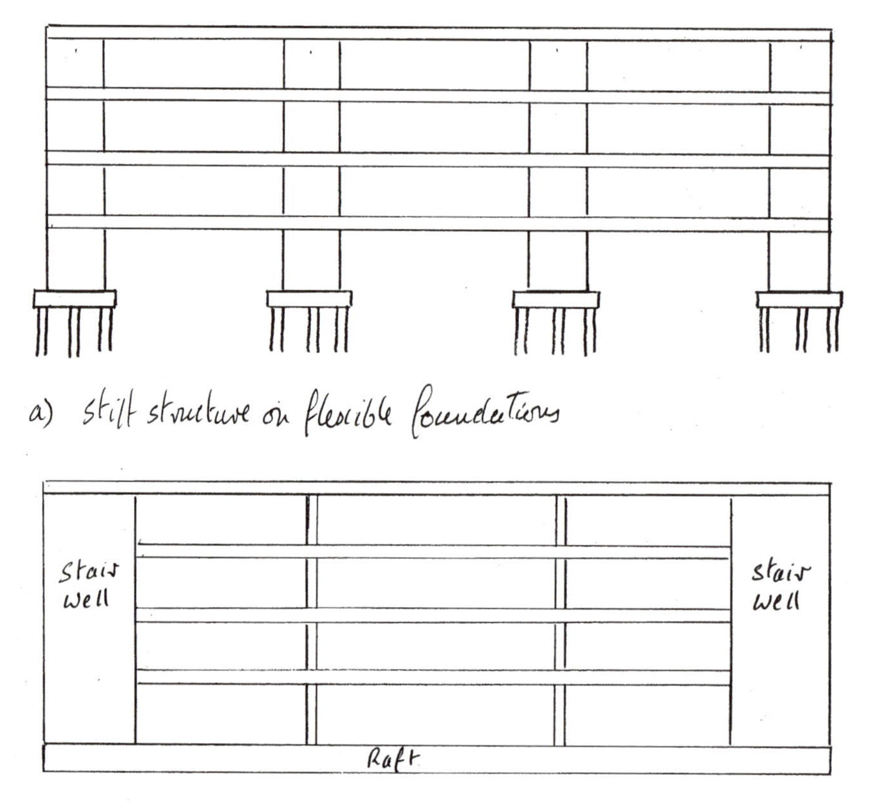

Figure 5.21 The structure must be free to shorten

in the transverse direction for bridges that are built into their piers, for some portal structures, and for buildings.

It is clear that if a single floor of the building shown in Figure 5.21 (a) were to be prestressed, most of the force would be dissipated in the stocky columns, with potential damage to the columns. However, if all the floors were to be prestressed in a co-ordinated sequence, and if the piled foundations were flexible enough to accept the deformations due to the shortening of the floors, prestressing would be possible. The building shown in Figure 5.21 (b), where stiff stairwells are placed at each end, and the whole is founded on a raft, could not be successfully prestressed.

5.24 Forces applied by prestress anchorages

5.24.1 General

There are generally three areas behind an anchor that need to be reinforced. Immediately behind the anchor are splitting forces due to the tendency of the anchor to be driven into the concrete member by the force of the tendon. Adjacent to the anchor, on the end face of the member, are zones of tensile stress known as spalling zones. Finally, as the prestress force disperses into the bridge deck, further zones of tensile stress are created.

5.24.2 Splitting forces behind anchorages

Most prestress anchorages are sized to apply a pressure on the concrete of about 37 MPa. The typical lines of force behind an anchor are shown in Figure 5.22 (a). The change of direction of these lines of force creates transverse stresses, compression if the lines are concave towards the centre line, and tensile when convex. The stresses perpendicular to the line of the anchor are shown in Figure 5.22 (b). These stresses are compressive directly behind the anchor and tensile further away. The primary splitting or bursting reinforcement is designed to resist these transverse tensile stresses. It usually consists of spirals or a series of mats. The design of primary bursting reinforcement is described in references [1, 2, 3 and 4].

However, the lines of force are an elastic concept. Design methods where the sizing of the bursting reinforcement is based on this elastic distribution of stresses produce safe structures, although in general the amount of reinforcement required is found to be excessive, and the design procedures generally introduce empirical methods to reduce it.

Once the concrete reaches its limit in tension and either yields or cracks, this elastic distribution of stresses is completely changed. At the ultimate limit state one may see the problem as one of mechanics. The concrete immediately beneath the anchor forms itself into a wedge, which tends to be driven into the beam member, splitting it, Figure 5.23. The wedge that is trying to force the concrete apart may be resisted by a tie or a series of ties placed closely behind the anchor, in the compressive zone of the elastic force diagram.

More research on anchor zones designed on the assumption that the concrete is cracked should allow lighter reinforcement to be used.

The design of this primary bursting reinforcement is one of the more critical activities in the design of a prestressed member. An unthinking or over-conservative application

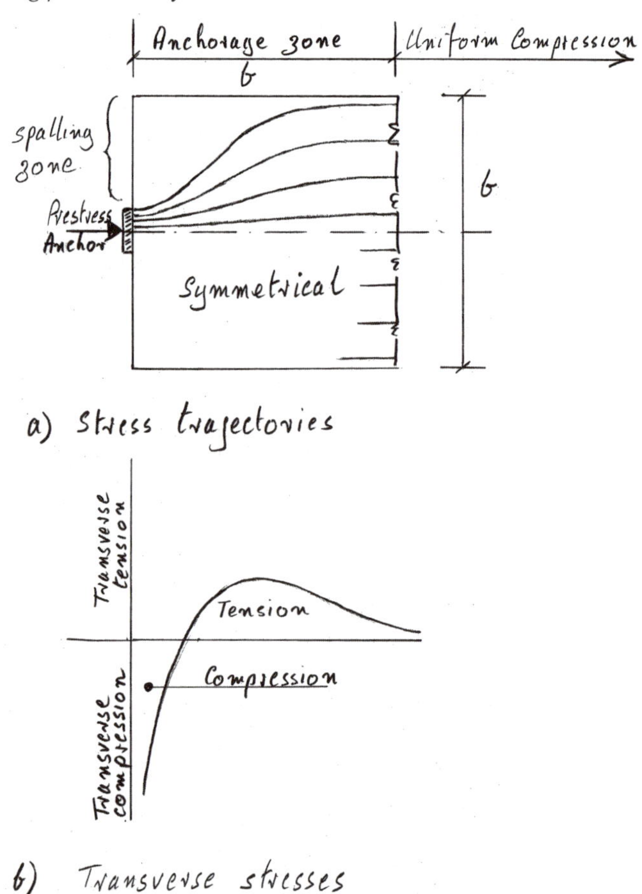

Figure 5.22 Stresses behind a prestress anchor (Source: Derived from Leonhardt)

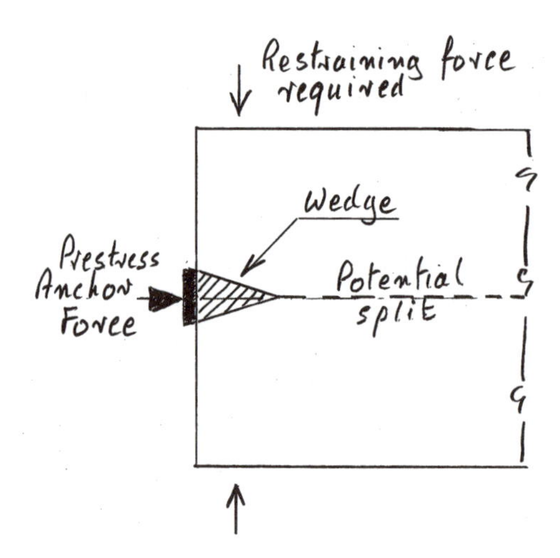

Figure 5.23 Post-cracking anchor behaviour

of design rules often leads to excessively heavy reinforcement. It is not uncommon to see square links of 300 mm side made of 25 mm bars. It needs to be remembered that the force is transferred from concrete to reinforcement by bond or by direct bearing in hooks and bends, and it is most improbable that such large bars can develop their full strength. A more thoughtful design and careful analysis would allow the use of smaller bars that bond better with the concrete.

Prestress anchors are frequently situated at the ends of a bridge deck, where there are several other sets of reinforcement, such as shear links for the beams, shear and bending reinforcement for the abutment diaphragms, bursting reinforcement over the bridge bearings and holding-down reinforcement for the expansion joint, as well as transverse deck reinforcement. This accumulation of sets of reinforcement frequently gives rise to excessive congestion of the section which makes it very difficult to cast and compact the concrete. However, the most important prerequisite for a successful anchor is the presence of sound concrete immediately behind the bearing surfaces that transfer the tendon force. It is thus very important that the design of the primary anchor reinforcement is carried out carefully, and that it is not over-conservative. It is also essential to prepare large-scale drawings of all the reinforcement in the anchor zone, where the bars are shown at their correct thickness, with two lines.

The designer should always avoid placing anchorages at the minimum distance from concrete edges, or adopting the minimum distance between anchors, unless it is essential to the project. The failure rate for anchors rises as edge distances are reduced. When it is essential to use minimum distances, extra care is required in the design and in the reinforced concrete detailing.

Clearly, the safe edge distance also depends on the strength of the concrete when the tendons are stressed. More caution is required for cast-in-situ structures, particularly if they are being stressed as early as possible, than for precast structures where the concrete is likely to be considerably stronger than its nominal cube strength when the tendons are stressed.

5.24.3 Spalling

Adjacent to the anchor are areas of so-called spalling tension, Figure 5.22. Although these areas of tension cannot in theory compromise the strength of the member, they may cause the concrete cover to break away exposing the bursting reinforcement, and they may delay the job as explanations are demanded and repairs planned. The cause of these areas of tension may be understood by referring again to Figure 5.22. The compressive force of the anchor is carried between the stress trajectories. The concrete within these trajectories shortens under the effects of the high compressive stresses that exist behind the anchor. However, the concrete immediately outside the outermost trajectory is not subjected to this compression. Consequently, there is a strain discontinuity between these two zones that sets up shearing and tensile stresses in the concrete adjacent to the anchor.

The forces involved in spalling are weak, and these areas need to be reinforced by small-diameter bars, typically not larger than 10 mm, which may be bent to tight radii and thus can be fixed into the corners of the concrete member. Using large-diameter bars, which have large bending radii and require considerable bond lengths to develop their working force, is pointless.

5.24.4 Dispersion of the prestress

The third area of reinforcement is often called equilibrium steel, or secondary bursting steel. As the prestressing forces spread out and adopt their elastic distribution, transverse tensile stresses are set up which may require substantial reinforcement.

The slab shown in Figure 5.24 (a) is stressed by two tendons located near its edges. At some distance from the anchor face, the force of the two anchors will be reacted by a uniform load spread across the width of the slab. Figure 5.24 (a) also shows the lines of force, and Figure 5.24 (b) shows the idealised strut-and-tie diagram. It is clear that a substantial tie is required at the end of the deck. The force in this tie depends on the angle assumed for the struts. Generally, it is satisfactory to assume that the angle of the steepest line of force is no greater than 30°. Steeper angles result in excessive reinforcement with no benefit. However, the designer must be alert to special situations where the angle of the struts is affected by the geometry of the member. If cracking is to be avoided, the tie should consist of reinforcing bars working at not more than 250 MPa or of prestressing tendons.

Another way of understanding the need for this tie is to imagine the slab cut down its axis. Each half slab then behaves like a column loaded eccentrically by the prestress anchor, Figure 5.24 (c). The deflection of the two columns away from each other is the mechanism causing the slab to crack and the tie force is that necessary to pull the two halves together again. This model also demonstrates that if the reinforcement was omitted, the structure would find an alternative equilibrium, albeit with wide cracks.

If the longitudinal prestressing cables are themselves angled in plan, they create additional transverse forces which must be considered. For instance, if the tendons were angled to follow the mean lines of force, Figure 5.24 (d), they would generate transverse forces which would cancel out the dispersion forces shown in Figure 5.24 (c).

Figure 5.25 shows the same slab with the two tendons anchored near the axis. Here the lines of force spread outwards from the anchors, creating a transverse compression between the anchors, and a zone of tension some distance into the slab. The main practical difference to the previous example is that here the tension zone may be assumed to be quite deep, while the tension in the previous example is concentrated at the slab end. Consequently, it is likely that existing transverse slab reinforcement that is underused, such as reinforcement on the compression face of the slab, may make up a considerable part of the tie force required.

From these simple examples, it should be clear that the arrangement of tendons at a beam end needs thought to minimise the additional reinforcement required, with its attendant costs and congestion.

Particular problems can occur when the transverse tensions caused by the spreading out of the prestress force combine with other tension forces. For instance, in a bridge built by free cantilever erection, it is common practice to anchor the prestressing tendons in the webs, Figure 5.26. These tendons give rise to an upwards shear force, which causes a principal tensile stress that is approximately perpendicular to the line of the prestressing ducts. As the prestress force spreads out from the anchor, it adds its transverse tensile stress to the principal tensile stress due to shear, and there is an enhanced risk of cracking along the weak plane constituted by the presence of the ducts. In this case, conservative assumptions need to be made in the design of the dispersion steel in the webs. Several bridges known by the author have cracked during construction in this way.

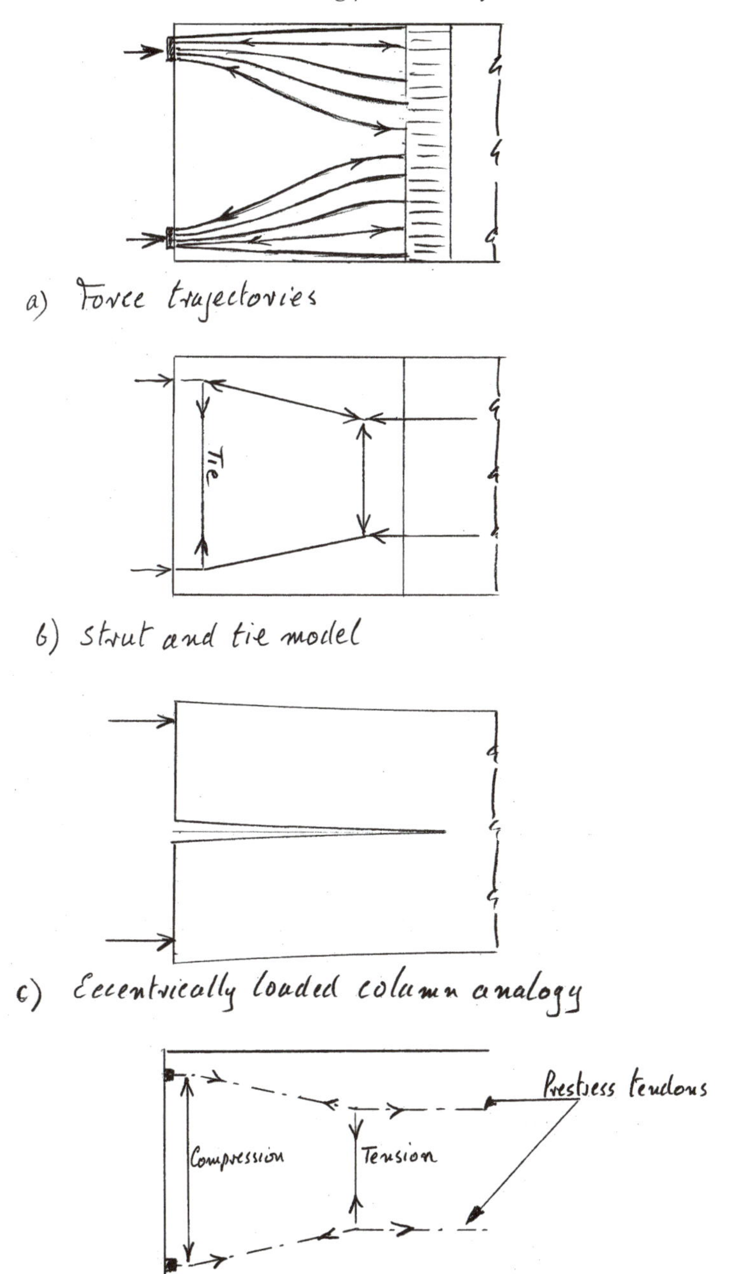

Figure 5.24 Equilibrium forces for slab loaded at its edges

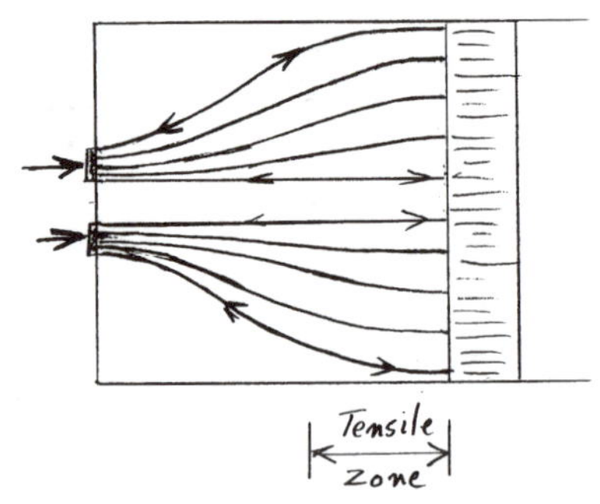

a) Force trajectories

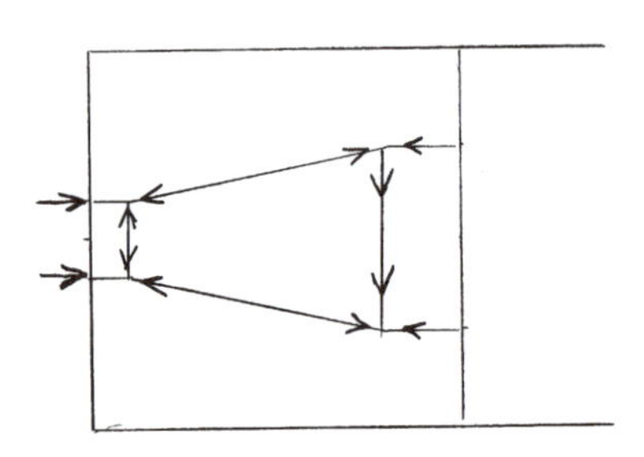

b) Strut and tie model

Figure 5.25 Equilibrium forces for slab loaded at the centre

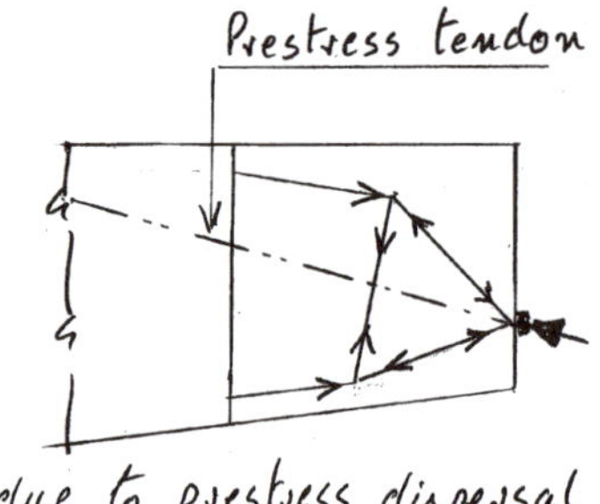

a) Tension due to prestress dispersal

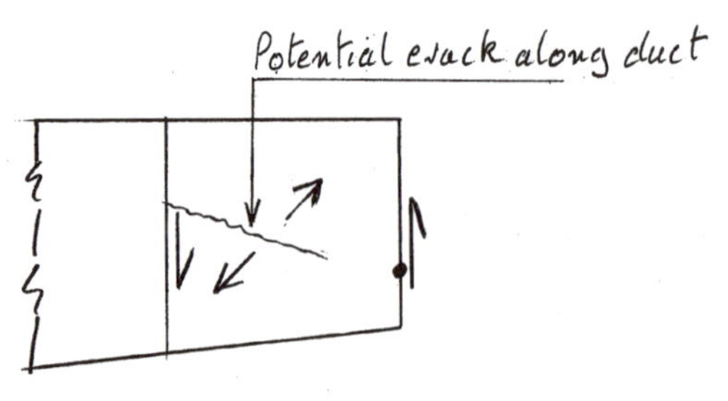

b) Principal tensile stress due to prestress shear

Figure 5.26 Potential web cracking due to prestress shear

A similar example occurred on one of the earlier projects in which the author was involved (but did not design!). A large number of 32 m long post-tensioned Tee beams (Chapter 10) were being prefabricated for a viaduct. In the casting yard, they were temporarily supported some distance in-board of their final bearing position. The tendons that were anchored in the end of the beams gave a very significant relieving shear force that, when the beams were on their final bearings, would partially counteract the working shear. However, in the temporary condition, in the absence of dead load shear in the length beyond the temporary bearing, they created a principal tensile stress that was orientated approximately perpendicular to the prestressing ducts, as described above. When the tendons were stressed, the beams cracked along the line of the ducts, and successive increases in shear links failed to cure the problem. In some beams, the experimental installation of vertical prestress did eliminate the cracking.

The designer must be wary of any cumulation of tensile stresses, which may be due, for example, to a temporary bending condition, to restrained heat of hydration shortening, to the presence of holes or other stress raisers, to temperature gradient effects, as well as to prestress dispersion.

The forces due to the dispersion of prestress, and the reinforcement required to resist them, are assessed at the SLS. In the example of the web of the deck shown in Figure 5.26, it is important that the concrete does not crack when these forces are combined with the working load shear forces. In making this assessment, it should be noted that when the tendons are stressed, the force applied may be up to 80 per cent of the strength of the tendon, but by the time the live load shear forces are applied, this will have fallen to about 60 per cent. The reinforcement to resist the dispersion forces should not be added to the shear reinforcement calculated at the ULS.

5.25 Following steel

If a prestress anchor were to be placed in an elastic medium, far from any free edge, the force of the anchor would be carried half in compression in front of the anchor, and half in tension behind. If the elastic medium was to be replaced by unreinforced concrete, as it is weak in tension, the concrete behind the anchor would crack and all the force would be carried in compression. Consequently, it is necessary to provide local reinforcement that controls this cracking, and which carries a proportion of the force in tension. Generally, it is adequate to provide following steel that can carry a maximum of one-third of the prestress force with the reinforcement working at a stress of 250 MPa, Figure 5.27.

If the concrete in which the anchor is situated is compressed due to prestress and to the overall bending of the deck, this compression will have to be overcome before the concrete behind the anchor may crack, and consequently the following steel may be reduced. Also, if there is a free concrete edge close behind the anchor, less of the anchor force will be carried in tension (to the limiting case when the anchor is at the end of the member and all the force is carried in compression), and less following steel is required.

A special case exists in precast segmental decks, which have frequent unreinforced transverse joints. If prestress anchors are placed closely in front of such joints, there is a possibility of the joint being opened locally by the following tensile strains. It is good practice to place anchors at least one metre in front of such joints, when some following steel needs to be provided.

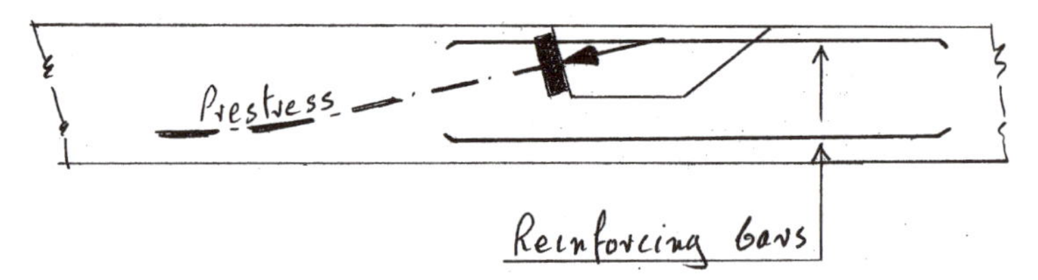

Figure 5.27 Following steel

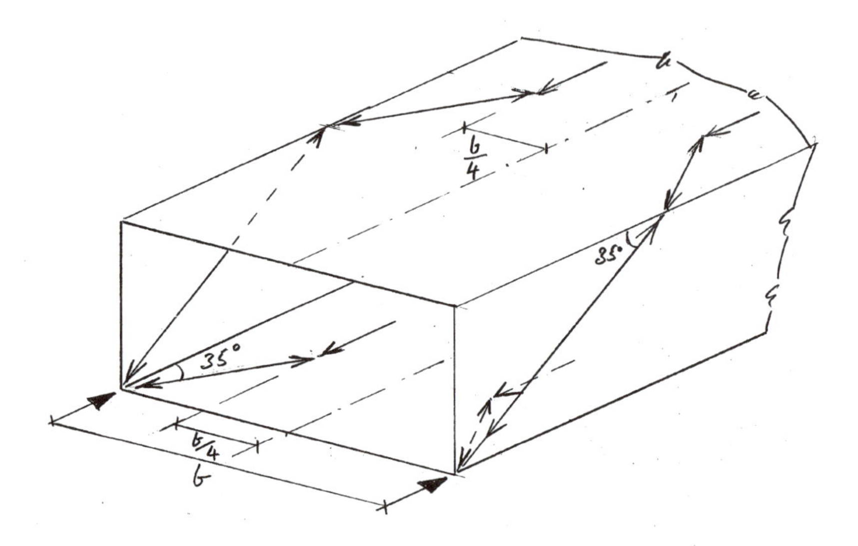

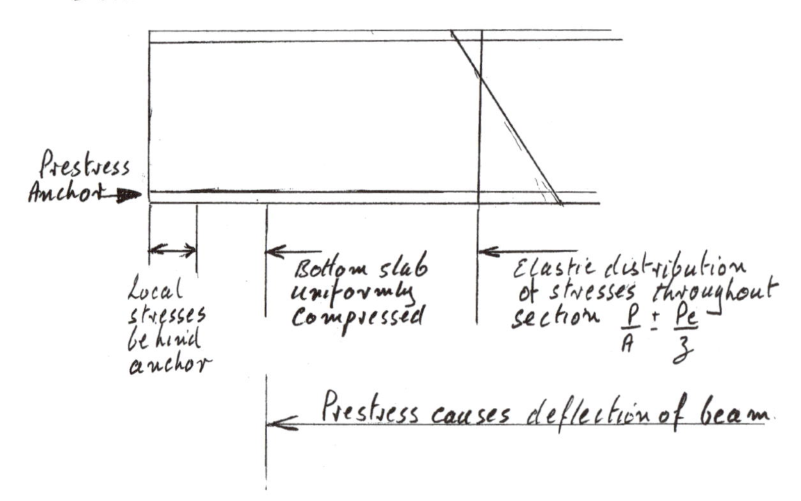

Figure 5.28 Dispersal of prestress

5.26 The introduction of prestress forces

In continuous bridge decks, it is often essential to stop off tendons at regular intervals. The tendon anchor applies a concentrated compressive force to the concrete. Immediately behind the anchor the elastic assumption that plane sections remain plane is not justified, and the tendon does not create a bending moment in the beam. The anchor force disperses through the concrete section, until, at some distance from the anchor, the normal assumption of elastic behaviour is re-established.

The angle at which the force disperses through the concrete lies between 30° and 45°. If the anchor is located at the end of a member, the anchor force can only be transmitted forwards by compression, and the angle of dispersion will tend to be flatter. If the anchor is placed midway along a member, and if the section is in compression or there is adequate following steel, some of the anchor force is carried backwards by tension, effectively shortening the dispersion length. However, the angle of dispersion of prestress is conventionally accepted as 35°.

Three separate effects of the prestress must be considered by the designer. First, the prestress causes local compressive stresses. These are present before the dispersion of the prestress is complete, and they may overstress the concrete locally if combined with other stresses due to bending of the deck. However these local compressive stresses may assist in cancelling out tensile bending stresses in the deck. In Figure 5.28 (a) for instance, if the bottom slab of the box shown was subjected to tensile bending stresses due to the bending action of the beam, the local compression due to the new anchors would be beneficial once it had spread over the full width of the bottom slab, well before it had spread throughout the box section.

Second, the prestress causes deflections of the beam even before the stresses have been uniformly dispersed throughout the section. In a continuous structure, these deflections will generate prestress parasitic moments.

Finally, once the dispersion of the prestress force is complete, it causes elastic bending and compressive stresses of $P/A \pm Pe/z$, as described in *5.8 et seq*. These three phases are summarised in Figure 5.28 (b).

5.27 Bonded and unbonded cables

Tendons may be housed in ducts that are buried in the concrete webs and slabs of the bridge deck. These internal tendons are protected from corrosion by adequate concrete cover, by the lack of cracking in the surrounding concrete under the action of permanent loads, and by filling the ducts with cement grout. If the grouting is well done, and if the deck has been competently designed and built, these grouted tendons will be durable and will need no maintenance. A correctly designed and built prestressed concrete deck with internal tendons will be more durable than an equivalent reinforced concrete structure, due principally to the much-reduced incidence of cracking. However, if the above conditions are not fulfilled, there is the possibility that the tendons will be subject to corrosion due to cracking under permanent loads or to voids in the grout. Such corrosion is likely to be difficult to repair, and may well require the demolition and reconstruction of the bridge deck. In fact many thousands of prestressed concrete bridge decks with internal cables have been built, and the incidence of failure due to corrosion of the tendons is extremely small.

In some types of bridge deck, the prestressing tendons may be housed in ducts that are external to the concrete, generally within the void of a box section deck, Figure 15.26. The ducts are usually made of high density polyethylene (HDPE). In order to achieve the desired profile of the prestress centroid, the ducts are deflected at pier diaphragms and intermediate deviators. Once the tendons have been stressed, the ducts are grouted with cement, or with a petroleum jelly.

The principal advantage of this arrangement is that the tendons may be detailed in such a way that they can be replaced in the event of their suffering corrosion. The corollary to this benefit is that as the tendons are external to the concrete, they are inherently less well protected than internal tendons. They must be subjected to an organised inspection regime that is capable of detecting corrosion occurring within the ducts, and in particular in the most vulnerable areas which are immediately behind the anchorage, at any junctions between plastic and steel duct and at the field joints in the ducts. This may impose a considerable burden on the maintaining authority. A further advantage of external tendons is that the webs of the bridge are not encumbered with the ducts and consequently are easier to cast and may be made thinner.

There are several practical disadvantages of external tendons. The need to avoid a confusion of ducts within the box section deck generally leads to the use of large prestress units. These large units require substantial deviators that generally take the form of internal ribs or frames, and also require, for tendons that cannot be anchored at pier diaphragms, large internal anchorage blisters. The volume of these deviators and anchorage blisters generally more than cancels out any saving in concrete weight achieved by making the webs thinner. Also, they generally require dense reinforcement that significantly increases the overall weight of passive reinforcing steel in the deck. Furthermore, the deviators may locally stiffen the deck cross section and act as diaphragms, attracting transverse bending moments.

The weight of prestressing steel is greater than for internal tendons. This is due to the reduced eccentricity of the prestress as the tendons have to lie between the top and bottom slabs of a box, and to the reduced flexibility in stopping off tendons. In order to provide adequate ultimate strength to the beam, the prestress may have to be significantly over-designed, or the shortfall in strength made up by additional passive reinforcing steel.

The installation of the external ducts has to be carried out once the concrete has been cast or erected, putting this activity on the critical path, and generally slowing construction. This can be a critical consideration when speed of construction is of the essence.

Finally, as the tendons are not bonded to the concrete, they do not significantly increase in stress when the deck is overloaded. This lack of composite action between the main reinforcement and the concrete makes such decks less able to cope with extreme situations.

External prestress is at its most cost effective when the prestress profile is simple and when the tendons can be anchored in the end diaphragms of a deck, such as in statically determinate spans. External prestress may also be successful as continuity prestress in bridges built by the balanced cantilever method, both cast in-situ and precast, as the lack of internal tendons in the webs facilitates construction, and the external tendons may be anchored in the pier diaphragms. However, as in such bridges the negative moment tendons are internal, the main reason for using external prestress, the facility to replace the tendons is not relevant.

6

Prestressing for continuous beams

6.1 General

In this chapter, the general principles of the design of continuous bridge decks will be explained by following the essential steps of the design of a typical box girder. However, before this can be carried out it is necessary to explain the concept of prestress parasitic moments.

6.2 The nature of prestress parasitic moments

Consider a three-span continuous beam that has been released by hinges over the supports to create three statically determinate spans. The three equal spans are subjected to uniform loads, Figure 6.1 (a). The free bending moment at the centre of each span, $M_{max} = wl^2/8$, where w is the load per metre and l the span, and the moment at any point may be called M_{iso}. The beams will deflect, and the end rotation of each beam $\theta = M_{max} \times l/3EI$, Figure 6.1(c). If the spans are to be made continuous, hogging moments must be applied at each internal support such that the rotations of the beam ends become compatible. The hogging moments for a three-span beam are as shown in Figure 6.1 (d), with, for this particular case of three equal spans, a value of $wl^2/10$ at each internal support; that is 80 per cent of the free bending moment. We may call these hogging moments the 'continuity moments', the values at any point being M_c. The total moment at any point along the continuous beam will be the sum of the free bending moment and the continuity moment, or $M_{iso} + M_c$, Figure 6.1 (e). This is basic theory of structures, which causes no problems to engineers or students. The hogging continuity moments in the beam are as easy to understand as the free bending moments, and there is no reason to label the free moments as primary and the hogging moments as secondary.

The uniform loads on the statically determinate spans may be replaced by parabolic prestressing cables anchored on the neutral axis at the beam ends, which apply a uniform upwards load on the beams, Figure 6.1 (f), (*5.18*). The prestress force P and the eccentricity at mid-span e_c may be chosen to give an upwards distributed force = w kN/m, and hence to produce exactly the same bending moments and beam end rotations, although of the opposite sign, as the externally loaded beams described above, Figure 6.1 (g) and (h). The free bending moment at the centre of each span is Pe_c and the moment at any point is Pe. In order to make the beams continuous, numerically the same moments as for the distributed loads, that is 80 per cent of Pe_c

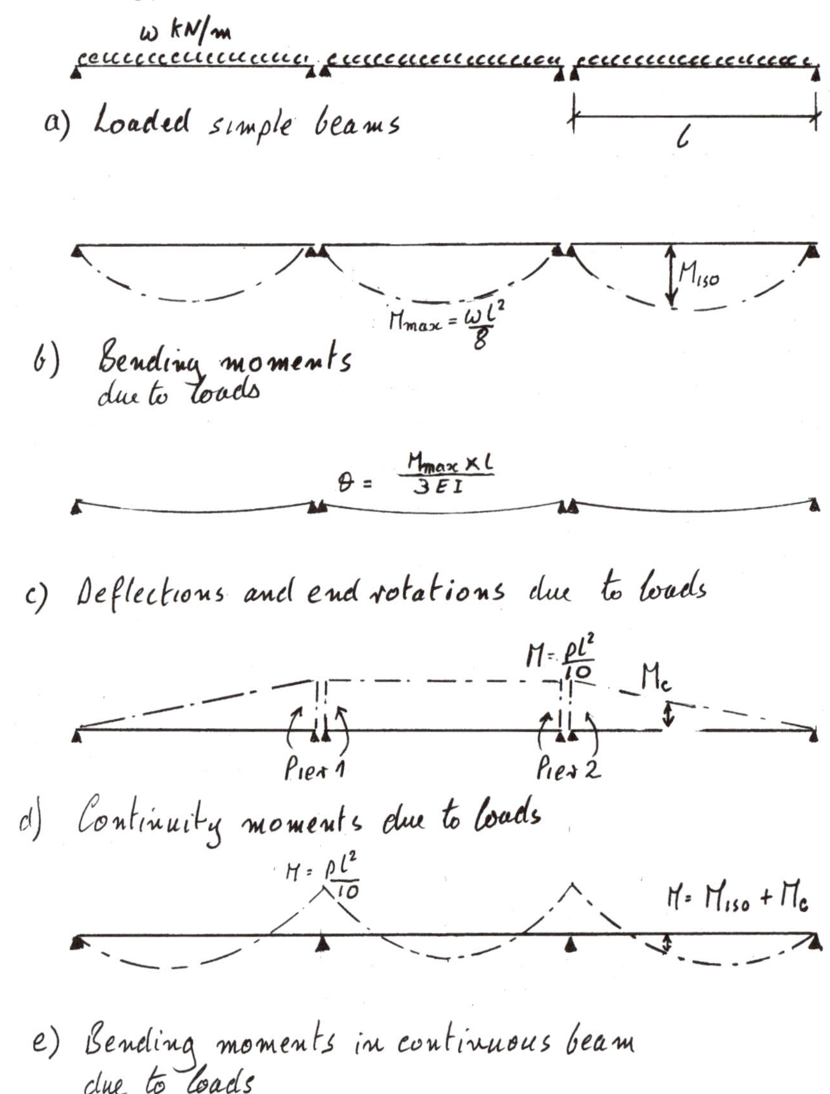

Figure 6.1 The principle of parasitic moments

but of opposite sign, must be applied at each internal support Figure 6.1 (i). These continuity moments are called the 'parasitic moments' or M_p. The prestress moment at any point in a continuous beam is then $Pe + M_p$, Figure 6.1 (j). Although these continuity moments due to prestress are of exactly the same nature as the continuity moments due to external loads, engineers and students appear to be baffled by them.

However, whereas the free bending moments and the continuity moments due to external loads are linked by the laws of statics and are consequently considered together as 'the bending moment', it is convenient to consider Pe and M_p separately. The eccentricity of the prestress is under the control of the designer, who may thus vary the resulting M_p between certain limits, as explained in 6.8.

Figure 6.1 continued...

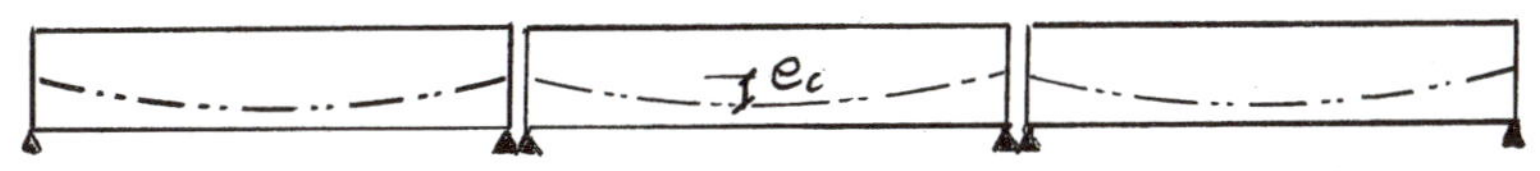

f) Simple beams with parabolic prestress

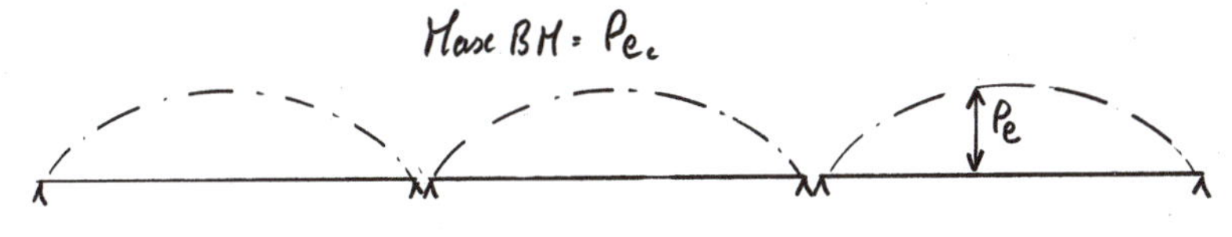

g) Prestress bending moments in simple beams

$$\theta = \frac{Pe_c \times L}{3EI}$$

h) Deflections and end rotations due to prestress

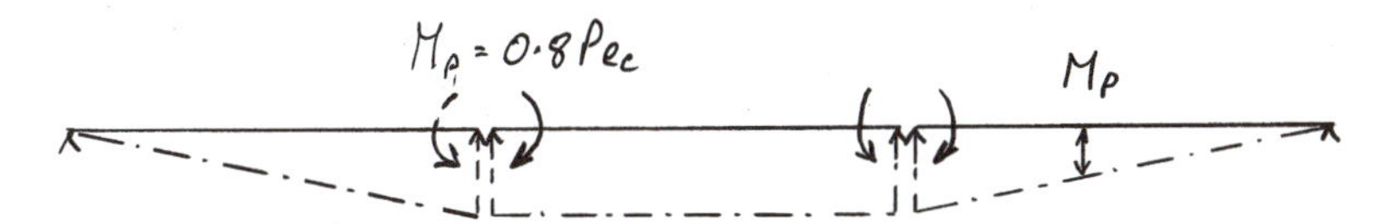

i) Continuity moments due to prestress
(Prestress parasitic moments)

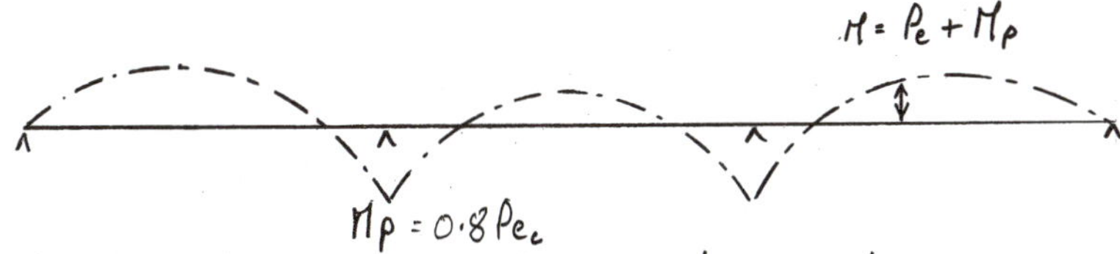

j) Prestress bending moments in continuous beams

This particular case of three equal spans with parabolic cables anchored at the neutral axis gives rise to parasitic moments equal to 80 per cent of the mid-span *Pe*. If the cable had the same mid-span *Pe*, but if its shape had been other than parabolic, or if it had been anchored above or below the neutral axis, the end rotations of the simple beams would have been different, and consequently the parasitic moments required to re-establish geometric compatibility would also have been different.

Prestressing cables that lie below the neutral axis of the beam tend to make it hog upwards, requiring sagging parasitic moments to restore compatibility of rotations, which are given a positive sign in accordance with the sign convention described in *5.12*. It should be noted that parasitic moments change the reactions on the supports, and create shear forces in the beams (as do the continuity moments in externally loaded continuous beams), see *6.4* below.

6.3 Parasitic moments at the ULS

If it is necessary to apply a load factor to the prestress at the ULS, this factor should be applied only to the force P. As the parasitic moment is proportional to P, it will be factored in the same way.

For instance, assume that the bending moments at an internal support of a continuous beam are:

- moment due to applied loads, M_L;
- prestress primary moment, Pe;
- prestress parasitic moment, M_p.

If, as an example, the ultimate load factors are 1.5 on loads and 0.9 on prestress, the ultimate moment

$$M_{ULS} = 1.5M_L + 0.9Pe + 0.9M_p.$$

In general M_p is sagging, while at supports M_L is hogging and Pe is sagging. Introducing the correct signs,

$$M_{ULS} = -1.5M_L + 0.9Pe + 0.9M_p,$$

thus the support moment is numerically reduced by the two components of the prestress. At mid-span, where M_L is sagging and Pe is hogging,

$$M_{ULS} = +1.5M_L - 0.9Pe + 0.9M_p.$$

M_p and Pe cannot rationally be factored differently.

Where the code of practice allows the redistribution of moments at the ULS, in reinforced concrete the support moments are usually greater than the span moments, and redistribution will be from support to span. In prestressed concrete, the total support moment may well be less than the total span moment, and such redistribution is either inappropriate, or may be in the reverse direction.

When such redistribution is appropriate, a proportion of the total moment M_{ULS} is redistributed, which we may call kM_{ULS}. Thus, if the redistribution is from the support towards the span, the total moment at the support becomes

$$M_{ULS} = -1.5M_L + 0.9Pe + 0.9\,M_p + kM_{ULS},$$

and the total moment in the span becomes

$$M_{ULS} = +1.5M_L - 0.9Pe + 0.9\,M_p + kM_{ULS}.$$

This redistribution has no effect on the value of M_p.

When the load on a continuous prestressed concrete beam is increased towards failure, an elasto-plastic hinge may form over a support. This hinge will put a ceiling on the total moment at the support. For instance, if it forms when the factor on the applied load is 1.7, the total moment at the support will be

$$M_{ULS} = -1.7\,M_L + 0.9Pe + 0.9\,M_p.$$

Any further increase in load will not increase the moment at the hinge (the increased moment being carried in the adjacent spans). However, the prestress moments, both primary and parasitic, are not affected by the formation of this elasto-plastic hinge. Of course, if a complete hinge were to be introduced at a support, all moments including M_P at the hinge would become zero, and the M_P at the other supports in the beam would be redefined.

6.4 The effect of parasitic moments on the beam reactions

The parasitic moments for a typical four-span beam are shown in Figure 6.2 and may be labelled M_{p1}, M_{p2} and M_{p3} at the three internal supports, with the parasitic moments being zero at the end supports. The variation of parasitic moment along the spans creates shear forces in the beam, and reactions at the end of each span. In general, the reactions and the shear forces are $\pm (M_{p\,right} - M_{p\,left})/L$, where L = span length.

For instance, in the side span the reaction at the end support is $(M_{p1} - 0)/L_1$, which is an upwards reaction if M_p is positive. There is a corresponding downward reaction on pier 1. The reactions at each end of span 2 are $\pm (M_{p2} - M_{p1})/L_2$.

These reactions must all sum to zero, as no external vertical loads have been applied to the beam.

The prestress primary moments, *Pe*, are internal forces, and although they apply shear forces to the beams, they do not create reactions; the shear forces at the beam ends are matched by equal and opposite vertical components of the prestress anchorage force (*5.19*).

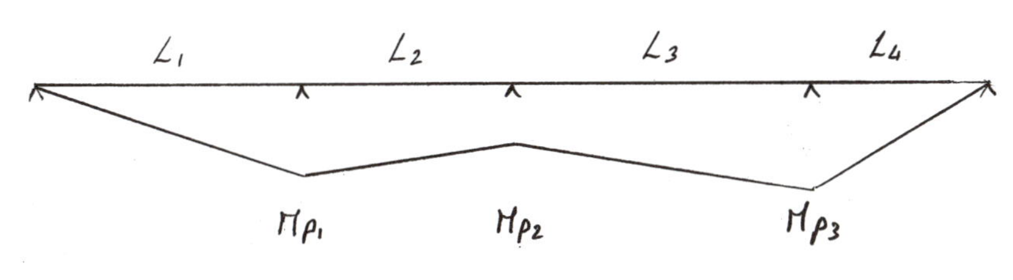

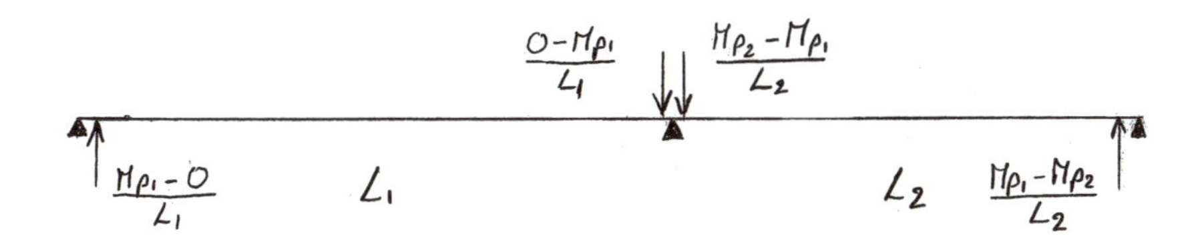

Figure 6.2 Reactions due to prestress parasitic moments

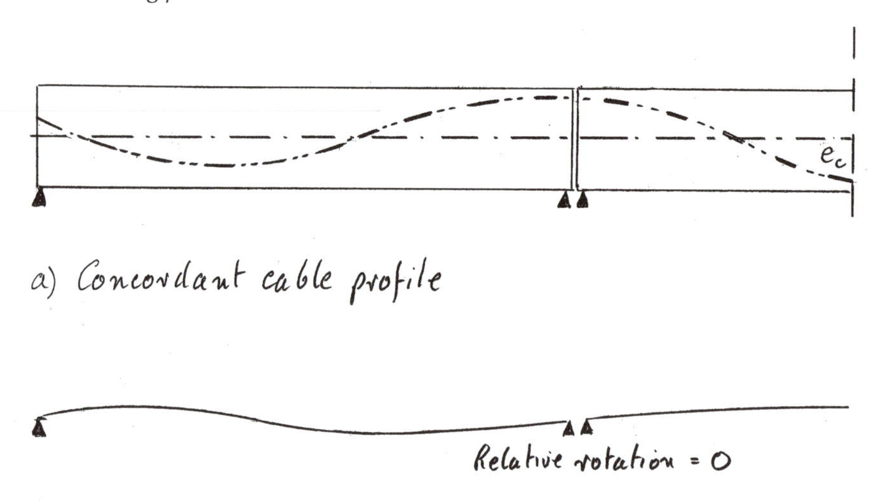

Figure 6.3 Concordant cables

6.5 Concordant cables

We have seen that the continuity moments due to prestress for a continuous beam are those moments that must be applied to the ends of the released beams to render compatible the beam end rotations, Figure 6.1. It is possible to arrange the prestressing tendons in such a way that they create zero relative rotations at the ends of the statically determinate beams. For instance, a prestress profile of the general form shown in Figure 6.3 (a) for the statically determinate beams shown in Figure 6.1 (f) may, by suitably arranging areas of positive and negative eccentricity, be designed to give zero relative rotations of the beam ends at the intermediate supports. Under these circumstances, the M_p would be zero. A cable profile that gives a zero parasitic moment is called a concordant cable.

In the early days of prestressing, when engineers had trouble understanding and manipulating prestress parasitic moments, it was considered desirable to achieve concordant cable profiles. In fact such profiles are almost universally uneconomical, frequently requiring some 50 per cent more prestressing steel than a properly designed profile. One of the great strengths of prestressing is the facility to manipulate M_p in order to optimise the bending moment distribution between pier and mid-span.

6.6 Straight cables in built-in beams

In a span that is effectively built-in at each end (which could be a single built-in span or the interior spans in a long continuous deck), any straight cable of constant eccentricity will generate a parasitic moment that is equal and of opposite sign to the Pe. That means that such a straight cable is always effectively centred, producing only a direct compression of P and no bending moment.

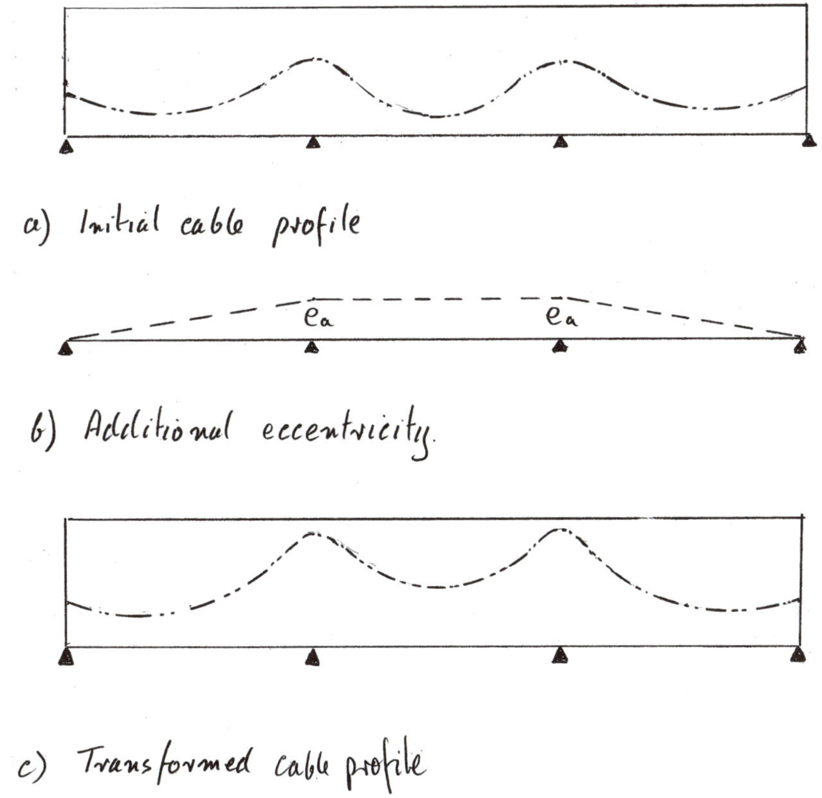

Figure 6.4 Cable transformations

6.7 Cable transformations

Consider the continuous beam stressed with a cable anchored at the ends of the side spans, as shown in Figure 6.4 (a). If the eccentricity of the cable along the beam is changed linearly but the point of attachment of the cables remains unaltered, Figure 6.4 (b), the total prestress moment $Pe + M_p$ at any point remains unchanged. The change in M_p exactly compensates the change in Pe.

6.8 Control of prestress parasitic moments

As we have seen in *5.14*, the width of the cable zone tends to be close to zero at mid-span of a simply supported beam, but widens out in areas of lower bending moment. In a continuous beam, the greatest moments exist close to mid-span and over the supports, and the prestress force is sized for these critical locations where consequently the cable zone is very narrow. At all other locations the prestress force tends to be greater than the minimum necessary, and consequently the cable zone is wider. This means that the designer has some latitude in where he draws the cable within the zone, Figure 6.5.

By adjusting the position of the centroid of the prestressing force in the non-critical locations, the designer has a degree of control over the size of M_p. The lower the

Figure 6.5 Typical cable zone

prestress profile, the more positive (sagging) becomes the parasitic moment. This is a vital tool in the hands of the engineer, but is neglected by many bridge designers.

6.9 Details of the sample bridge deck

The sample bridge has spans of 36 m, 40 m, 40 m, 36 m; Figure 6.6 (a). Note that the side span is 90 per cent of the main span, which is the appropriate proportion for a prestressed deck, as the sagging parasitic moment at the critical design section of the side span is only 40 per cent of the value in a typical internal span. For the design of continuous prestressed concrete bridge decks, it is normal practice to check stresses at least at every tenth point of each span. The numbering of the design sections we will adopt is shown in Figure 6.6 (b).

The deck consists of a 12 m wide box section bridge deck as shown in Figure 6.7. The depth is 2.2 m, giving an economical span/depth ratio of 18, and providing sufficient headroom for work inside the box. The web thickness in the spans is 350 mm, which is close to the practical minimum for a cast-in-situ deck with internal prestressing tendons. At the piers, the webs thicken to 600 mm to provide sufficient width to carry the shear force.

The top slab is 200 mm thick which is a practical minimum for a slab carrying traffic (*9.1*). Also, for the 5.3 m clear span it provides a span/depth ratio of 26.5, which is within the guidelines given in Chapter 9. The bottom slab is 200 mm thick, which is also close to the practical minimum for cast-in-situ construction.

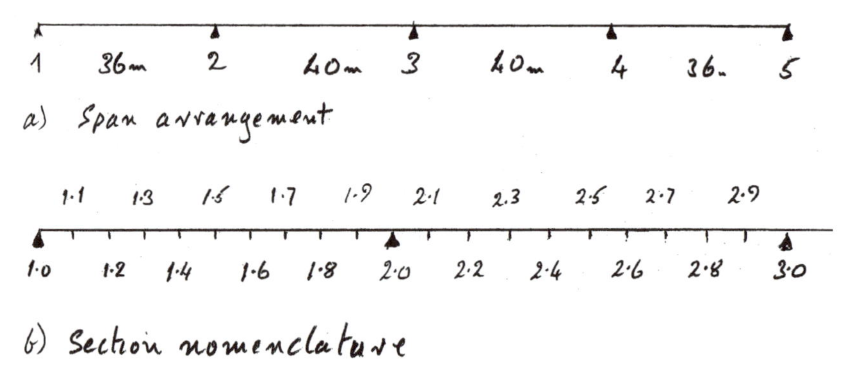

Figure 6.6 Span arrangement of sample bridge

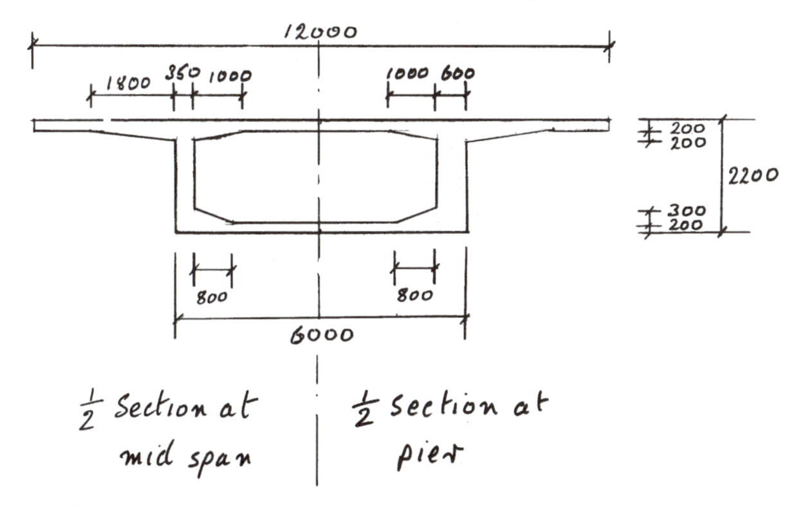

Figure 6.7 Deck cross section

6.10 Section properties

6.10.1 Section properties for the sample box

Table 6.1

	At each mid-span	*At each support*
Area m²	5.66	6.56
Inertia m⁴	4.115	4.411
y_t m	0.839	0.875
y_b m	1.361	1.325
a_t m	0.534	0.507
a_b m	0.867	0.768
z_t m³	4.905	5.041
z_b m³	3.024	3.329
η	0.637	0.580

Note: See 5.3 for the definition of the above symbols.

6.10.2 The effects of shear lag on section properties

When a beam consists of a web associated with top and or bottom slabs, shear stresses exist at the junction of the web and slabs. These shear stresses are essential to induce the slabs to participate in the overall bending of the beam. For a box section they fall to zero at the free end of cantilever slabs and at the axis of symmetry of the box for the top and bottom intermediate slabs, Figure 6.8 (a). The slabs deform under the effect of the shear stresses, and as a consequence, the compressive and tensile stresses in the slabs due to the overall bending of the beam fall off from a maximum at their junction

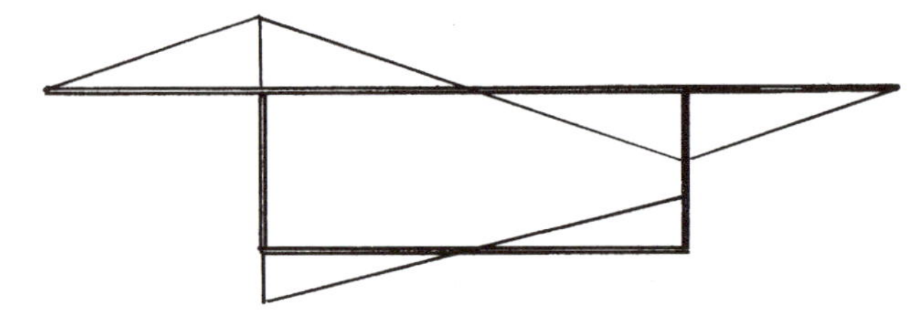

a) Shear stresses in slabs.

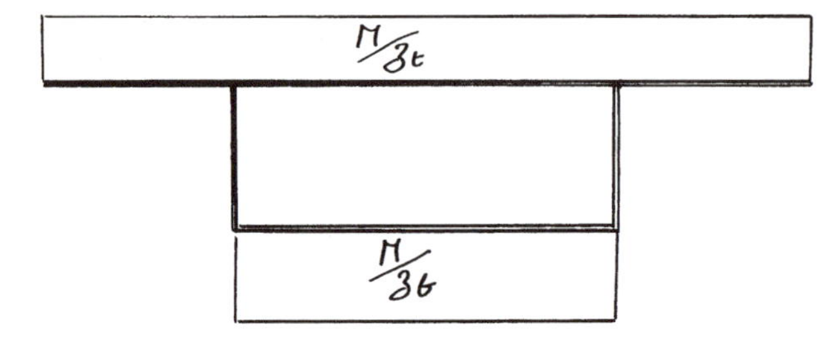

b) Longitudinal bending stresses in slabs calculated using full section properties

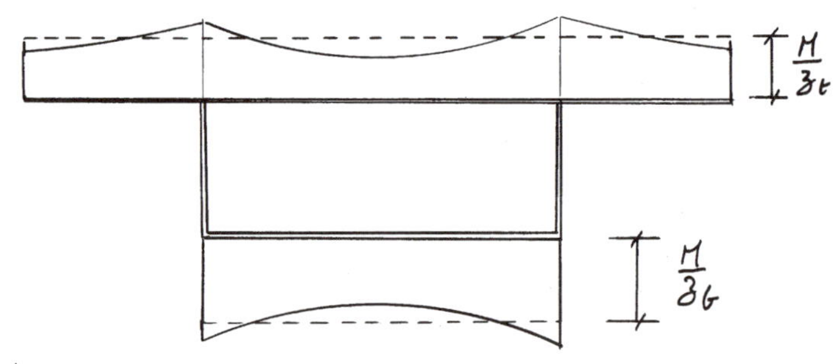

c) Longitudinal bending stresses in slabs affected by shear lag

Figure 6.8 Effect on bending stresses of shear lag

with the web, to a minimum at the points of zero shear described above. As the slabs are not fully effective, the maximum overall bending stresses in the box will be greater than those calculated using the section properties in Table 6.1; Figure 6.8 (b) and (c). This is an elastic effect that only concerns working loads and the calculation of stresses; at ultimate loads, the full width of the slab is effective.

The exact calculation of the effects of shear lag is tedious, as the shear stress in the slabs is dependent on the shear force in the webs of the beam, which itself depends on the loading that is applied to the beam, and on the location along the beam. The effect of shear lag thus varies with each load case, and with the location along the beam.

Most codes of practice cover the matter by proposing reductions in the width of slab that should be assumed to be associated with each web for the calculation of bending stresses. The effect on the calculation of the stresses due to prestress is that the compressive component P/A is assumed to spread over the full area of the cross section, while the stresses due to the bending component Pe are calculated with the section properties corresponding to the modified cross section. Thus in the expression $P/A \pm Pe/z$, P/A is calculated using the complete cross section, while Pe/z uses the section properties of the shear lagged cross section. Similarly, the stresses due to bending moments caused by applied loads are calculated with the shear lagged section properties.

It should be clearly understood that the codified calculations are approximate, and that the real bending stresses in the section are not identical to those calculated. It is also clear that the effects of shear lag will be reduced by the presence of slab haunches, as they will reduce the shear stress in the slabs, and hence their shear deformations.

In the interests of simplicity and clarity, the effects of shear lag have been ignored in the current example.

6.11 Comment on the accuracy of calculations

Designers should be aware of the approximate nature of the calculations required for the detailed design of concrete bridge decks.

All codes of practice known to the author assume that the bending stresses will be calculated by the 'engineer's bending theory', which only considers uni-axial stresses and which assumes that plane sections remain plain, and that consequently stresses vary linearly. The structure is deemed to be acceptable at working load if the limiting stresses, calculated by this theory, remain below certain values.

The designer must be careful not to apply the 'deemed to satisfy' provisions of codes of practice to sophisticated analyses that they were not intended to cover. If a structure that was satisfactory under this simplification was to be analysed in three dimensions by finite elements it is likely that local areas of stress will be found that exceed the permitted limits. For instance, above the bridge bearings, which have contact stresses of the order of 20 MPa, the lateral Poisson's ratio effects will combine with the overall bending stresses to produce local areas of compressive stress that may exceed the limiting values defined by the code of practice. In general these excess stresses may be safely ignored; the strength of concrete is increased when it is subjected to bi-axial or tri-axial states of compressive stress. However, the designer must use his judgement and experience to decide if high stresses defined by elastic analysis are in fact acceptable.

The approximations used to cater for the effect of shear lag, as defined in *6.10.2* above, for the effect of creep on bending moments, described in *6.21* and for the effect of creep on the section properties described in *6.22*, should make clear that there is no point in carrying out a very sophisticated analysis for code compliance, when the basic assumptions themselves are approximate.

However, code compliance is only one component of a designer's task. He must work to understand how the structure behaves in complex areas, and sophisticated elastic analysis can be extremely useful in some circumstances.

6.12 Dead and live loads

Dead loads include the self weight of the deck and any deck finishes such as road surfacing, footpaths and parapets.

For this exercise the density of concrete is assumed to be 25 kN/m^3; consequently the self weight is $25 \times 5.66 = 141.5$ kN/m.

The finishes may be assumed for the first stages of a preliminary design to be represented by a uniform thickness of concrete of 150 mm, and hence their dead load is $25 \times 0.15 \times 12 = 45$ kN/m.

Live loads are principally the weight of traffic which must include the appropriate dynamic enhancement. Other effects of traffic loading are centrifugal force on curved decks, braking and traction forces that are particularly significant on railway bridges, and lateral nosing on railway bridges.

For this example, the live load is assumed to be a distributed load of 100 kN/m, of any length, which is assumed to include the dynamic enhancement.

6.13 Bending moments

6.13.1 Bending moments due to live loads

The 100 kN/m live load is arranged to maximise the effect on the design section being considered by loading only areas of the influence lines that are of the same sign. For mid-span and support sections, the loaded lengths are always complete spans. For some other design sections, part spans may be loaded.

6.13.2 Bending moments due to differential settlement

The bending moments in continuous bridges are affected by differential settlement of the foundations. The settlement of a foundation is made up of several components.

a) The first is the purely geotechnical component of settlement due to the response of the ground beneath the foundation. This not only concerns the soil or rock immediately beneath the foundation, but also may involve deeper layers.

   In principle, a reasonable calculation of the total and differential settlement due to the response of the ground beneath a foundation may be made, albeit with a degree of uncertainty inherent in all geotechnical calculations. If the foundation conditions are extremely variable, for instance when adjacent piers are founded on different strata, or foundations alternate between piles and pads, the uncertainties in the calculation of differential settlement will be much greater.

   For instance, the River Nene Bridge on the Nene Valley Way in Northampton was founded on pads resting on a layer of gravel, Figure 6.9 and Figures 11.16 and 11.17. (The bridge is further described in *11.6*.) Although the gravel was relatively consistent, and would not have given rise to significant differential settlement, it was underlain by a thick layer of clay. The pressure bulb beneath the relatively small pier foundations was confined to the gravel, giving rise to only small settlements. However, the pressure bulb of the 30 m wide approach embankments extended deep into the clay, causing substantial settlement of the bridge abutments. The design of the bridge was dominated by the differential

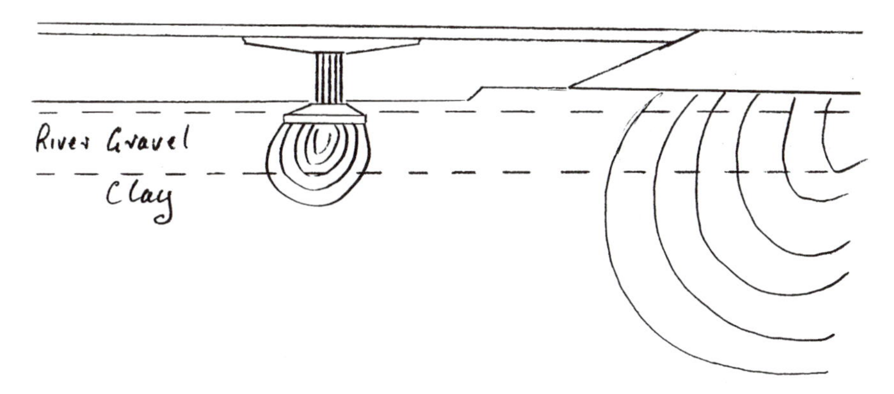

Figure 6.9 River Nene Bridge: differential settlement

settlement between the abutments and the adjacent piers. A very slender flexible deck was adopted, with provision left for jacking up the abutment bearings if the settlements proved to be greater than those calculated.

b) Second, there is the structural component of differential settlement consisting principally of the compression of the bearings and the shortening of piers and piles under load, elastic under live loading and including creep under permanent loads. Whereas mechanical bearings do not compress significantly, the compression of elastomeric bearings may need to be considered.

   If the piers and piles are of significantly different length, they will shorten differentially. If the foundation piles are end bearing, their shortening is simple to calculate. However, if the piles work in friction, or if the load is carried both in friction and end bearing, some care needs to be exercised in estimating their response under service loads. Piles are normally designed to carry an ultimate load that is 2.5 or 3 times greater than the working load. Consequently, the working load may often be carried in friction entirely in the upper layers of the bearing stratum, giving a short effective length of pile. As the load is increased towards ultimate, so greater lengths of the pile are mobilised in friction until the end bearing comes into play. In some cases, this reasoning leads to much less pile shortening under working loads than would be calculated from an oversimplified assumption that the full length of the pile shortens under load.

c) Finally, there are differential settlements due to inconsistencies in construction technique. For instance, end-bearing piles may be cleaned out more or less well, bored friction piles may be left open for different periods of time etc. The differential settlements due to these effects generally defy calculation, but may be very significant. If there is reason to doubt the care with which piles are being installed, a conservative view of differential settlement should be taken, the foundation levels should be monitored, and in extreme cases, provision made for re-levelling the bearings.

The timing of the development of differential settlements is very important in the analysis of their effects on the deck. In most cases, it may reasonably be assumed that the differential settlements are due solely to permanent loads, of which the self weight of the deck is the greater part. In some forms of construction, the self weight of the deck is mobilised before the deck is made continuous. An obvious example is balanced cantilever construction, when it is likely that the greater part of any settlement, and hence of any differential settlement, will have taken place before the deck is made continuous.

Another example is span-by-span construction, where the weight of the deck is applied when it is continuous in one direction only, generally reducing the effect of a given settlement. In this form of construction, it is necessary to distinguish between cases when the falsework rests on the ground between piers, or when it spans from pier to pier, resting on the pile caps. In the former case the weight of the wet concrete is carried by the ground between piers, and the forward foundation is only loaded when the falsework is struck, once the span is structurally continuous. Thus all the instantaneous settlement of the forward foundation will be differential with respect to the previous pier. In the latter case, the weight of the wet concrete is carried by the forward foundation before the span is structurally viable, thus its short-term settlement does not cause any bending moments.

Differential settlement is an applied deformation, not an applied load. Consequently relaxation of the deck concrete will reduce its effect. It is normal practice to adopt a Young's modulus of $^1/_2$ to $^1/_3$ of the short-term modulus for the deck concrete in estimating the effect of differential settlement (see *3.9.2*). Also, as long as the structure is adequately ductile, such an imposed deformation may usually be ignored at the Ultimate Limit State.

If a deck rests on piers of similar length, is founded on a consistent medium and the construction of the foundations is well controlled, it is generally acceptable to assume that differential settlements will not exceed ±5 mm. These settlements should be applied to the supports in the manner that maximises the resulting bending moments in the deck. If the bridge is being built in such a way that the self weight is applied before it is continuous, consideration should be given to ignoring differential settlements completely. In the author's experience, the amount of differential settlement and its effect on the bridge are often over-simplified and exaggerated by inexperienced engineers. However there are cases in which differential settlement is a very significant problem, when a careful assessment should be made.

The moments due to differential settlement should be added to the table of moments due to loads.

6.13.3 Bending moments and stresses due to temperature change and temperature gradients

Children used to be taught that in the desert, the sudden drop in temperature at night splits rocks and helps create sand. They may not know that the reason is that as the surface of the rock cools more quickly than the core, so it is stressed in tension and cracks. Temperature gradients may have similarly important effects on concrete bridges if they are not correctly considered in the design. In fact, a concrete bridge deck is much more likely to be damaged by temperature gradients than by traffic loading. It is thus essential that a designer really understands the behaviour of the deck under such gradients, and does not just blindly apply the appropriate formulae.

As the sun shines on a deck, it heats the top surface, creating temperature gradients through the deck and raising its average temperature. At night, the deck re-radiates its heat, cooling and creating different gradients.

There are two distinct effects of temperature on a bridge deck:

- The bridge deck will expand or contract in response to changes in its average temperature. Most bridges of modest length that are carried on sliding or elastomeric bearings are free to change length without creating significant normal stresses in the deck. Long decks will experience substantial normal forces due to the cumulated effect of friction or shear in the bearings, known as bearing drag. Some decks, such as portals, arches or decks built into piers will experience both normal stresses and bending moments due to their change in length.
- The deck will deflect up or down due to the temperature gradients. When the top surface of a beam is heated by the sun for instance, it expands relative to the bottom fibre and the beam tends to hog up, acting as the familiar bimetallic strip. If the rotations of the beam ends are restrained by continuity, bending moments are set up in the beam.

The changes of bridge deck temperature may be converted into strains by multiplying them by the coefficient of expansion of concrete, α, which is generally considered to lie between $12 \times 10^{-6}/°C$ for normal concrete, and $7 \times 10^{-6}/°C$ for concrete made with limestone aggregate [1] (7.2.2). If the strains are restrained, the stresses caused are found by multiplying them by the Young's modulus of concrete, E, so that $\sigma = t\alpha E$ (where σ = stress and t = temperature change). For daily fluctuations in temperature, the short-term modulus measured at 28 days should be used. For seasonal changes, a lower modulus is appropriate; 75 per cent of the short-term 28-day modulus is a reasonable assumption, which takes into account the increase in strength and modulus with time.

The effect of temperature changes and gradients on a beam may best be understood by considering initially a single span beam of rectangular cross section. Assume that

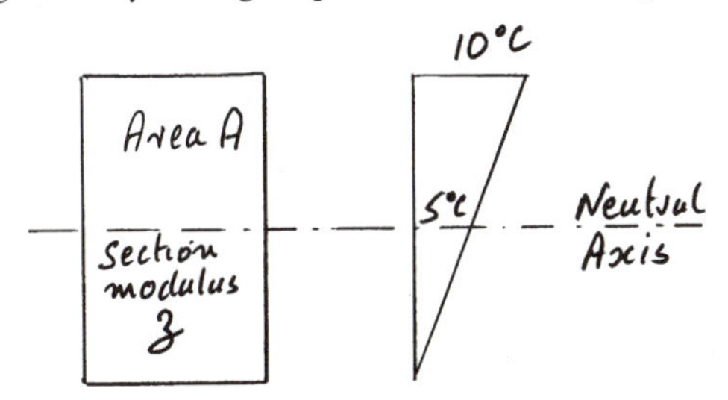

a) Beam cross section and temperature gradient

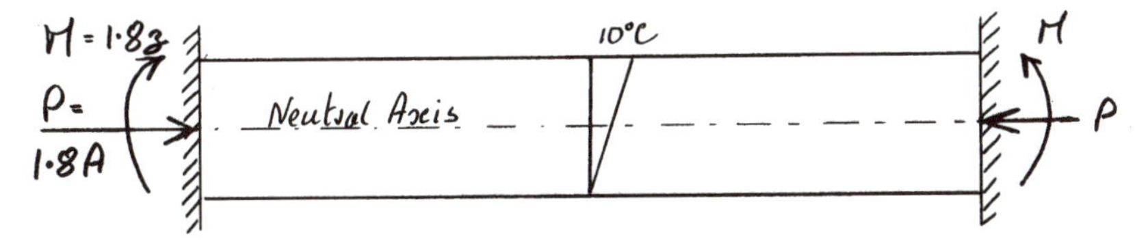

b) Elevation of built in beam

Figure 6.10 Rectangular beam subject to linear temperature gradient (figure continues overleaf)

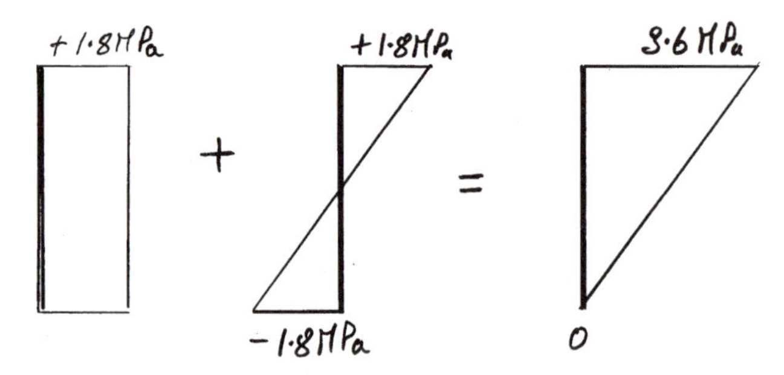

c) Stresses on built in beam

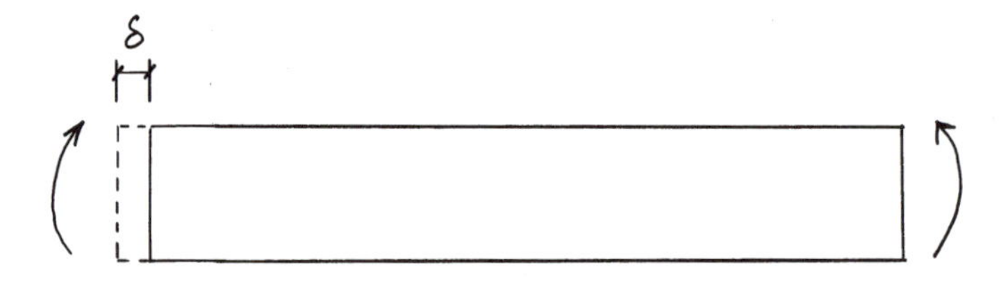

d) Beam released for normal force; built in for moment

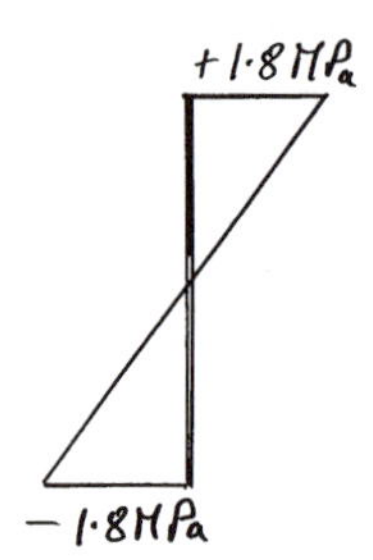

e) Stresses in beam built in for moment only

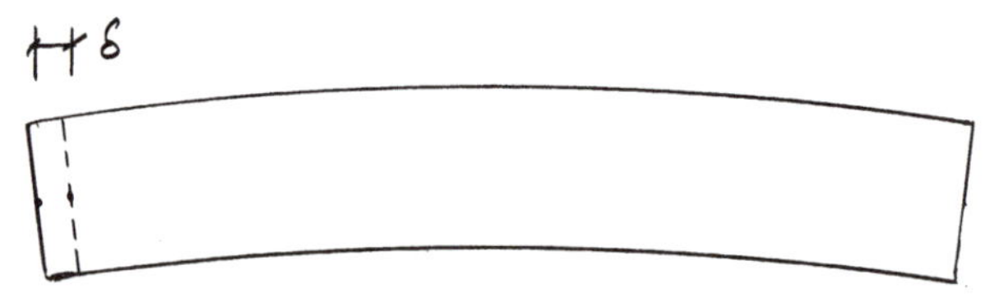

f) Beam released for normal force and moment

Figure 6.10 Rectangular beam subject to linear temperature gradient (continued)

the temperature increases by 10°C on the top surface, that the gradient is linear, that $E = 30{,}000$ N/mm^2, that $\alpha = 12 \times 10^{-6}$/°C, that the cross-section area is A and the section modulus is z, Figure 6.10 (a). If the span were built in at its ends such that it could neither expand nor deflect, the stresses in the beam would be zero on the bottom fibre, $10 \times 30{,}000 \times 12 \times 10^{-6} = +3.6$ MPa on the top fibre and +1.8 MPa at the neutral axis, Figure 6.10 (b). This state corresponds with an axial compressive force of $P = +1.8 \times A$, necessary to maintain its length, together with a sagging bending moment of $+M = 1.8z$ MNm necessary to maintain zero deflection, Figure 6.10 (c).

If the deck is now allowed to expand, but the ends are still restrained for rotation, the compression will fall to zero while the moment will be unchanged; the top and bottom fibre stresses will become ±1.8 MPa, Figure 6.10 (d) and (e). Effectively a tensile force of $1.8 \times A$ MN has been subtracted from the stresses in the restrained beam.

If the rotational restraint were now removed, effectively by applying hogging moments $-M$ to each end, the simply supported deck would arch up into a circular profile (as the stresses were constant along the span), Figure 6.10 (f), and be completely unstressed.

However, as usual the reality is more complicated, as the temperature gradient through bridge decks is far from linear. It can be taken from codes of practice, or calculated from the first principles of thermodynamics. The temperature distributions typical of concrete box girders in the UK, shown in Figure 6.11, are taken from the British code BD 37/88 [2], which is based on [3]. The actual temperatures to be used will vary according to the climate and the applicable code of practice. The distribution through a deck is sensitive to the presence, and thickness of, surfacing. Consequently, the effects may be most severe when a bridge deck is under construction.

For a typical box section highway bridge the top slab, which is generally between 200 mm and 400 mm thick, constitutes about half the total concrete section of the deck. This top slab heats up quickly under the effect of solar radiation during a daily cycle while the webs and bottom slab remain relatively cool.

As before, the span is at first considered restrained both in length and rotation, subjected to a daytime temperature gradient. The temperature may be converted into stresses by using the appropriate coefficient of expansion and Young's modulus,

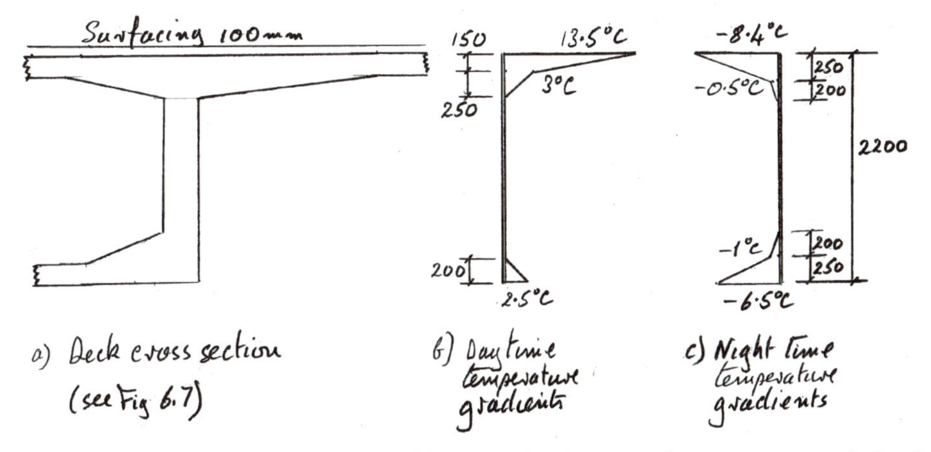

Figure 6.11 Temperature gradients defined by UK code of practice drawn to same vertical scale as sample box girder

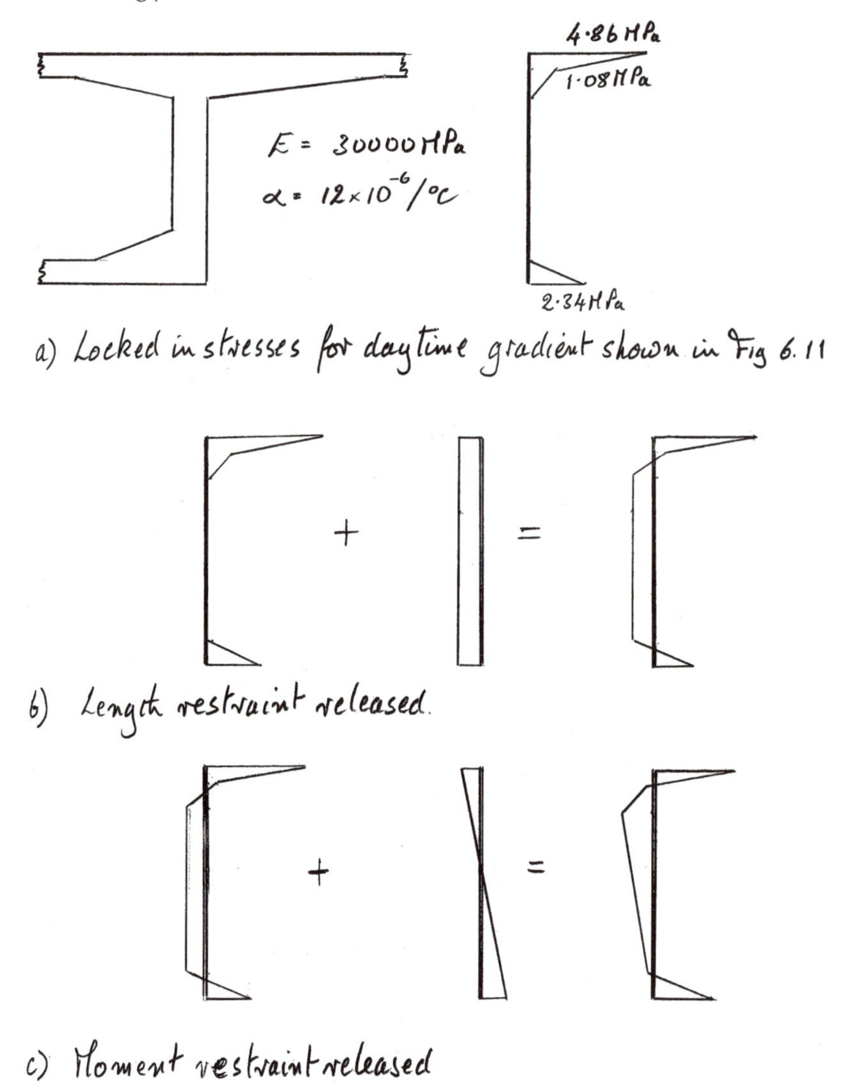

Figure 6.12 Temperature stresses in box girder

Figure 6.12 (a). It will be subjected to an overall compression corresponding to its average rise in temperature, and to a sagging moment. Both the compression and the bending moment may be computed by considering the cross section divided into appropriate strips, and by summing the force in each strip and its moment about the neutral axis.

If the length restraint is released, allowing the beam to expand, effectively a tensile force is added to the beam, cancelling the overall compression and creating a new set of stresses that have no net normal force, Figure 6.12 (b); that is the compressive areas of the cross section balance the tensile areas. Now rotational releases may be applied to the beam ends allowing it to hog upwards. Effectively, a hogging moment has been applied to the beam, modifying the stresses so that there is no overall bending moment,

Figure 6.12 (c). However, as the temperature gradient is not linear through the depth of the deck, even when fully released, stresses remain locked into the section. These stresses must, by definition, sum to give zero longitudinal force and zero bending moment. They may be termed the 'internally balanced stresses'.

a) Effect of temperature gradients on the design of statically determinate beams

In statically determinate beams both the longitudinal and rotational restraints have been released, leaving only the internally balanced stresses. The compressive stress on the top of the top flange due to solar heating may attain 3 or 4 MPa, a significant proportion of the allowable stress at the serviceability limit state. However, it falls off very quickly within the thickness of the top slab, generally reducing to zero within 150 mm. Such shallow compressive stresses clearly have little effect on the reserves of strength of the beam. At the ultimate limit state these compressive stresses that are due to locked-in temperature strains, disappear. Consequently, they can do no damage, and it is suggested that they should be ignored in design. (However, some codes of practice do include such temperature effects at the ULS.)

The tensile stresses due to night-time re-radiation generally do not exceed –2 MPa. When they are combined with stresses due to prestress, these tensile zones are shallow. The spacing, and hence the width of any potential cracks is related to the depth of the tensile zone. A tensile zone 200 mm deep will have a crack spacing of less than 200 mm. If all the temperature strain were to be concentrated into a series of cracks at this spacing, their width would be only $200 \times 2/30{,}000 = 0.013$ mm. This crack width is quite insignificant, either on its own or when added to any cracks due to other causes. The minimum crack width that can be seen by the naked eye is reported by Neville to be 0.13 mm (*3.5.5*), while the crack width considered acceptable for considerations of durability lies between 0.15 mm for the most aggressive conditions, and 0.35 mm for internal members.

Concrete subjected to tensile stresses of low intensity caused by applied strains may yield and does not necessarily crack. The limiting tensile strain of concrete lies between 1×10^{-4} and 2×10^{-4} [5], while the tensile strains due to these temperature effects are of the order of 0.7×10^{-4}. Furthermore, if the concrete were to crack or yield, the stresses would disappear. Consequently, it is suggested that tensile stresses due to the non-linear temperature distribution in statically determinate beams be ignored for the purposes of design of the prestress, a position that is supported by most codes of practice. However, the surface strains due to temperature gradients are a reality, even if the tensile stresses they give rise to may be ignored. Consequently, any surface that is subjected to tensile stresses due to temperature gradients should be provided with a mesh of reinforcement. If no design steel is present for other purposes, this reinforcement should consist of small-diameter bars (12 mm or less) at a spacing of not greater than 200 mm.

In bridge decks with a wide top slab and thin webs, the heating of the top slab will give rise to tensile stresses in the webs that may attain 3 MPa or more. These will increase significantly the principal tensile stresses due to shear force, and could cause the web to crack under working loads. Although this will not affect the ultimate strength of the web in shear, it should be borne in mind by the designer, particularly in tropical climates, when the deck is unsurfaced, or even worse, surfaced with a very thin black layer, such as a waterproof membrane. In general, it is adequate to ensure that there is an appropriate mesh of anti-crack steel on both surfaces of the web.

The temperature gradients cause significant deflections and beam end rotations, and this is likely to have an effect on the dimensioning of bearings and expansion joints. When building decks in balanced cantilever, the free end of the cantilever will move up and down significantly on a daily cycle, affecting the surveys required to control the shape of the bridge, and the procedures for casting the mid-span stitch.

b) Effect of temperature gradients on the design of continuous beams

In typical continuous beams the longitudinal restraint has been released, while the rotational restraint has only been released at the end supports. Consequently the temperature gradients cause bending moments in the beam, Figure 6.13. At any section of the continuous beam there will be a bending moment together with the internally balanced stresses. At the SLS, these bending moments should be added to the table of moments due to loads, and the prestress designed in consequence. As the moments are due to locked-in strains, they should not be added to the moments due to external loads and prestress at the ULS.

In the author's opinion, the internally balanced stresses on the extreme fibres are harmless for the reasons explained in (a) above, and should be ignored in the sizing of the prestress. An additional reason for this is the disproportionate effect the inclusion of these shallow surface tensile stresses would have on the amount of prestress required. For instance, in a deck with an average prestress *P*/*A* of 5 MPa, which is typical of bridge decks, the need to provide typically an additional compressive stress of 1.5 MPa on both extreme fibres would increase the prestress force required by 30 per cent. This is not only uneconomical, but increases the permanent compressive stress in the concrete, the congestion of the cross section for concreting, the movement at the expansion joints and bearings, and in general deteriorates the quality of the design. The best prestress scheme is that which needs the least pre-compression in order to fulfil the design objectives!

Consequently, in the author's opinion, the prestress for continuous bridges subject to non-linear temperature gradients should be designed to accommodate the bending moments caused by the gradients, but not the internally balanced stresses. These are adequately catered for by ensuring that the top and bottom surfaces of the bridge deck are reinforced with a mesh of bars as described above. Where the deck is made of precast segments, clearly the tensile stresses are released harmlessly at the segment joints.

Some authorities do insist that these internally balanced stresses should be considered for the design of the prestress, although this position would appear to be inconsistent with ignoring these same effects in statically determinate beams. Under

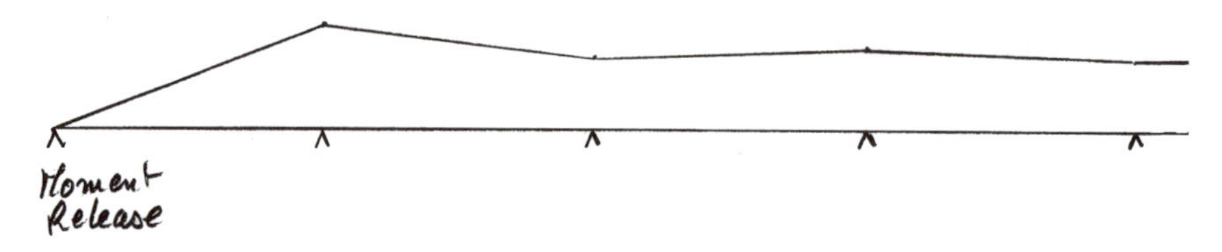

Figure 6.13 Bending moments caused by temperature gradients

these circumstances, either fictitious bending moments that give the appropriate stresses on the extreme fibres should be added to the table of moments, or the prestress should be designed to leave residual compressive stresses on the extreme fibres that are equal to the tensile stresses due to temperature.

However, there are some circumstances in which these internally balanced tensile stresses may act as a trigger that sets off cracking due to other, locked-in stresses. For instance, the bottom fibre of a continuous deck of short or medium span (less than about 50 m) at locations either side of an internal pier is particularly susceptible to cracking during construction, for the following reasons, Figure 6.14:

- Generally, this location is permanently compressed by hogging moments when in service, and consequently may be virtually devoid of design longitudinal reinforcing bars or of bonded prestressing cables.
- The prestress centroid is generally above the upper limit of the central kern, and thus there are tensile stresses due to prestress on the bottom fibre during construction. For bridges of modest span, before the prestress has suffered its time-dependent losses and before the deck finishes have been added, the self weight bending moment may not be enough to overcome these tensile stresses.
- Immediately adjacent to the pier, the convex shape of the prestressing profile fits badly to the cusp-shaped bending moment. As design sections are generally located at the support and at the first tenth point of the span, it is not uncommon to find that between these two sections the prestress centroid is above the cable zone, and that the tensile stresses on the bottom fibre are significantly higher than assumed.
- Tensile stresses due to moments caused by differential settlement may well be present at this location.
- In cast-in-situ decks where the webs are thick compared with the slabs, heat of hydration effects due to the webs cooling more slowly than the slabs are likely to

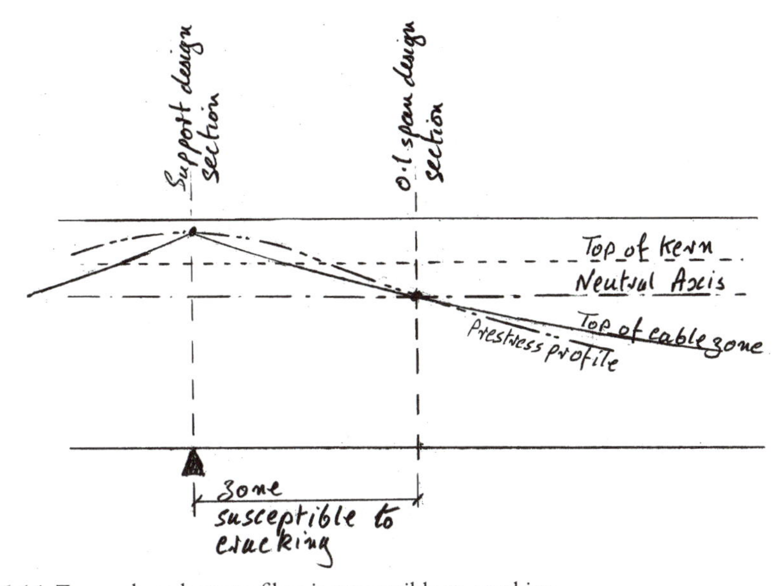

Figure 6.14 Zone where bottom fibre is susceptible to cracking

stress the bottom of the webs in tension. In box sections with thick webs, or in double Tee type decks (*12.6*), these stresses may be substantial.

All these actions may be evaluated by careful design. However, in the experience of the author, as this problem is not raised specifically by the codes of practice, it is neglected by designers and bridges crack as a result. In the absence of adequate passive reinforcement the section would be brittle and cracks are likely to be wide and extensive.

It is recommended to proceed as follows:

The stresses should be checked at the first 20th point of the span, either side of all internal supports.

For a cast-in-situ deck, the designer should ensure that the bottom fibre for a distance of 10 per cent of the span either side of an internal support (15 per cent for twin rib bridges) is subject to a residual compression of at least 1.5 MPa under the effects of initial prestress plus self weight. Alternatively it should be reinforced with an area of passive steel that corresponds to that required to render the section ductile (*3.7*). Twin rib type decks should always be so reinforced.

For a precast segmental deck, heat of hydration stresses will be absent. However as it is not possible to reinforce across the joints, either the section should be fully analysed at every joint close to the pier to check there are no tensile stresses, or a residual compression of 1.5 MPa should be respected.

c) Effect on railway viaducts and non-standard decks

On ballasted railway viaducts, the top slab of the deck will be insulated from direct solar radiation and from night-time re-radiation. Clearly this greatly reduces the temperature gradients affecting the deck, to the point where they can probably be ignored for the conditions in service. Conditions during construction or during refurbishment of the deck when the ballast may be removed should still be considered.

When the rails are carried on plinths or slabs, the situation is more complicated. The plinths and any raised footpaths will partially protect the deck from solar radiation. It is necessary either to carry out a heat flow calculation from first principles, or to adapt the normal temperature distribution calculations to fit the reality.

For non-standard cross sections, such as through bridges, the designer has no option but to return to first principles, and either to adapt the original fieldwork described in [3], or to carry out heat flow calculations.

d) Bridge decks that are restrained longitudinally

For bridge decks that are restrained longitudinally by built-in piers, the expansion or contraction of the deck will give rise to small longitudinal compressive or tensile stresses. The tensile stresses must be added to the flexural stresses at the SLS, and the prestress designed in consequence. The compressive stresses due to expansion may generally be ignored. Some codes substitute an effective bridge deck temperature for that derived from the temperature gradients for this calculation.

For a more detailed treatment of the calculation of stresses caused by temperature effects, see [4].

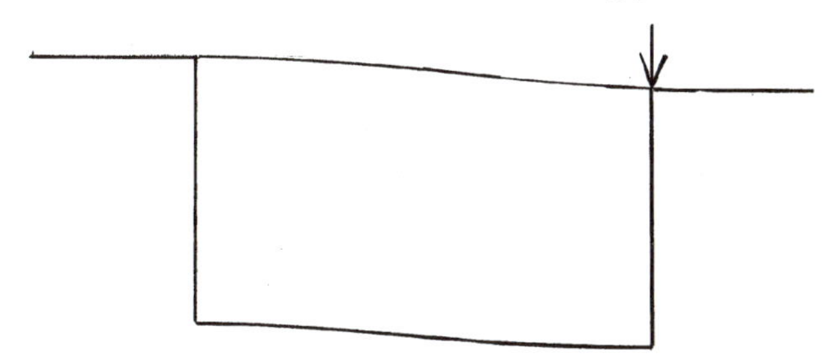

Figure 6.15 Distortion of box girder

6.13.4 Stresses due to distortion of the cross section and torsional warping

Box sections under the effect of eccentric live loads will distort as shown in Figure 6.15. This distortion causes longitudinal stresses on the extreme fibres of the webs, as well as transverse bending moments in the box.

The longitudinal stresses due to distortion are greater in a rectangular box than in a trapezoidal box, and disappear in a box of triangular cross section. If the box is very wide for its depth, the effect becomes greater. At the limit, the two webs become virtually independent of each other, and the deck behaves more like a twin rib deck than a box. Longitudinal stresses are also caused by torsional warping of the cross section.

At preliminary design, the longitudinal stresses due to these effects may be taken into account most easily by maintaining a margin with respect to the limiting stresses. Thus if the limiting stress is zero, the prestress would be sized for an appropriate compressive residual. For decks up to about 60 m span and of normal proportions, subjected to full British HB loading, it is usually conservative to reserve a residual of 1.5 MPa on the bottom fibre at mid-span, and 0.8 MPa on the top fibre at the supports. Where the governing loads are distributed rather than concentrated, half of these values is typical.

Alternatively, the residual stresses may be converted into equivalent moments and added to the table of bending moments.

It is of course essential at some stage to carry out an accurate calculation of these effects [6, 7 and 8].

6.13.5 Rounding of support moments

If a continuous beam is supported on a bearing of finite width d, the moment on the axis of the support will be less than the theoretical value by a parabolic decrement $Rd/8$, where R is the beam reaction, Figure 6.16 (a). In addition, if the height from the bottom fibre to the neutral axis is y_b, the effective width of the bearing is $d+2y_b \times \tan 35°$, and the reduction in the peak moment is $R(d+2y_b \times \tan 35°)/8$, Figure 6.16 (b). Usually, this reduction is arbitrarily limited to 10 per cent of the peak moment.

The reduction of moment due to the physical width of the bearing is purely mechanical, and is effective both at working and ultimate loads. The spread through the height of the beam is an elastic effect, and should not be used at the ULS.

This rounding is important in the design of prestressed concrete beams. If the prestressing cable is designed to cater precisely for the peak, unrounded moment, it

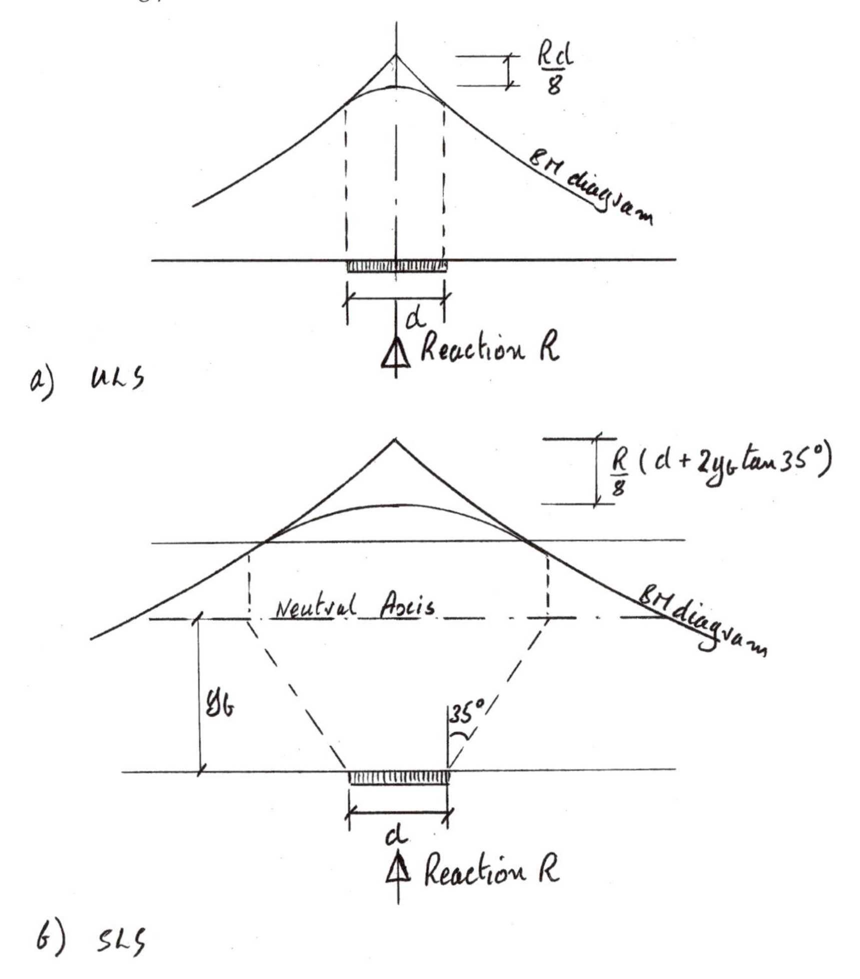

Figure 6.16 Rounding of support moments

will fit badly to the moment diagram, and is likely to be too high immediately adjacent to the pier section, enhancing the risk of the cracking described in *6.13.3* [9 and 10].

6.13.6 Summary of bending moments

The preceding sections have described the principal actions for which the prestress has to be designed. However, in order to describe the process of sizing the prestress, it would be needlessly complicated to include all these effects. Consequently, a much simplified table of moments will be used, due only to self weight, finishes and live load. The moments over the supports have not been rounded.

The bending moments, in MNm, due to applied loads are as shown in Table 6.2.

In this table, ΔM represents $M_{max} - M_{min}$ (sagging moments are positive, hogging negative), which describes the variation of bending moment at each section. In the side span, the maximum sagging moment is at Section 1.4 rather than at mid-span. The bending moments are shown in Figure 6.17. Note that the discontinuous nature of

Table 6.2

	Section 1.4	*Section 2.0*	*Section 2.5*	*Section 3.0*
Self weight	+13.6	−21.0	+8.9	−17.8
Finishes	+4.3	−6.7	+2.8	−5.7
Live load +	+13.2	+2.2	+11.6	+4.3
Live load −	−3.5	−17.0	−6.2	−16.9
M_{max}	+31.1	−25.5	+23.3	−19.2
M_{min}	+14.4	−44.7	+5.5	−40.4
ΔM	16.7	19.2	17.8	21.2

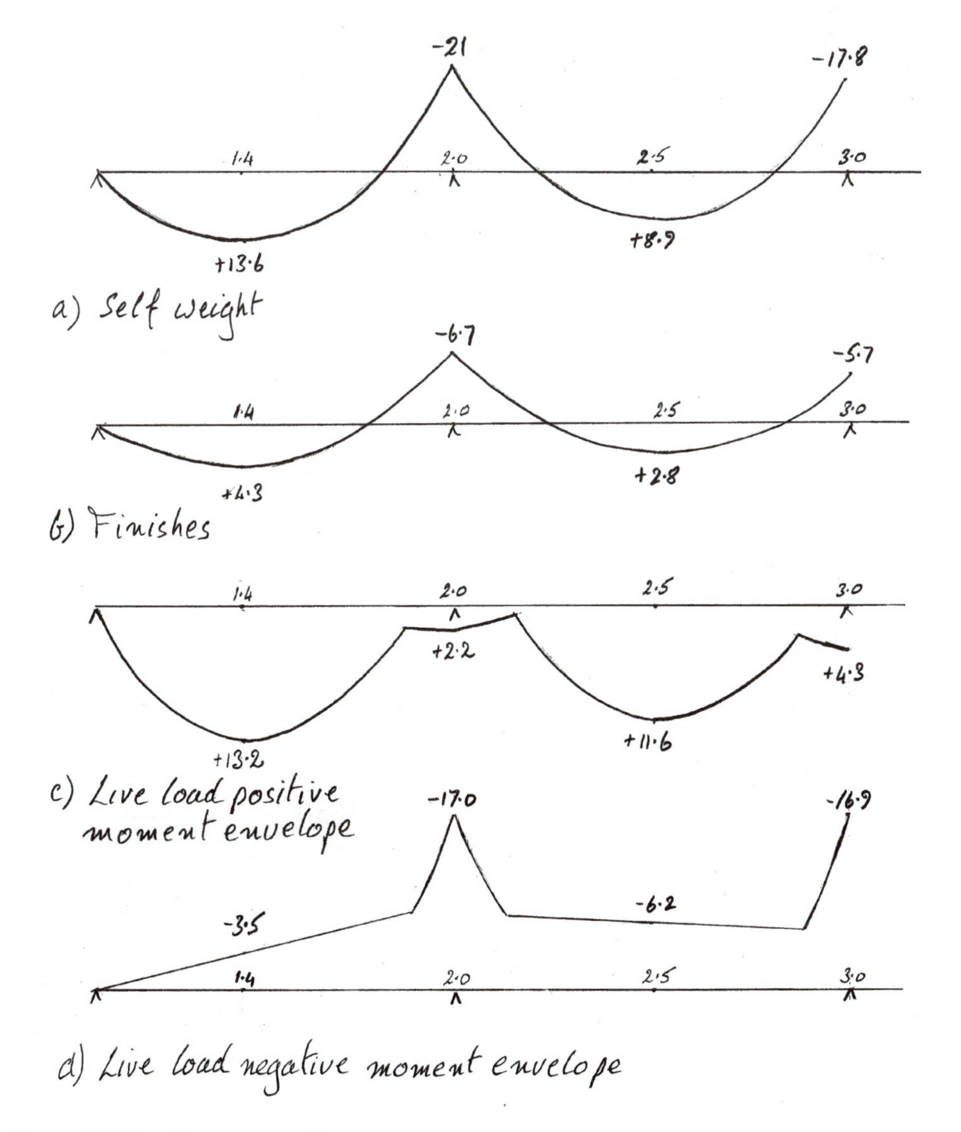

Figure 6.17 Bending moment diagrams

the live load bending moment envelopes is due to their being the result of several load cases as spans are loaded alternately.

It is most important that designers draw out the bending moment (and shear force, torsion and normal force as appropriate) diagram corresponding to each load case. All designers, however conscientious, make mistakes, and an essential element of their training, and of the growth of their experience, is to develop techniques that make it easier to spot them. Grouping together the results of several actions is one of the best ways of hiding mistakes.

During the design of a major urban viaduct with many spans, the engineers developed, as one would expect, productive methods of carrying out the analysis. The bridge was divided by expansion joints into sub-viaducts of approximately 10 spans each. The span lengths were very variable, so virtually each sub-viaduct was different. The bending moments were plotted out for the total moments in each sub-viaduct, grouping together the effects of self weight, finishes, differential settlement, temperature gradients and live loads. In one sub-viaduct with a particularly short end span, the design check picked up the fact that the analysis had omitted to include the live loading on the end span; the other effects were dominant, and the shape of the moment envelope was convincing. This could not have happened if each effect had been plotted out separately.

6.14 Considerations on the choice of tendon size

Before calculating the prestress force, some considerations on choosing the appropriate tendon size are relevant. The breaking force of tendons normally used in bridge decks ranges from about 1.2 MN to 10 MN (120 tons to 1,000 tons). Even larger tendons, with a breaking force of 15 MN are used in the construction of nuclear pressure vessels and other special structures. The designer's choice of size suitable for a particular scheme should be informed by a variety of considerations:

- the number of webs in the deck; in general the tendons are distributed evenly among the webs;
- less labour is required to place, stress and grout a few large tendons than more smaller ones;
- an economical prestress scheme often requires tendons to be subdivided into groups that are stressed at different times, or have different lengths; smaller tendons give more flexibility in such sub-divisions;
- powerful tendons require large ducts, and this may determine the minimum thickness of the webs, which will have implications on the weight and economy of the bridge deck;
- in general, it is easier to fit into a confined space more small ducts than a few large ducts;
- the same consideration applies to finding suitable space for large or small tendon anchorages;
- when tendons are anchored in the webs, such as in balanced cantilever construction, the thickness of the web may be determined by the size of the prestress anchor;
- large tendons require stressing jacks that weigh up to 1,000 kg, which are difficult to manoeuvre in a confined space;

- if large tendons are to be anchored in blisters within a box section deck, they will require a substantial volume of concrete and weight of reinforcement;
- powerful anchors require a considerable weight of additional equilibrium steel in the deck behind the anchor;
- small anchors may need very little equilibrium steel, if any, as existing reinforcement may prove adequate;
- the webs and slabs of the concrete section must be thick enough to resist the forces transmitted by large anchors;
- large anchorage blisters may significantly complicate the internal shutter, particularly if it is mechanised for rapid construction, as in the precast segmental system;
- for tendons that are external to the concrete, the placing of ducts is likely to be a critical activity for the construction programme, leading to the use of fewer, more powerful tendons.

In general, for internally prestressed decks, inexperienced designers tend to use tendons that are too powerful, attracted by the simplicity of the profiles. The difficulties in housing and anchoring these tendons, with the attendant increase in thickness of members and in the weight of reinforcement, come as a surprise later in the design process. As a general rule, for internally prestressed decks, it is advisable to consider tendons of modest size, with 19 No 15 mm or 27 No 13 mm strands as an upper limit. There are of course exceptions, when the use of more powerful tendons may solve problems and simplify construction.

For externally prestressed decks, where the tendons are anchored on pier diaphragms, tendons up to 37 No 15 mm strands are used frequently. If the tendons are anchored on blisters, smaller sizes will generally be chosen to limit the cost of the blisters and of their associated reinforcement.

6.15 Calculating the prestress force

6.15.1 General

With the statically determinate beam, it was possible to calculate the prestress force directly at any location along the beam; the only data required were the total bending moment, the eccentricity of the prestress centroid and the geometric properties of the cross section. For continuous beams, one cannot calculate the prestress force directly, as the total bending moment includes the prestress parasitic moment. This parasitic moment depends on the prestress force and on the locus of the prestress centroid along the beam. Consequently, one has to proceed by trial and error. At each design section there are three unknowns, the prestress force P, the eccentricity e and the parasitic moment M_p. The following techniques allow the designer to control the trial and error process.

In a continuous beam, there are several options for the arrangement of the prestress. Two schemes will be discussed in *6.16* and *6.17*.

6.15.2 Prestress eccentricity and lever arm

It is assumed for the purpose of the preliminary design that the maximum prestress eccentricity is achieved when the centroid of the prestress force is at 200 mm from the extreme fibres. Thus the eccentricity e at supports is 0.875 – 0.2 = 0.675 m, and at the critical span sections is 1.361 – 0.2 = 1.161 m (refer to *6.10.1*).

In the following calculation, it will be assumed initially that the bridge is designed to Class 1 prestressing, that is stresses must not fall below zero in service. Consequently, the lever arm for the prestress is the distance from the centroid of the tendons to the opposite limit of the central kern, as described in *5.8.2* and in Figure *5.5*. At mid-span the lever arm,

$$l_{a\,sp} = y_b - 0.2\text{ m} + a_t = 1.695\text{ m}$$

while at the support

$$l_{a\,sup}\; y_t - 0.2\text{ m} + a_b = 1.443\text{ m}.$$

6.15.3 Maximum moment and variation of moment criteria for defining the prestress force

There are two principal criteria that determine the prestress force required at a section, and its eccentricity. The first is the 'maximum moment criterion'. Under the effect of the maximum moment applied at a section, including the prestress parasitic moment ($M_{max} + M_p$), the centre of pressure must remain within the prescribed limit (the upper or lower limit of the kern for Class 1 prestressing), as described in *5.4*, *5.7* and *5.8*. The second is the 'variation of moment criterion'. Under the effect of the full range of moment applied at a section (ΔM in the table of moments) the centre of pressure must remain within the central kern for Class 1 prestressing.

There is in fact a third criterion which only applies to a relatively small number of bridge decks, which is discussed in *6.19*.

a) Maximum moment criterion

At each principal design section, generally mid-spans and supports, it is initially assumed that the eccentricity of the prestress is at its maximum value, and thus is defined. The two remaining unknowns are the prestress force, P, and the parasitic moment M_p. Pairs of adjacent principal design sections are considered, and initially it is assumed that both P and M_p are equal, or related, at both sections. One may then set up and solve pairs of simultaneous equations, relating P and M_p for each pair of sections. Although the results will be inconsistent, in that each pair of equations will give different values for the two unknowns, the designer will have a good idea of the pattern of prestress forces and parasitic moments that are required along the full length of the beam. The equations take the form:

Support section: $P \geq (M_{min} - M_p)/l_{a\,sup}$

Span section: $P \geq (M_{max} + M_p)/l_{a\,span}$

Note that the M_{max} and M_{min} are given their numerical values (ignoring their sign). Also note that the parasitic moments have been assumed to be of positive sign, meaning that they are added to the sagging moment in the span (M_{max}) and deducted from the hogging moments at the support (M_{min}). This is immaterial, as the sign of M_p will be corrected as the equations are solved.

b) Variation of moment criterion

This criterion applies at any point along the beam. The variation of moment may not be at its maximum at the mid-span and support sections, although it is usually close to maximum at these sections.

The equation takes the form

$$P \geq \Delta M/(a_t + a_b).$$

This equation yields the absolute minimum force required at each section.

6.16 Prestress scheme 1

Scheme 1 assumes that the prestress will be constant over the full length of the continuous beam.

6.16.1 Maximum moment criterion

The first pair of equations considers the critical section of the side span, section 1.4, and the first internal support, section 2.0.

Over the length of the side span the parasitic moment falls from its value M_p at the first internal support to zero at the free end of the beam. Consequently the parasitic moment will be 0.4 M_p at section 1.4.

Equation 1 At section 1.4 $P = (M_{max} + 0.4\ M_p)/l_{a\ sp} = (31.1 + 0.4\ M_p)/1.695$

Equation 2 At section 2.0 $P = (M_{min} - M_p)/l_{a\ sup} = (44.7 - M_p)/1.443$

Hence $P = 21.6$ MN and $M_p = 13.6$ MNm

Considering sections 2.0 and 2.5

Equation 3 At section 2.0 $P = (44.7 - M_p)/1.443$

Equation 4 At section 2.5 $P = (23.3 + M_p)/1.695$

Hence $P = 21.7$ MN and $M_p = 13.4$ MNm

Considering sections 2.5 and 3.0

Equation 5 At section 2.5 $P = (23.3 + M_p)/1.695$

Equation 6 At section 3.0 $P = (40.4 - M_p)/1.443$

Hence $P = 20.3$ MN and $M_p = 11.1$ MNm

The designer now has a good idea of the prestress force and parasitic moment required all along the bridge deck. Of course, the prestress force derived from these equations relies on the parasitic moment calculated being achievable. In most slab and box section bridges of constant depth with normal span ratios, a parasitic moment up to about 50 per cent of the mid-span Pe is achievable, and the results generally fall within this range, except for very unequal spans, where some of the results may be anomalous. The designer will, with experience, appreciate which results are to be treated with diffidence or ignored. A particular problem arises with very unsymmetrical cross sections, where the ratio y_t/y_b is below some critical figure. This will be discussed in 6.*19* below.

6.16.2 *Variation of moment criterion*

This criterion requires $P \geq \Delta M/(a_t + a_b)$. This gives the following minimum values for P:

Section 1.4 $P \geq 11.9$ MN
Section 2.0 $P \geq 15.1$ MN
Section 2.5 $P \geq 12.7$ MN
Section 3.0 $P \geq 16.6$ MN.

It is clear that in this example, P is controlled by the maximum moment criterion at all sections. It follows that at the design sections the maximum available eccentricity will be used.

6.16.3 *Choice of prestress force and parasitic moments*

The simultaneous equations above can also be used to check the range of M_p that may be accommodated by a chosen P, or if M_p is defined, the limiting values of P.

For instance, if we choose a prestress force of $P = 22$ MN for all sections, one may calculate from equation 1 that M_p at section 1.4 must not exceed 0.4×15.5 MNm = 6.2 MNm, and from equation 2 that M_p at section 2.0 must not be less than 13.0 MNm. From equations 3 and 4, M_p must not be less than 13.0 MNm at section 2.0 and not greater than 14.0 MNm at section 2.5. From equation 6, M_p must not be less than 8.7 MNm at section 3.0.

Thus with $P = 22$ MN we may choose parasitic moment values of 13 MNm at section 2.0, and 10 MNm at section 3.0, yielding an M_p of $13 \times 0.4 = 5.2$ MNm at section 1.4, and 11.5 MNm at section 2.5, Figure 6.18.

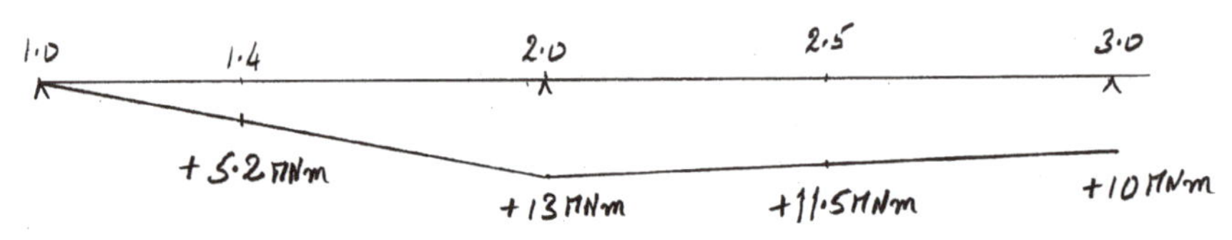

Figure 6.18 Prestress parasitic moments, scheme 1

6.16.4 Check on bending stresses

As stated previously, it is essential to tabulate the bending stresses yielded by the P and M_p chosen. This is to check that maximum allowable stresses are not exceeded, whether there are likely to be potential problems due to the higher initial prestress, and more generally that no mistakes have been made.

Section 1.4

P = 22 MN; e = –1.161 m; M_p = +5.2 MNm

Table 6.3

	Top fibre		*Bottom fibre*	
	σ partial	*σ cumulated*	*σ partial*	*σ cumulated*
Self weight	+2.77		–4.50	
Prestress	–1.32		+12.33	
Mp	+1.06	+2.51	–1.72	+6.11
Finishes	+0.88	+3.39	–1.42	+4.69
Live load $+^{ve}$	+2.69	+6.08	–4.37	+0.32
Live load $-^{ve}$	–0.71	+2.68	+1.16	+5.85

From Table 6.3 it may be seen that compressive stresses are not a problem, and that on the bottom fibre there is a residual of +0.32 MPa under maximum live load, confirming that the section had the capacity to carry an additional M_p of approximately 1 MNm if required.

Section 2.0

P = 22 MN; e = 0.675 m; M_p = +13 MNm

Table 6.4

	Top fibre		*Bottom fibre*	
	σ partial	*σ cumulated*	*σ partial*	*σ cumulated*
Self weight	–4.17		+6.31	
Prestress	+6.30		–1.11	
Mp	+2.58	+4.71	–3.91	+1.29
Finishes	–1.33	+3.38	+2.01	+3.30
Live load $+^{ve}$	+0.44	+3.82	–0.66	+2.64
Live load $-^{ve}$	–3.37	+0.01	+5.11	+8.41

It may be seen that the stress on the top fibre under the full live load falls to zero, as the M_p adopted was equal to the minimum acceptable value. It may also be seen that the stress falls to only 1.29 MPa on the bottom fibre under the effect of self weight plus prestress. If the prestress were factored up by 15 per cent to represent the initial prestress force, this stress would fall to 0.54 MPa. This illustrates the vulnerability of the bottom fibre to cracking, as discussed in *6.13.3*.

Section 2.5

$P = 22$ MN; $e = -1.161$ m; $M_p = +11.5$ MNm

Table 6.5

	Top fibre		*Bottom fibre*	
	σ partial	*σ cumulated*	*σ partial*	*σ cumulated*
Self weight	+1.81		−2.94	
Prestress	−1.32		+12.33	
Mp	+2.34	+2.83	−3.80	+5.59
Finishes	+0.57	+3.40	−0.93	+4.66
Live load $+^{ve}$	+2.36	+5.76	−3.84	+0.82
Live load $-^{ve}$	−1.26	+2.14	+2.05	+6.71

Section 3.0

$P = 22$ MN; $e = 0.675$; $M_p = +10$ MNm

Table 6.6

	Top fibre		*Bottom fibre*	
	σ partial	*σ cumulated*	*σ partial*	*σ cumulated*
Self weight	−3.53		+5.35	
Prestress	+6.30		−1.11	
Mp	+1.98	+4.75	−3.00	+1.24
Finishes	−1.13	+3.62	+1.71	+2.95
Live load $+^{ve}$	+0.85	+4.47	−1.29	+1.66
Live load $-^{ve}$	−3.35	+0.27	+5.08	+8.03

It may be seen that the stress on the top fibre under the full live load has a residual of 0.27 MPa. This is because the M_p at 10 MNm is greater than the minimum acceptable figure of 8.7 MNm. The margin is 1.3 MNm. With $z_t = 5.041$, this margin gives rise to a top fibre stress of 0.26 MPa, confirming that if M_p had in fact been 8.7 MNm, the stress on the top fibre would have fallen to zero. This is the type of simple check that designers should be carrying out all the time to confirm the accuracy and consistency of their calculations.

The M_p at section 2.0 is 13 MNm, slightly greater than 50 per cent of Pe at the adjacent section 2.5. It would be possible to redefine M_p as say 12 MNm and calculate the required P. However, the 13 MNm will be accepted for the purpose of this example.

6.16.5 Choice of tendons and tendon arrangement

We have seen in *5.13* that strand is available in nominal 13 mm and 15 mm sizes, and that for preliminary design purposes, the working force for each size is 100 kN and

150 kN respectively. Thus for a prestress force of 22 MN, we would need 220 of the 13 mm strand, and 147 of the larger size.

We will adopt 9 tendons of 12 No 13 mm strands in each web, although this falls marginally short of the required 220 strand, providing 21.6 MN of force. However, there is no point in seeking too much precision, as the working prestress force per strand will only be finalised once a calculation of prestress losses has been carried out. The following calculations will be carried out with $P = 22$ MN.

These tendons may be arranged in a variety of ways. Two possibilities will be illustrated:

a) The tendons are all located inside the web reinforcement, Figure 6.19 (a). This arrangement is simpler to build, but it would not be possible to achieve the eccentricity adopted in the calculation. The prestress sizing exercise would need to be redone, and a greater prestress force adopted.
b) The tendons are located partially outside the web reinforcement, Figure 6.19 (b). This arrangement allows the designer to maximise the prestress eccentricity and has generally been adopted by the author in his practice. It requires more careful design and detailing, but is more economical.

The tendon arrangements shown are based on UK practice where they are spaced at two times their internal duct diameters.

6.16.6 *Plotting the cable zone*

With the assumed P and M_p, and with design sections at tenth points of each span, the upper and lower limits of the cable zone may be plotted, using the techniques explained in *5.14*. It will initially be assumed that the limiting stresses are zero on both extreme fibres.

For instance, at section 1.4:

$P = 22$ MN
$M_p = +5.2$ MNm
$M_{max} = +31.1$ MNm
$M_{min} = +14.4$ MNm.

Then, under the maximum moment the centre of pressure must not rise above the top of the kern, while under minimum moment it must not fall below the bottom of the kern. Consequently the upper limit of the cable zone is plotted down from the top of the kern by a distance $(M_{max}+M_p)/P = (+31.1 + 5.2)/22 = 1.65$ m, and the lower limit of the zone is plotted down from the bottom of the kern by $(M_{min} + M_p)/P = (+14.4 + 5.2)/22 = 0.891$ m. The sign convention is that sagging moments are positive, and positive dimensions are plotted down from the relative kern limit.

For section 2.0:

$P = 22$ MN
$M_p = +13$ MNm
$M_{max} = -25.5$ MNm
$M_{min} = -44.7$ MNm.

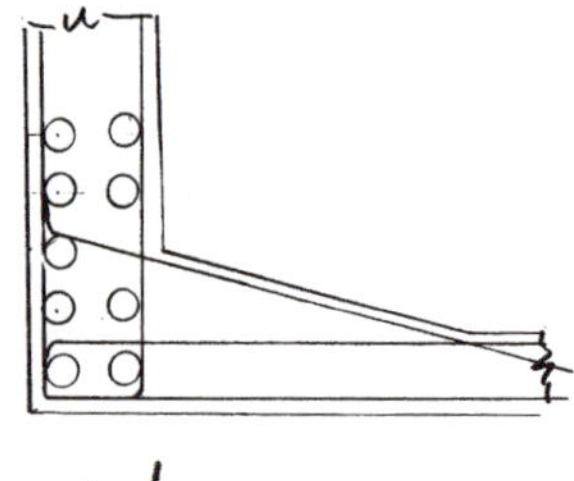

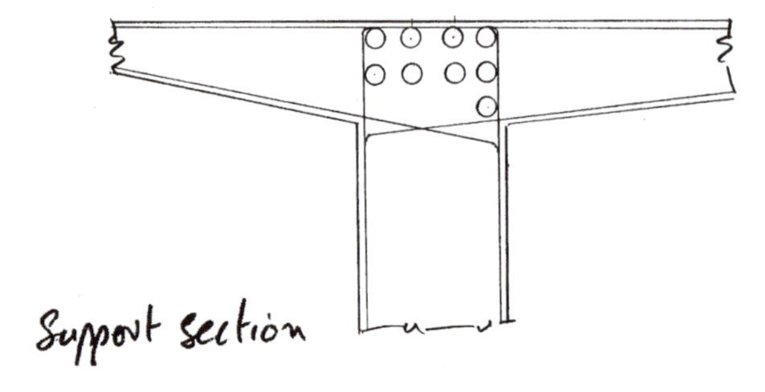

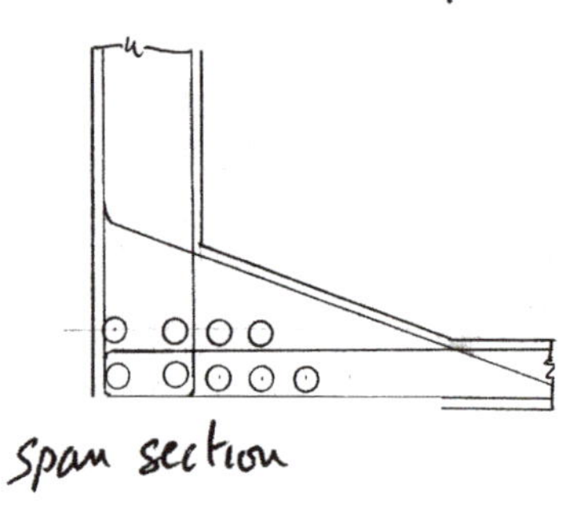

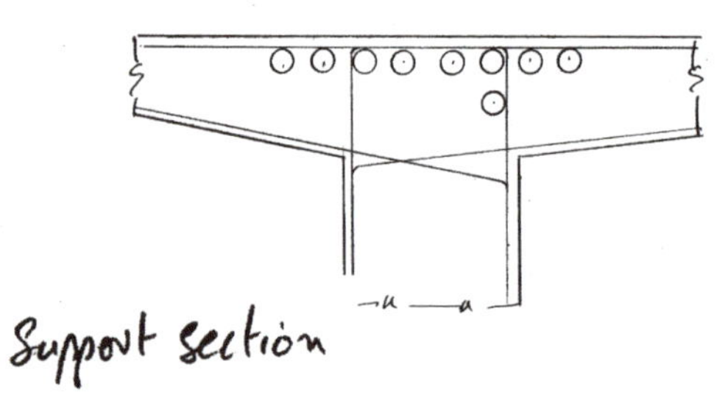

Figure 6.19 Arrangement of tendons

Then, under the maximum moment (the smallest hogging moment) the centre of pressure must remain below the top of the kern, while under minimum moment, it must not fall below the bottom of the kern. Consequently, the upper and lower limits of the cable zone are (–25.5 +13)/22 = –0.568 m, plotted up from the top of the kern, and (–44.7 + 13)/22 = –1.441 m plotted up from the bottom of the kern.

For section 2.5 the cable zone is +1.582 m below the top of the kern and +0.773 m below the bottom of the kern. For section 3.0 the cable zone is –0.418 m above the top of the kern and –1.382 m above the bottom of the kern.

The relevant diagrams for sections 1.4 and 2.0 have been drawn in Figure 6.20. It is strongly advised to draw such diagrams for each design section where these calculations

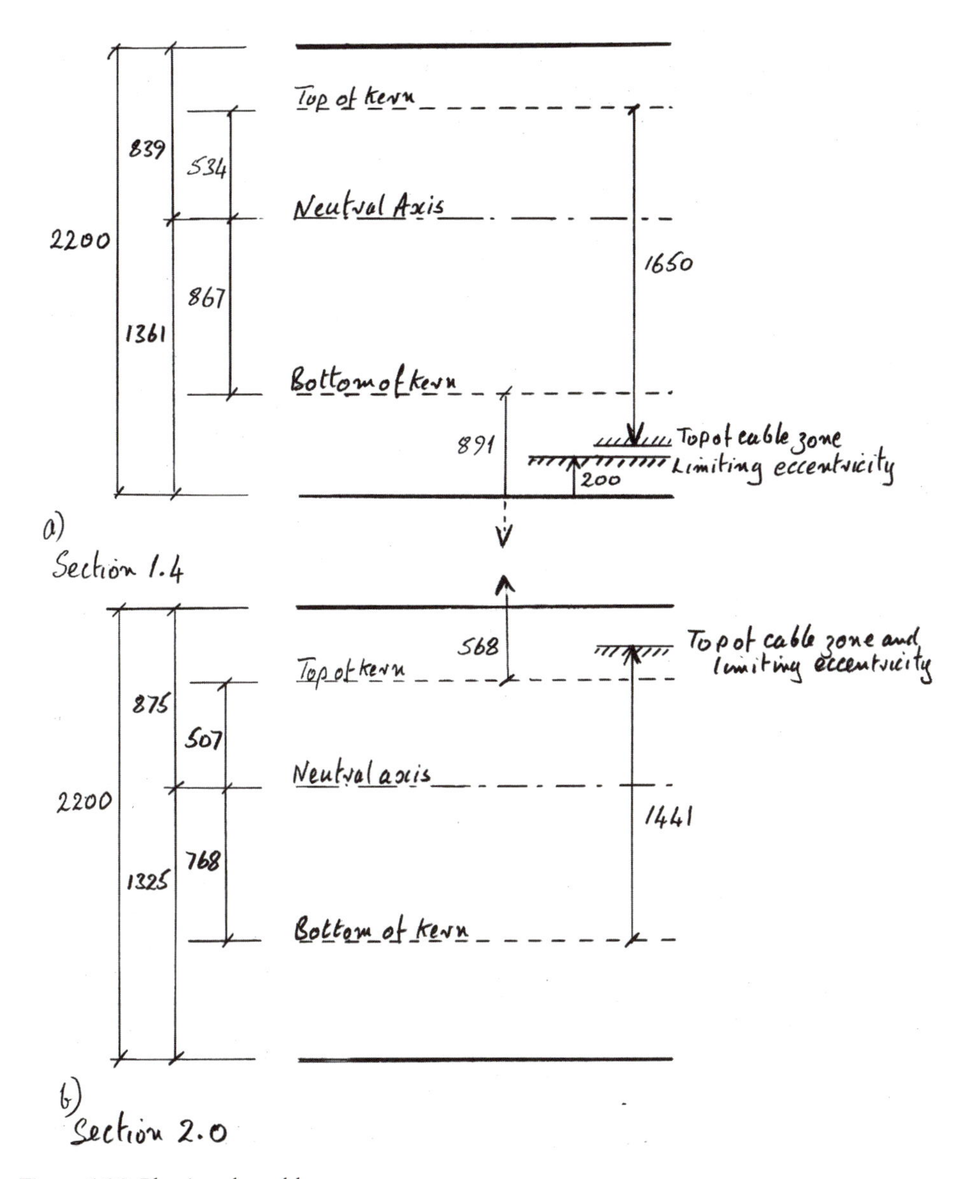

Figure 6.20 Plotting the cable zone

are carried out. It will be seen that the outer limits of the cable zone at positions of maximum and minimum moment are often defined by the limiting eccentricity.

The upper and lower limits of the cable zone should be plotted at every tenth point along the spans, giving a zone of the general shape shown in Figure 6.5.

As described for the statically determinate beam, the cable zone will be very narrow at design sections which determine the prestress force, generally supports and mid-span, and wider in between. The prestress centroid must lie between the two limits defined by this cable zone.

However, the position of this cable zone within the depth of the beam is dependent on the value of M_p chosen, and the designer is not sure whether this M_p is in fact achievable. This may be checked by calculating the M_p that results from a cable profile placed firstly along the top of the zone, and then from a profile placed along the bottom of the zone. If these two profiles give rise to values of M_p that bracket those chosen, then it is clear that there exists a cable profile between the two limits that will give rise to parasitic moments equal to those chosen. The designer then has to place the cable centroid within the cable zone so that he achieves the desired parasitic moments, using the techniques described in *6.20*.

6.17 Prestress scheme 2

Scheme 1 involves 9 tendons in each web, extending over the full length of the deck. Although it is possible to use tendons of 152 m length, the friction losses are likely to make the option uneconomical. It would be necessary either to couple tendons or to overlap them to reduce the maximum length. An alternative scheme would be as shown in Figure 6.21 (a). Here half the tendons are overlapped over piers 2.0 and 4.0, where the design was the tightest, and where the M_p required was rather close to the maximum that is likely to be achievable. The new arrangement provides 8 tendons per web, yielding a total of 19.2 MN at sections 1.4, 2.5, 3.0, 3.5 and 4.6, and 12 tendons yielding 28.8 MN at sections 2.0 and 4.0. Furthermore, assuming that the anchorage points of the cables that are stopped off are 5 m either side of the penultimate piers, the total length of prestressing tendon has been reduced by some 5 per cent.

The equations of *6.16.1* may be used to discover the acceptable ranges of M_p with the newly defined forces.

- From equation 1, M_p at section 1.4 must not exceed 1.44 MNm
- From equation 2, M_p at section 2.0 must not be less than 3.14 MNm
- From equation 4, M_p at section 2.5 must not exceed 9.24 MNm
- From equation 6, M_p at section 3.0 must not be less than 12.69 MNm

Thus a scheme with M_p of 3.5 MNm at section 2.0 and 12.7 MNm at section 3.0, yielding 1.4 MNm at section 1.4 and 8.1 MNm at section 2.5, satisfies all criteria, although the M_p required at section 3.0 is almost as large as that previously found at pier 2.0, Figure 6.21 (b).

These two examples illustrate how the basic equations allow the designer to refine his prestressing scheme, and define the parasitic moments that are required to make it work.

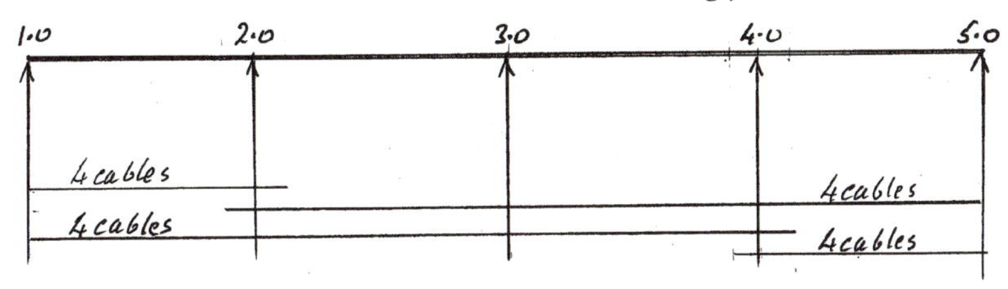

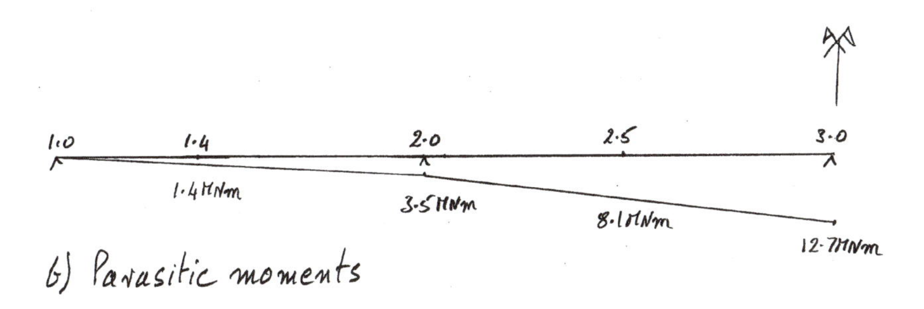

Figure 6.21 Prestress scheme 2

6.18 Non-zero stress limits

For continuous beams, it is usually the case that the designer is working to non-zero limits of stress on the extreme fibres. This is because either the code of practice allows tensile stresses under some load combinations, or it is necessary to maintain a compressive residual to cope with the stresses due to some of the effects described in *6.13*.

As was explained in *5.10*, this may be easily carried out by changing the notional boundaries of the central kern. For instance, if at section 2.5 it is necessary to maintain a residual of 1.5 MPa on the bottom fibre, under maximum sagging moments the centre of pressure may only move up from the prestress centroid to the limit of the reduced top of the kern, a_{t1}, Figure 6.22 (a). a_{t1} may be calculated by assuming that the stress on the bottom fibre reduces linearly from P/A, when the centre of pressure is at the neutral axis, to zero when it is at the top of the kern a_t.

Assume that $P/A = 4$ MPa, then

$$a_{t1} = a_t \times (4 - 1.5)/4 = 0.625\ a_t.$$

As $a_t = 0.534$ m, $a_{t1} = 0.334$ m. Consequently, the new prestress lever arm for use in equations 4 and 5 (*6.16.1*),

$$l_a = 1.361 - 0.2 + 0.334 = 1.495 \text{ m}$$

reduced from its original value of 1.695 m.

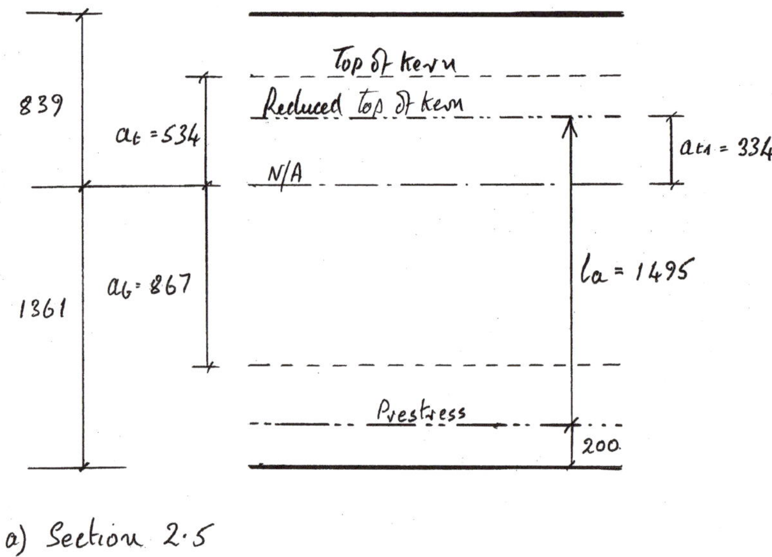

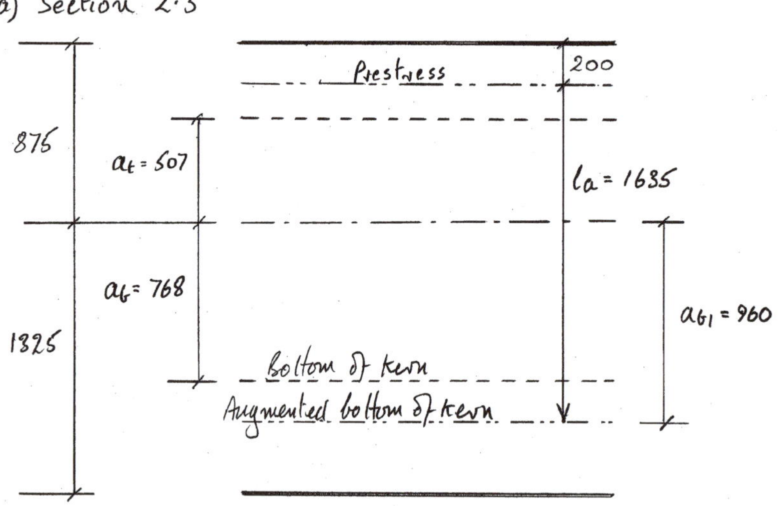

Figure 6.22 Non-zero stress limits

Similarly, if the stress limit on the top fibre at section 2.0 is –1 MPa instead of zero, under the minimum moment (greatest hogging moment) the centre of pressure may move down from the prestress tendon centroid to the limit of the augmented bottom of the kern, a_{b1}. Using the same assumptions as above,

$$a_{b1} = a_b \times (4 + 1)/4 = 1.25\, a_b = 1.25 \times 0.768 = 0.96 \text{ m.}$$

Consequently, the new prestress lever arm

$$l_a = 0.875 - 0.2 + 0.96 = 1.635 \text{ m,}$$

increased from 1.443 m.

6.19 Very eccentric cross sections

In a few cross sections, such as twin rib type decks (Chapter 12) or box sections with an unusually wide top slab, where the neutral axis is high and the ratio of y_t/y_b is smaller than some critical figure, the simultaneous equations for finding the prestress force and M_p under the maximum moment criterion described above, yield an M_p that cannot be achieved.

In such decks, when the prestress centroid is placed along the top of the cable zone to give the lowest possible value of M_p, as described in *6.16.6*, it is found that the parasitic moment calculated is greater than that assumed in the definition of the cable zone. The practical solution is to increase the force in the cables and reduce their eccentricity at mid-span. The simultaneous equations in *6.16.1* are then used to define a new set of prestress forces and parasitic moments, defining a new cable zone. The prestress centroid is then placed along the top of the new zone, and a new lower bound M_p calculated. This procedure is repeated until the lower bound M_p is smaller than that used to define the zone. This demonstrates that a cable profile exists that will deliver the required parasitic moments. Alternatively, the prestress force and eccentricity may be found by adopting the calculation method described in [11] and [12]. In some cases of very eccentric cross sections, it is necessary to increase the prestress force by up to 20 per cent.

A much better solution than increasing the prestress force and reducing eccentricity would be to adopt a tendon profile which uses the maximum available eccentricity at supports and mid-span, to accept the tensile stresses on the bottom fibre caused by the excess M_p, and to reinforce the bottom fibre to control the cracking. However, this is likely to be categorised as partial prestressing, and would not be allowed by all codes of practice.

It is important when calculating the section properties for such eccentric cross sections that due account is taken of the effects of shear lag, as this may reduce the effective width of the top flange, and hence render the section better balanced. Unfortunately, the ratio of y_t/y_b at which this problem arises has not, to the author's knowledge, been identified. It would be a useful research project.

6.20 Design of the parasitic moments

Computer programs are available to calculate the effect of a particular prestress cable profile on a structure including the value of prestress parasitic moments. However, as they generally need to be re-run in their entirety each time a local amendment to the tendon profile is made, they are not suitable for the exploratory phases of the design of the profile. Attempting to optimise a cable profile by trial and error in this way is slow and cumbersome, even with the most well-designed program, and does not favour clear thinking, initiative and innovation.

For these exploratory phases, it is essential that the method used allows the designer to predict the effect any changes he plans to make to the prestress force or eccentricity will have on the M_p.

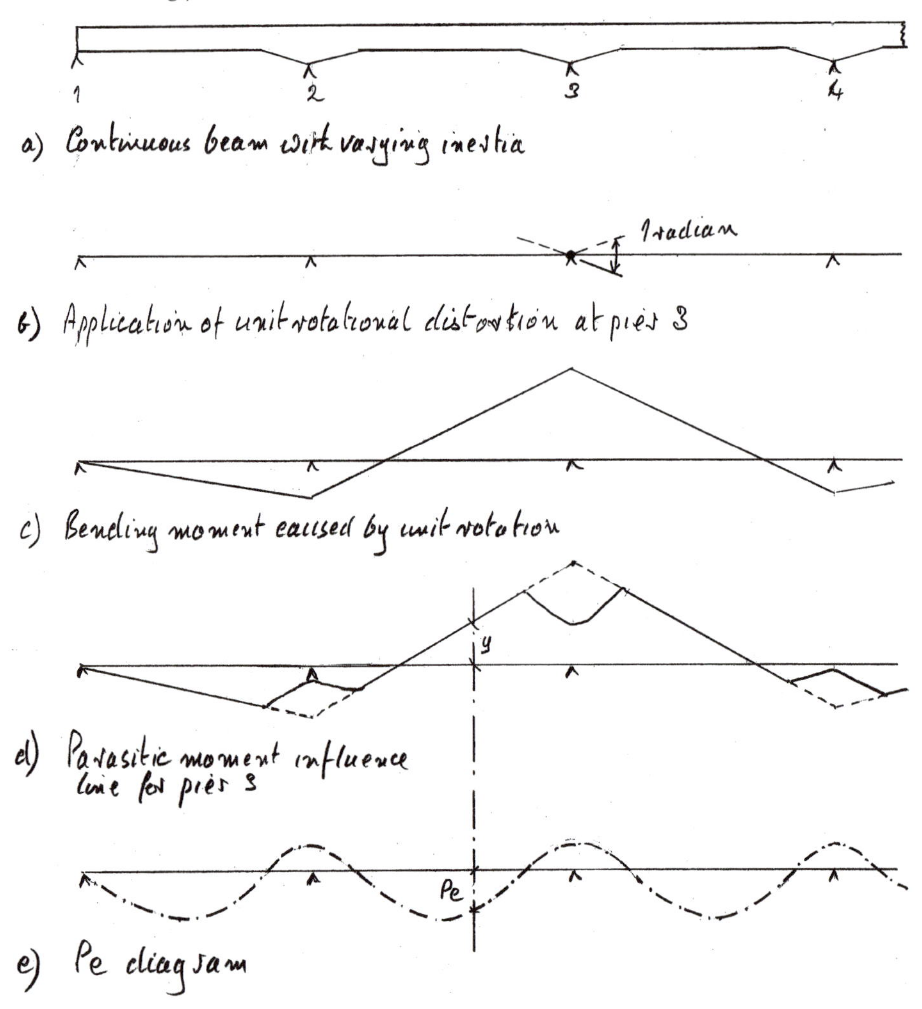

Figure 6.23 Typical influence line for parasitic moment

6.20.1 *Influence lines for parasitic moment*

In the author's experience, the best method for adjusting the cable profile to give a specific M_p is the use of influence lines for parasitic moment, Figure 6.23. The influence line for parasitic moment at a pier of a continuous beam is generated by introducing a hinge at that pier and applying a unit rotational distortion. The bending moment caused by the distortion at any point in the beam, divided by the local *EI* (Young's Modulus × Moment of Inertia), is the value at that point of the influence line for parasitic moment at the pier. For a continuous beam of varying inertia, the influence line for parasitic moment at support 3 is of the form shown in Figure 6.23 (d). A unit *Pe* (prestress force × eccentricity) applied over a unit length at the position shown would produce a parasitic moment at support 3 equal to the ordinate of the influence line *y*. Consequently, the product integral of *yPe* over the length of the deck will give the parasitic moment at the support. Using these influence lines, it is possible to predict the effect on the parasitic moment of any local adjustment of the cable profile by simple hand calculation; it is not necessary to re-run the whole structure. This is a great improvement over adjusting the cable profile by trial and error. It is even possible to make minor adjustments to the parasitic moment at one support by changing the cable profile in a remote span.

As parasitic moments vary linearly between points of support, influence lines are required at support positions only. These influence lines may be used for more complex structures such as beams built in to their supports, or portal frames, with the bending moment due to the unit distortion calculated by simple computer programs.

6.21 Modification of bending moments due to creep

6.21.1 General

Creep or the deferred deformation of concrete, is described in *3.9.1*. If the statical definition of a structure is changed during construction, then creep will change the distribution of bending moments. Most concrete structures are in fact built in phases, whether it is a building where adjacent floor slabs are built successively, or a bridge deck being built span by span, and consequently creep changes their self weight bending moments.

The simplest demonstration of this behaviour of concrete is illustrated in Figure 6.24. A reinforced concrete beam is built as a single span, and the falsework struck, allowing the beam to deflect elastically under its self weight, Figure 6.24 (a). Then the statical definition is changed by placing an additional support beneath the beam at mid-span, just in contact, Figure 6.24 (b). As the beam tries to continue its creep deflection as a single span, it gradually loads up this additional support, Figure 6.24 (c), and the bending moment is progressively modified. The final shape of the bending moment depends on the relative magnitudes of the elastic deflection and the creep deflection. If the creep deflection were very large compared with the elastic value, the final moment diagram would be similar to that of a two-span beam, which is the limit towards which the bending moment is tending. If the beam had been made of well-matured precast segments that creep little, the bending moment would remain closer to the statically determinate value.

The bending moment diagram for the simply supported beam shown in Figure 6.24 (a) is known as the 'as-built moment diagram', or M_{ab}. The bending moment diagram for a continuous beam built ab-initio as two spans shown in Figure 6.24 (e), which is the diagram to which the structure is tending as it creeps, is known as the 'monolithic moment diagram' or M_{mono}. The final bending moment in the structure, shown in Figure 6.24 (d) is the 'design moment diagram' or M_{des}.

Then $M_{des} = (M_{ab} + \phi M_{mono})/(1 + \phi)$.

The above formula is the simplest definition of creep; other, more complicated formulae exist. However, it should be borne in mind that the value of the creep coefficient ϕ is not known with any precision, and it is illusory to seek too much sophistication in the calculation of creep. Consequently, it is good practice to design concrete bridges to limit the difference between the as-built moment diagram and the monolithic diagram, so that the amount of creep, and hence the uncertainty on the

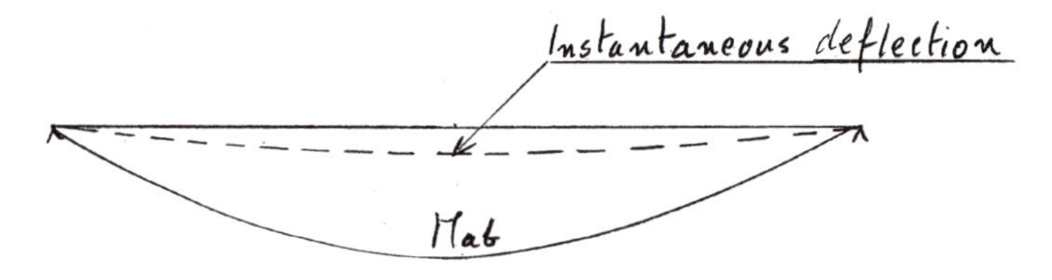

b) Support placed just in contact with deflected beam

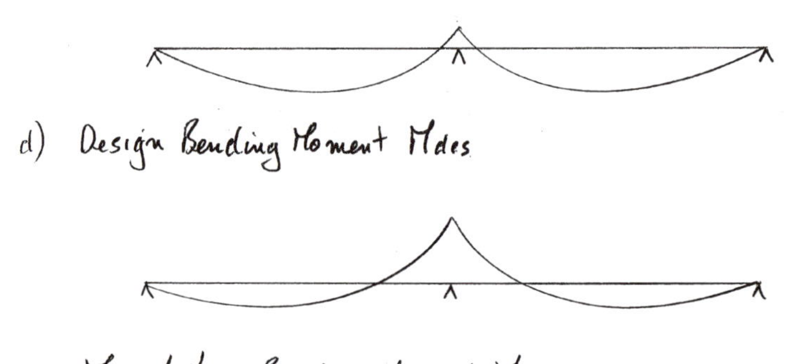

Figure 6.24 Modification of bending moment due to creep

distribution of the self weight bending moments, is minimised. The use of prestressing, as described below, helps in that respect.

If the beam had been made of prestressed concrete rather than reinforced concrete, the principles would be identical, but the results would be quite different. For instance, if the statically determinate beam had been prestressed with a parabolic cable that applied an upward uniformly distributed load to the beam that was equal to its dead weight, the beam would not deflect. Consequently, no load would be shed onto the central support, and no change in the moments would take place. Analytically, the bending moments under self weight and prestress would be considered separately. Under self weight acting alone the beam would behave as described above for the reinforced concrete beam, with a hogging moment developing above the central support. Under prestress acting alone the beam (assumed to be weightless), would arch upwards, and if it were tied to the support, a sagging moment would develop above the central support. These two equal and opposite effects, when added together, cancel each other.

Although the example above where the self-weight deflection is completely cancelled by prestress gives a simple understanding of the problem, it is not a common design situation. In the design of real bridges, the bending moments due to self weight and prestress do not balance exactly and are consequently modified by creep due to any changes in their statical definition during construction.

6.21.2 Bridges built in free balanced cantilever

Bridges built in balanced cantilever start life as statically determinate cantilevers, and then are stitched together at mid-span, changing the statical definition and creating a redundant structure. Thus they fulfil the criteria for structures where the bending moments will be modified by creep.

Consider a pair of reinforced concrete cantilevers, as shown in Figure 6.25 (a). The as-built bending moments under self weight are 0.125 wl^2 at the piers and zero at the free ends of the cantilevers, where w is the uniform distributed self weight and l is the length of the complete span. The deflection of the structure causes a downward deflection and a rotation at the ends of the cantilevers. The mid-span stitch is cast to make the structure redundant, and the cantilever deflection continues to increase due to creep. However, now this creep deflection causes sagging bending moments to develop at the mid-span. In fact, the self-weight moment diagram is being lowered, reducing hogging moments over the supports.

The as-built moment is shown in Figure 6.25 (a), and the deflections in Figure 6.25 (b). The monolithic bending moments are obtained by applying the self weight to the redundant structure, and are –0.0833 wl^2 at the supports and +0.0417 wl^2 at mid-span, Figure 6.25 (c). The design moment, which lies between the two, is shown in Figure 6.25 (d), and is found using the equation shown in *16.21.1* above. Thus if the creep factor $\phi = 2$, the design moment at the support would be

$$M_{des} = -wl^2(0.125 + 2 \times 0.0833)/3 = -0.0972\, wl^2,$$

and at mid-span

$$M_{des} = +wl^2(0 + 2 \times 0.0417)/3 = +\,0.0278\, wl^2.$$

The prestressing cables are shown in Figure 6.26 (a). They produce an as-built bending moment on the statically determinate structure of the form shown in Figure 6.26 (b) with a moment at the support of $Pe_{support}$, zero parasitic moments as the structure is determinate, and an instantaneous deflection of the form shown in Figure 6.26 (c). If these same prestressing cables had been applied to the continuous

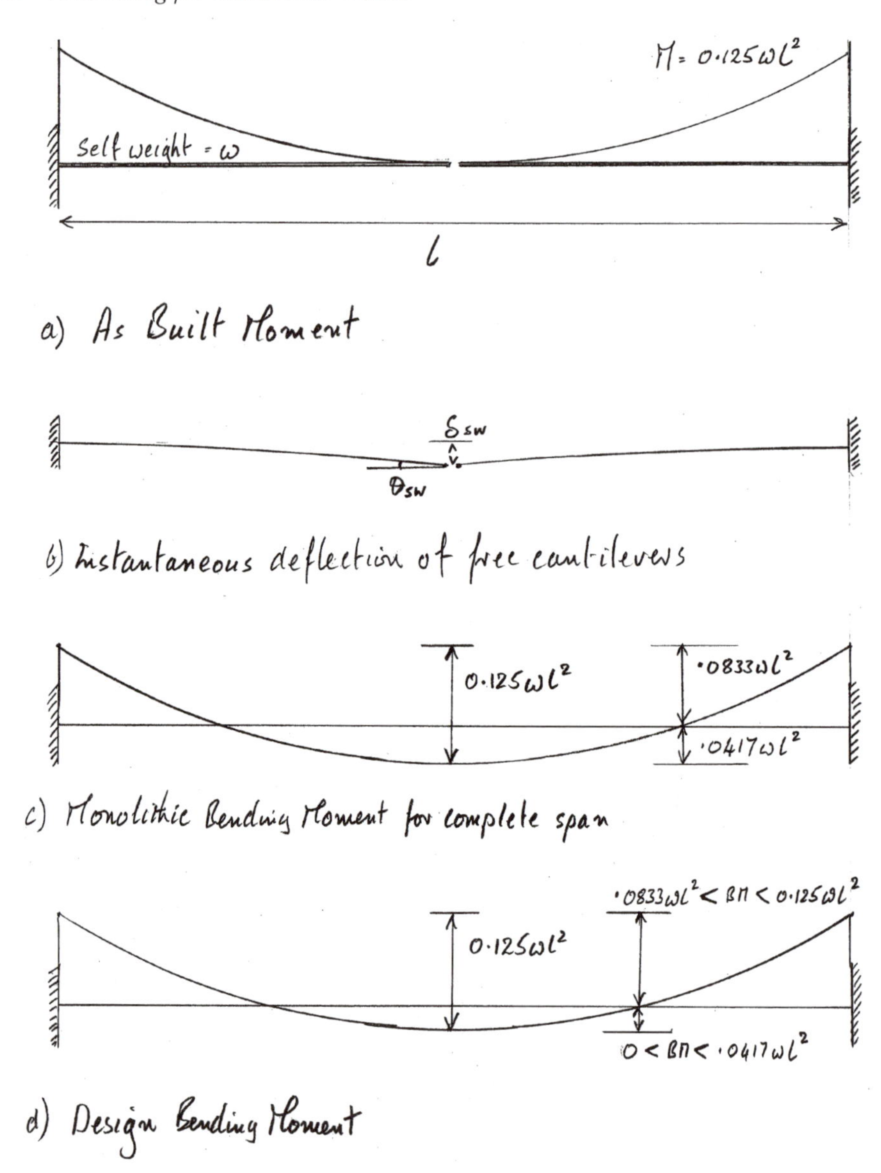

Figure 6.25 Free cantilever construction: self-weight effects

structure, the *Pe* would of course have been the same, but in addition there would have been a hogging (negative) parasitic moment (as all the cables are above the neutral axis). The monolithic prestress moments are shown in Figure 6.26 (d).

The structure is now made continuous by the mid-span stitch, and the prestress deflections increase due to creep. As the structure creeps, a parasitic moment will be created that reduces the effectiveness of the prestress at the supports. This parasitic moment gradually tends towards the monolithic value. The design prestress parasitic moment lies between the as-built and monolithic extremes, and is obtained using the equation shown in *16.21.1* above.

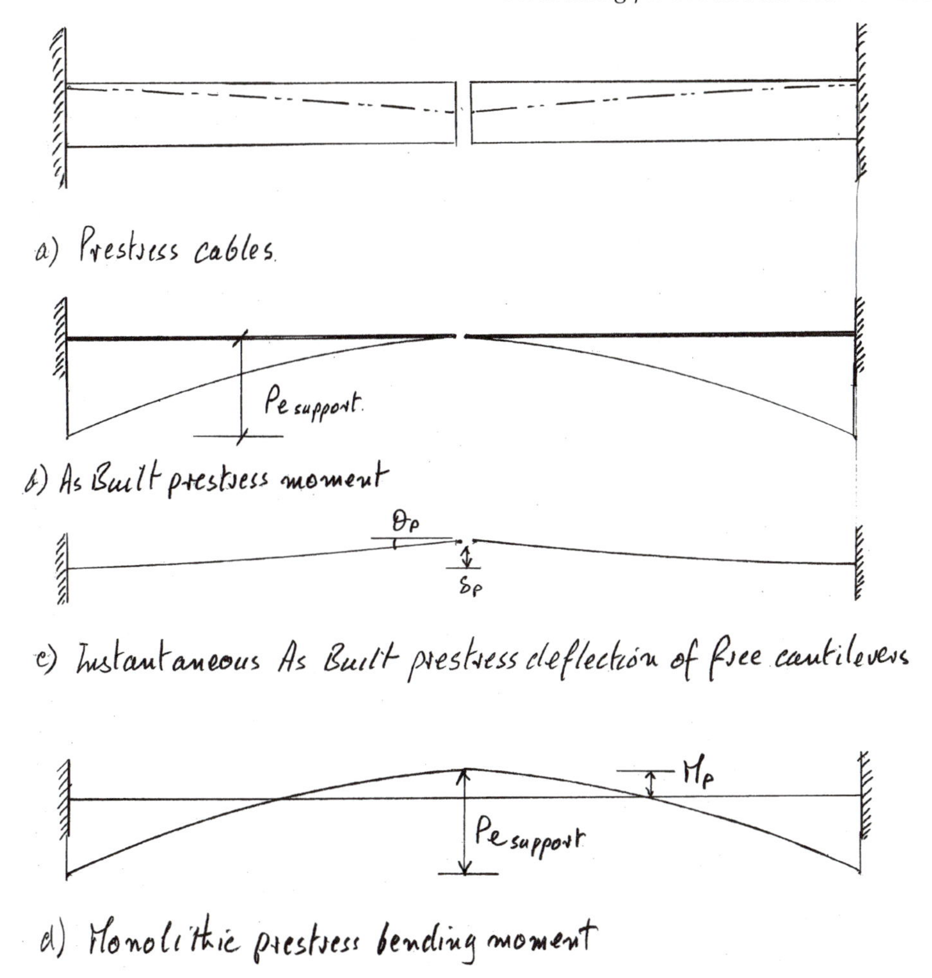

Figure 6.26 Free cantilever construction: prestress effects

Strictly, it is not the deflection of the cantilever under self weight or prestress that causes the introduction of continuity moments, but the rotation of the cantilever ends. Whereas under self weight the deflection and rotation of the cantilever tip are indissolubly linked by the laws of statics, that is not the case for prestress, where the profile of the tendons, and hence the deflected shape of the beam, including the end rotation, is under the control of the designer. Thus the prestress designer may control the parasitic moment that is introduced by creep in a number of ways. In particular, he can reduce or increase the rotation of the cantilever ends due to prestress, and he can postpone the stressing of some tendons until the structure has been made continuous, and lengthen these cables beyond strict structural necessity to increase the parasitic moment they generate.

6.22 Modification of bending stresses due to creep following change of cross section

When the cross section of a bridge deck is changed during the construction sequence, the bending stresses due to self weight and prestress are affected by creep. For instance, post-tensioned statically determinate beams of the type described in Chapter 10 are precast, prestressed, and launched into place. The beams are then joined together by a reinforced concrete slab, Figure 6.27. The weight of this slab, in the form of fluid concrete, is applied to the beams as a superimposed dead load. However, once this second phase concrete has hardened, the section properties of the beams change substantially, due to the much increased area of top flange. Initially, the second phase concrete slab will be unstressed. However, as the beams continue to shorten and to deflect under the effect of self weight and prestress, this new concrete will gradually pick up stress and the bending stresses in the first phase concrete will be modified. The stresses in the beam will tend towards the monolithic situation, where the full section of the beam including the additional top slab, had been cast in one phase.

The stresses in the beam due to the initial self weight, the prestress and the weight of the wet second phase slab are the 'as-built stresses'. The stresses in the beam that would have arisen if the complete cross section had been built and stressed in one phase are the 'monolithic stresses'. The 'design stresses' lie between the two extremes and depend on the creep factor in the same way as the bending moments described above.

The creep affects all the section properties of the beam. Thus the as-built stresses would be

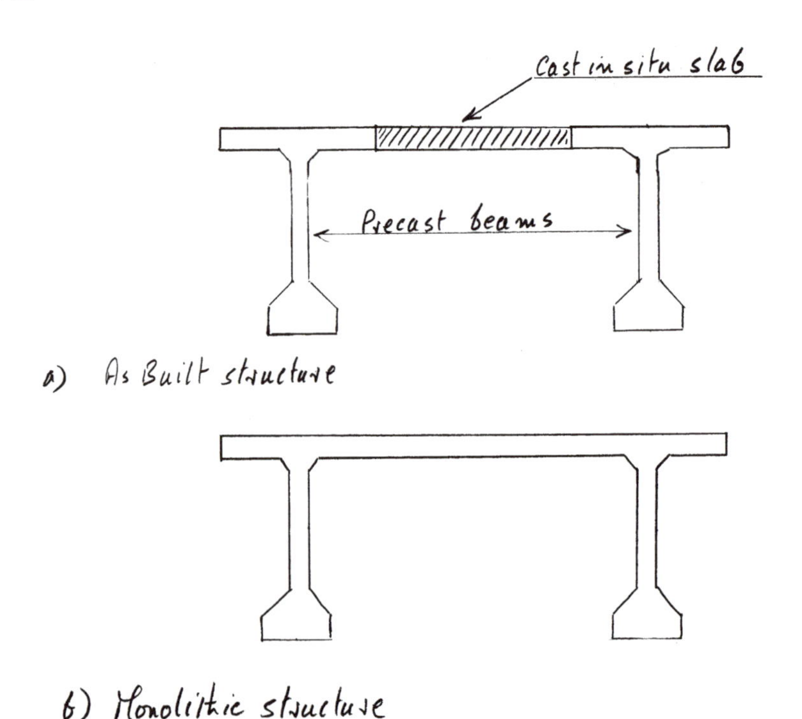

Figure 6.27 Creep due to change in cross section

$$\sigma_{ab} = M_{self\ weight}/z_{ab} + P/A_{ab} \pm Pe_{ab}/z_{ab},$$

the monolithic stresses would be

$$\sigma_{mono} = M_{self\ weight}/z_{mono} + P/A_{mono} \pm Pe_{mono}/z_{mono}$$

and the design stresses

$$\sigma_{des} = (\sigma_{ab} + \phi\sigma_{monov})/(1+\phi),$$

where $M_{self\ weight}$ is the moment due to the weight of the complete cross section.

The bending stresses due to subsequent loads applied to the beams after the second phase concrete has hardened, such as finishes, live loads and prestressing tendons stressed after the change in section will be calculated using the monolithic section properties.

Another common occurrence of this problem occurs when a bridge deck consists of two parallel box girders that are built and fully prestressed, and subsequently stitched together with a cast-in-situ slab. The bending stresses in the deck may be calculated ignoring the structural contribution of the slab (considering its self weight as a superimposed dead load), and then again assuming that the slab was present as the deck was erected and prestressed. The bending stresses for the two limits may then be compared. A best estimate of the actual stresses may be made as described in the paragraph above.

In a prestressed concrete bridge deck, the self weight and the prestress generally act in opposition to each other. Consequently the uncertainties in the value of moments and stresses are much reduced with respect to a reinforced concrete deck. Furthermore, the self weight and prestress constitute only a proportion of the total bending moments. Consequently the overall effect of creep on the suitability of bridge decks of medium span to carry their specified loads is generally of limited significance; excessive accuracy in the calculation of moments and stresses is illusory and should not be sought. For bridge decks with a span in excess of about 120 m, where the self weight and prestress moments predominate, more care must be taken to establish where, between the as-built and monolithic limits, the reality lies.

6.23 Bursting out of tendons

A curved tendon applies pressure to the concrete on the inside of the curve. The pressure Q KN/m $= P/r$, where P is the force in the cable and r is its radius of curvature. For a duct radius of 7 m, which is typical of many schemes, a tendon consisting of 12 strand of 13 mm diameter, with a force at stressing of approximately 1,600 kN, the pressure applied to the concrete would be 230 kN/m, while for a tendon consisting of 27 strand of 15 mm, with a stressing force of 5,300 kN, the pressure would be 760 kN/m. Where the curve of the tendon is concave towards a free concrete surface, such that the cover to the tendon may be pushed off by this force, it is necessary to detail reinforcement that carries the force back, behind the tendon, Figure 6.28 (a). Examples of such curved tendons may be found for instance in the blisters for internal tendons described in 6.24, at rapid changes in thickness of bridge deck webs and in the curved bottom slabs of variable depth bridges.

The pressure on the inside of the curve may also cause curved tendons that are stacked in a web to collapse onto the tendon above or below, if the spacing is inadequate,

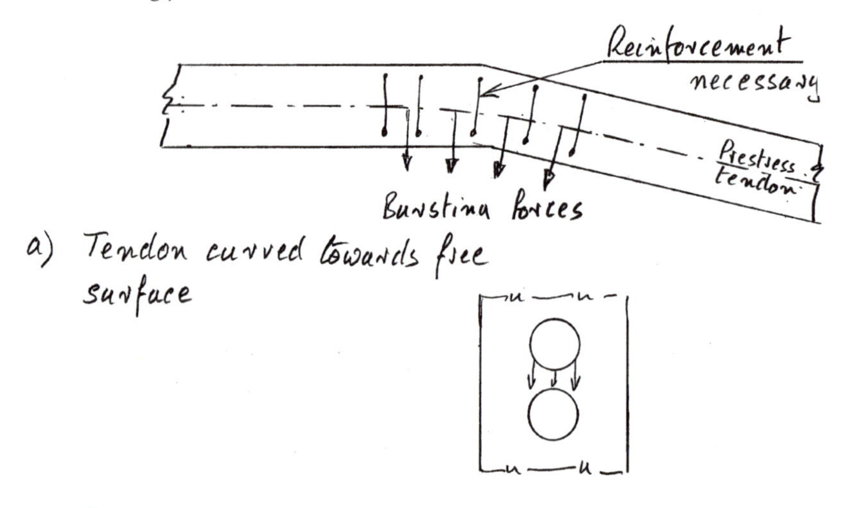

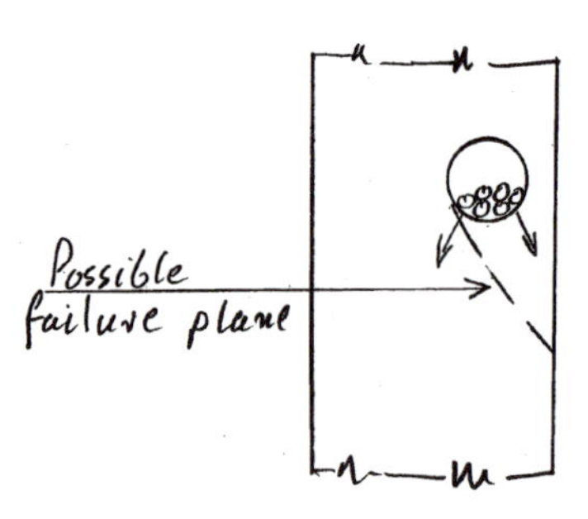

c) Curved tendon parallel to free surface

Figure 6.28 Forces exerted by curved tendons

Figure 6.28 (b). However, for tendons spaced vertically at two internal duct diameters there can never be a problem.

When tendons are curved in a plane parallel to a concrete surface, and located close to that surface, the pressure on the inside of the curve can cause spalling of the concrete in the absence of appropriate reinforcement. This tendency is enhanced by the fact that the strands within the curved duct create splitting forces along the plane of the duct, Figure 6.28 (c). However, with the cover to prestressing ducts and the reinforcement normally provided in prestressed concrete bridges, this is only likely to be a problem for unusual situations, involving large prestress units on tight radii very close to lightly reinforced surfaces. Common sense would normally lead one to avoid such a situation.

In the author's opinion, the guidance given on the subject of the spacing and cover appropriate for curved tendons in Appendix D to BS 5400: 1990: Part 4, Tables 36 and 37 should not be followed, as they are conservative to the point of uselessness.

In situations where the designer considers that the curvature of the tendons gives rise to a risk of failure, reinforcement may be detailed to carry the forces safely. In some situations, when the bursting forces are particularly large, the stress in the reinforcement carrying them back may have to be limited to a relatively low figure, say 150 MPa, to reduce the strain in the bars to a value that will not crack the concrete cover.

6.24 The anchorage of tendons in blisters

Blisters for external tendons take the form shown in Figure 6.29 (a) and in Figure 15.26. In addition to the primary and secondary effects, described in *5.24*, such anchorages also apply very significant bending moments and shearing forces to the parent concrete to which they are attached, as well as compression immediately in front of the blister. The thickness of the concrete member carrying the blister must be adequate for these forces, which can be very substantial for the large anchorage units often adopted for external tendons. When possible, the blister should be located in a corner of the box, so that two approximately orthogonal slabs can carry these forces. If the anchorage is remote from a free end of the concrete deck, it will also require following steel, as described in *5.25*.

Blisters for internal tendons take the form shown in Figure 6.29 (b). Such anchorages apply compressive forces to the slab in front of the anchorage, and require primary and secondary bursting reinforcement, as described in *5.24*, as well as following steel. However, although they do not apply such large moments or shearing forces to the slabs, care must be taken to ensure that the slabs carrying the blisters are thick enough to resist the concentrated forces and local eccentricities involved, as well as the local bursting forces due to curvature of the tendons. Such blisters are best located in the corners of the box if possible, Figure 6.29 (c) [13].

6.25 Checks at the ULS

For continuous prestressed concrete bridge decks, it is essential to carry out checks at the Ultimate Limit State (ULS). The need for such checks may be simply demonstrated by considering the two-span continuous beam illustrated in Figure 6.30 (a). In this beam, the uniformly distributed loads are exactly balanced by parabolic cables in each span. The cables are discontinuous over the central pier. The moment over this pier due to the applied load, that is equal to $-wl^2/8$, is cancelled out by a parasitic moment due to the cables, equal to $+wl^2/8$. Consequently at working load the bending moments in the beam are zero at all sections, and no reinforcement, prestressed or passive, is required over the supports. However, once the applied loads are factored to arrive at the ULS, it is clear that the unreinforced pier section would fail.

More generally, the presence of parasitic moments makes it essential to check the strength of prestressed beams at the ULS. It has become customary to check even statically determinate beams in the same way, although for designs with internal tendons that are familiar to the author, this is a formality as the ratio between the working and the ultimate stress in the steel is greater than the ratio between the working and ultimate loads. However, for statically determinate and continuous beams with external tendons, where the force in the tendons between the working and ultimate conditions does not increase much, the ULS is likely to be the critical case.

A second reason to check prestressed beams at the ULS is to ensure that there is adequate ductility. In particular for twin rib type bridges, or box sections with very small

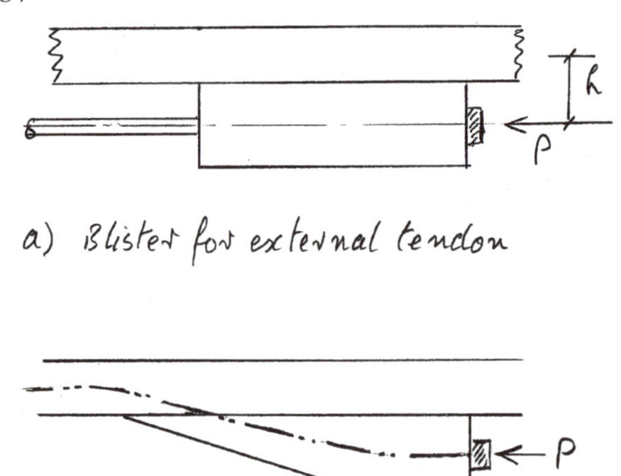

Figure 6.29a and b Anchorage blisters for tendons

Figure 6.29c Anchorage blisters for tendons: internal blister (Photo: Benaim)

bottom slabs, the neutral axis at the ULS at support sections may be uncomfortably high, indicating a lack of ductility. It is conventional wisdom that the compressed zone of concrete should not exceed 40 per cent of the depth of the section. If this is found to be a problem, measures that can be taken to increase the ductility of the section are to increase the strength of the concrete, locally to increase the thickness of the bottom flange of a box, or to introduce compression reinforcement close to the bottom fibre.

For twin rib type bridges, there should be reinforcement in this location for the reason described in *6.13.3* and *12.6*. This reinforcement, if linked as compression steel, may suffice, or may need to be locally increased. As the bending moment falls off very quickly, these bars may be short.

A further reason for checking at the ultimate limit state is shown in Figure 6.30 (b). This diagram shows the support section of a typical bridge deck built by the balanced cantilever method. As the bending moments fall off very quickly near the support, it is normal practice to stop off prestressing cables in the first segment faces adjacent to the support. These short tendons are effective in compressing the top slab of the box section and controlling the stresses under service loading. However, at the ULS, the first bending/shear crack will be as shown, and is likely to miss some of these short cables, which thus must be discounted at the ULS.

It is customary for the ULS bending moments and shear forces to be derived from factored up SLS bending moments resulting from an elastic analysis. However, in some cases, it may be more rational to adopt a plastic analysis, where moments are redistributed, generally from the pier towards the spans, as long as the deck section is adequately ductile. This is subject to the comments made in *6.3*, and to the requirements of the relevant code of practice.

In a prestressed beam at the SLS, the lever arm of a section designed to Class 1 or 2 is smaller than that of a similar reinforced section, Figure 6.31 (a) and (b). However at the ULS, the lever arm of a prestressed section increases substantially, becoming very similar to that of the reinforced section, Figure 6.31 (c). The procedure for checking

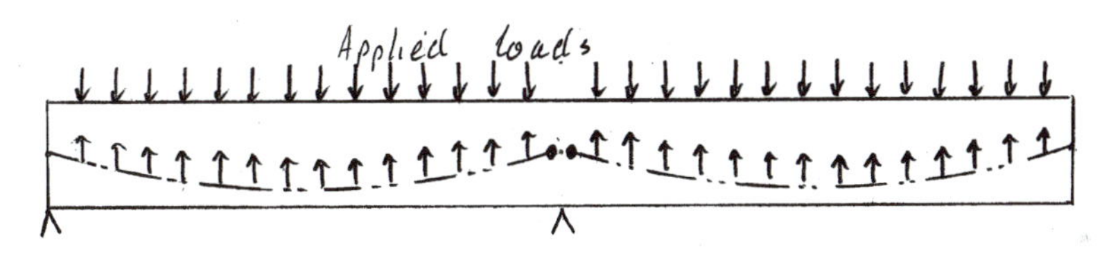

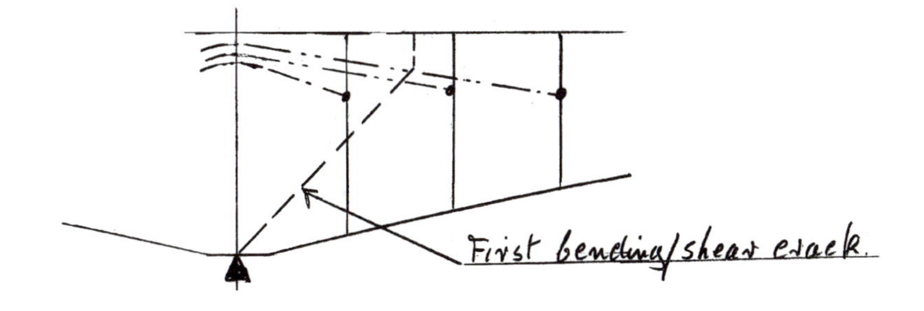

Figure 6.30 Checks at the ULS

the section is as described for reinforced concrete in *3.4.3*. This rough check assumes that the section is under-reinforced and ductile, so that the prestressing steel may reach its ultimate stress.

For bridges with external prestressing, the section check at the ULS is more complicated. If the tendons are not bonded to the concrete, overloading the deck will not cause the tendons to yield. In fact, there may be very little increase in stress in the tendons, the only increase in the moment of resistance of the section coming from the greater lever arm of the internal couple. For cast-in-situ bridge decks it is possible to augment the ultimate strength by including bonded reinforcing bars. However, if the deck consists of precast segments, it is necessary to over-design the prestressing to provide an adequate tensile force at ULS.

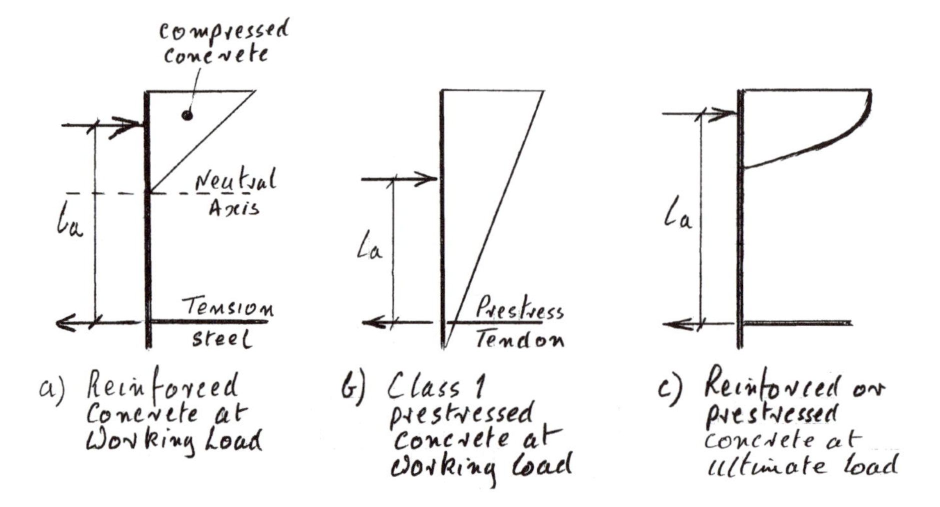

Figure 6.31 Increase of prestress lever arm at the ULS

7

Articulation of bridges and the design of substructure

7.1 General

Concrete bridges expand and contract with temperature changes, they shorten under the effects of concrete shrinkage and creep and they deflect under the effect of applied loads, prestress and temperature gradients. The deck must be held in place when subjected to wind, seismic activity and to forces applied by the traffic carried. If the deck is long, it may be necessary to divide it into several expansion lengths by intermediate joints. The articulation of a bridge may be defined as all the measures taken to hold it firmly in position while allowing it to change in length and width and to rotate at supports.

The articulation interacts closely with the type of bridge deck chosen, with the method of construction of the deck and with the design of the foundations. It cannot be considered in isolation; it is part of the overall concept of the bridge and should be present in the first 'back of the envelope' schemes.

7.2 Design parameters

7.2.1 Temperature change

Concrete bridge decks have considerable thermal inertia, and so do not follow daily extremes of temperature. The bridge deck temperatures which should be considered are defined by the local codes of practice. For instance, in the London area of the UK it is accepted that the temperature of a concrete bridge may vary from −7°C in winter to +36°C in summer, while in an equatorial climate such as that relevant to Kuala Lumpur, for instance, there are no significant seasons, and designers use bridge temperatures of 30 ± 10°C.

One cannot of course expect that the bridge will be built at the mean temperature; consequently it is necessary to consider carefully the actual range for which the articulation should be designed. This may not be the same for all parts of the structure.

For instance, consider a bridge built in the London area consisting of several continuous spans, with the deck built into the piers, and erected by balanced cantilever. While the balanced cantilever is under construction, it is free to expand and contract with changing temperature without any bending moments being created. Once it is connected to the adjacent balanced cantilever it becomes a portal and expansion and contraction of the spans due to temperature change will cause bending moments in both the deck and the piers of the completed bridge. The construction is likely to span several months, covering a considerable temperature range. The critical temperature that affects each span is that on the day the structure is made continuous by casting the mid-span stitch. This is certain to be different for each span, and may cover the seasonal range from summer to winter. For a span closed in summer at say a bridge temperature of 25°C, the effective temperature increase causing expansion is only 11°C, while the effective temperature decrease is 32°C. Conversely a span closed in winter will have to be designed to accommodate a large expansion and a small contraction. Unless the bridge has exceptionally long spans, it is generally quite impractical to design each span differently, and consequently a substantial temperature overlap must be used, designing the whole deck for, say, expansion from +10°C to +36°C and contraction from +25°C to −7°C.

An alternative strategy would be to adjust the length of the mid-span closure according to the bridge temperature at the time by jacking apart the ends of the cantilevers. The jacking stroke and thrust would be adjusted for each span according to the bridge temperature on the day. However, as always it is necessary to balance the costs of this operation against the cost of adopting the temperature overlap described above.

The situation is quite different when the roadway expansion joints are designed. When they are due to be fitted to the completed bridge deck, the bridge temperature may be assessed or measured quite accurately, and the joint openings may be set for the temperature of the day of installation. Consequently the design temperature overlap may be quite small. For instance, if on the day of installation the bridge temperature is actually 15°C the joint may be set to accommodate a temperature fall from +16°C to −7°C and a temperature rise from +14°C to +36°C.

In temperate climates, it is possible to distinguish between daily, weekly and seasonal fluctuations of temperature. This can be important in designing members that are stressed by temperature variations, as the effective Young's modulus of concrete is dependent on the duration of the load (*6.13.3*). However, in equatorial climates where there are no seasons, it is necessary to take a local view on how quickly the temperature may oscillate between extremes.

7.2.2 Coefficient of expansion

The coefficient of expansion of concrete varies principally with the type of aggregate used, from about 12×10^{-6} per °C for gravel to about 7×10^{-6} per °C for limestone. However, these are not the extreme limits, and when it really matters, for instance for very long expansion lengths, the designer should consult specialist literature, or carry out trials on the concrete mix to be used. The coefficient of expansion of concrete may also be prescribed by the various national codes of practice.

For bridges in a climate similar to that of the UK, a coefficient of expansion of 12×10^{-6} defines a movement which is typically a 26 mm expansion or contraction from the mean temperature for a 100 m length. As a design overlap is required, the actual design movements will be greater.

7.2.3 Temperature gradients through decks

Temperature gradients cause rotations of the unrestrained ends of decks, and thus must be taken into account in the design of the bearings. The rotation of the deck end about the bearing can cause additional movements at the level of the expansion joint. This can also affect the design of the measures to inhibit cracking of the road surfacing at the fixed end of short spans. See *6.13.3* for a full discussion of the importance of temperature gradients in the design of bridge decks.

7.2.4 Shrinkage of concrete

As concrete matures, it gradually reduces in volume. The amount and rate of shrinkage depends on the type of cement, the aggregate, the mix proportions, the size and shape of the specimen, and on the relative humidity of the air. The shrinkage to be considered when designing the articulation of a bridge may be defined by the local code of practice, or it may be necessary to refer to specialist literature [1]. There are large variations in the shrinkage values called for by the different national codes of practice.

Shrinkage is restrained by the presence of bonded reinforcement, which is stressed in compression by the shortening of the concrete. This is graphically illustrated by the early prestressed concrete balanced cantilever bridges which had a hinge at mid-span. Some of these bridges exhibited unexpected downward deflection of the mid-span hinge over a period of years. Part of the explanation for this behaviour was differential shrinkage between the top and bottom flanges of the box section. In the top flange, the shrinkage was restrained by the large number of bonded prestressing cables, while the bottom flange was only very lightly reinforced in the longitudinal direction.

BS5400: Part 4: 1990, Appendix C gives values of shrinkage for a typical bridge deck in UK conditions, of some 180×10^{-6}, or 18 mm per 100 m length, of which 10 per cent is completed in 10 days, and 35 per cent in 100 days. The actual values depend on the criteria mentioned above, and should be calculated for each case. If the section is significantly reinforced with bonded steel the shrinkage will be less.

The amount of shrinkage that affects the design of the articulation of any bridge will depend on the method and programme of construction. For instance, when a single-span bridge is cast in-situ in one pour the bearings will have to cope with 100 per cent of the shrinkage. However, if the bridge consists of several continuous spans, each being cast at three-weekly intervals, a considerable portion of the shrinkage of earlier spans will have taken place before the last span is cast and consequently the bearings may be designed for reduced movements. When a bridge is made of precast components which will be several weeks old when they are incorporated into the deck, the shrinkage to be used for the design of the substructure is significantly reduced.

The amount of shrinkage to be considered in the design of the expansion joints will depend on the average age of the deck when they are installed.

The codes of practice may give useful guidance for typical structures. However for very long structures where movements of expansion joints and bearings must be known with some certainty, they should not be relied upon.

7.2.5 Elastic shortening under compression

Prestressed concrete decks shorten during the tensioning of the cables, and this is known as elastic shortening. The elastic modulus for concrete to be used to calculate this shortening depends principally on the strength of the concrete, its age when the prestress is applied and also on the aggregate type. Generally the modulus to use is defined by the appropriate code of practice. However it should not be forgotten that this is a conventional value, and real values may well be significantly different.

For a typical bridge deck concrete with a cube strength of 50 MPa, the elastic modulus given by the British code for concrete that is at least 28 days old is 34,000 MPa. For a bridge with an average prestress $P/A = 4$ MPa, this gives rise to a shortening of 12 mm per 100 m.

7.2.6 Creep of concrete

When concrete is loaded in compression, it suffers an immediate elastic shortening, and then, if the load is maintained, continues to shorten for a considerable time. This deferred shortening is known as creep, and is described in *3.9*. The creep coefficient ϕ depends on the same criteria as shrinkage, plus the very important additional criterion of age at first loading; the older the concrete at first loading the smaller the creep coefficient.

For a typical cast-in-situ bridge deck in UK conditions, prestressed to an average compressive stress of $P/A = 4$ MPa with $\phi = 2$, the shortening due to creep is 24 mm per 100 m length. For a typical precast segmental deck similarly compressed, $\phi = 1$ and the creep shortening is 12 mm per 100 m. These figures are of course purely for the purpose of illustrating the order of magnitude of the effect. In reality, creep coefficients need to be considered carefully. There is some evidence that conventional calculations under-estimate the amount and the time to completion of concrete creep.

7.3 Bearings: general design considerations

This is not the place to give detailed descriptions of the bearings available on the market for consideration by the bridge designer. However, a general survey of the characteristics of the different types of bearing is necessary to allow the development of the rest of the chapter. There are three generic types of bearing in general use: mechanical bearings, elastomeric bearings and concrete hinges. Their purpose is to allow the bridge to change in length and width and to deflect freely.

The design of the bridge must make provision for the eventual replacement of mechanical or elastomeric bearings. This is most conveniently done by arranging space for jacks on the pier head, and such a measure would be essential on very high or inaccessible piers. However, on many viaducts, jacking off falsework resting on the ground or on the pile caps is an acceptable option.

It is necessary to inspect all mechanical and elastomeric bearings regularly. However, this is particularly important for sliding bearings, as described in *7.4.3* below.

7.4 Mechanical bearings

7.4.1 General description

There exist a variety of rocker, roller, spherical and other mechanical bearing types. However, the commonest type of mechanical bearing adopted for concrete bridges, and the only one to be described here, is the pot bearing, Figure 7.1. It consists of a steel cylinder containing a rubber disk. The disk is compressed by a steel piston. The rubber acts rather as a contained liquid, giving very little rotational resistance

Figure 7.1 Sliding pot bearing (Image: CCL Stressing Systems Ltd)

to the piston. The piston and the cylinder are attached to base plates that connect to the concrete of the deck and pier respectively. Such a basic bearing allows rotation but no displacement. Standard pot bearings can carry a limited amount of horizontal load, usually about 10 per cent of the rated vertical load. Special pot bearings may be manufactured where larger horizontal loads must be carried. Pot bearings are normally sized such that the pressure on the concrete of the deck and pier does not exceed 20 MPa.

It is very important that the designer does not accept proprietary items such as bearings uncritically. There is competition between manufacturers, and not all bearings are made to the same standards. By studying the catalogues, one can deduce, for instance, the working stress assumed on the rubber disc and on the concrete. The author has had the experience of a bearing supplier being undercut by a competitor who was simply rating his bearings more highly, and thus supplying a smaller product for the same load. Also, mechanical standards differ; for instance the pot may be simply welded to the base plate, or may be machined into it; corrosion protection systems may not be to the same standard.

7.4.2 *Guided and free sliding bearings*

If the bearing is to allow free sliding in any direction, the top surface of the piston base-plate is covered by a PTFE coating, and a further plate, equipped with a highly polished stainless steel sheet slides over it. The sliding friction of such bearings depends on the contact pressure and on the cleanliness of the sliding surface. In reality, it may vary from 0.5 per cent up to 5 per cent.

If the movement of the bearing is to be guided, a keyway and corresponding key are machined into the top two plates, or side guides are attached to the top plate.

7.4.3 *The life of PTFE*

The PTFE coating of sliding bearings wears with use, and it is not clear how long it will last [2]. Once the PTFE starts to wear out, the friction coefficient will increase, until it attains the value that corresponds to the contact between the stainless steel and the carbon steel backing to the PTFE, which may be 20 per cent or more. As the stainless steel is only a very thin sheet, it may well ruck under the effect of a much higher friction coefficient, further increasing the sliding friction or even jamming the bearing.

Typical piers and foundations are designed to withstand longitudinal forces applied by the bearings that are in general 5 per cent of the dead weight of the bridge deck. A significant increase in friction due to bearing wear could seriously damage the bearing holding down bolts, the piers or the foundations. Fixed bearings on piers or abutments would experience very greatly increased forces due to differential friction.

There are two principal components to the movement of sliding bearings:

- irreversible shortening of the deck due to elastic shortening, shrinkage and creep;
- repetitive temperature cycling.

It is most improbable that the PTFE would not outlast the phase of permanent shortening of the deck. The risk of wear is thus limited to the daily and seasonal temperature cycling.

Piers or abutments that are within say 20 m of the fixed point for deck movement would be subjected to no more than ±5 mm of temperature movement for a range of ±20°C. Even if the bearings were totally jammed, this amount of movement is unlikely to cause major structural damage in most cases, although it could cause progressive deterioration of the substructure. However for longer viaducts, the risks are much greater. For instance, for a pier that is 500 m from the fixed point of the deck, the yearly temperature movement would be of the order of ±120 mm, and even a daily variation of ±5°C would cause ±30 mm of movement. Such movements could cause very serious structural damage, compromising the safety of the bridge.

All sliding bearings should be subjected to a regular visual inspection that includes checking the pier stems for cracking. All but the shortest bridges should in addition have an occasional in-depth inspection of the bearings, involving sample dismantling and measurement of the thickness of the PTFE layer. This in-depth inspection could, for lack of better advice, be undertaken 10 years after commissioning of the bridge and then at 5-yearly intervals; more frequently if movements are due to live load.

If it is considered that such inspections cannot be guaranteed, on short bridges the piers and abutments may be designed such that they could accommodate the temperature cycling without significant damage, even if the bearing locked. For bridge abutments and piers of longer bridges where the magnitude of the movements would make this approach impossible, each bearing, pier and foundation should be designed with a hierarchy of strength, such that the bearing itself, or its connection to the deck or to the substructure will fail before the pier or its foundation is seriously damaged. It is likely this would entail some increase in strength, and hence in cost, of the piers and of any piled foundations.

Alternatively, the bridge should be designed with elastomeric bearings as described below, as they do not suffer from this particular risk.

In lens type bearings, where rotations due to live loads and temperature gradients involve sliding movements, the report [2] suggests the life of PTFE may be as short as 5 years.

7.4.4 Cast-in items

Although bearings are designed to be replaceable, some items such as dowels are cast permanently into the concrete, and in some installations upper or lower plates may also be cast in. Many bearings on box section bridges are protected by long overhanging side cantilevers when there is little chance of moisture being present regularly. In other situations, where deck expansion joints may allow water onto the bearing shelf, where bearings are not protected from rain by side cantilevers, in situations where the bearing would be subject to salt-laden spray from the sea, or from an adjacent carriageway, or if it is impossible to repaint regularly, stainless steel should be employed for such cast-in items. Electrolytic corrosion will occur at the interface between carbon steel and stainless steel in the presence of water. If water will be present, the two steels should be separated by suitable washers.

7.5 Elastomeric bearings

These consist of natural rubber or Neoprene slabs vulcanised between steel plates, Figure 7.2. They cater for rotation by differential compression across the width of the slabs, and for displacement by shearing of the slabs. The bearing can be designed to suit the particular bridge by providing the appropriate plan area to carry the load, an adequate total thickness of elastomer to cater for the displacement, and by assembling the appropriate number of layers of a specified thickness to cater for the required rotation. Rubber bearings are regularly made in sizes up to 700 mm square, when they may carry loads up to 8 MN. Such bearings may be assembled in series to carry greater loads.

The allowable working compressive stress of the bearings lies generally in the range 12–17 MPa with bearings of larger plan area working at the higher stresses. The allowable shear displacement of a bearing is approximately 70 per cent of the thickness of rubber. However, the design of these bearings is complex, as the capacity

Figure 7.2 Elastomeric bearing (Image: CCL Stressing Systems Ltd)

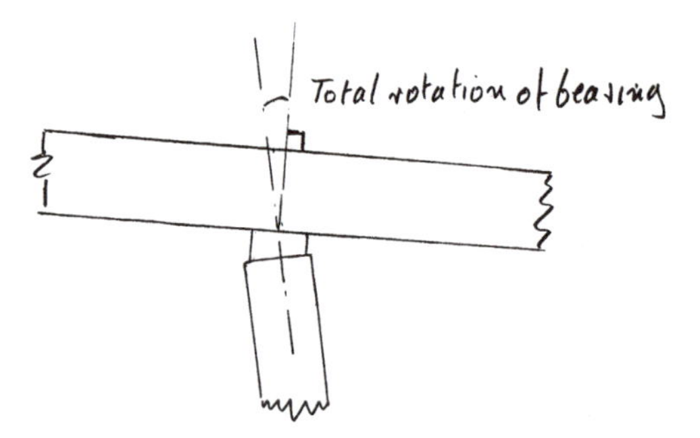

Figure 7.3 Design rotations for elastomeric bearings

of the bearing in direct load, shear and rotation are inter-related. Furthermore, if the total thickness of rubber in the bearing is greater than between one-third and one-half of its smallest plan dimension, it may suffer a form of elastic instability. Although the bearing manufacturers provide a design service, the bridge designer should understand the basis of this design.

Rubber bearings may be locked in position, by placing a steel dowel through them. They may also be associated with a PTFE/stainless steel sliding surface to increase their movement capability. However, rubber bearings are best kept simple, and when it is necessary to increase the displacement or to provide locked bearings, they may be used in conjunction with mechanical bearings.

If precast beams are placed onto rubber bearings, the bearing design must take into account the rotation caused by any pre-camber of the beams or tolerance in flatness of their bearing surface. Alternatively, the beam may be propped clear of the bearing, and the void filled with mortar.

The design rotations of bearings must include the rotation of the pier head caused by its longitudinal deflection, as well as the rotations of the deck, Figure 7.3.

7.6 Concrete hinges

A very simple, cheap and reliable rotational bearing may be made in plain concrete. This was invented by Eugene Freyssinet, and is used more extensively in France than elsewhere. The former French rules [3] specify that the concrete throat should be sized so that under working loads it is stressed at no more than twice the 28-day cylinder strength of the parent concrete, and that the rotation should not exceed typically 1/40, that is 25 mille-radians.

Typical French practice was to make the hinge very simply, by casting the deck concrete directly onto the pier, with polystyrene sheet or other thin formwork defining the throat. The lines of stress in the hinge are shown diagrammatically in Figure 7.4, and make it clear that the concrete in the throat is tri-axially compressed. Although not called for in the code, it would be sensible to ensure that the minimum plan dimension of a hinge built in this way is not less than 100 mm, with a height of 20 mm. The hinge may be linear, rectangular or circular.

The concrete above and below the hinge needs to be reinforced with mats of bursting steel. The hinge itself does not need reinforcement through it, although if there is a risk of it being displaced before it is substantially loaded, light vertical steel bars may

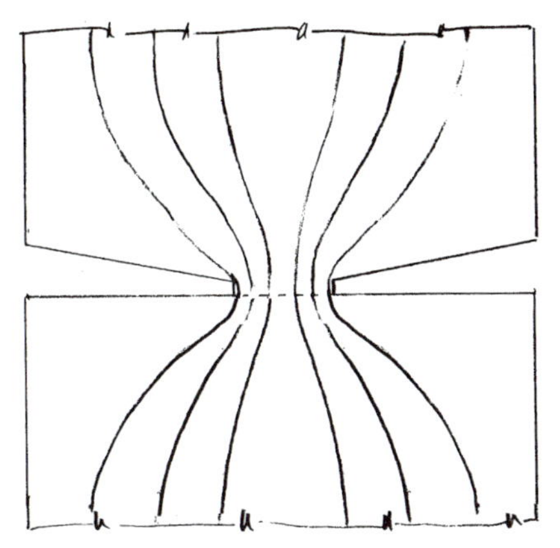

Figure 7.4 Concrete hinge

be used. The unreinforced hinge may carry horizontal loads up to one-quarter of the co-existing vertical load.

Concrete hinges are not appropriate if the minimum load is too light, as they are then vulnerable to displacement. As they rely on the tri-axial containment of the concrete, the hinge must be suitably set back from all edges of the concrete. Finally, in order to make it possible to inspect the hinge in service, the concrete above or below (or both) should be gently chamfered.

The use of concrete hinges in the UK has been seriously inhibited by an ill-considered, over-theoretical attempt at regulation [4]. This imposed drastic limits on the rotation that the hinge could accept, and also required fussy details, insisting for instance that the hinge and the zones immediately above and below it should be made in small aggregate concrete, that there must not be a construction joint through the hinge, which means it cannot be cast together with the parent bridge deck, and that the throat should have a parabolic shape.

7.7 Design of foundations

7.7.1 General

It is not the intention of this book to cover in detail either the geotechnical or the structural aspects of the design of foundations. However, the choice of foundation type affects fundamentally the articulation of the bridge. This section will discuss bridge foundations in the context of the design of the articulation.

It is very surprising the number of times one comes across well-designed prestressed concrete decks placed on over-designed foundations. It sometimes appears that bridge designers are ready to delegate foundation design to soil specialists, as if it was some incomprehensible black art. In particular, it is frequently not appreciated that the most economical pile, recommended by the specialists, may not provide the most economical foundation. The cost of the foundation usually includes a pile cap and generally a substantial excavation in addition to the piles. Furthermore the elastic properties of the foundation may affect the design of the piers and bearings and influence the provision of expansion joints, and hence the overall cost and maintenance of the bridge. Consequently foundation type should not be determined by soil specialists, but by the designer, with the benefit of expert advice.

7.7.2 *Pad foundations*

Usually, if it is possible to adopt a pad, this will provide the most economical foundation. An additional advantage is that the general bridge contractor does not need to bring onto site a specialist foundation sub-contractor. There are times when, for instance, all the piers except one can be founded on pads at a reasonable depth. It is worth finding ways of avoiding mobilising a piling rig for a few piles under just one pier, such as using cofferdams or mining techniques. However, if there is no alternative to piling, once the cost of mobilising a rig has been incurred piles may be found to be economical at other pier positions.

Pads give rise to stiff foundations that do not move under the effect of horizontal loads, and have very limited rotation under overturning moments. This rigidity can affect the articulation, as described later in this chapter.

There is no reason why pads and piles should not be mixed beneath the same deck, as long as the relative settlements and differential stiffness of the foundations are properly considered in the design.

7.7.3 *Driven piles*

Most driven piles used in the UK have a relatively small capacity, generally 1.5 MN or less. A typical medium span concrete highway bridge weighs, including finishes, live load and substructure, between 0.025 MN/m^2 and 0.035 MN/m^2. Thus a continuous bridge 13 m wide with spans of 52.5 m will give rise to pier reactions of some 18 MN, and would require 12 such piles at each pier to carry the vertical loads alone. However, foundations are subjected not only to vertical loads, but also to longitudinal bending moments caused by bearing friction and wind forces on the pier stem as well as transverse moments due to live load eccentricity and wind forces on the deck and on the pier stem. In a rectangular array of piles subjected to bi-axial bending and vertical load, only the corner piles will be stressed to the maximum. Consequently, the average pile load will be significantly less than the rated load, increasing the number of piles required.

Driven piles in steel or concrete are normally considered in design as pin-ended, incapable of carrying bending moments. The consequence of this assumption is that any horizontal loads have to be carried by a system of triangulation, using at least three rows of piles, in which at least one row is inclined, Figure 7.5. This further increases the number of piles required, and leads to a larger pile cap which, together with the earth carried, increases the vertical loads and leads to the need for yet more piles. The total number of piles required to carry a typical pile cap is likely to be two to three times the number required for centred vertical loads alone.

Significant savings can be made in number of piles and size of pile cap if it is possible to use only vertical piles, with the horizontal forces carried in bending. However, there are complex group effects, which may spread the bending unequally among piles, and specialist literature should be consulted or expert advice sought. If the bridge is subject to very high wind loading such as typhoons, to railway loading, to hydraulic loading,

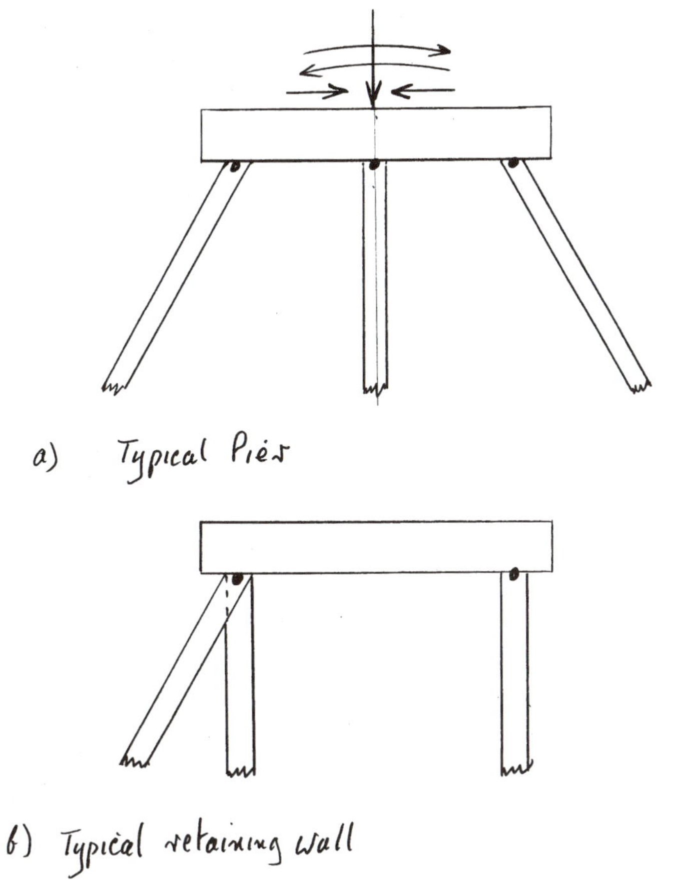

Figure 7.5 Statically determinate arrangements for driven piles

to ship impact or to seismic loading, or if the driven piles are of small capacity, usually it is not possible to dispense with raking driven piles.

In some countries, in particular South East Asia, prestressed piles of up to 800 mm diameter are used, which have much greater capacity to resist bending moments. Some specialised companies have prefabricated concrete piles up to 1.2 m diameter, and steel driven piles may have diameters up to 2 m, using technology developed for offshore rigs. The bending strength of such large-diameter driven piles may be similar to that of bored piles, when they will share their methods of design.

Foundations with a triangulated arrangement of piles will be very stiff under the effect of horizontal loads, deflecting little. Foundations that rely on the bending of the piles to resist horizontal forces will be much more flexible. These differences can be very important in the design of the articulation of the deck.

If piles can be inclined so that the projection of their centre lines meets at the point of application of the horizontal force, there will be no significant bending in the pile, Figure 7.6. This is particularly useful where there is one dominant horizontal force with a fixed location, such as the centrifugal force of a train on curved track, or the friction force applied by a sliding bearing. It is not particularly useful where the force may vary in point of application, such as ship impact on a foundation in tidal water, or where there are several components to the horizontal force applied at different levels.

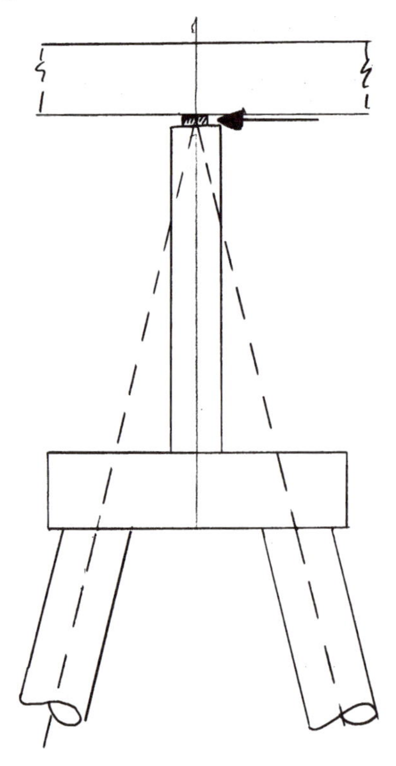

Figure 7.6 Triangulated piles

7.7.4 Bored piles

Bored piles may be made with diameters from 600 mm up to 2500 mm, with some rigs making even larger sizes; the Kincardine Bridge across the Forth, where Benaim is carrying out the detailed design, is using bored piles up to 3.85 m in diameter. Bored piles can carry substantial bending moments, and generally foundations may be designed without the need for raking piles. The piles and pile caps are analysed as portals, making use of the bending strength of the piles. Abutments and retaining walls which have a high ratio of horizontal to vertical forces may require raking piles. It should be noted that the use of raking bored piles at inclinations greater than about 1/10 requires specialist advice to confirm that their construction in the specific soil is feasible. In particular, it needs to be confirmed that the 'roof' of the inclined bore will not collapse, and that the boring tool will maintain its inclination, and not gradually curve towards the vertical.

When it is possible to aim the inclined piles at the centroid of the horizontal forces, as described for driven piles, the bending moments on the piles will be minimised.

a) Site investigation

It cannot be emphasised too strongly that the design of bored piles depends on a well-specified and carefully carried out and supervised site investigation. Most bored piles derive their carrying capacity from friction, either in soft materials over long lengths of shaft, or over relatively short rock sockets. They also rely on the excavation remaining open long enough to lower the reinforcing cage and to fill it with concrete. This may require temporary casing or the use of bentonite or other drilling mud. In order to choose the method of construction, and to calculate the capacity of the pile, the properties of the various strata all the way down the length of the pile must be known. Also, whereas triangulated driven pile systems are generally analysed as if the soil around the piles did not exist, vertical bored piles rely on the ground for support. When horizontal loads are applied to the foundation, it tries to sway and the soil around the piles resists the movement. Thus accurate knowledge of the properties of the upper levels of the soil is required to analyse correctly this soil/structure interaction.

The site investigation should start with a desk study based on geological maps, and on previous site investigations or foundation construction in the vicinity. It should also look at any archaeological history that may be relevant, so that the presence of man-made objects may be predicted. The study of aerial photographs, particularly stereoscopic pairs when available, can be very revealing. For instance if valleys between adjacent hills are seen disappearing under the sediments of the plain, one may expect to meet the complications caused by these buried valleys when piling. When a bridge designer interprets the results of a site investigation, he is tempted to extrapolate the information obtained from isolated boreholes to the remainder of the site. The knowledge obtained from the desk study helps dispel such complacency.

The investigation should then continue with a first stage campaign of boreholes at, say 100 m centres, to establish the general succession of strata, to plan which type of testing will be required, and to allow a specification for the second stage investigation to be written. The second stage investigation should consist of at least one borehole at each pier position. If the pile caps are large, say longer than 10 m, and if the strata are known to be sloping, faulted or discontinuous, more boreholes per pier position will be justified. It is very rare to see an over-specified site investigation, and it is also rare that additional knowledge of the sub-soil will not result in cost savings that far exceed the cost of the investigation.

There are, of course, some sites where the sub-soil is very uniform over large distances, when all that is needed is confirmation that there are no unexpected changes. In these circumstances, the above warnings are less relevant.

However, the experience of one section of the investigation for the STAR railway viaduct foundations will act as a useful antidote to the complacency or lack of imagination that appears to afflict designers when dealing with the sub-soil. These foundations consisted of a single 1.8 m diameter pile per pier, as described in *7.15.3* below. For a particular length of the viaduct, the ground consisted of soft strata overlying limestone. Although it is well known that the limestone in Kuala Lumpur may be heavily eroded, a borehole sunk at each pile position was interpreted as regular rock head profile shown in Figure 7.7 (a).

When the first pile was sunk, it reached bedrock at a much greater depth than expected. Additional boreholes showed that the pile was located at the edge of a near-vertical 20 m deep cliff face in the limestone. Thereafter, three additional boreholes were sunk in a triangular pattern at each pile position, demonstrating that the rock head profile, far from being regular was tortuous in the extreme, with the construction of each pile becoming a major engineering feat, Figure 7.7 (b).

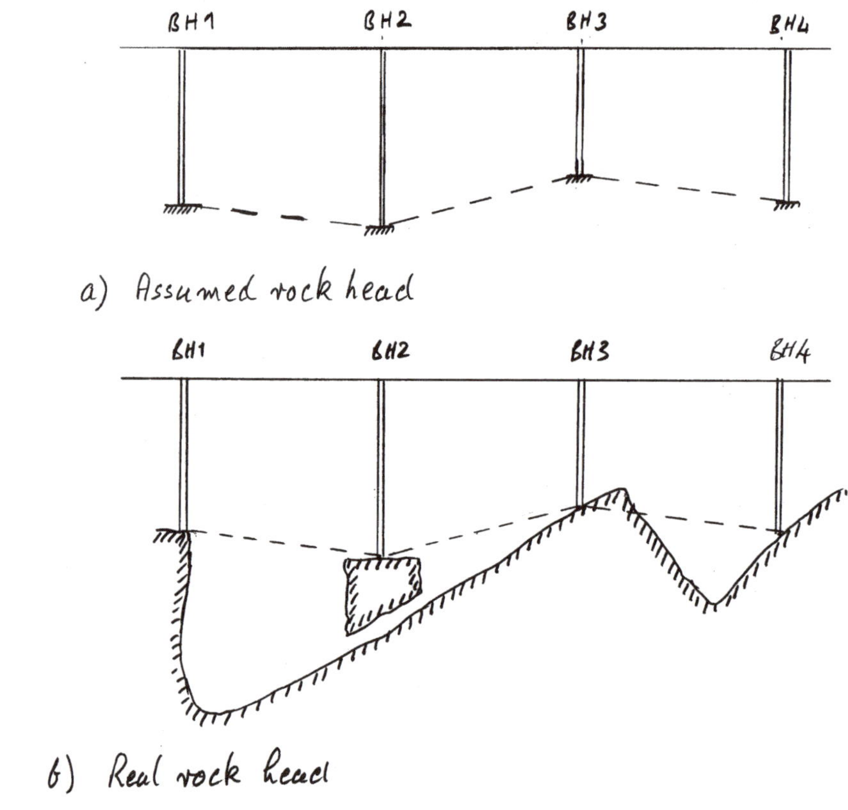

Figure 7.7 The perils of site investigation

b) Pile design

The subject of the design of bored piles is of course vast, and well beyond the ambitions and scope of this book. The suitability of a site for bored piles, and the determination of the safe load for any particular pile are subjects for geotechnical specialists, unless the bridge designer knows very well both the ground and the method of construction of the piles. However there are a few items of practical experience which may be useful to bridge designers.

The load capacity of the pile will depend on the combined resistance of end bearing and side friction. Clearly it is inevitable that debris will fall to the bottom of the pile during its excavation. An essential part of the process of making the pile will be to attempt to clean the base to allow close contact between the pile concrete and the foundation stratum. However, it is very difficult to ensure that this has been done properly, particularly on a site where several piles may be under construction simultaneously, 24 hours per day. Furthermore, the hole will be left open and vulnerable to debris falling in for several hours after cleaning, before the reinforcing cage is lowered in. This action itself may dislodge further soil from the walls of the excavation. Thus the designer cannot count on a clean base for the piles.

A layer of loose material in the toe of an end bearing pile would cause it to settle before stiffening up and carrying its design load. Consequently, even when a pile is founded in rock, it is advisable to penetrate the rock by 2–3 pile diameters to create a rock socket, where at least the working load is carried by side friction.

The frictional resistance of a pile depends on the condition of the walls of the bore, which in turn depends principally on how the pile was made. If the pile passes through clays into firmer strata, it is important that the tool used for excavation does not smear the walls at depth with a layer of soft clay. Also, if the clays are being relied on for part of the frictional strength, the design should take into account that the boring may rework the surface layer, reducing its strength, while if the bore is left open for too long, the clay surface may deteriorate. If the pile is being bored under water or bentonite, some boring tools that are insufficiently vented may cause suction as they are being drawn out, loosening or even collapsing the side walls.

If the pile is bored through sand or gravel it will usually be under bentonite. The bentonite limits the loss of drilling water through the soil by building up an impermeable layer, or filter cake, on the pile wall. Thus the concrete of the pile walls will not be in direct contact with the ground, and the frictional load capacity of the pile will be defined by the lower of the internal friction of the bentonite filter cake or of the soil; it is usually the former that governs.

When a pile is loaded, some settlement takes place as the load transfers to the soil. As friction is a stiffer action than end bearing, initially the load is all carried in friction. When the limiting friction is reached continuing settlement brings the end bearing into play. Consequently it is often considered good practice to design bored piles such that they carry their working load with a small factor of safety in friction alone. For instance, the design could be based on friction alone carrying 1.5 × working load, while the required factor of safety of 2.5 or 3 is provided by the combined action of friction and end bearing. Clearly, the detailed definition of these factors of safety needs to be established for each case.

c) Pile sizing

A foundation consisting of numerous bored piles beneath a pile cap suffers from the economical disadvantage described for vertical driven piles, namely that only the corner piles work at their maximum load, and the average load in the piles is thus significantly less than their rated load. Corner piles work hardest, followed by edge piles, with piles in interior rows carrying the least load. The design logic that follows is to attempt to limit the number of piles to a maximum of four.

If a single pile is used, it only needs to be designed to carry the actual vertical load. The horizontal loads and moments in both directions are carried in bending in the pile. (Considerations of flexibility of the pier/pile or the amount of bending reinforcement required may require the diameter of the pile to be increased beyond that required for vertical load.) If two piles are used side by side, longitudinal forces and moments are carried in bending, while transverse loads and moments are carried in push/pull and in portal action, increasing the load on the piles and requiring greater pile cross-sectional area. If three or more piles are used, moments in both directions will be carried in push pull. If more than four piles are used, non-corner piles will be under-used.

It should be noted that the vertical load capacity of piles is usually assessed at the SLS, while the bending strength of piles is assessed as for any other reinforced concrete structure at the ULS. The initial sizing of bored piles may be based on a compressive stress at the SLS of 5 MPa (generally a 25 per cent over-stress may be appropriate for load cases that include wind). This is appropriate for most soft rocks. In hard rocks the stress may be increased, while in very weathered rocks or hard clays the length of pile necessary to achieve this stress may be excessive and a lower value may be appropriate. Clearly, expert advice must be sought before finalising the design of the piles.

Consider the following example of a bridge deck carried on two bearings on each pier. Assume a vertical SLS load of 18 MN at each pier, and longitudinal and transverse SLS moments on the pile group of 12 MN-m and 10 MN-m, Figure 7.8 (a). Longitudinal moments on the foundations are due to horizontal forces applied on the bearings or on the pier. Transverse moments on the foundations will be due to horizontal forces on the bearings and pier as well as to push-pull effects on the bearings due to the eccentricity of live load, and to lateral wind and traffic forces on the deck. A maximum pile stress of 5 MPa is used for all the pile arrangements.

The pile area required for a single pile is 18 MN / 5 = 3.6 m^2, requiring a pile diameter of 2.1 m, Figure 7.8 (b). A pile of this diameter suitably reinforced may readily carry the applied moments, factored up to the ULS. If two or more piles are adopted, a pile cap is necessary, which, together with the earth carried, will increase the vertical load. With two piles spaced laterally at 3 pile diameters, Figure 7.8 (c), the maximum pile load is 11 MN, requiring a diameter of 1.67 m, with a total pile area of 4.38 m^2, and an average pile stress of 4.1 MPa. If four piles are adopted, Figure 7.8 (d) again at a spacing of 3 pile diameters, the maximum pile load is 7.22 MN, requiring piles of 1.35 m diameter, with a total area of 5.72 m^2 and an average pile stress of 3.15 MPa. If 16 piles are adopted in a square pattern at 3 diameter spacing, Figure 7.8 (e), the maximum load on a corner pile is 1.91 MN, requiring piles of 0.7 m diameter, with a total pile area of 6.16 m^2 and an average pile stress of 2.92 MPa.

This example corresponds loosely to the East Moors Viaduct designed by Benaim, which was founded on four piles of 1.35 m diameter at each pier. It is approximate, as it has ignored the weight of the pile cap. It has also ignored the fact that the moment on the foundation that is due to horizontal loads must be calculated at the level where the bending moment in the piles is zero, Figure 7.9. This is between 3 and 6 pile diameters below the base of the pile cap, depending on the strength of the ground. Despite these approximations, the example illustrates how economy is served by limiting the number of piles.

On the basis of a working stress of 5 MPa, the total cross-sectional area of piles required for a four-pile group under UK highway loading (including 45 units of HB), may be estimated at A/120m^2, where A is the area of deck carried by the pile group. For a pile group including more than four piles, the pile area required will be increased by about 10 per cent. The cross-sectional area required for a single pile may be estimated as A/180m^2, although its diameter may need to be increased in the interests of limiting the deflection of the pier, or to reduce the density of reinforcement required to carry the applied bending moments.

Clearly, if a bridge is to be founded on only one or two piles beneath each pier, the stability of the pier will rely principally on the support offered by the ground in the upper areas of the pile. This requires a thorough understanding of the ground and of the soil/structure interaction, and excellent geotechnical advice. In particular, the following considerations must be addressed:

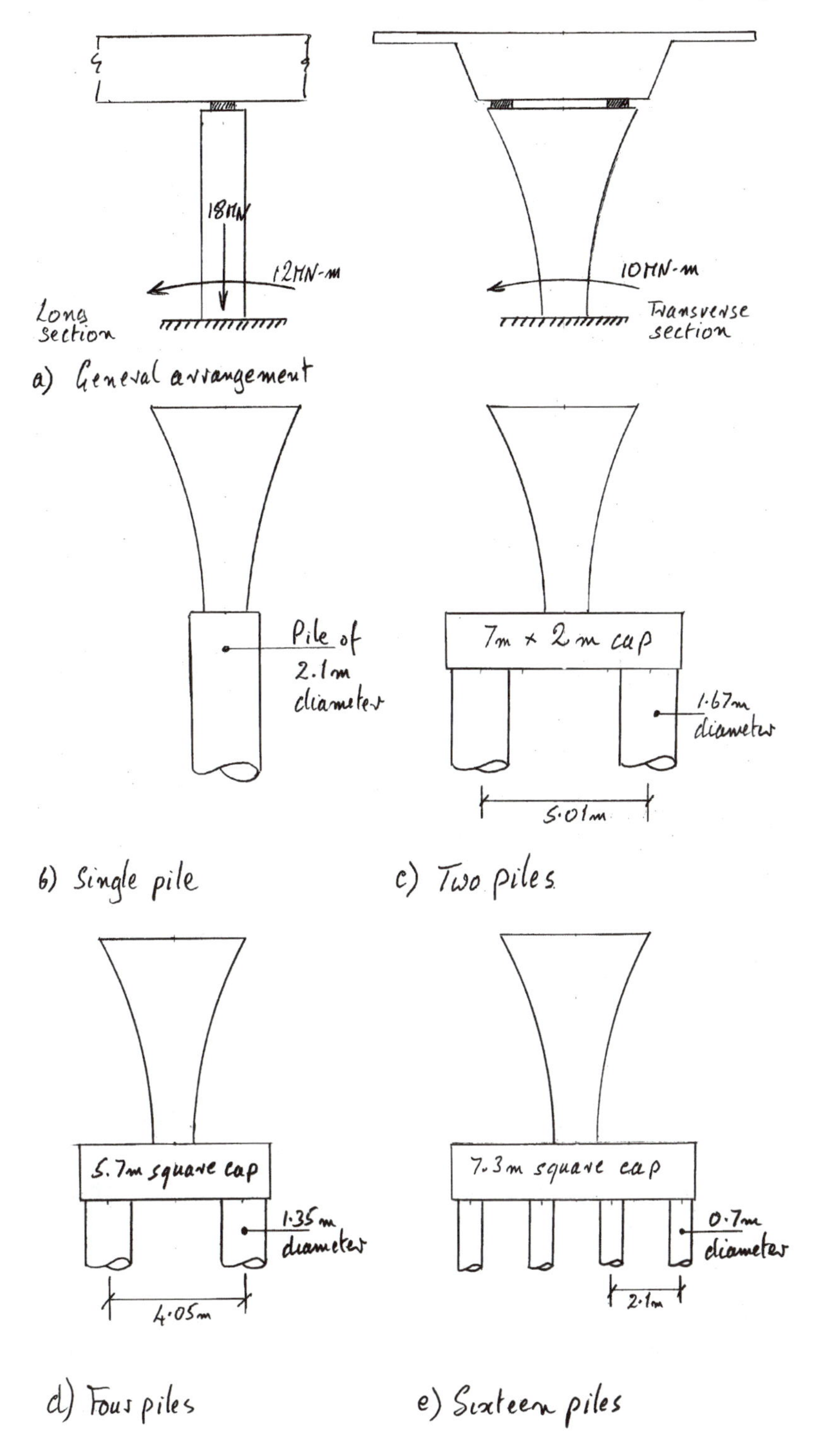

Figure 7.8 Bored pile options

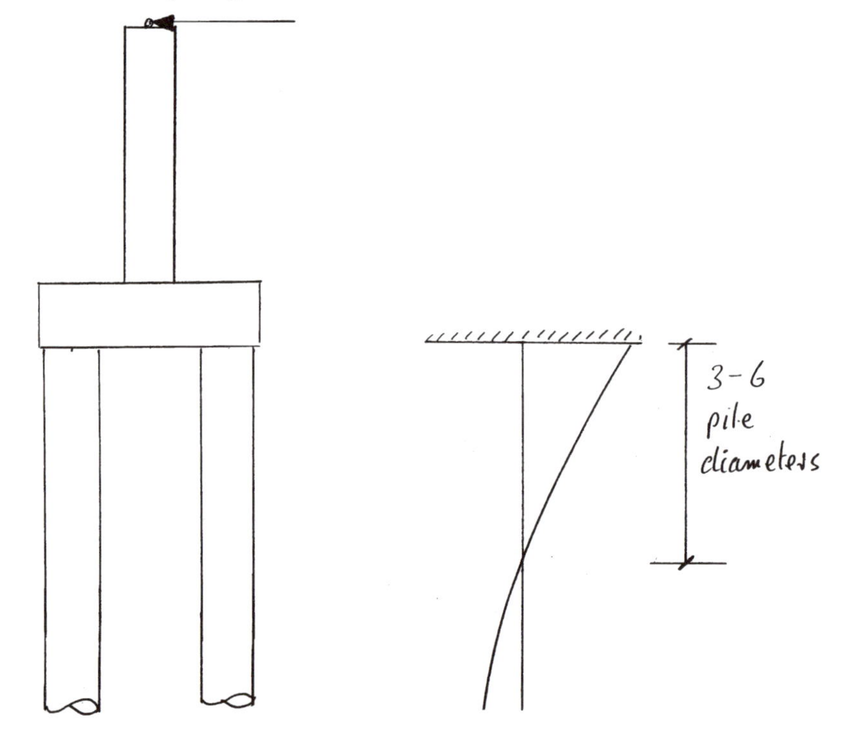

Figure 7.9 Pile bending moment under the effect of a horizontal load

- The pile may not be vertical; a tolerance on verticality of 1/75 is common. Consequently, there will be a kink where the pier joins the pile, giving rise to horizontal forces and additional bending moments.
- The pile may not be precisely centred beneath the pier, giving rise to additional bending moments at their point of junction; a positional tolerance of 75 mm is typically used.
- If the horizontal forces are repeated often, the pile could gradually loosen the ground around it.
- It may be necessary to carry out a preliminary pile test under horizontal forces, to check the deflection and rotation of the pile head at ground level.
- The pile must be sound both structurally and geotechnically, and appropriate testing regimes must be devised to ensure this.
- The pile must be long enough so that it is effectively encastré in the ground, and will not fail by rotating about its base.

7.8 The design of piers

Piers have a relatively simple engineering role, carrying the vertical loads from the deck to the foundation, and resisting the various horizontal loads that can be applied to the deck. However, they are very important visually. The simplicity of their function, which makes them easy to understand by non-engineers, combined with their visual importance, appears to exercise an irresistible fascination for some architects. The number of simple, clean-lined bridge decks that have been ruined by neo-classical

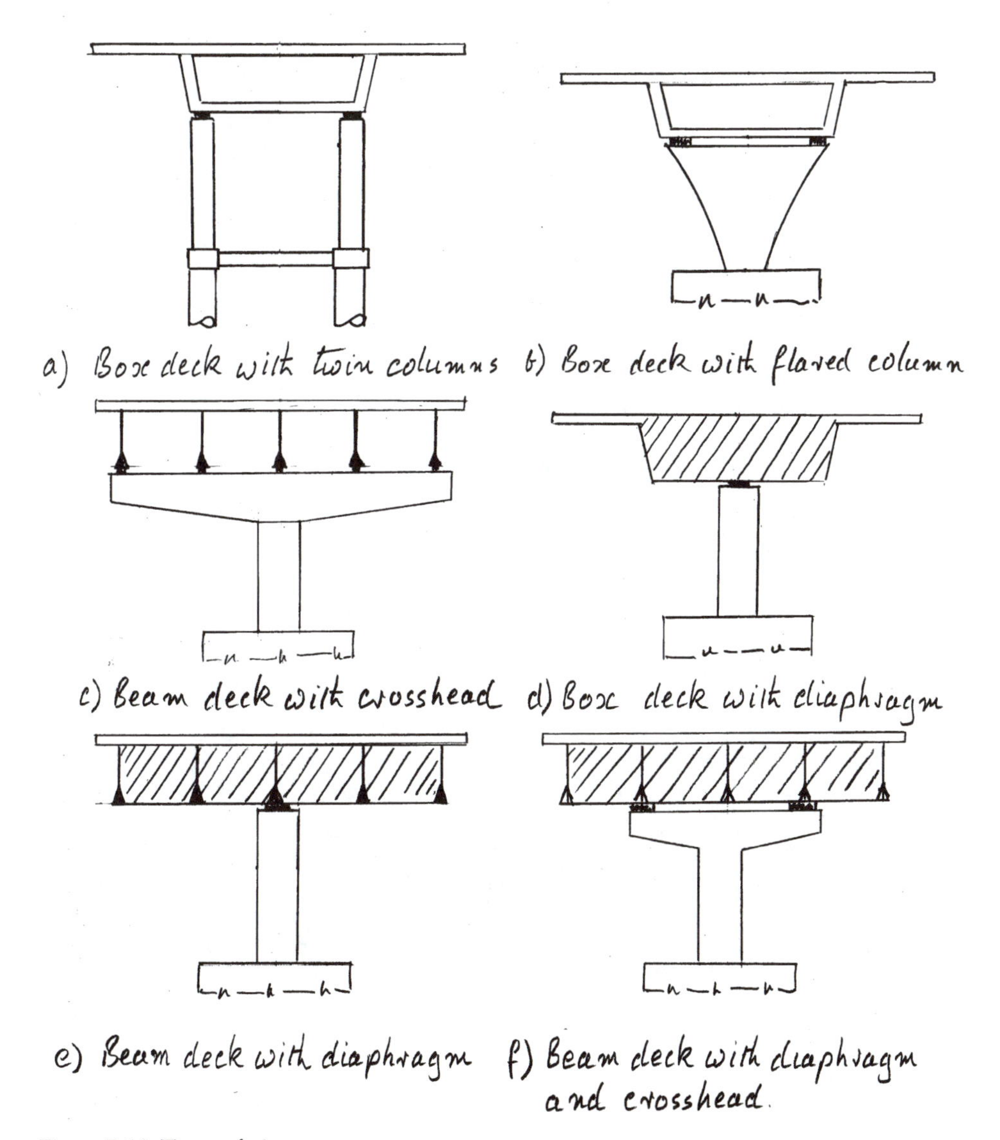

Figure 7.10 Types of pier

columns or other such fantasies is legion. Also, for some reason, there is a fashion for inclining piers, usually for no engineering reason.

Despite their simplicity, there is an engineering and functional logic that should act as a starting point for the aesthetic design of piers.

Concrete is at its most economical when it is in direct compression, needing only the bare minimum of reinforcement. When it is subjected to significant bending, heavier steel reinforcement is needed which increases its cost, and slows construction. Consequently, the simplest and cheapest way to carry loads from the deck to the foundations is to place a vertical column between the bearing and the foundation. If the foundation then consists of a single pile beneath each column, the vertical

Figure 7.11 Pier of STAR Viaduct (Photo: Benaim)

deck reaction is transmitted to the foundation stratum as economically as possible without bending moments, Figure 7.10 (a). Eccentric live load moments, or dead load torque due to plan curvature of the deck, are also carried to the foundations in direct compression.

However, the pier is also subjected to lateral forces applied at the bearing, both longitudinally and transversally. These create bending moments that are zero at the top of the pier, and a maximum at the base. There is thus logic in widening the columns near the base. This is shown clearly on the piers of the East Moors Viaduct, whose articulation is described in *7.15.5* and Figure 7.26.

Piers consisting of one column per bearing are not always possible, or desirable. When it is necessary to carry the deck on a pier consisting of only a single column, the loads must be gathered in from the bearings by a crosshead. At one limit, this crosshead may consist of a gradually flared column, Figure 7.10 (b), while at the other limit it may consist of twin cantilevers, Figure 7.10 (c). Alternatively, the crosshead may be incorporated into the deck as a diaphragm, Figure 7.10 (d) (the arrangement

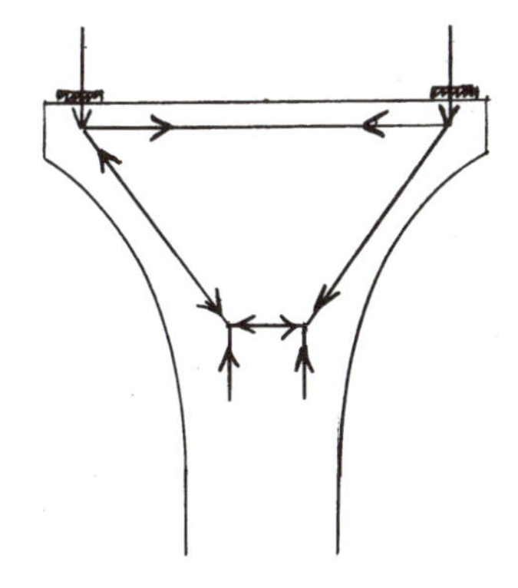

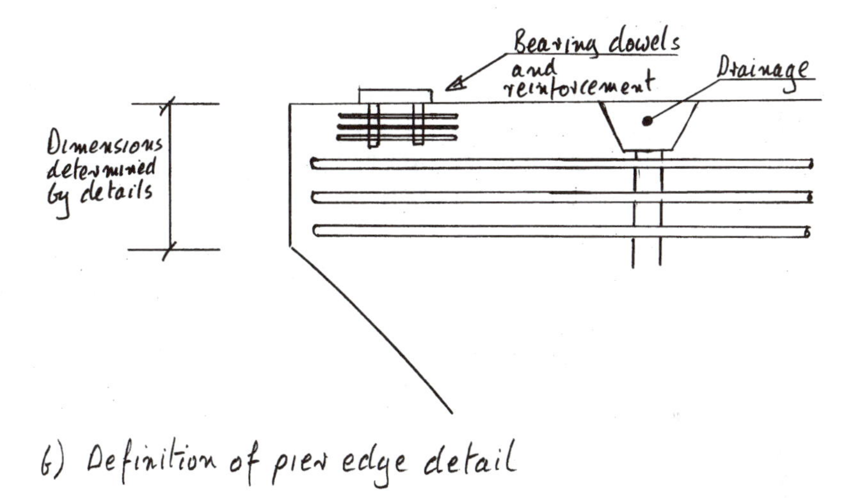

Figure 7.12 Flared pier details

used on the Stansted Abbotts Bypass, shown in Figure 10.4) and Figure 7.10 (e). It is also possible to share the bending moment between a deck diaphragm and a pier crosshead, Figure 7.10 (f). Crossheads, whether they form part of the deck or the pier are costly, and should only be used when strictly necessary.

When different types of bridge deck are compared, it is often forgotten that the deck type may require expensive crossheads, and that this may have a very considerable influence on the cost of the bridge as a whole.

If a box section deck carried by a flared column has a trapezoidal cross section, such as in the STAR project, the bearings will be closer together, reducing the necessary flare. This flare may be further reduced by placing the bearings in-board of the webs, beneath a deck diaphragm, Figure 7.11, when a price will be paid in reinforcement in the diaphragm.

Truss analogy may be used to determine the tension force that must be resisted across the top of a flared pier, Figure 7.12 (a). This tensile force may be resisted by passive reinforcement, or by prestressing. Whichever system is used, considerable care must be taken in the detailing, to ensure that the bearing reactions are effectively enclosed by the reinforcing bars or tendon anchorages, and frequently the pier head must be widened for this reason. Also, the main mat of reinforcement may have to be lowered to miss the bearing dowels, the bursting steel beneath the bearings, or drainage details in the pier, affecting the appearance of the pier, Figure 7.12 (b). Clearly, the tensile force at the top of the pier will be minimised if the flare is made as deep as possible.

When mechanical bearings are used and the deck is pinned to one or more piers, these special piers are subjected to additional forces due to acceleration/braking of traffic, longitudinal wind and to differential bearing friction (unequal coefficients of friction either side of the points of fixity). If the deck is fixed to only one pier, it is likely that this pier will need to be larger than typical piers, which creates a problem both visually and functionally as it will require special formwork. If the deck is fixed on several piers as described in 7.9, it is usually the case that the standard pier may be used, albeit with increased reinforcement.

In a long or complex viaduct, which may include a considerable variation of deck height above ground level, of span lengths, of deck widths and include slip roads, it is important design to create a family of piers that suits all situations with elegance, economy of materials and with logical re-use of formwork.

The finish that should be given to the surface of piers depends on how closely they may be seen by the public. For instance, it is not worth paying for an expensive ribbed finish for piers that are isolated off shore. However, if the piers will be seen close to, it is necessary to avoid large flat concrete surfaces that are inevitably ugly. This may be achieved by breaking up the surface with facets which change the way light reflects and hence the perceived shade of grey, or by applying a decorative finish. Examples are the piers of the Byker Viaduct that were located in a public park, Figure 1.5, and of the STAR urban viaduct, Figure 1.4.

7.9 The articulation of decks with mechanical bearings

7.9.1 General

Mechanical bearings are either pinned, guided or free sliding. One or more pinned bearings hold the deck in place, with guided or free sliding bearings at all other supports. This causes large concentrations of force on the pinned bearings.

Design strategies are aimed at reducing the cost and visual implications of this concentration of fixing force. Some of these strategies are examined in the following sections.

One of the aims of the bridge designer should be to minimise the number of maintenance-intensive mechanical devices, such as bearings and expansion joints that are incorporated in a bridge deck. When they are necessary, the designer should choose them to minimise the risk of malfunction and the cost of maintenance. Pinned bearings are the simplest of the mechanical bearings, and should be used where possible; free sliding bearings are simpler than guided bearings.

7.9.2 Typical arrangement of mechanical bearings

Typically the deck is fixed longitudinally on one abutment, on a single pier or on a series of piers, and guided longitudinally on all other piers and abutments.

On a pier providing longitudinal fixity, there would normally be one pinned bearing, with the other bearings guided transversally, such that they share in carrying the longitudinal loads but allow transverse expansion and contraction. On all other piers or abutments, there is normally one bearing guided longitudinally, which carries the transverse wind and traffic loads, with the remainder being free sliding. However, if a pier consists of a row of transversally flexible columns, it may well be possible to place longitudinally guided bearings on more than one column, distributing the transverse load more evenly on columns and foundations. If the longitudinal movements are large the bearing guides must be exactly parallel to each other.

7.9.3 Longitudinal fixity at the abutment

Many continuous bridges in the UK are designed with the assumption that the deck must be held rigidly in place longitudinally. As piers are more or less flexible, this requires the deck to be pinned to one abutment. However, this is the very worst place to fix a deck, and should be a last resort, for the following reasons:

- When the deck expands or contracts, the sliding friction of all the bearings reacts against the fixed point. This bearing drag amounts to some 4–5 per cent of the dead weight of the deck, which can be a very substantial force.
- For most bridge decks, the vertical reaction on the abutment is much smaller than on intermediate piers, particularly when the penultimate span is loaded. Bearings that have to resist high horizontal loads combined with low vertical loads are non-standard and are consequently expensive, may be slow to deliver, and may require a special testing programme.
- The abutment is usually subjected to horizontal forces from earth pressure; the addition of longitudinal forces from the deck will usually result in a considerably more expensive structure. The increase in cost is likely to be most significant if the abutment is carried by piles, where increasing the ratio of horizontal to vertical forces is particularly onerous.

7.9.4 Longitudinal fixity on a single pier

It is much better where possible to fix the bridge deck on a pier, which has a greater vertical reaction. Figure 7.13 (a) shows an example of a five-span bridge carried on four rows of columns. Bearing drag will be minimised if the fixed pier is situated close to the centre of the viaduct. However, it must be assumed that the bearings on either side of the fixed pier will have different friction coefficients. The differential friction coefficient assumed should be greater as the number of bearings is smaller. Thus in a bridge with only two spans, it should be assumed that the bearing friction coefficient is 1 per cent on one abutment and 5 per cent on the other. When there are, say, ten or more bearings either side of the point of fixity, it would be unreasonable to expect that all those on one side are very different to those on the other, and a differential friction

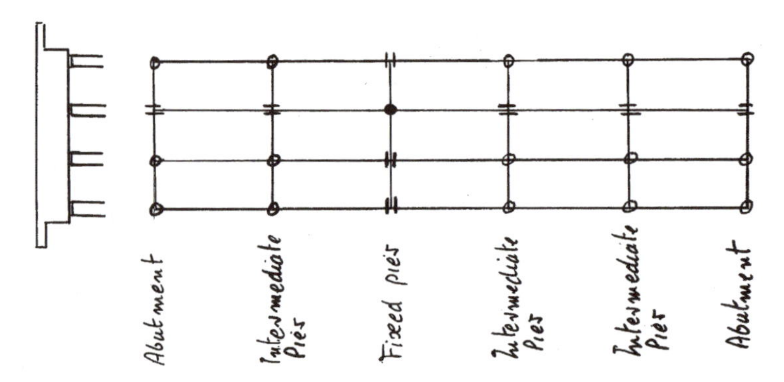

Legend

o Free sliding bearing

• Pinned bearing

= Longitudinally guided bearing

|| Transversally guided bearing

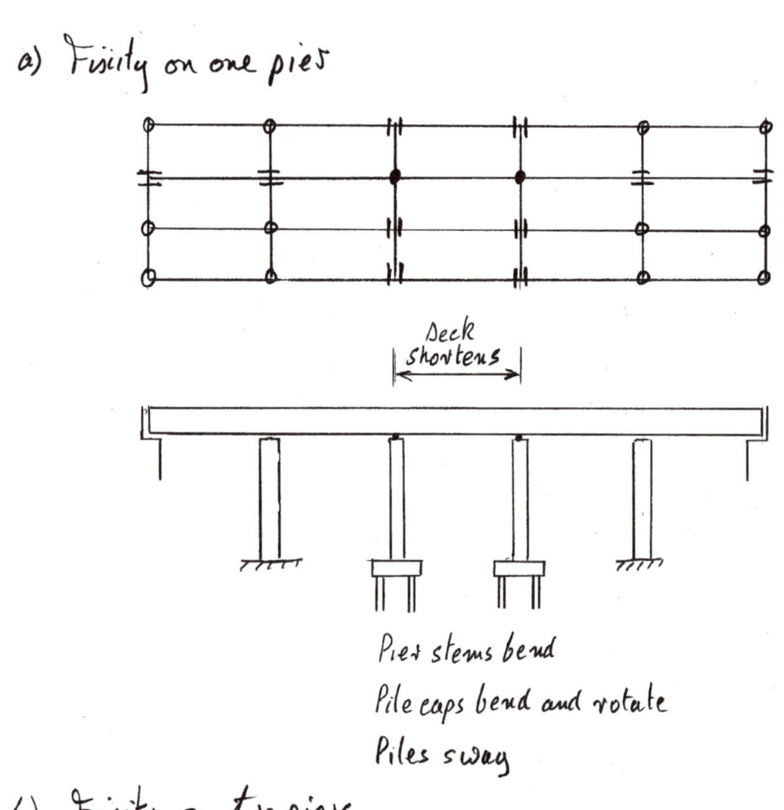

Figure 7.13 Typical articulation with mechanical bearings

coefficient of only 0.5 per cent would be reasonable. Codes of practice may impose other ways of defining differential friction, for instance [5].

Clearly, once the deck is pinned to a pier which is to some extent flexible, it is no longer rigidly held in position. It is necessary to ensure that the movement under longitudinal forces is not excessive, and can be accommodated by the expansion joints. If this movement is too great, it is generally better to stiffen up the anchor pier, or to anchor on a stiffer pier even if it is off centre, rather than to revert to fixing the deck on the abutment.

If the bridge is subjected to seismic forces, it may not be cost effective to design the expansion joint to survive the deck movements caused by a severe earthquake. However, the designer should ensure that vital parts of the deck, such as the prestress anchors in the end face, would not be damaged by collision with the abutment.

External longitudinal forces applied to the deck may, as a first approximation, be assumed to be carried solely by the fixed pier. However, in a long viaduct, the contribution of all the other piers should be considered. As the fixed pier deflects longitudinally the sliding bearings on other piers and abutments are displaced. Using a conservative low friction coefficient of 1 per cent, the sliding bearings will carry a significant proportion of the longitudinal load, relieving the fixed pier of stress and making it more similar to typical piers. This applies also to decks that are pinned to several piers, as described below.

Generally, for a deck of any length, a single fixed pier will need to be thicker than standard piers with sliding bearings.

7.9.5 Longitudinal fixity on two piers

It is frequently possible to fix the deck on two piers near the centre of the bridge, sharing the longitudinal forces, Figure 7.13 (b). If the piers and their foundations are identical, the longitudinal force will be shared equally. However, if the two piers are different, the longitudinal force will be carried in proportion to their stiffness. Generally, if the deck may be fixed on two piers, the longitudinal forces can be carried by increasing their reinforcement but without increasing their size, although their foundations are likely to be larger than those of typical piers.

The effect of the length change of the span between the two fixed bearings must be added to the effects of the external longitudinal forces and differential bearing friction. If the piers were too stocky, the forces engendered by the deck shortening may overcome the benefits of sharing the external longitudinal loads. The bending of the piers will be caused by an initial rapid elastic shortening of the deck as the prestress is applied, by the continued shortening due to creep and shrinkage, and by daily and seasonal temperature movements.

At the SLS it is necessary to check that the deck shortening will not cause excessive cracking of the piers. The flexibility of the piers should be assessed taking into account the bending of the foundation pad or pile cap, the sway of the piles and the rotation of the foundation, as well as the bending of the pier stem. All the calculations of flexibility should use the cracked inertia of concrete sections and due account should be taken of the relaxation of concrete under sustained load, as described in *3.9.2*.

In general, if this calculation is done carefully, most bridge piers will be found to be suitable. It is important to remember that the aim of the design of the anchor piers is to maintain their flexibility so they may accommodate the length changes of the deck, while retaining adequate strength to resist the longitudinal external forces applied to the deck, and stiffness to control its longitudinal displacement. This is an excellent test of the skill of a designer.

As the bending moments in the two anchor piers due to the deck shortening are of opposite sign, at the SLS external longitudinal loads applied to the deck will increase the bending on one of the two anchor piers and relieve it on the other.

At the ultimate limit state it must be remembered that the deck shortening consists of imposed strains, not loads. Consequently if the anchor piers become over-stressed by the deck shortening at the ULS, they would release the bending moments, either by hinging at their base, or by rotation and sway of the foundation. For this to be a safe assumption, the substructure must have a ductile mode of failure. The piers and any piles must be adequately reinforced in shear. Piles must not be allowed to fail in direct compression; either the pile shafts should be stronger than the ground, so over-stressing would cause settlement rather than concrete failure, or the foundation should be designed to withstand the ultimate vertical loads.

Under normal circumstances, foundation systems are adequately ductile. However, if necessary, the ductility may be improved, for instance by designing the base of the pier or the connection of the piles with the pile cap for plastic rotation. If the piers/foundations are ductile, at the ULS both piers may be assumed to resist the external longitudinal forces and the internal forces resulting from deck shortening may be ignored.

7.9.6 Longitudinal fixity on more than two piers

If the bridge deck rests on tall slender piers, it is quite possible to fix the deck on several of them. In fact, with design friction coefficients of the order of 5 per cent, sliding bearings on piers near the null point may, in theory, never slide. Fixed bearings are cheaper than sliding bearings and need less maintenance, so there is every incentive to use them on as many piers as possible.

7.9.7 Null point

When a bridge deck is pinned to a single pier, it provides the fixed point about which the deck changes length. However, when the deck is pinned to two or more piers, the position of the point of zero movement, or null point, needs to be calculated in order to estimate the forces on the piers, the sliding movement at each of the bearings, and the travel of the roadway expansion joints.

For a symmetrical deck with a symmetrical substructure pinned to two identical piers, the null point for expansion and contraction will be mid-way between them, Figure 7.14 (a). On the other hand, if for instance one pier is longer than the other, the null point will be displaced towards the stiffer pier, Figure 7.14 (b). If the substructure constituted an elastic system, as described in *7.10*, the position of the null point could be readily calculated, as the sums of the horizontal bearing forces either side of it must be equal and opposite.

Unfortunately, a substructure that includes sliding bearings is not an elastic system. The exact friction coefficient in each bearing is not known. Furthermore, sliding bearings will not slide until the horizontal force overcomes the static friction. Thus as the deck

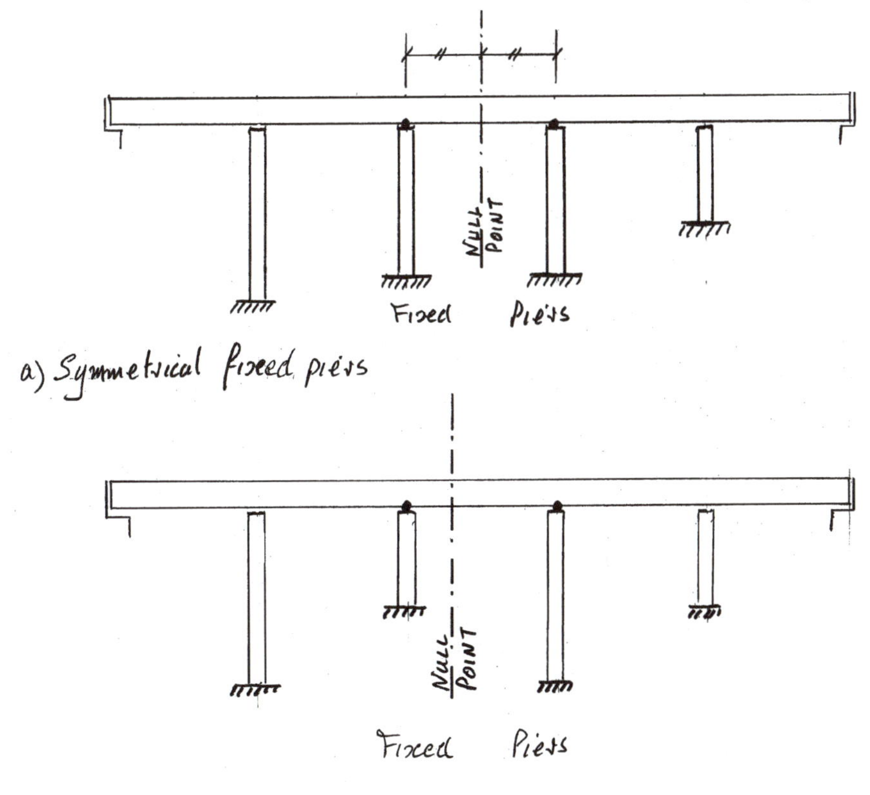

Figure 7.14 Null point

changes in length, a pier will initially deflect elastically. As the deflection increases and the shear force in the bearing builds up, the static friction will be overcome and the bearing will slide, giving elasto-plastic load deflection behaviour.

In calculating the null point, it is usually adequate to assume that the friction coefficient of all bearings is some realistic mean figure, such as 2.5 per cent. The position of the null point is then guessed. By calculating the movement of the bridge deck at each pier position and comparing it with the flexibility of the pier, it is then possible to define which bearings will remain locked, and which will slide. The shear force in the piers with locked bearings can be calculated elastically, while the shear force in piers where the bearings have slid is 2.5 per cent of the dead load reaction. The total force each side of the null point may be calculated. If the position of the null point used in this calculation was correct, the forces either side of it would be equal and opposite. If the forces are not equal, the position of the null point may be corrected, and the calculation repeated until equilibrium is achieved.

However, as the friction coefficient on each bearing is not known, there remains some uncertainty about this calculation, which should be kept in mind when making design decisions based on the position of the null point. In some situations, such as for very long or highly unsymmetrical decks, it may be necessary to reduce the uncertainty by repeating the calculation with different values of friction coefficient, or with some variability of coefficient between bearings. This would give a range of positions for the null point, which would allow a more realistic design of the expansion joints, for instance.

7.9.8 Curved decks

As a curved deck expands, both its length and its radius increase. Thus any point on the deck has a component of movement along the bridge centre line, and a component at right angles to it. Theoretically, all the guided sliding bearings should be aimed at the point of fixity of the deck, Figure 7.15 (a). This is not very convenient, as the orientation of each bearing must be individually set. Moreover the lateral component of the bearing movement that occurs at each pier may have aesthetic consequences, and, at abutments, may compromise the operation of the expansion joints.

For most curved viaducts it is possible to align the bearing slide tracks with the local axis of the deck, Figure 7.15 (b), and to constrain the deck to retain approximately its nominal radius. The lateral forces on the bearings and the lateral bending moments

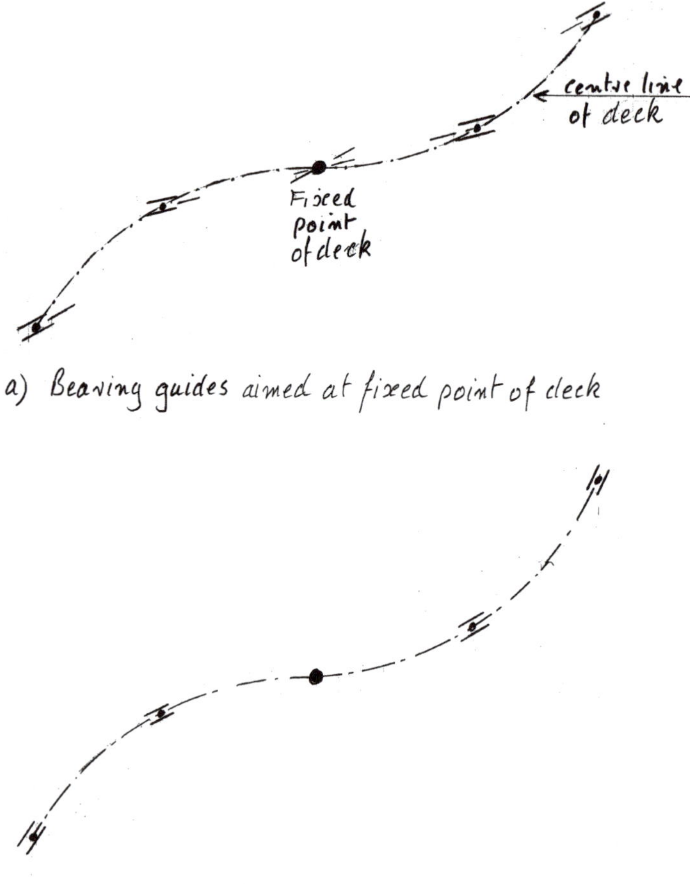

Figure 7.15 Orientation of bearings for curved decks

on the deck caused by constraining it in this way are simply calculated using a space frame, and in the great majority of cases are small compared with the other design actions. In such a calculation, it is important to appreciate accurately the transverse flexibility of the piers, including their foundations, as very small lateral deflections of the piers will greatly reduce the transverse forces calculated on the assumption of rigid supports. In some cases, the lateral forces may be found to be larger than desirable on the abutments, which are usually rigid and cannot deflect sideways. Omitting the bearing guides on the penultimate pier, leaving the bearings as free sliding, gives the deck extra freedom and generally reduces these forces to manageable proportions. Of course, it is then necessary to ensure that the guided bearings on the second pier and on the abutment are designed to carry the increased lateral wind loads.

7.9.9 Temporary fixity during construction

When building a long viaduct span-by-span, it is necessary to ensure that the deck is safely held in place at each stage of the construction. The extreme lower bound friction coefficient of sliding bearings is so low, less than 0.5 per cent, that an unsecured bridge could be displaced by accidental loads or even by exceptional winds.

In general, the completed viaduct would be held in place by a series of pinned bearings near the centre, with all the other bearings being free sliding or longitudinally guided. As construction proceeds from one end, the first spans will be carried only by sliding bearings. Some of these bearings need to be designed with locking bolts that temporarily stop them sliding. Alternatively, the deck may be held in place during construction by temporary locking devices that are independent of the bearings. The load capacity of these temporary fixings will be determined by the differential bearing friction that may be applied at a particular stage of construction. If the temporary fixed point is moved forwards frequently as the deck erection progresses, the forces in the temporary fixing may be kept low.

When construction reaches the first bearing that is pinned in the final configuration, if no precautions were taken, it could be grossly overloaded by all the bearing friction forces being applied from one side only. Consequently, it may be necessary to temporarily free this bearing so that it can slide in response to a temporary point of fixity situated in the completed portion of the deck. Once sufficient spans have been erected beyond the permanent fixed bearing to bring the differential friction loads within its design capacity, it may be permanently locked. A simple example of this procedure is given in Figure 7.16.

It is very important to consider these aspects of temporary fixity and temporary release of bearings sufficiently early in the design. Bearings require long lead times for ordering, and if these considerations are approached only at the last minute, they can lead to expensive modifications to bearings that have already been fabricated.

7.9.10 Use of shock absorbers

High capacity shock absorbers that allow a slow extension of the piston but lock up under rapid movements, can be fixed between a bridge deck and a fixed point such as a pier or an abutment, or between successive sections of a bridge deck in a long viaduct. Thus they allow the bridge to expand and contract freely under the effects of

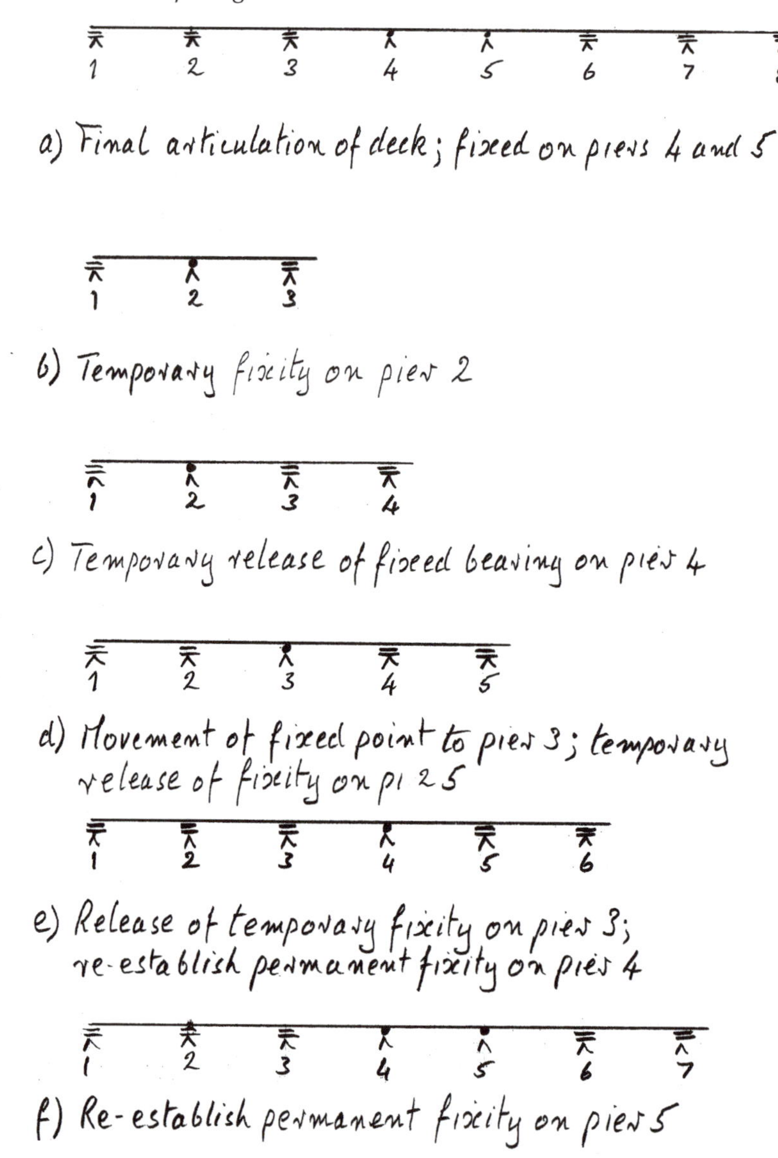

1 2 3 4 5 6 7 8
a) Final articulation of deck; fixed on piers 4 and 5
1 2 3
b) Temporary fixity on pier 2
1 2 3 4
c) Temporary release of fixed bearing on pier 4
1 2 3 4 5
d) Movement of fixed point to pier 3; temporary release of fixity on pi 2 5
1 2 3 4 5 6
e) Release of temporary fixity on pier 3; re-establish permanent fixity on pier 4
1 2 3 4 5 6 7
f) Re-establish permanent fixity on pier 5

Figure 7.16 Temporary fixity of deck built span-by-span

creep, shrinkage and temperature change, but lock up under the effects of acceleration/braking or seismic forces.

An example would be a railway bridge on very tall piers. The deck may well be fixed onto some of the piers, with sliding bearings on the others and on the abutments. Thus effects of differential friction would be shared between several piers, as described above. However, under the effect of a train accelerating on one track and braking on the other, the columns providing deck fixity may be too flexible, allowing excessive longitudinal movements. If shock absorbers were to be attached between the bridge deck and one of the abutments (or both abutments), these sudden forces would be carried rigidly by the abutment, Figure 7.17 (a).

Another example is a very long railway viaduct, split up into expansion lengths. If the piers were flexible, they could be all pinned to the deck. The acceleration and braking forces are applied by trains of finite length. If the whole viaduct is considerably longer than one train length for a single track railway, or two train lengths for twin tracks, such forces may be distributed to all the piers of the viaduct by placing shock absorbers between each of the expansion lengths. Thus for all slow movements the expansion lengths behave as separate entities, while for sudden loads the whole viaduct acts as one, Figure 7.17 (b).

An alternative solution to the above example would be for each expansion length to have at its centre an anchor pier (or piers) that was stiffer than the others, and strong enough to carry the acceleration/braking forces on that expansion length. This would probably have a higher first cost, but would not have the maintenance liability of the locking devices, Figure 7.17 (c).

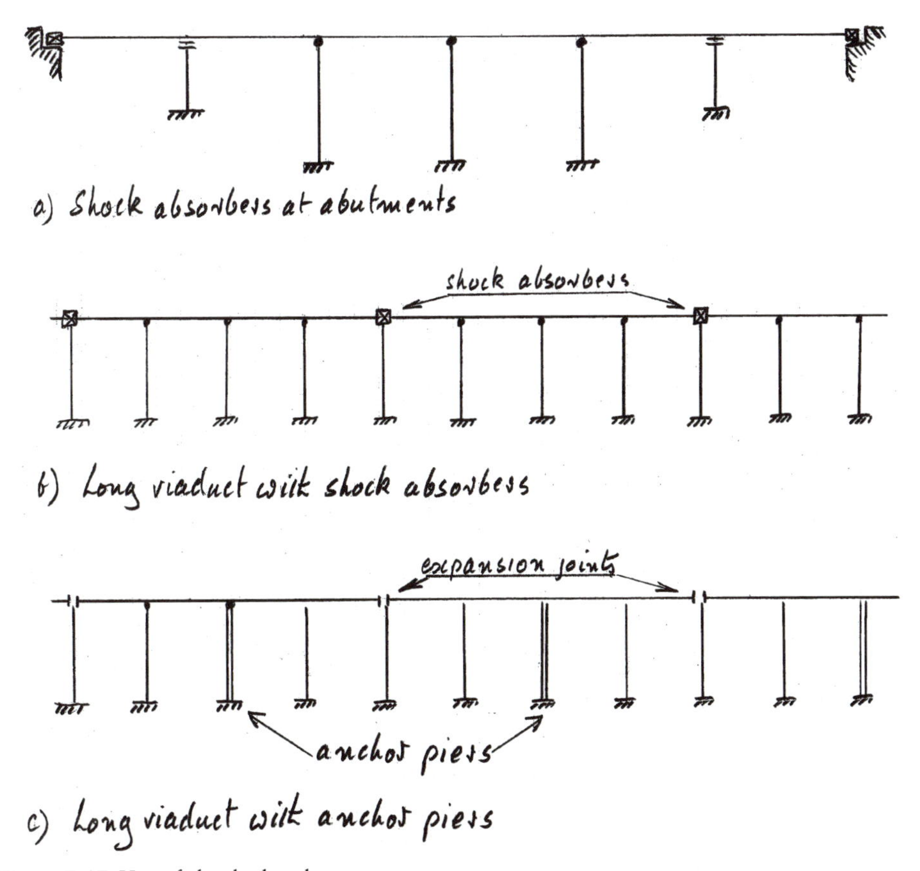

Figure 7.17 Use of shock absorbers

7.10 Deck on laminated rubber bearings

The use of laminated rubber bearings provides the designer with a very useful additional tool for controlling the articulation of his bridge. In particular, the shear stiffness of the bearings can be adjusted so that the distribution of longitudinal forces between the piers is under his control.

The plan area of a bearing is determined by the design load. The capacity of a bearing to accept rotations depends on its length in the direction of the rotation, on the total thickness of rubber and on the thickness of the individual layers; the thicker the layers, the more rotationally flexible is the bearing. The shear stiffness of a bearing depends on its plan area and on the total thickness of rubber.

When designing the articulation for a long viaduct the engineer would normally start by defining bearings that had the minimum thickness of elastomer required to accept the design rotations. The central piers of the viaduct would be equipped with such bearings that maximise shear stiffness. Thus the piers that are the least loaded by length changes of the deck are provided with the stiffest bearings to attract a greater proportion of the external longitudinal applied loads. For piers further from the centre of the viaduct that are more affected by the length changes of the deck, the bearings would be made progressively thicker, the aim being to equalise as far as possible the longitudinal loads carried by each pier, Figure 7.18. In this calculation, the flexibility of the piers must be added to that of the bearing; the pier and bearing constitute a single elastic system. Clearly, the choice of the shear stiffness of the bearings is a cyclical calculation. The abutments, which cannot deflect, must be equipped with thicker bearings, with bearings that can be reset, or with mechanical sliding bearings.

A substructure equipped with rubber bearings is an elastic system, and the null point may be calculated accurately, as can the share of longitudinal load carried by each pier. For a substructure where the piers vary in height and thus in stiffness, the bearing thickness can be tuned to control the distribution of longitudinal force on the piers. For example, if one pier were excessively stiff due to it being founded on high ground, it would tend to pick up a greater share of longitudinal loads. However, the rubber bearing on that pier may be made thicker and hence more flexible in shear, to bring the overall flexibility of the pier into line with the others.

A deck carried on rubber bearings does not have any 'fixed point'. However, it is a very safe, highly redundant system, as the longitudinal stability of the deck does not depend on any one pier.

Laminated rubber bearings need no maintenance, although they should be inspected regularly, and provision must be made to replace them, usually after a life of several decades.

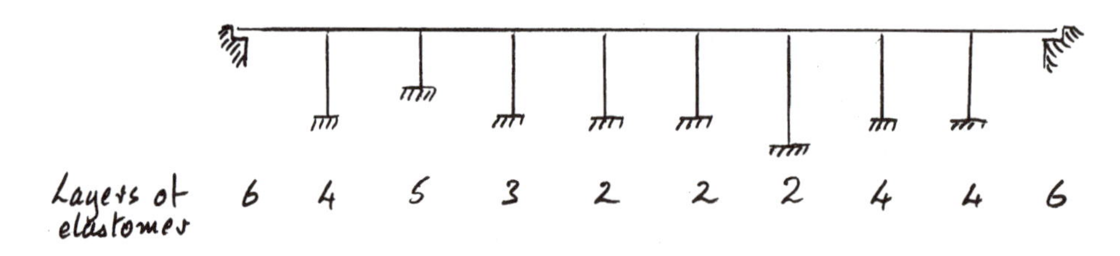

Figure 7.18 Typical example of the use of rubber bearings

Generally it is not necessary to organise temporary fixity for decks carried by rubber bearings during span-by-span construction, as there is always a degree of restraint of the bridge deck. However, the sequence of application of shear deformation to the bearings during construction should be thought through, as some bearings may become temporarily or permanently over-stressed in shear. Rubber bearings can in general accept a substantial degree of temporary over-stressing, in shear, rotation and direct load. However, the manufacturers should be consulted.

A typical example of this form of articulation is given by the GSZ project in China which is described in *7.15.2* below.

7.11 Piers built into the deck

The lowest maintenance solution of all is to eliminate the bearings altogether, and build the pier into the deck. This creates a portal structure, and live loads on the deck generate bending moments and shears in the piers, which are also transmitted to the foundations. The bending moments in the deck are substantially changed, with an increase in hogging moments over the supports, a decrease in sagging moments at mid-span, together with a decrease in the variation of live load bending moments. These changes to deck moments generally significantly reduce the amount of prestress required, particularly if the deck has variable depth, allowing the greater support hogging moments to be carried economically by the increased depth. However, this cost saving in the deck may be at the expense of increased cost in the piers and more particularly in the foundations.

The feasibility of fixing the deck to the piers in this way depends principally on whether the structure can cope economically with the forces generated by the shortening of the bridge deck. If the deck is long, and not very high above the ground, it may only be possible to build in the two central piers, with those further from the centre of the bridge requiring pinned or sliding bearings to accommodate the length changes. If the deck is high above the ground, more piers may be built in. However, for most bridges, particularly those that cross valleys, the end piers are likely to be shorter than those near the centre, which of course makes them less able to accommodate the deck shortening.

Once the building in of the piers is accepted as a design objective, simplifying the bridge mechanically and reducing first cost and maintenance, it may be used far more widely than is currently the case. For instance, the decks of most motorway over-bridges could be built into their piers, increasing at the same time the resistance of piers to vehicle impact.

The limit on this form of construction is usually the degree of cracking of the pier stem under service conditions that is considered acceptable. More research is required on the extent of the relaxation of reinforced concrete piers subjected to permanent shortening due to creep and shrinkage of the deck, and on the extent and harmfulness of the cracking due to these effects when added to the cyclic temperature movements of the deck.

7.12 Split piers

When a pier is built into the deck, its bending stiffness modifies the bending moments in the deck, with economic benefits. On the other hand its shear stiffness is a disadvantage, as it resists the length changes of the deck. If the pier is divided into two flexible leaves, the bending stiffness can be retained, while the shear stiffness is greatly diminished. This may best be understood by considering a pier consisting of an 'I' beam, Figure 7.19. The bending strength of the beam is provided by the two flanges, while the shear strength is provided by the web. If the web is removed, the bending strength remains, but the shear strength is destroyed. Thus the benefits to the deck bending moments are retained, while the length changes of the deck are less restrained, allowing a longer length of the deck to be built-in in this way.

The spacing of the two leaves has a marked influence on the design. Live load on one span will cause uplift on the remote column of the pair. If the leaves are close together, the uplift will overcome the dead load reaction on the remote column and put it into tension. The degree of moment fixity of the deck gradually reduces as the columns are brought closer together. This is due to the shortening of the near column and the extension of the remote column, which allows the node between

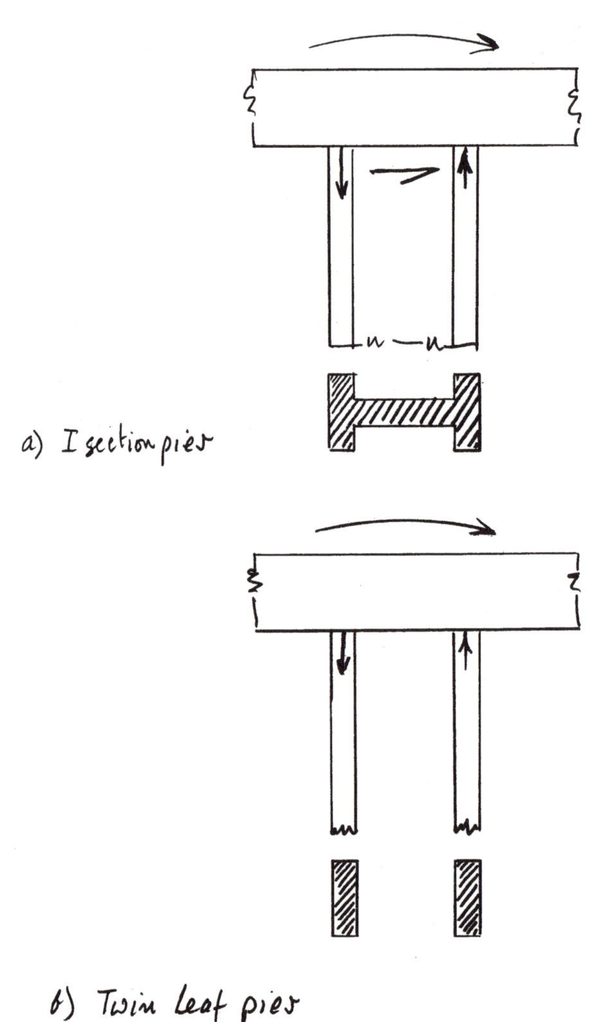

Figure 7.19 Development of twin leaf piers

deck and pier to rotate. Also, as this rotation becomes significant, it introduces further bending moments into each leaf of the columns, which may determine their minimum thickness.

If the leaves are sufficiently far apart, the live load uplift will not overcome the dead load on the remote column, leaving both compressed under all load cases. The columns may then be equipped with pinned bearings rather than being built into the deck, increasing their flexibility, and allowing them to accommodate greater deck movement. The optimum spacing of twin columns lies between 1/25 and 1/13 of the adjacent span. At the closer spacing the columns will be built into the deck, while at the greater they may be pinned.

The slenderness and hence the flexibility of the columns may well be limited by considerations of elastic stability. Although columns that are built in to the deck are inherently stiffer than pinned columns, they are less susceptible to elastic instability,

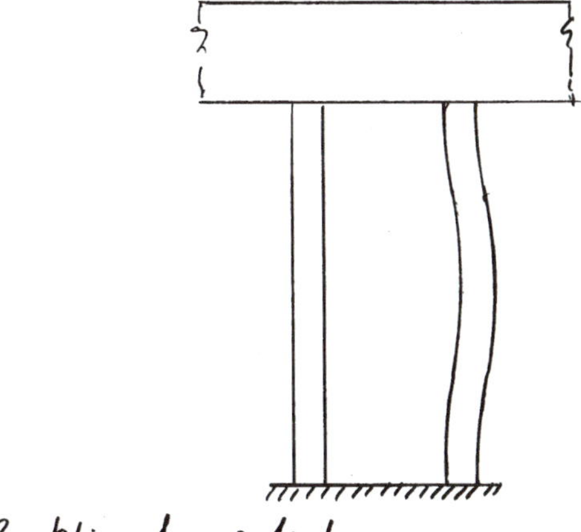

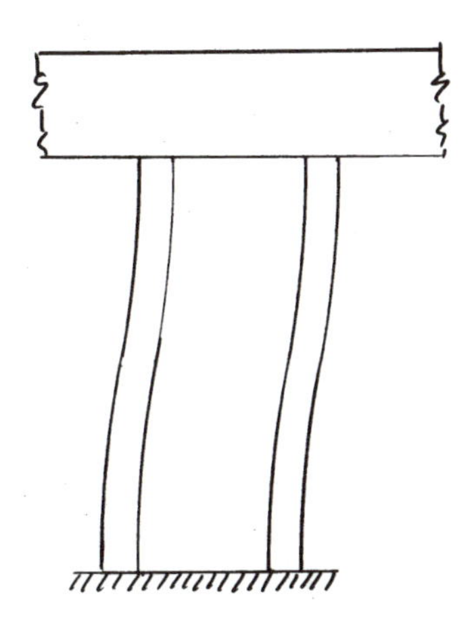

b) Buckling of both leaves by translation of deck

Figure 7.20 Buckling of split piers

and thus may be made thinner, recovering flexibility. The essence of the design of such split piers is to find the correct compromise between the flexibility and the elastic stability of the leaves. In considering stability when the deck is in service, it is relevant that usually only one leaf of a pair is highly compressed under any load case, with the less compressed leaf providing stability by resisting translation of the deck. Also the columns are usually only highly stressed in compression under transient loading, when the higher, short-term Young's modulus is used to check buckling.

As well as considering the buckling of the individual leaves of the split piers, it is also necessary to check the buckling mode where the whole deck translates longitudinally, Figure 7.20. This will be controlled by the shortest piers, and may also be inhibited if the deck is curved in plan.

A good example of a deck supported on split piers is the valley section of the Byker Viaduct, Figure 1.5, described in *7.15.4*.

If the piers are very short, it may be necessary to further subdivide them in order to reduce the shear stiffness enough, while maintaining adequate moment capacity. This is demonstrated on the River Nene Bridge, described in *11.6* and in Figures 11.16 and 11.17. It was necessary to make this bridge deck as slender as possible so that it would fit between an imposed highway alignment and river clearance requirements. Thus a 700 mm thick prestressed concrete voided slab was allowed to span 39.9 m by providing local haunches that were built into the piers. The piers, which were divided into five 200 mm thick precast concrete planks, provided a very high degree of moment fixity to the deck combined with very low shear stiffness.

7.13 Integral bridges

One of the aims of a designer of concrete bridges should be to reduce maintenance by minimising the number and sophistication of mechanical engineering devices required for its operation. Thus, wherever possible, piers should be built into the deck. Where this is not possible, the next best option is to adopt concrete hinges (codes allowing), then rubber bearings, then fixed mechanical bearings, and last of all, sliding bearings.

However, the greatest single item of maintenance expenditure on highway bridges is the expansion joint. Not only do such joints need regular repair and renewal, but they also allow salt-laden water to attack and corrode the substructure. Most attempts at creating waterproof joints fail after a number of years of service. Expansion joints also need to be inspected from beneath, which greatly increases the cost and complexity of abutments.

Integral bridges take the philosophy of mechanical simplicity to its logical conclusion by either pinning the deck to the abutment or building it in. This eliminates the expansion joint and greatly simplifies the abutment structure that becomes more like a pier.

The consequence of this type of design is that the abutment either rocks or slides back and forth as the bridge expands and contracts, causing settlement of the backfill behind the abutment, and disrupting the road surface. This is overcome either by regular maintenance of the road surface, or by bridging the disrupted area with a short transition slab. The slab is attached to the abutment, and so follows its movement, sliding on the substrate. Consequently, a flexible mastic type joint is required in the blacktop at the end of the transition slab. Some authorities adopt transition slabs, while others prefer to maintain the road immediately behind the abutment [6, 7, 8].

A very economical form of abutment is achieved by separating the functions of soil retention and support of the deck. The soil is retained by a reinforced earth wall, while the deck is carried on piles that are allowed to rock within the fill. The piles may be concrete or steel, and may be encased in pipes to give them the freedom to rock. It needs to be demonstrated that steel piles would be adequately protected from corrosion; they would appear to be more at risk than similar piles driven into the embankment.

Integral bridges have been designed in the USA with lengths up to 352 m according to [5]. However, current UK guidelines in BA42/96 suggest a very conservative limit of 60 m. The same standard also suggests limiting the skew to 30°.

Building in piers and abutments and eliminating bearings and expansion joints should be a major aim of the bridge designer.

7.14 Continuity versus statical determinacy

7.14.1 General

The designer of small to medium span bridges often has the choice between decks that are continuous and those that are statically determinate. The choice is not straightforward, and depends on the consideration of many factors, which may be grouped under the headings:

- economy of materials;
- safety;
- method of construction;
- maintenance and use.

7.14.2 Economy of materials

Theoretically, the weight of reinforcement required for a reinforced concrete beam is proportional to the area beneath the bending moment diagram. Figure 7.21 (a) shows a simply supported beam of span l subjected to a uniformly distributed load. The maximum bending moment is 0.125 wl^2, where w is the unit load. The area beneath the diagram is 0.0833 wl^3. Figure 7.21 (b) shows a built-in beam of identical span hinged at mid-span, subject to the same load. Although the maximum bending moment is also 0.125 wl^2, the area beneath the diagram is 0.0417 wl^3, half that of the simply supported beam. Figure 7.21 (c) shows a built-in beam without the mid-span hinge, where the bending moment at the supports is 0.0833 wl^2 and at mid-span is 0.0417 wl^2, and the area beneath the diagram is 0.0278 wl^3, one-third that of the statically determinate span. This simple demonstration shows that hogging moments are cheaper to reinforce than sagging moments, because the bars can be shorter, and that a continuous beam should be much more economical than a statically determinate beam.

However, for bridge decks the theoretical advantage of the continuous beam is eroded in several ways. Live loading may be applied to individual spans of a continuous beam, giving rise to an envelope of moments that has a much larger requirement in reinforcement, Figure 7.21 (d). Furthermore, a continuous bridge deck must be designed for additional bending moments due to differential settlement of foundations,

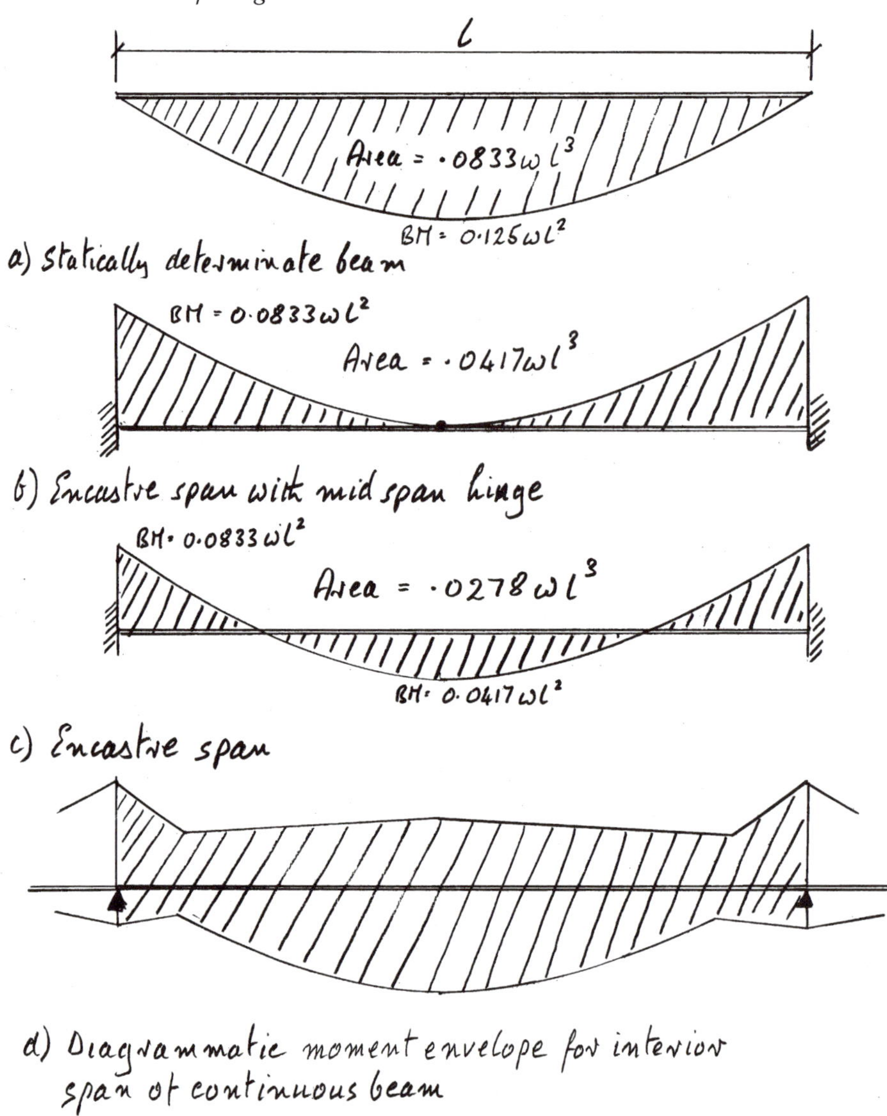

Figure 7.21 Areas beneath bending moment diagrams

and to temperature gradients through the deck, both of which are absent in a simply supported deck. On the other hand, statically determinate beams show some small savings in concrete and simplifications in construction due to the lack of anchorage blisters within the span and to the use of thinner webs near the supports, as the prestress may be effectively tuned to counteract shear force.

In practice, for prestressed concrete decks below about 45 m span, continuity saves about 20 per cent of the prestress. Although this saving is substantial, for these short-span bridges the most appropriate method of construction will generally dictate the form of bridge. For longer spans, where self weight predominates, the savings offered by continuity are greater; it is most unusual to see statically determinate beams with spans exceeding 45 m.

7.14.3 Safety

Indeterminate structures have more reserves of strength than statically determinate structures. For instance, if a mid-span section of a continuous bridge deck lost some of its reinforcement due to corrosion, bending moments may be redistributed to supports, giving warning deflections and delaying collapse. Also, in the event of extreme accidental damage, such as the collision of a train with the supports of an overbridge, a continuous structure may avoid complete collapse and reduce the severity of the accident. On the other hand, if the accident is so severe that it completely removes a pier, a statically determinate deck would limit the damage to the bridge to just two spans, where a continuous bridge is likely to be more extensively damaged.

In severely seismic regions, continuity reduces the risk of decks being shaken off their bearings. However similar benefits can be obtained by linking determinate spans together as described in *7.15.2*.

Continuous decks are more vulnerable to damage due to unforeseen or under-estimated effects, such as settlement of foundations, temperature gradients or heat of hydration effects if cast in-situ.

In general, although continuity does give a greater resilience to a structure, this is less important for a bridge than for a building. Bridges are more highly analysed, are more easily and regularly inspected, and are subjected to a more determinate variety of risks than buildings.

7.14.4 Method of construction

Some forms of construction lend themselves more naturally to determinate or indeterminate decks. Decks formed of precast, prestressed beams are naturally simply supported, and measures taken to make them continuous are time-consuming and generally uneconomical.

In general, very long viaducts with regular spans consisting of precast elements, erected by highly mechanised gantries, where speed of construction is essential for economy, are designed as statically determinate. The time needed to create continuity between decks usually rules out this option. For instance, the first scheme designed by Benaim for the 17 km long GSZ Pearl River Delta Viaducts (*15.9.2*) used prefabricated decks each consisting of a 32.5 m long, 15.4 m wide box girder weighing 620 tons. These decks were due to be placed by gantries at the rate of up to three per day per gantry. For such a rate of construction, creating continuity would have caused unacceptable delays, more than doubling the duration of the erection programme.

On the other hand, when decks are cast in-situ span-by-span, either on gantries or on staging, it is more difficult to make them statically determinate than to make them continuous. Consequently such a form of construction leads more naturally to continuous beams. Similarly, decks built in balanced cantilever, both precast and cast in-situ, are usually made continuous; mid-span hinges and drop in spans are no longer considered economical or desirable. Bridges built using the method of incremental launching are naturally continuous, although expansion joints may be introduced if necessary, at additional cost due to the disruption to the flow of construction.

7.14.5 Maintenance and use

There are many poor existing examples of statically determinate prestressed concrete highway bridges. Precast beams were frequently built without negative pre-cambers, and as a result hog up under permanent loads and give a poor ride, which sometimes excites the resonant frequency of car suspension. Expansion joints at each pier require excessive maintenance and further reduce the quality of the ride, as well as providing paths for aggressive run-off water to attack deck ends and piers. Such bridges have created a prejudice against statically determinate decks.

However, these defects were the result of poor design, and are not inherent in the lack of continuity. For instance, the GSZ viaducts (*7.15.2* and *10.3.8*) consist of 32.5 m span statically determinate precast, post-tensioned Tee beams built with the correct pre-camber, and linked through their top slabs in expansion lengths of nine spans. Thus expansion joints only occur at some 300 m intervals. The ride and the maintenance requirements are as good as on the equivalent continuous deck.

A similar solution was adopted on the many kilometres of the Bangkok Second Stage Expressway, where statically determinate precast segmental box section decks, erected on gantries, were linked through their top slabs in groups of four spans.

Railway viaducts have a different logic. For ballasted track it is advantageous to limit the bridge deck expansion lengths to about 100 m to avoid the need for expensive and maintenance-intensive rail expansion joints. Also, the deck expansion joints required for rail bridges are very simple, and as the decks are not salted, water run-off onto the substructure is not as corrosive as on highway bridges. Thus for ballasted track, statically determinate spans are frequently preferred. However, for city centre locations where deck slenderness is important and where span lengths may be variable, continuous decks ballasted or not, are frequently used, with rail expansion joints every 400 m or less.

If paved track is used, deck end rotations and deck length changes are transmitted directly to the rails, as are any vertical deflections of the bearings or pier crossheads. Short statically determinate spans may still be used, but it is necessary to check that the rails and their fasteners can accommodate these deck end movements, and that the relevant railway authority accepts this solution. Otherwise continuous decks should be adopted.

7.14.6 Conclusion

For spans below about 45 m, there is no overwhelming economic, safety or maintenance case for either continuous or statically determinate decks. By and large, continuous beams are more suited to balanced cantilever erection and to all forms of cast-in-situ construction, while their decks and piers may be more slender and elegant. Statically determinate decks are well suited to long precast concrete viaducts with regular spans, where speed of erection is of the essence, and to ballasted railway viaducts. For spans above about 45 m, statically determinate spans are not competitive. Both continuous and statically determinate decks have their place, and it is up to the designer to weigh the options.

7.15 Examples of bridge articulation

7.15.1 General

The descriptions of the preceding sections of this chapter may best be illustrated by giving some typical examples of the articulation of bridges designed by the author. In accordance with the general theme of this book, these examples are not given as models to be followed, or as examples of 'best practice'. If the author were to design these bridges again he may do them differently. These examples just show how the principles described above work out in practice.

7.15.2 GSZ Pearl River Delta Viaducts

The viaducts extended for 17 km across the Pearl River Delta in South China, and carry the dual four lane carriageways of the Guangzhou–Shenzhen–Zuhai Superhighway on two adjacent decks, each 15.4 m wide, with spans of 32.5 m. Each deck consisted of five precast, post-tensioned beams, joined by a cast-in-situ top slab (*10.3.8*). Groups of nine statically determinate spans were linked by their top slabs into 290 m long expansion units, Figure 10.18.

The ground was soft in the upper layers, with sedimentary rocks at a variable depth below, generally between 10 m and 20 m. The foundations were made with bored piles that penetrated 2.5 diameters into the hard stratum. This rock socket allowed the piles to rely for their working load strength on side friction in the rock, rather than on end bearing. With such a short penetration into rock, the piles were dependent on the support of the ground for their stability.

Each expansion unit was stabilised by two central anchor piers. The anchor piers were founded on a group of four piles of 1.35 m diameter while the six intermediate piers of each expansion unit were founded on pairs of 1.5 m diameter piles, one per column. Each deck beam rested on a laminated rubber bearing 400 mm square. The bearings consisted of 12 mm thick rubber leaves alternating with 3 mm steel plates, with 6 mm of rubber on the top and bottom surfaces. At the anchor piers the bearings had five internal rubber layers, which was the minimum necessary to cope with the deck and pier head rotations. The joint piers, which could not deflect with the length changes of the deck, had up to 11 rubber layers. At the intermediate piers the bearing thickness depended on the distance from the null point and on the flexibility of the pier/foundation. These piers were stabilised against cantilever mode buckling by their elastic connection to the deck. As it was considered that the thicker bearings on the joint piers did not give them adequate stability, they were founded on a group of four 1.35 m piles.

7.15.3 STAR Viaduct in Kuala Lumpur

This viaduct extends for 6 km through the centre of Kuala Lumpur, and carries the twin tracks of the STAR Light Rapid Transit System, Figures 7.22 and 7.11 (*15.5.11*). The rails are carried on concrete plinths. It was built by Taylor Woodrow International to a preliminary and detailed design by Benaim [7]. The continuous precast segmental box girder deck was divided into expansion lengths of approximately 300 m, which is close to the limit for which a simple scarf expansion joint may be adopted for the

Figure 7.22 STAR Viaduct (Photo: Benaim)

rails. The bridge had a sinuous alignment, with horizontal curves with radii as tight as 130 m. The spans of the deck varied considerably due to the constraints of the city centre location. Typically, interior spans were 35.1 m or 51.3 m. End spans were 22.95 m when associated with shorter interior spans and 33.75 m adjacent to the longer spans. The bridge deck also carried the stations, when the spans were normally 40.5 m. For the 35.1 m spans, the deck had a constant depth of 2.2 m, while for the 51.3 m spans it was haunched to 2.7 m at supports. For the spans carrying the stations, the depth was constant at 2.7 m.

The soil consisted generally of 15–25 m of soft materials overlying marl type soft rocks for the greater part of the length. On some sections the bedrock consisted of heavily eroded limestone. Each pier for the typical 35.1 m spans was founded on a single bored pile of 1.8 m diameter, generally over 30 m long, founded in a rock socket, Figure 7.23. The piles were designed with a factor of safety of 1.5 on friction alone, disregarding end bearing, and with a factor of safety of 2 on ultimate shaft friction and 3 on ultimate end bearing.

As the tightly curved railway alignment applied centrifugal forces to the foundations, as well as moments due to transverse eccentricity of load when only one track was loaded, full-scale trial piles were subjected to horizontal load tests, using a second reaction pile. The load was cycled to check that the horizontal stiffness of the ground did not deteriorate unacceptably with repeated loading.

One of the principal advantages of using a single-pile foundation for this town centre location was that no pile cap was necessary, only a junction section hardly larger than the pile itself, minimising the displacement of services and traffic disruption. However, in any context it is an economical option as long as the ground conditions are suitable.

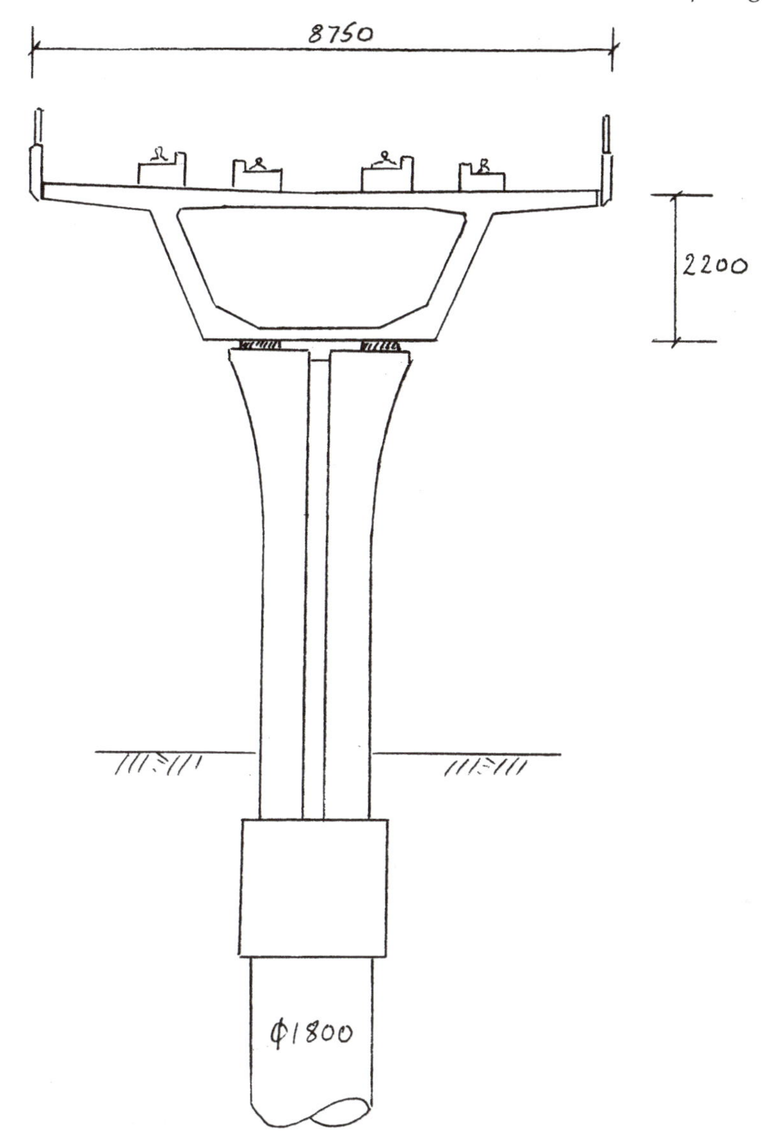

Figure 7.23 STAR Viaduct: typical cross section

Under typical 52.5 m spans, the foundation consisted of two 1.5 m diameter piles, arranged laterally and inclined to reduce the bending moments due to centrifugal forces. This arrangement of piles left the pier flexible in the longitudinal direction. The spans carrying the stations were founded on pile caps supported by four piles.

Each expansion length was anchored on a group of central piers equipped with fixed pot bearings. Joint piers and intermediate piers were equipped with one guided and one free-sliding pot bearing. The number of fixed piers was maximised, taking advantage of the flexibility of the single-pile foundation. As the two bearings on each fixed pier were less than 3 m apart and as the deck was made of precast segments, the transverse length changes between bearings were insignificant, and two pinned bearings were used on each pier.

7.15.4 The Byker Viaduct, Valley section

The Byker Viaduct was designed by the author when at Arup, with the architectural assistance of Humphrey Wood of Renton Howard Wood Levine. This section of the viaduct was approximately 380 m long and was aligned on a 390 m radius curve, Figure 1.5. It carried the twin railway tracks of the Tyne & Wear Metro on paved track and consisted of three main spans of 69 m, flanked by side spans of 52.8 m, and at one end two short approach spans. The main spans crossed a deep valley, with the tallest piers being some 30 m high to deck level [8].

The four main piers which carried the three longest spans and their side spans consisted of two leaves built into the deck and to the pile caps, Figure 7.24. Each leaf was 'I' shaped with 1.1 m thick ends and a 400 mm thick web. The two leaves are spaced at 2.9 m centres, Figure 7.25. The end thickenings flare outwards near the base of the columns to increase stability under the centrifugal forces, and to allow the web to be cut away. These piers are stiff for moments and flexible for shear, absorbing the length changes of the deck. They also provide a stable base for free cantilever erection, obviating the need for temporary supports. The buckling stability of the longest piers was enhanced by their attachment to shorter piers, and by the plan curvature of the deck.

Such slender columns are particularly vulnerable to buckling during construction, as it is possible for the incomplete deck to sway longitudinally. The stiffness of

Figure 7.24 Byker Viaduct: valley piers (Photo: Robert Benaim)

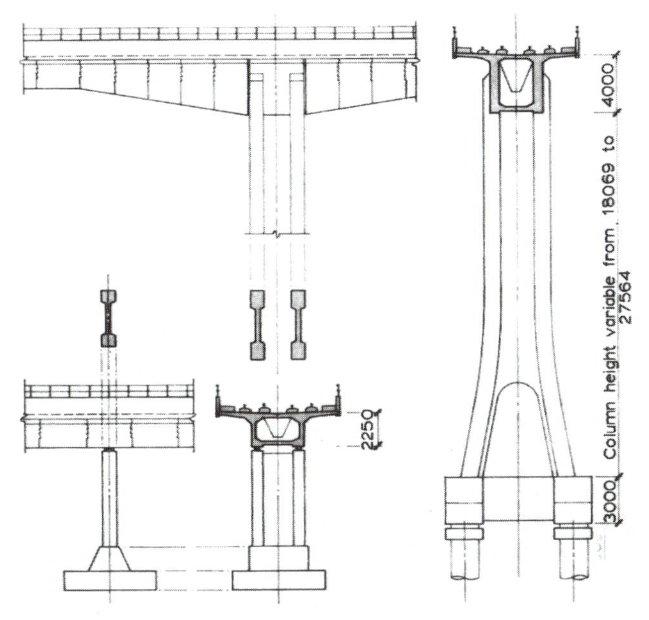

Figure 7.25 Byker Viaduct (Based on drawing by Arup)

the columns was enhanced during construction by stressing a temporary concrete diaphragm between them. This was sufficiently long (vertically) to provide rotational fixity to the columns at their mid-height, halving their effective length, Figure 15.31. The abutment and the short piers carrying the end spans were equipped with sliding bearings.

Each main pier is founded on four bored piles of 2.1 m diameter. As the foundations are in a coal-mining area which may be subject to settlement, the piles were not built into the pile caps, and were provided with the facility to be re-levelled if necessary.

7.15.5 East Moors Viaduct

The viaduct was 915 m long with expansion joints only at the ends. The prestressed concrete single-cell box section deck was on an 'S' shaped alignment, and had typical spans of 52.5 m. At each pier the deck was carried by two pot bearings that each rested directly on a reinforced concrete column, Figure 7.26. The deck was fixed onto the two central piers, each equipped with one pinned and one transversally guided bearing. These fixed piers were the same size as the other piers, but were more highly reinforced. All other piers had a guided sliding bearing on one column, with the slides orientated along the deck axis, and a free sliding bearing on the other column. The two columns of each pier were linked by a cross-beam that shared the lateral load between them, and flared transversally at their base to resist the lateral and longitudinal moments. Each pier was carried by four 1.35 m diameter piles beneath a pile cap. The abutments were carried by one row of vertical and one row of raking 1.35 m piles.

Figure 7.26 East Moors Viaduct (Photo: Benaim)

7.15.6 Project for the Dong Po South Bridge

The bridge was designed to cross a major affluent of the Pearl River in Guangdong Province, South China, Figures 9.32 to 9.36 (*9.5.7*), and references [9] and [10] to Chapter 18. It comprised two adjacent double deck bridges with trussed concrete webs, each accommodating four motorway lanes on its top deck, and three lanes inside the box structure. The three main spans were each 160 m, with end spans of 110 m. The shipping clearance beneath the bridge was 20 m. Each intermediate pier consisted of two rows of two columns built into the deck, and spaced at 12 m along the bridge axis. These twin piers would provide a stable base for balanced cantilever erection and provide moment fixity to the deck, while the columns were sufficiently tall and flexible to accept the length changes of the deck. At each end, the deck rested on a single flexible column, also built into the deck. Thus the 700 m long bridge had no bearings. As the bridge was prolonged at each end by approach viaducts, there were no abutments. Roadway expansion joints were provided at each end of the bridge, where it connected to the approach viaducts.

7.15.7 Project for the Ah Kai Sha Bridge

This double deck, cable-stayed bridge was designed to cross the Pearl River in South China with a shipping clearance of 34 m. It was 42 m wide and carried 16 lanes of traffic on two levels. It had a main span of 360 m and side spans of 172 m, Figures 18.18 to 18.21 (*18.4.11*) and references [9] and [10] to Chapter 18.

The two 114 m high reinforced concrete towers of each pylon were split longitudinally into two leaves that were built into the deck. Each leaf was dumb-bell shaped, and was

2.5 m thick and 7 m wide at the base. The leaves were spaced at 7.5 m centres at the base, reducing to 5 m at a point beneath the cable stay anchorages. Above this point the leaves were joined together to form a single box section. This arrangement gave the pylons the strength necessary to resist lateral and longitudinal forces on the deck while maintaining the flexibility necessary to accommodate the length changes. It also provided a stable base for the erection of the deck in balanced cantilever. The end supports consisted of flexible concrete columns, built into the deck. Thus, the bridge had no bearings, minimising maintenance.

8
The general principles of concrete deck design

8.1 General

A prestressed concrete bridge deck consists principally of a top slab, webs and a bottom flange. Each of these elements can be rationally sized, with reasons for the choice of each dimension. The dimensions may be the result of analysis, or governed by considerations of buildability; they should never be the result of preconceived ideas.

For bridges other than those with through type decks, the width of the top flange is generally defined by the road or railway carried. Consequently, for most bridges with spans less than about 120 m, the top flange compressive stresses due to longitudinal bending are never critical, and top slab thickness is governed by transverse bending requirements, by considerations of longitudinal shear, by the cover required to protect reinforcing steel from corrosion and by the thickness needed to house longitudinal prestressing tendons.

The bottom flange area must be adequate to provide ULS hogging moment capacity at supports, to limit bottom fibre SLS compressive stresses due to longitudinal bending at supports, and to limit the range of bottom fibre stresses at mid-span. The bottom flange area required is critically dependent on the span/depth ratio of the deck. As the depth selected increases for a given span, the bottom flange area required decreases very rapidly.

The web area required is defined by shear and torsion forces. These forces reach maxima at supports, and logically webs should vary in thickness along the span. The web area required is not dependent on span/depth ratio.

Hence bridge decks follow the notional pattern shown in Figure 8.1, with increasing depth for a given span.

8.2 Transverse bending

The action of the top slab of a bridge deck spanning transversally between webs can be considered largely in isolation from its role as a longitudinal compression member. (This is not the case of course for a top slab spanning longitudinally between cross-beams.) Thus it can be dimensioned as a conventional reinforced concrete plate, subject to distributed and point loads, either spanning transversally between supports or cantilevering beyond the last support.

One finds that cantilever slabs work, under UK loading, at a span/depth ratio at the root of about 1/7 to 1/9 (depending principally on the width of the footpath),

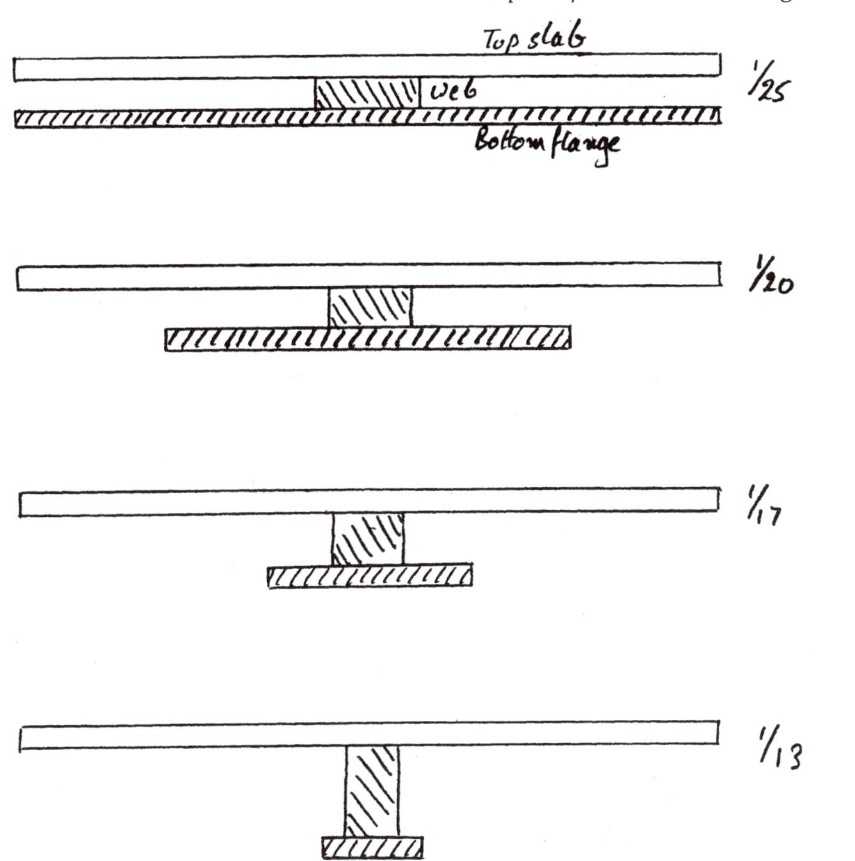

Figure 8.1 Variation of bottom flange width with span/depth ratio

while simply supported slabs are comfortable at about 1/15 to 1/20, and continuous, haunched slabs at about 1/25 to 1/30.

As deck slabs vary generally between 200 mm to 300 mm thick, they need support at between 3 m and 9 m centres, depending on the conditions of end fixity.

One is now approaching deck dimensioning from two different angles:

- material required for longitudinal bending;
- support required for the top slab.

The schematic diagrams of Figure 8.1 can now be modified, splitting up the web into discrete points of support to the slab, and logically these take a piece of bottom flange with each of them, Figure 8.2.

Web concrete is expensive as two planes of formwork and two planes of reinforcement are required, and the concrete can be labour intensive to place and compact, particularly if the web is thin and houses prestressing ducts. Also, near mid-span, where weight is most expensive to carry, the web is generally lightly stressed in shear, and its thickness is usually the minimum buildable. Consequently the designer should use as few webs as possible which will then be thicker and easier to build in

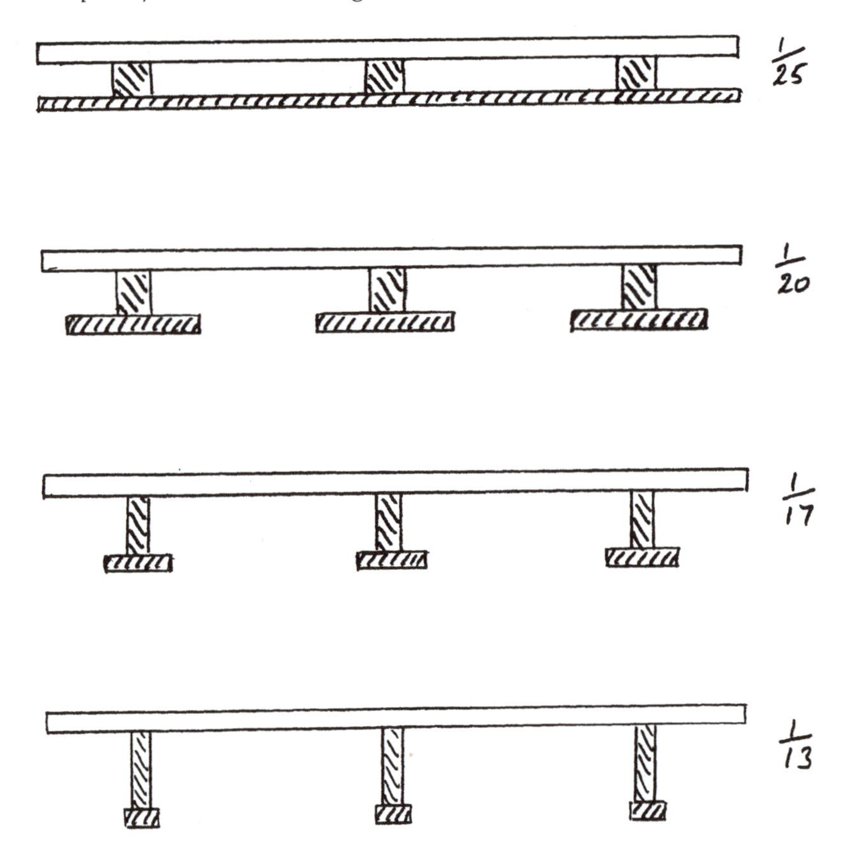

Figure 8.2 Distribution of longitudinal stiffness to provide support for top slab

areas of high shear, and will minimise redundant concrete near mid-span. Clearly this concept will place more demands on the transverse spanning action of the top slab.

Traditional bridges had their longitudinal bending strength distributed evenly across the width of the deck, with closely spaced beams. This arrangement lead to much redundant concrete in the webs and bottom flanges and virtually eliminated the transverse bending role of the top slab. Many pseudo-slabs made of adjacent precast, prestressed beams are still conceived in this way. Rationally designed decks minimise the number of webs and the bottom flange area, and use the transverse strength of the slab to distribute concentrated live loads. A comparison between the two approaches can be seen in the example of Figure 8.3 which shows the alternative design proposed by the author, and eventually built, for a bridge crossing the Sungai Kelantan in Malaysia.

8.3 Transverse distribution of live loads

It is essential for the bridge designer to understand how bridge decks cope with the transverse distribution of concentrated live loads. Where the specified highway loading includes a single indivisible load, for instance the HB load in the UK, this distribution has a major effect on the sizing and the economy of decks. Under such concentrated

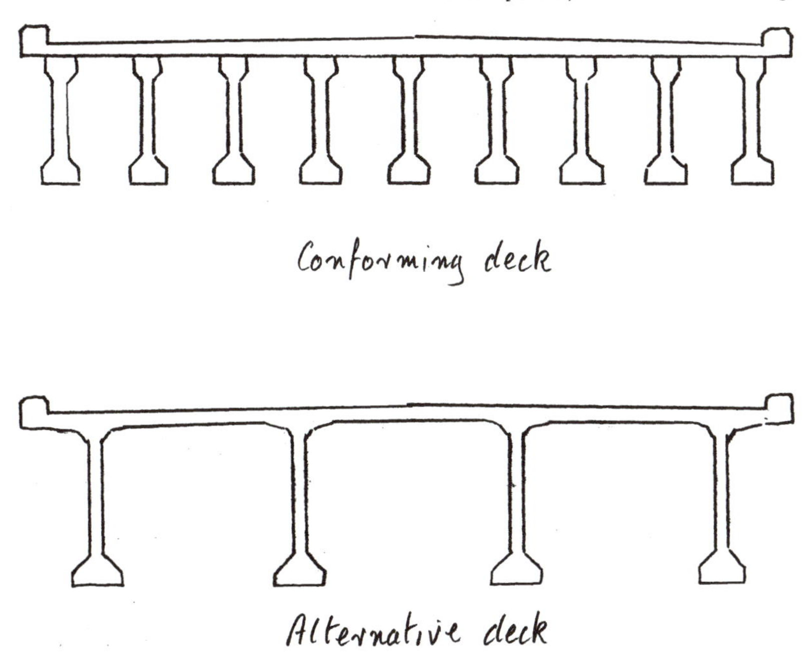

Figure 8.3 Alternative deck for the Sungai Kelantan Bridge

loading the effective total live load bending moment and shear force carried by the deck is governed by the capacity of the deck to re-centre eccentric loads.

For instance, consider initially a deck consisting of two torsionless beams carrying a top slab. As a point load is placed on each beam in turn, it deflects and rotates, and no transverse distribution of the load takes place. As each beam must be designed to carry 100 per cent of the load, it is carried twice. For a span L and a load W the total moment carried by the deck is $2\times PL/4$. Joining the beams with diaphragms makes no difference to this result. Figure 8.4 shows such a deck that was designed as part of a jetty, carrying heavy dump trucks.

This conclusion may be extended to a multi-beam deck such as the notional torsionless three-beam system with a hinged slab shown in Figure 8.5 (a); no load is distributed from the edge beam to the centre beam, as illustrated by the deflection diagram. As the live load may be placed over each beam in turn, it is effectively carried three times. Therefore, the logic of dimensioning beam type decks is to minimise the number of beams.

If the deck is given a finite transverse bending stiffness, loads will be distributed transversally to some degree. This may be achieved by using diaphragms, Figure 8.5 (b), or more simply by giving the deck slab an adequate transverse bending stiffness, Figure 8.5 (c). Analysis and experience have shown that intermediate diaphragms are costly and unnecessary (see 9.6), as an appropriately dimensioned slab provides adequate transverse strength. Real, virtually torsionless systems without intermediate diaphragms are described in Chapter 10, and typical examples are shown in Figure 10.14 and Figure 10.17. A concentrated load on one beam will now be shared to a small degree by adjacent beams, due to the transverse bending strength of the slab. A consideration of the deflected shapes shown in Figure 8.5 helps to clarify this behaviour.

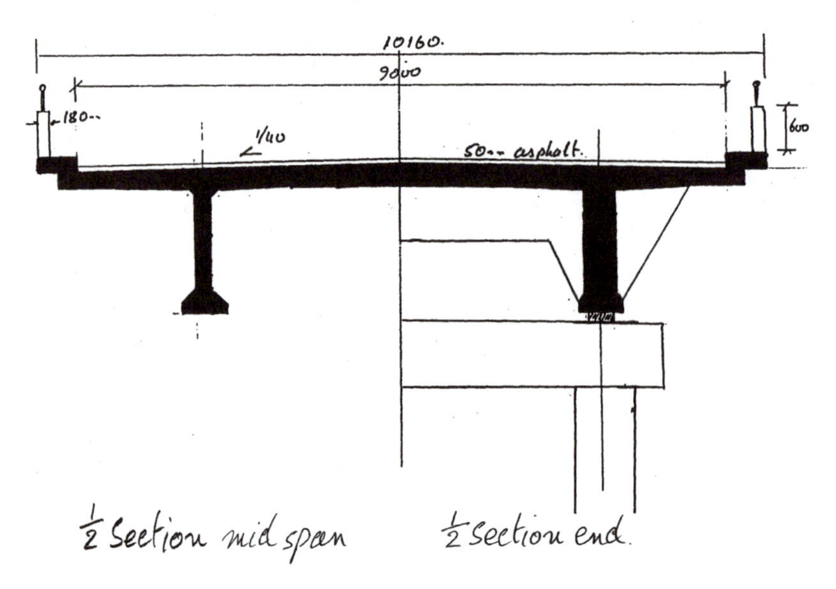

Figure 8.4 Two-beam torsionless deck

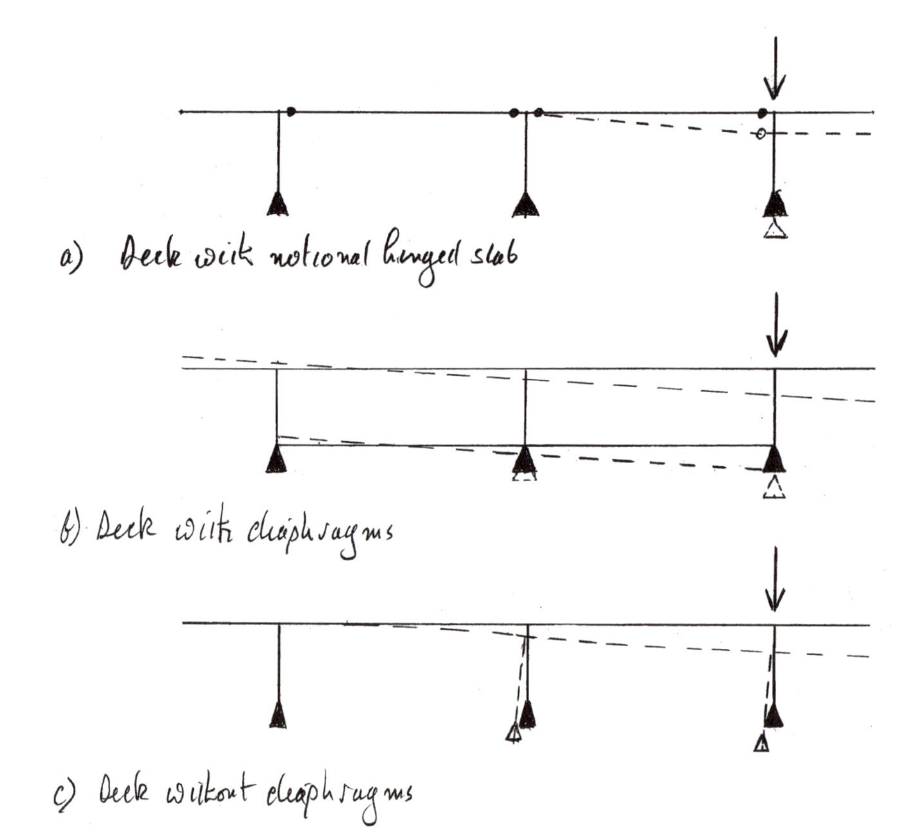

Figure 8.5 Transverse distribution of loads on beam type decks

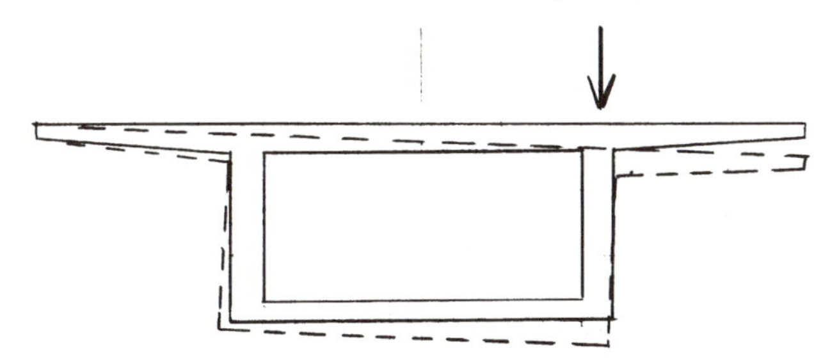

Figure 8.6 Single-cell box girder subjected to an eccentric point load

The capacity of the deck to re-centre loads is improved if the beams have significant torsional stiffness. The beams resist being twisted by local loading, and their relative deflections generate bending moments and vertical shears in the slab, transferring load to the adjacent beam. This type of deck, known as a ribbed slab, may be designed without any diaphragms, except at the abutments where support for the end of the deck slab is essential. When the slab between the ribs is subjected to a concentrated load, it becomes effectively built in, rather than supported on knife edges, as is the case for torsionless beams. It can thus span further, allowing the number of ribs to be reduced. Most typical carriageways can be carried by two such torsionally stiff ribs, giving rise to the name 'twin rib decks'. This type of deck is described in Chapter 12, and typical examples are shown in Figures 12.1 and 12.12. A very useful concept for assessing the ability of a deck to re-centre loads is the transverse influence line for load carried by each beam. For these twin rib type decks, this is shown in Figure 12.6.

The other extreme from the torsionless deck is a deck with a single central torsionally stiff member such as a single-cell box girder, Figure 8.6. The concentrated live load is only carried once, but at the expense of torsionally induced shear stresses in the webs and slabs, which may require thicker members and more reinforcement. This latter concept uses less material for longitudinal bending. This does not necessarily mean that it is more economical as economy depends also on the quantity of materials required for torsion and transverse bending and on the unit costs of materials, which is itself dependent on the methods of construction.

8.4 Material quantities and costs

8.4.1 General

There are three principal components to the cost of a bridge deck:

- the quantity of materials
- the unit cost of the materials
- the cost of the erection method.

These three components are all within the control of the designer, although consultation with the contractor, when this is contractually possible, should be encouraged. Very approximately, the materials and the erection method each account for half the cost of the deck.

The powerful logic behind the concept of prestressed concrete decks described above, combined with the sizing rules for bridge deck elements described in Chapter 9, give rise to predictable quantities for correctly designed decks. The author has collected statistics over a considerable period, giving rise to the benchmarks described hereunder. They are particularly relevant to bridge decks designed to the UK code of practice, where the live loading consists of 45 units of HB associated with HA. Most other codes of practice adopt lighter live loads, and consequently bridges designed to them will require less material quantities.

The benchmarks do not represent the minimum possible quantities of materials. If concrete sections are made too thin, leading to more difficult casting and higher reinforcement densities, or as prestressing cables are subdivided to save steel, the unit cost of the principal materials will rise, defeating the object of the exercise. The benchmarks proposed have been found to be compatible with 'construction friendly' schemes. They are useful in making a preliminary estimate of the cost of an option and in assessing whether a design has been competently done. The author has seen many projects where the quantities proposed for a bridge deck have been up to twice those that are necessary.

Several authors have published charts giving the relationship between the quantities of the principal materials and span [1, 2, 3], some of which are particularly relevant to different national codes.

8.4.2 Concrete

Inexperienced designers tend to consider concrete decks as inherently heavy, and to adopt thick sections. Nevertheless, self weight is the dominant bending moment in concrete decks and controls the weight of prestressing steel required. Furthermore, if the cross-section area of a deck is greater than necessary extra prestress is required

Figure 8.7 East Moors Viaduct box section (Photo: Robert Benaim)

to compress it. Concrete self weight also tends to dominate the foundation loads. Consequently, it is very important to refine the design of the deck, and indeed of the substructure, to remove excess weight. In the East Moors Viaduct, Figure 8.7, where a prestressed concrete deck replaced a steel composite design, refining the design of the deck, removing the crosshead that carried the steel beams and rationalising the pile cap resulted in a foundation load that was no greater than that of the steel deck.

The author has found a less clear relationship between concrete quantity and span than the other references given, above all at small to medium spans. The thickness of the top slab, which accounts for about half the total concrete in the deck for short spans, and a third for spans greater than 100 m, depends principally on the web spacing and on the intensity of the live loading, and hence is not related to the span. For short and medium spans the bottom flange area depends principally on span/depth ratio, while for spans above about 60 m, when it is usual for the deck to be of box section and of variable depth, the bottom slab close to the supports thickens with increasing span. It is principally the webs that increase in depth and volume as the span increases.

The equivalent concrete thickness of a deck is defined as the volume of deck concrete (excluding parapets, but including diaphragms and internal blisters) divided by the gross plan area of the deck (width including parapets). This thickness plotted against span is shown in Figure 8.8. Up to a span of about 35 m, the mean equivalent concrete thickness is found to remain substantially constant at 0.48 m. However, the thickness of carefully designed box section decks remains below 0.5 m up to 50 m span, as shown by the –10 per cent line. After 35 m and up to about 150 m, the mean

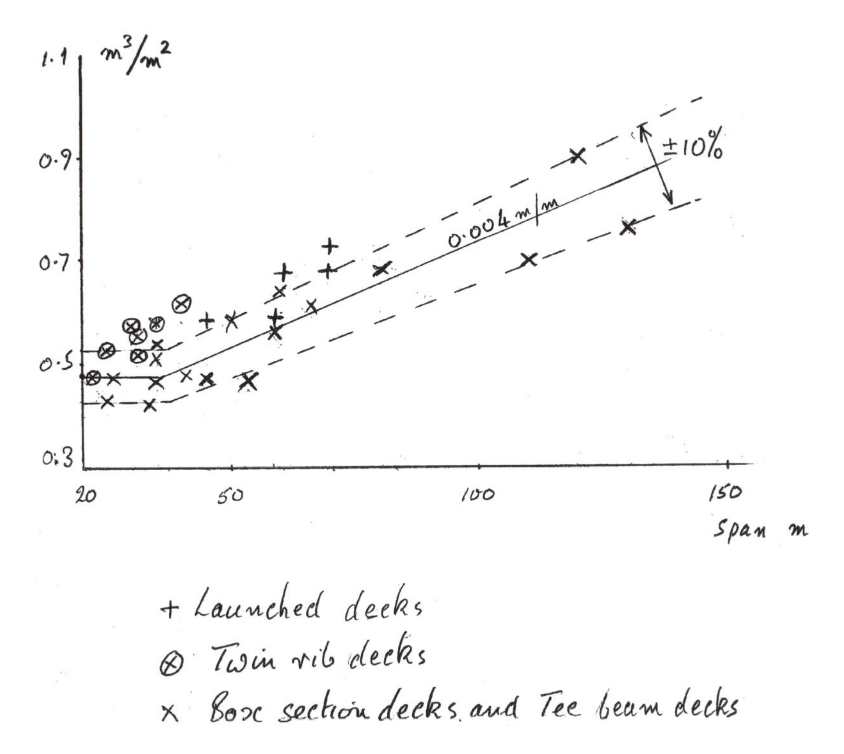

Figure 8.8 Equivalent thickness of concrete decks for increasing span

thickness increases at a rate of approximately 0.004 m per metre of span. One would expect a variation either side of this mean line of about 10 per cent.

Bridges built by the method of incremental launching tend to require some 10–15 per cent more concrete than those built by other methods, principally because they have thicker and deeper webs, although if built using intermediate props to cut down the launching span, this is less marked.

Twin rib type bridges are generally up to 20 per cent thicker than the mean, and follow a steeper increase in thickness with increasing span, due to the preponderance of their webs in their cross-section area. They follow a different logic, with a thickness of 480 mm at 20 m, increasing at a rate of approximately 0.006 m/m.

Figure 8.8 includes only bridges designed by Benaim to British codes of practice, or to very closely related codes such as those used in Hong Kong and Malaysia. This historical record may need adjusting slightly where new regulations increase the cover required for reinforcement to values greater than those described in *9.1*.

It should not be forgotten that certain types of deck are generally associated with specific designs of pier, and these may affect the overall economy of the project. Tee beam decks (Chapter 10) which are very economical require substantial crossheads on their columns, Figure 10.17, while twin rib decks (Chapter 12) which are heavier than the norm, may be carried by very economical simple columns, Figure 12.1.

8.4.3 Reinforcement

The quantity of reinforcement in a fully prestressed concrete deck is not closely related to the span. The bending reinforcement of the top slab, which accounts typically for some half of the total, depends on its transverse span between webs and on the length of the side cantilevers and is hence independent of the span of the deck, while the bottom slab of a box girder is generally reinforced at nominal rates over most of its length. Only the reinforcement of the webs, which increase in depth with span, is clearly span related. Consequently at the preliminary design stage, the quantity of reinforcement is better estimated as a rate per m^3 of concrete.

The rate of reinforcement achieved in a design is an accurate measure of the skill and care with which the deck has been designed. Inexperienced designers will over-reinforce a deck, often to the point of rendering it uncompetitive, and difficult to build. Designers too concerned about their legal risks will over-reinforce in the hope of eliminating the possibility of concrete cracking. Such designers often waste much steel by over-providing the nominal reinforcement in concrete members. For instance, reinforcing a 200 mm slab with four layers of T10 bars at 150 mm spacing gives rise to a reinforcement rate of approximately 80 kg/m^3, while adopting 16 mm bars at the same spacing, which is not uncommon, gives a rate of over 200 kg/m^3. The 10 mm bars provide more than the minimum reinforcement required by the UK code for slabs up to 350 mm thick, and in the right circumstances, are quite appropriate to a bridge deck.

Even when a deck is designed expertly, minimising the reinforcement is time consuming, and may be difficult to achieve within limited design fees or to a tight deadline. For instance, at a particular design section of a web, the live load cases that give the maximum shear and the maximum transverse bending are usually not coincident, and consequently some of the shear reinforcement can double as bending steel. It is clearly more economical of design time to assume that the two actions are at a maximum simultaneously, although this will over-design the reinforcement.

Minimising reinforcement offers diminishing returns as one approaches the absolute minimum that is compatible with the code of practice. As the designer carries out this exercise, he must keep in mind that the cost per ton of reinforcement will rise as the bars become shorter and of smaller diameter.

Each type of deck has a characteristic rate of reinforcement per m^3 of concrete, which is dependent on the code of practice being followed. Typical rates of reinforcement for the various prestressed concrete deck types subjected to 45 units of HB loading in accordance with the UK code of practice are as follows:

- Solid slab — 45–60 kg/m^3
- Voided slab — 110 kg/m^3
- Ribbed slab — 120 kg/m^3
- Precast Tee beams — 110–130 kg/m^3
- Concrete box girders; span < 80 m — 150–180 kg/m^3
- Concrete box girders; span > 80 m — 120–160 kg/m^3

These figures assume that the deck is conventionally reinforced transversally. For transversally prestressed decks the reinforcement will be reduced by some 20–25 per cent. These reinforcement rates are heavily influenced by the rate of reinforcement required in the top slab, which is very dependent on the relevant loading code. Many national codes, such as AASHTO (American Association of State Highway and Transportation Officials), adopt lighter vehicles than the UK, and these will give rise to significantly lower rates of reinforcement.

There is some evidence that externally prestressed decks with the tendons anchored in blisters require significantly more reinforcement than decks with internal prestress. This is due to the large size of prestress units that are generally used for such decks, and the large amount of reinforcement necessary to distribute the anchor forces into the cross section.

8.4.4 Prestress

The prestress compression that is necessary for a deck is more dependent on the span/depth ratio than on the length of the span and consequently it is more useful to relate the weight of prestressing steel to a rate per m^3 of concrete. For decks of constant depth up to 50 m span, the average prestress compression, that is the compressive stress at the neutral axis, will vary from about 3.5 MPa for a span/depth ratio of 14, to about 8–10 MPa for a ratio of 25. For prestressing strand with an ultimate stress of 1,860 MPa, where the working stress may be assumed to be 1,000 MPa, it follows that $W \approx \sigma_p \times 8$, where W is the rate of prestress in kg per m^3 of concrete and σ_p is the average prestress compression in MPa. Thus the rate of prestress will vary from about 30 kg/m^3 to 80 kg/m^3, with most projects falling between 35 kg/m^3 and 55 kg/m^3. Variable depth decks with spans up to about 120 m which respect the typical depths at mid-span and support given in *15.4.2* would be expected to have prestressing rates of between 45 kg/m^3 and 55 kg/m^3. For longer spans the rate of prestressing will rise, and it would be unsafe to quote typical values.

If wire or bar tendons are used, these rates of prestressing steel should be factored inversely with respect to their working stress. For twin rib type decks, statically determinate decks and incrementally launched bridges built without intermediate supports, add 15–20 per cent.

Clearly, for very slender decks or decks carrying heavy rail or other exceptional loads, these rates should be increased, while for very deep or very lightly loaded decks the figures will be over-conservative. A quick preliminary design, using the techniques described in this book, will allow the prestress quantity to be refined.

8.5 Choice of most economical span

When the span length is not dictated by the obstacles to be crossed, the choice of the most economical span is not straightforward. Any analytical approach is likely to discover that the graph of cost against span length does not show a distinct minimum. Clearly one must consider the costs of both the deck and the substructure.

As described above and shown in Figure 8.8, the quantities of materials in a well-designed deck are more or less constant up to a span of about 50 m. This refers to a mean of all types of deck; if one were to study each deck type, more distinct trends would appear. For instance, consider twin rib decks built span-by-span. On the one hand, the average deck thickness, and hence the quantity and cost of materials in the deck rises with increasing span. As the deck becomes heavier, the falsework necessary for its construction will also become more costly as will the cost of the foundations necessary to carry the self weight of the deck. On the other hand, as the span increases and the number of spans reduces, the labour required to strip the shutter and advance the falsework will diminish. For the British live loading code, which is dominated by a single heavy vehicle, the fewer the foundations the less often this vehicle is carried, and consequently the foundations required to carry the live loading become cheaper as the spans increase. Putting all these trends together would probably show a minimum cost lying somewhere at a span of about 25 m, although the curve of cost against span is likely to be quite flat. As the span increases beyond 40 m, the weight increase of the deck will tend to dominate and costs will rise more steeply.

If one were to apply a similar analysis to decks made of precast Tee beams, the determining factor is likely to be the increasing cost of moving and launching the beams as they become longer and heavier, with costs rising steeply once they exceed a span of about 45 m and a weight in excess of 130 tons. Thus, each deck type and method of construction has its own internal logic.

As long as the cost of the foundations is proportional to the weight of the deck, they will not have a disproportionate influence on the choice of the most economical span. This would be the case for foundations consisting of pads or of short driven piles. On the other hand, the cost of bored piles per ton of load carried typically reduces as the pile size is increased, favouring longer spans. This will be most marked where foundation conditions are difficult, requiring for instance very deep piles or piles in the sea. Where a bridge is founded on large-diameter bored piles, it may well be that the optimum span is influenced by the maximum utilisation of a single pile or of a two-pile group of a particular diameter which is competitively priced by local contractors.

The author's experience is that the height of the pier shafts does not influence the most economical span until it exceeds about 30 m, when it starts to become economical to lengthen the spans, above all if this can lead to shorter columns, such as when crossing deep V-shaped valleys.

Finally, the cost of a bridge will depend on how each contractor prices the various components, which will be coloured by his past experience. Four contractors pricing the same bridge are likely to show a range of at least 15 per cent in their conclusions. Consequently, it is vain to attempt to be either too scientific or too dogmatic about the most economical span for any particular situation. It is very much a matter of experience and intuition, and once the choice is made, of designing as tightly as possible. A well-designed bridge of the 'wrong' span is likely to be considerably more economical than a poorly designed bridge of the 'correct' span.

In general, for a viaduct that is crossing reasonably level ground with unexceptional foundation conditions, the most economical span length for a twin rib deck is likely to fall between 25 m and 35 m. For a box section deck of constant depth it is likely to lie between 30 m and 45 m, while if the deck is provided with a haunch, the most economical span will lie between 40 m and 60 m.

9

The design of bridge deck components

9.1 General

The purpose of this chapter is to describe the logic that controls the dimensioning of each component of a bridge deck. Clearly, this dimensioning will depend to some degree on the code of practice to which the bridge is being designed. In particular, the minimum thickness of members will depend on the current regulations on the cover required to protect reinforcement from corrosion. For instance, in the 1960s, the top slabs of precast Tee beam bridges (Chapter 10) on which the author worked in France typically had a minimum thickness of 160 mm for a beam spacing of 1.87 m (Pont du Saut du Loup) or 170 mm for a spacing of 3.35 m (Viaduc de la Porte de Versailles). The greater cover requirements of the British code, together with heavier loading lead Benaim to use a minimum thickness of 200 mm for the top slabs of precast Tee beam and box section decks. Recent changes in the British code of practice have further increased the required cover, and have led to another incremental increase in thickness in the UK. In other countries, where the environmental conditions and the rules are different, other minimum thicknesses will apply. The minimum thicknesses quoted in this chapter correspond generally to a concrete cover of 35 mm for the components of bridge decks in contact with the weather, and 30 mm for protected surfaces, such as the inside faces of concrete boxes.

9.2 Side cantilevers

9.2.1 General

The side cantilevers of a bridge deck carry the highway, the footpath and the bridge parapets, and they may act as storage for prestressing tendons. In addition they may house the bridge deck drainage and other services, and have an important effect on the appearance of the bridge.

9.2.2 Geometry of cantilevers

As explained in Chapter 8, one major design objective is to minimise the number of webs in the bridge deck. This concern inevitably leads the designer to contemplate long side cantilevers if the deck is wide. It is important for the economy of the deck that these cantilevers should not be thicker than necessary.

In the event of a collision between a vehicle and the parapet, the parapet itself or its connection to the deck should fail before the deck slab, thus localising the subsequent repairs and guaranteeing that the integrity of the deck is not compromised. This requires the upper bound strength of the parapet to be less than the lower bound strength of the slab. It is difficult to provide such strength in a slab thickness of less than 200 mm. Where high impact parapets are specified, this minimum thickness may increase to 220 mm or 250 mm, depending on the local rules. Thinner reinforced concrete slabs may be envisaged where the cantilever is used only for pedestrians and cyclists, and where a roadside barrier makes it impossible for traffic to encroach. However, it is unlikely that a thickness less than 160 mm will be viable even in these unusual circumstances.

A 200 mm thickness can span as a free cantilever at least 1.4 m and up to 1.7 m, depending on the relevant loading code, on whether the slab is carrying footpath or highway and on the weight of the parapet. Beyond this span, the cantilever must be thickened with a haunch. For the UK loading code, the root of the haunch should respect a thickness of between 1/7 and 1/9 of the cantilever length. Other codes of practice with lighter loading allow more slender cantilevers.

In general, the deflection of the side cantilever is not a consideration in deciding on its thickness. An exception could be where the cantilever carries an exceptionally heavy footpath or parapet when the cantilever may be prestressed to control dead load deflections. However, caution on the slenderness of cantilevers is required if the bridge is being built by the counter-cast method of prefabrication (*14.3.8*).

Some examples of cantilever geometry are shown in Figure 9.1.

Where side cantilevers carry footpaths that are used for services, the footpath structure can be made as a series of accessible chambers filled with sand or mass concrete, depending on the degree of accessibility required.

Drainage of the bridge deck must be considered early in the preliminary design. For bridge decks with side cantilevers, the rainwater gullies usually fall within the cantilever, and may well affect its thickness, and hence the section properties and self weight of the deck as a whole. For a discussion on deck drainage, see *9.7*.

A particularly efficient form of side cantilever which eliminates the additional dead weight of a raised footpath structure was used on the East Moors Viaduct and is shown in Figure 9.2. This type of cantilever is applicable where the footpath carries no significant services. Tubes for electrical cables, together with pulling pits, may be housed within the precast verge.

Once side cantilevers exceed a length of about 4 m, it becomes worthwhile to prestress them, or to adopt ribbed or propped slabs. The viability of these options depends fundamentally on the applicable code of practice, see *9.2.5 et seq*.

9.2.3 Longitudinal shear

For the cantilever to contribute to the longitudinal bending inertia, it has to be attached to the web by adequate shear reinforcement. The shear force at the root of the cantilever may be understood as a force tending to make the cantilever slide along the web which reaches its maximum adjacent to the piers and abutments, and is closely related to the vertical shear force in the webs of the deck. Most codes of practice require calculation of this effect at the ultimate limit state, which logically should consider the average effect over a length of the deck, rather than the peak effect. The

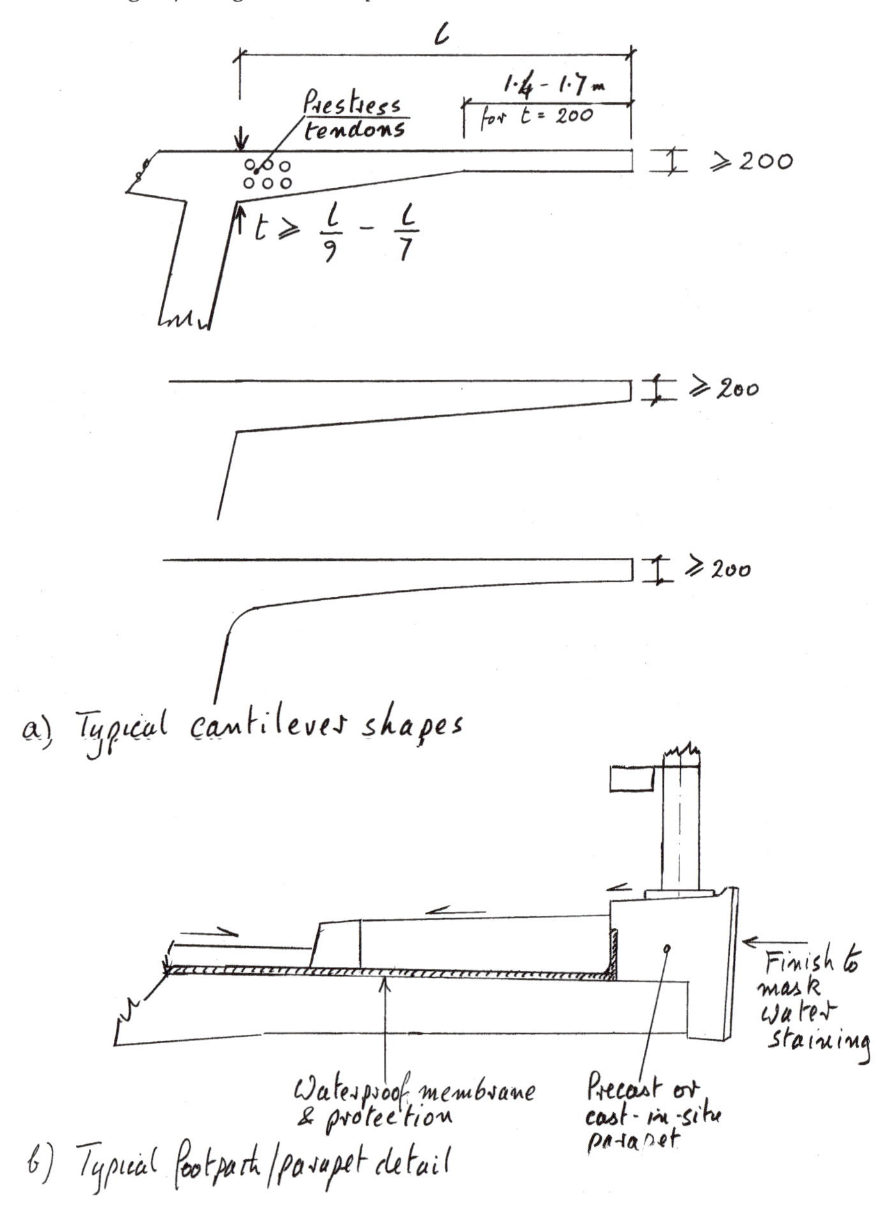

Figure 9.1 Geometry of side cantilevers

shear reinforcement is placed perpendicular to the axis of the deck, and resists the sliding force in shear friction (*3.11.4*).

As the cantilever is also subjected to a hogging bending moment due to its self weight and to superimposed loads, the root experiences a couple, with top fibre reinforcement in tension, and bottom fibre concrete in compression. It should be noted that in general, the maximum transverse bending of the cantilever root under local live loading is not compatible with the maximum sliding shear force which is due to general loading on the span. Also, while the ultimate shear at the cantilever root may be averaged over a

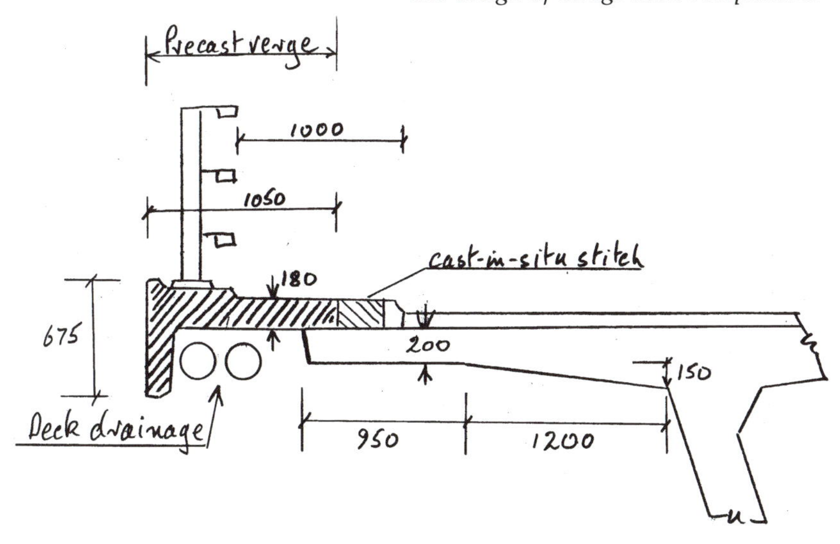

Figure 9.2 East Moors Viaduct: precast verge

significant length of deck, the maximum bending moment on the root is likely to occur over a relatively short length. This means that the shear may be resisted by underused bending reinforcement in areas that are not subjected to maximum bending moment. However, there is no consensus on how the reinforcement for combined shear and bending should be calculated.

A simple method of combining the reinforcement for shear and bending is to calculate the reinforcement required for both actions. One-third of the reinforcement required for shear is placed near the bottom fibre of the cantilever, while the greater of the bending reinforcement or two-thirds of the shear reinforcement is placed near the top fibre of the cantilever. For this approach to be valid, neither the shear nor the bending must be exhausting the capacity of the concrete in compression.

For further consideration of how to combine the reinforcement for shear and bending moment, see [1, 2 and 4].

The reinforcement required for shear should not be confused with that needed for the dispersion of the prestress force into the cantilevers at the end of a deck. This is a separate effect, requiring additional reinforcement (*5.24.4*).

If a direct tension was applied to the cantilever, additional reinforcement would be required. This can occur due to impact on the parapet, although such tension is unlikely to be compatible with the load cases giving maximum shear and bending. Tension will be applied when the cantilever is supported by inclined struts (*9.2.11*). On the other hand, compression due to transverse prestress would participate in resisting both the bending moment and the shear friction.

9.2.4 Shear lag

Shear lag must be considered when assessing how much of the side cantilever participates in the longitudinal bending of the deck at the SLS (*6.10.2*). The usual rule of thumb is that any part of the cantilever beyond a width of one-tenth of the span of a simple beam, or one-tenth of the distance between points of inflection for a continuous beam, should be ignored in calculating the bridge deck section properties. As the top flange of a typical highway bridge will not be very highly stressed in compression (except for very long span bridges or through girders), the effect of shear lag generally does not result in overstressing of the concrete in compression due to longitudinal bending.

Side cantilever slabs more than 2 m long usually have a variable thickness. This reduces the intensity of the longitudinal shear stress and logically should thus increase the length of cantilever that participates in the longitudinal bending. Conversely, where the side cantilever is carried by transverse ribs, it can be thinner than a cantilevering slab, reducing the ratio of the root thickness to the area of the slab, leading it to be more highly stressed in shear and increasing the effect of shear lag. These factors are not recognised by the simplified rules of thumb, or by many codes of practice.

9.2.5 Prestressed cantilever slabs

In the author's experience post-tensioned cantilever slabs designed to Class 1 or Class 2 (*4.2.3* and *4.2.4*) are rarely cost effective for the following reasons. Prestressing tendons usually need to be anchored at or near the ends of the cantilevers and so cannot be effectively curtailed, while reinforcing bars may be tailored to the local bending moment. Also, due to the relatively large size of prestressing ducts as compared with reinforcing bars, to the fact that tendons must be inside the surface grillage of distribution bars and the fact that prestress is designed at the SLS while reinforced concrete is generally designed at the ULS, the prestress lever arm is significantly less than that for reinforced concrete, Figure 9.3 and Figure 6.31.

As the deck becomes wider and the slabs become thicker, the deficit in the lever arm reduces and prestressing gains in competitiveness. Also, in the thicker slabs, it may be possible to achieve a degree of curtailment of the prestress and save some steel by providing intermediate buried dead anchors. If the bridge deck is made of precast segments, it may well be possible to use pre-tensioning for the transverse reinforcement, which is likely to be more economical than post-tensioning.

The former French code for transverse prestressing of decks allowed tensile stresses to penetrate to the surface of the duct for normal live loading, and to the centre of the duct for exceptional deck loading, with no tensions under permanent loads. This is a sensible method which increases the prestress lever arm. However, the most rational and competitive way of sizing a prestressed concrete slab is to design the section at the ULS, and to check at the SLS that no tensions occur under dead load, and that the crack widths under live loading are acceptable. A slab designed in this way will be uncracked over the greater part of its life, and will be both economical and more durable than reinforced concrete. Unfortunately, some codes do not accept this approach.

If the slab is subjected to loading that is highly repetitive and could constitute a fatigue risk, such as mass transit trains, or a proportion of standard highway loading, it would be sensible to ensure that the tendons are in uncracked concrete under these loads.

9.2.6 Precautions at expansion joints

Particular care must be taken in the dimensioning of side cantilevers at the ends of bridges. Here, the total deflection of the cantilever under combined dead and live loading relative to the fixed abutment wall may adversely affect the roadway expansion joint. It is usually necessary to carry the cantilever by a transverse rib that provides the

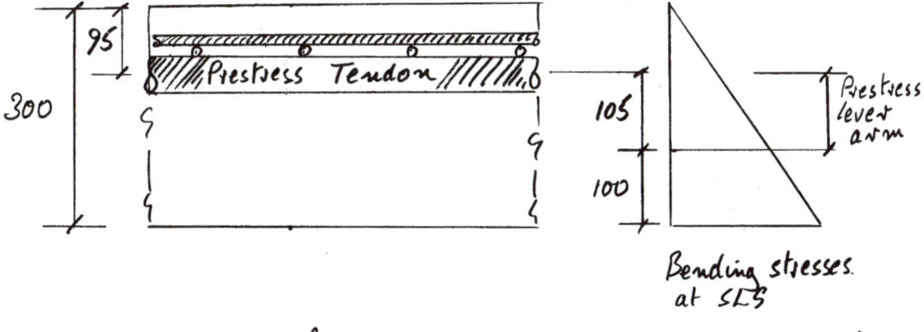

a) Typical lever arm for 300mm slab designed to Class 1 prestress

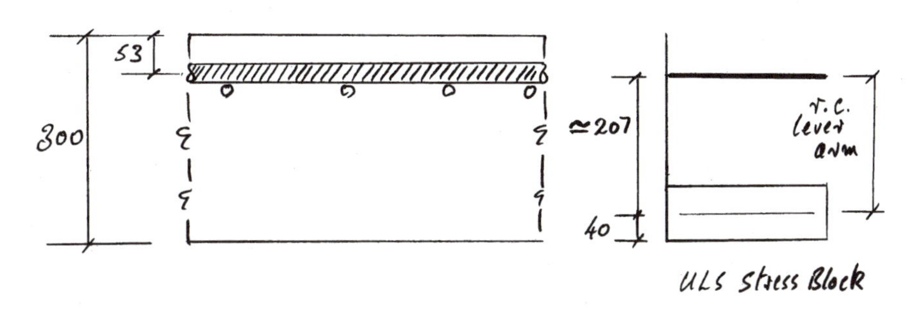

b) Typical lever arm for 300mm reinforced concrete slab at the ULS.

Figure 9.3 Lever arms for prestressed and reinforced solid slabs

necessary stiffness. The problem is particularly significant in the acute angle of skew bridges. It should not be overlooked that longitudinal hogging bending moments will exist in the cantilever slab at its junction with this rib.

The expansion joints are rarely perfectly installed, and they create an irregularity in the carriageway. As a result, there are very real dynamic impacts on the bridge deck as heavy axles cross the joint. In the absence of detailed data on these effects, it is good practice to provide a conservative design for this area of the cantilever slab.

Although railway bridges do not generally have expansion joints that are sensitive to relative deflections, considerations of bending and fatigue in the rails may also impose limits on the flexibility of the cantilever adjacent to the abutment.

9.2.7 Longitudinal moments in the slabs; cast-in-situ construction

In addition to the principal transverse bending moments in a side cantilever, longitudinal bending moments also occur under concentrated wheel loads. The largest moments are sagging immediately beneath the wheel, with smaller hogging moments either side, Figure 9.4. In a cast-in-situ deck there is no problem in carrying these moments with longitudinal distribution steel.

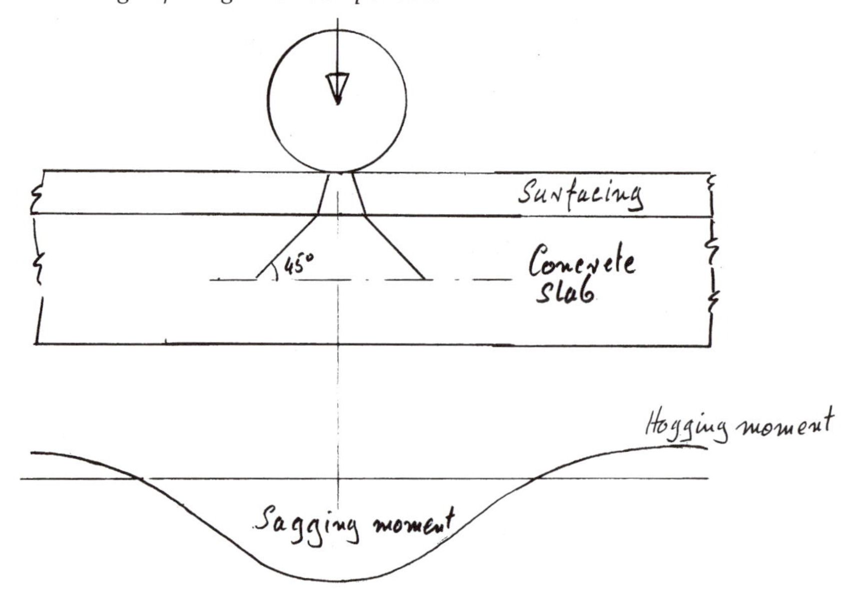

Figure 9.4 Longitudinal moments in cantilever slabs

9.2.8 Longitudinal moments in the slabs; precast segmental construction

If the deck is made of precast segments (see Chapter 14), at each transverse segment joint there is no longitudinal reinforcement to carry these bending moments. In a prestressed deck there will be some longitudinal compression across the joint that will allow a proportion of the moment to be carried. This compression varies with the live load cases and with the position along the deck and is unlikely to be adequate to resist the full amount of the longitudinal moment at all design sections. Under these circumstances it is necessary to re-analyse the transverse bending of the cantilever under localised loads, taking into account the local reduction in longitudinal moment capacity of the slab. This will lead to an increase in transverse reinforcement either side of the joint.

A simple empirical analysis at the ULS for the longitudinal sagging moment due to a heavy axle located over the transverse segment joint would be as follows. Assume that the compression force in the slab due to the overall bending of the deck is P kN per metre of deck, and that the centroid of this force is at a depth h below the top surface of the slab. The joint can thus resist a moment of approximately $P \times h$ kNm before its strength is exhausted, Figure 9.5. The proportion of the axle load that causes this limiting longitudinal sagging bending moment is calculated, as is the cantilever bending moment for this reduced load. Any further load would cause the joint to rotate. For the remainder of the axle load, calculate the cantilever bending moments assuming the cantilever to be hinged along the transverse joint.

9.2.9 Ribbed cantilever slabs

For side cantilevers longer than about 4 m, it is worth considering the use of a ribbed slab that increases the reinforced concrete lever arm while reducing the dead weight of the deck, Figure 9.6.

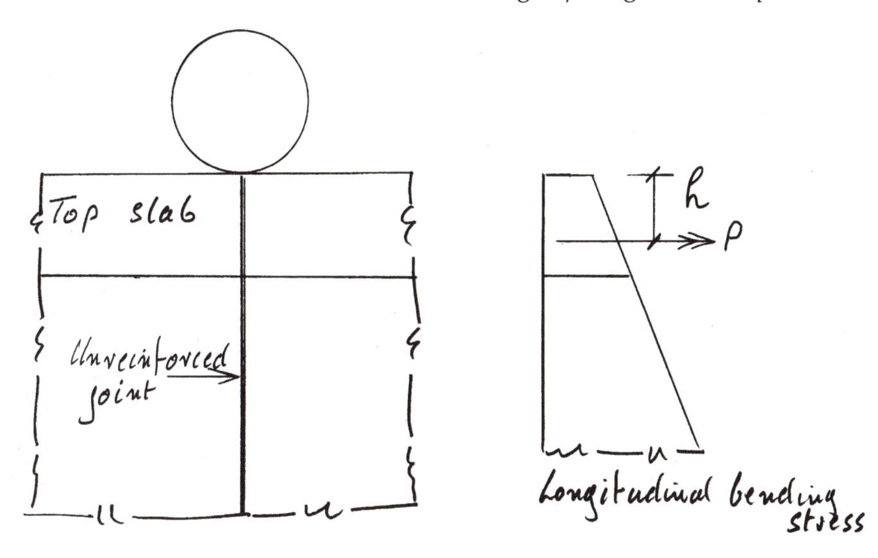

Figure 9.5 Transverse slab joint subjected to longitudinal bending moments

The deck slab for a ribbed side cantilever will generally be 200 mm thick, requiring ribs at centres of approximately 4 m under UK loading. Spans up to 5 m may be possible under other codes that use lighter loading. As the slab spans longitudinally between the ribs, its local bending moments combine with the longitudinal stresses in the slab due to the overall bending of the deck. This will affect the design of the longitudinal bending reinforcement in the slab, and may also reduce its ultimate bending capacity under local loads, requiring greater thickness or a reduced span. It is beneficial to haunch the slab longitudinally over each rib.

If the deck is built by precast segments a transverse cast-in-situ stitch is necessary, usually located half way between the ribs, to create continuity of the longitudinal reinforcement, Figure 9.6 (b).

Another consideration when using a ribbed cantilever slab is to ensure that the concentrated moment in the rib can be accepted by the web of the deck, or by the slab beyond the web. The most logical arrangement is to provide ribs across the full width of the deck, although this is not essential as long as the webs and slabs are designed correctly. The rib may also be deepened so that it applies its bottom fibre compression directly to the bottom flange of the box, Figure 9.6 (c).

The economy of ribbed cantilever slabs depends principally on the loading code. Where the design live load consists of an isolated load or vehicle, it has been the author's experience that ribbed slabs are not economical. A solid slab has great capacity to spread the effects of a concentrated load over a considerable length at the cantilever root, while a ribbed slab concentrates the bending moment on to a small number of ribs, Figure 9.6 (d) and (e). Thus the benefits of the greater reinforced concrete lever arm are lost, as the moment to be carried is more concentrated.

However, when the loading is principally a uniform load, the bending moment at the root of the cantilever is not subject to dispersion and the greater lever arm will reduce the reinforcement required. Transverse prestress is more likely to be economical in a ribbed slab than in a solid slab.

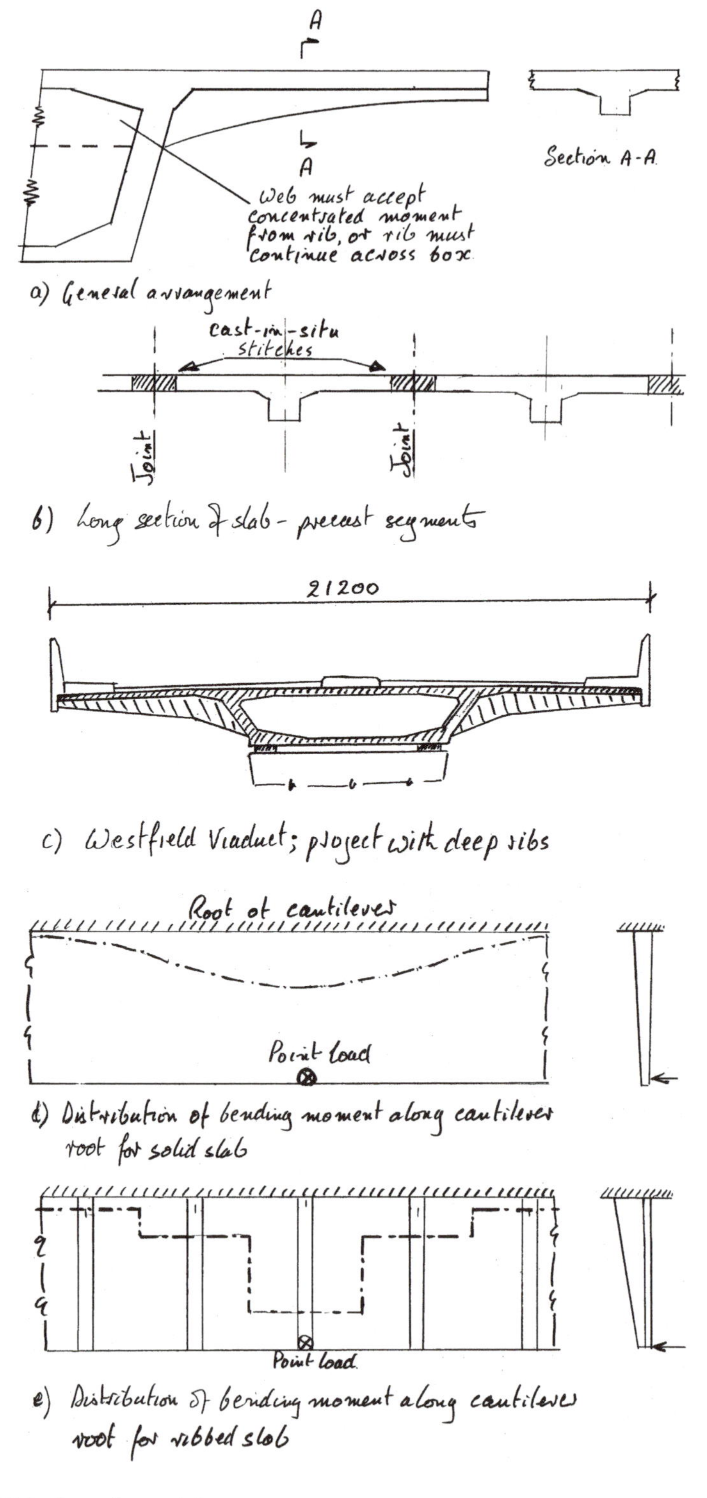
A
A
Section A-A
Web must accept concentrated moment from rib, or rib must continue across box
a) General arrangement
cast-in-situ stitches
Joint
Joint
b) Long section of slab - precast segments
21200
c) Westfield Viaduct; project with deep ribs
Root of cantilever
Point load
d) Distribution of bending moment along cantilever root for solid slab
Point load
e) Distribution of bending moment along cantilever root for ribbed slab

Figure 9.6 Ribbed cantilever slab

The depth of the rib depends on the bending moment carried and on the need to limit the reinforcement over the webs of the box to avoid congestion, while its width is governed by the shear force carried.

Ribs have additional costs in formwork, although it is generally not difficult to strike the shutter downwards to clear the ribs for the side cantilever.

9.2.10 Coffered cantilever slabs

For long-span cantilever slabs subjected to concentrated live loads, where solid slabs become uneconomically heavy, it is worth exploring the use of a coffered slab. This type of ribbed slab reduces the disparity in the stiffness in the transverse and longitudinal directions, and thus improves the spread of the concentrated load. The ribs may be at approximately 1.5 m centres in both directions, with a top slab 170–180 mm thick, Figure 9.7. The transverse ribs may need to increase in depth very locally at the junction with the deck to control shear stresses and to reduce the maximum density of reinforcement over the top of the web. A grillage or finite element analysis which includes the torsional stiffness of the ribs would be necessary to gain the full benefits of the load spreading characteristics of such a coffered slab. The author has not tried this arrangement, but it has potential.

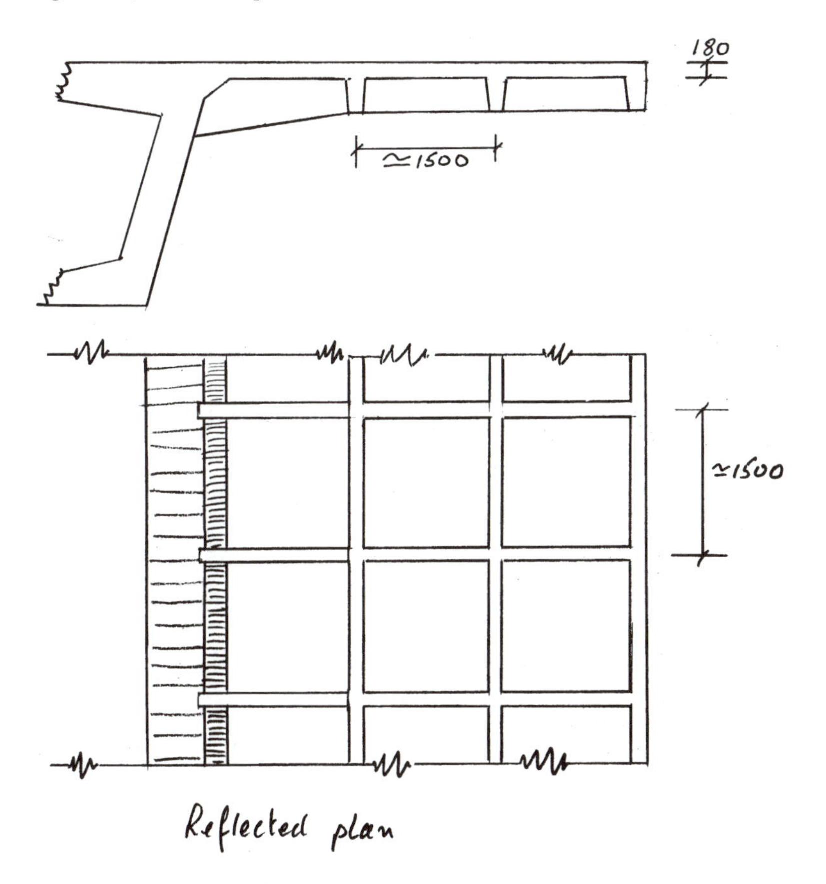

Figure 9.7 Coffered cantilever slab

9.2.11 *Propped cantilever slabs*

An alternative to ribs to support a long cantilever is the provision of inclined props. When used in conjunction with a box section deck, they have the advantage that they carry the compression of the cantilever moment directly to the bottom slab of the box, Figure 9.8. They also have the advantage that the cantilever slab retains its transverse bending action, reducing the problems at the joints for precast segmental decks.

Struts are usually made of precast concrete or occasionally of steel. However steel creates an unwelcome maintenance liability if the rest of the deck is in concrete. The props may be arranged perpendicular to the deck axis, or in a truss pattern.

a) Perpendicular struts

An example of the use of perpendicular struts is the deck of the STAR light railway viaduct in Kuala Lumpur. It consists of a single-cell box girder with reinforced concrete side cantilevers, built by the precast segmental method. Some of the stations are carried on the bridge, when it was necessary to widen the deck by extending the length

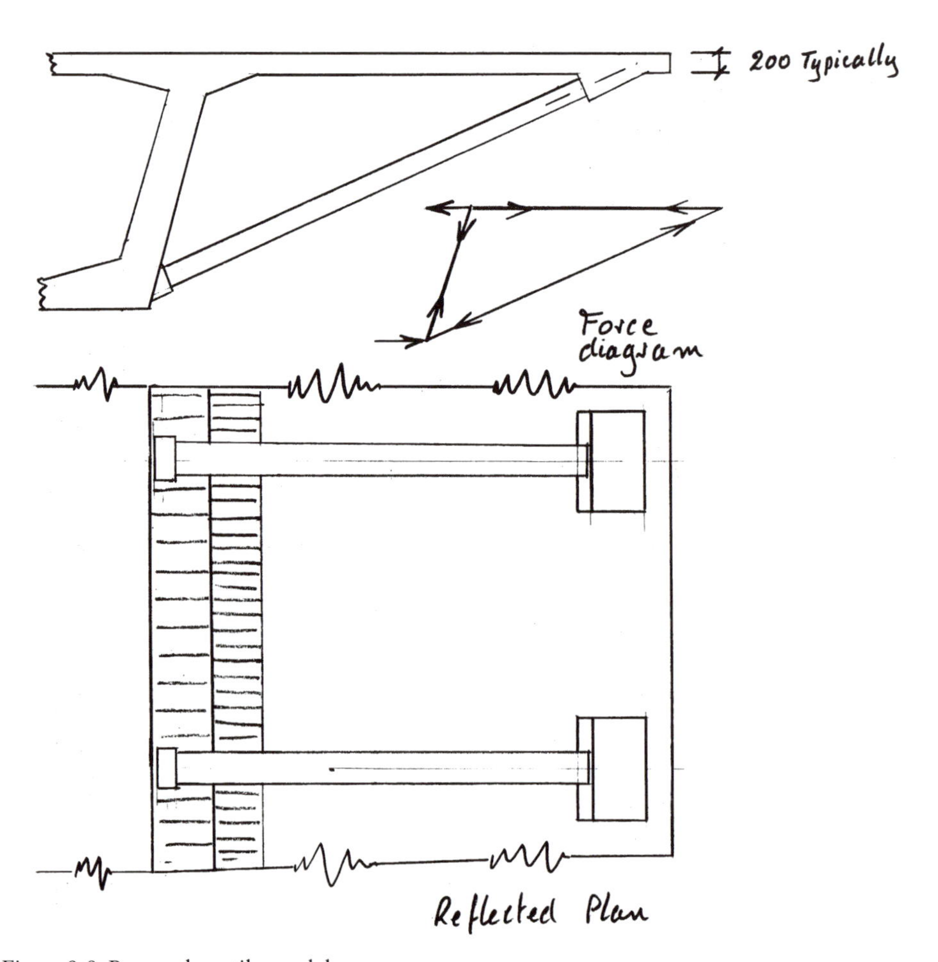

Figure 9.8 Propped cantilever slab

of the side cantilevers and to design them to carry much heavier loads, Figure 9.9. The cantilevers were strengthened by the addition of struts, which were added after precasting the deck segments, Figure 9.10.

Figure 9.9 STAR Viaduct: elevated station (Photo: Benaim)

Figure 9.10 STAR Viaduct: propped cantilever slabs (Photo: Benaim)

The spacing of the struts is typically between 2.5 m and 4 m. Although they act principally in compression under the self weight of the side cantilevers, it is necessary to check that live loading on the adjacent top slab or seismic accelerations cannot unload them. Even if the calculations show that there is no possibility of the struts being stressed in direct tension, it is better to have some nominal reinforcement linking them to the deck at their ends.

Where the struts meet the cantilever slab, they apply the horizontal and vertical components of their compression. The horizontal component puts the cantilever slab into tension and must be resisted by reinforcement or prestress. Room must be left beyond the strut to anchor this steel. The vertical component of the strut force will cause longitudinal bending moments in the slab. The thickness of the slab may need to be increased to cope with these moments, either by creating local drops around the point of support or a longitudinal beam. A longitudinal beam would give problems of creating continuity in precast segmental construction, requiring cast-in-situ connections between the segments, Figure 9.11. The struts apply compression across the bottom slab, and tension in the webs. This tension requires web hanging steel in addition to the normal shear reinforcement.

The struts must be designed to resist buckling under direct compression combined with their self weight bending. If they do not have reinforcement linking them to parent concrete at their ends, they should be considered as only partly fixed or pinned at their ends when calculating their effective length. If they are built in at their ends, the reinforcement should be designed to carry the fixed end bending moments, including any additional moments due to buckling calculations.

Figure 9.11 STAR Viaduct: stitches between precast segments (Photo: Benaim)

If the deck is cast in-situ, the struts would be placed in the formwork, and the joints at each end created by casting the deck, when there is no difficulty in tying in both ends with starter bars. If the deck is made of precast segments, the struts may be incorporated into the mould in the same way as for cast-in-situ construction, with the external shutter divided into two pieces, located either side of the strut.

However, the deck segments may be precast first and the struts placed afterwards, when it is more difficult to create a tensile connection at the ends of the struts. If the deck is wide, propped side cantilevers are likely to be too slender to carry their own dead weight without the presence of the struts. In this case, the first phase of prefabrication could include only stub cantilevers, the second phase including both the struts and the outer section of cantilever, as shown in Figure 9.10.

b) Trussed struts

When the struts are arranged as a truss, superimposed on their propping role is their action as an inclined web. Although these inclined webs will contribute relatively little to the shear strength of the deck, they may add very substantially to its torsional strength and stiffness. Shear and torque on the deck will put into the struts a pattern of tension and compression which is superimposed on the compression due to the weight of the cantilever. In common with any other fixed ended truss, secondary bending moments will be generated in the plane of the truss, and the struts should be kept slender in their plan dimensions to limit such moments. Bending moments will also occur in a vertical plane due to the weight of the struts. It is most likely that they will need to be linked to the parent concrete at each end with reinforcement or bolts, designed to carry the biaxial bending moments combined with direct tension, or at least with much reduced compression.

If the deck is curved, the length along one cantilever edge will be greater than the other, in theory making it necessary to vary the length of the struts. In practice, if the curvature is not too tight, the use of constant length struts may be achieved by varying the spacing of the point of contact of the bottom of the struts with the bottom flange of the box girder. The box webs and bottom slab easily carry the secondary bending moments due to the lack of conjunction of the truss. However, there may be a problem of appearance.

When the deck is built from precast segments, it would be normal for the top end of the struts to meet at the centre-line of the segment. This arrangement keeps the most intense local longitudinal bending moments in the slab away from the match-cast joints.

The trussed struts impose a relatively inflexible visual discipline on the deck. It is difficult to adopt a short pier segment, for instance, and if the spans vary, they need to respect a module dictated by the truss. Mid-span stitches need to be of one segment length, or to be very short to avoid a visual hiatus.

c) Conclusion on struts

Struts arranged as a truss constitute a satisfying engineering solution. They may contribute very substantially to the torsional strength of a box girder, and look attractive. However, they are more difficult to design than perpendicular struts, both technically and visually.

9.3 Top slabs

9.3.1 General

The top slab of the deck acts transversally to carry loads to the webs, it may house the longitudinal prestressing tendons in the vicinity of the piers and, if the deck is a box section, it forms one member of a torsion box.

The slab must of course be attached to the webs by adequate reinforcement or by compression due to transverse prestress. The discussion in 9.2 of the shear force between cantilever slab and web, and the interaction of this shear force with slab bending moments, is equally applicable to the connection between the top slab and the webs.

The dimensioning of the top slab depends on the loading code used, and on the degree of end fixity provided by the webs. Three different situations are considered, where the top slab is carried by precast beams, where the top slab is within a box girder, and where the top slab spans between torsionally stiff ribs or box girders.

9.3.2 Slab on precast beams

Where the slab is carried by the thin webs of precast beams which are not connected either by intermediate diaphragms or by a bottom slab, Figure 10.15, little end fixity is available. For UK loads, such slabs need a thickness of not less than span/20. However, the minimum thickness is normally accepted as 200 mm, corresponding to a span of up to 4 m. Although for such slabs the support moments may not be much different from the span moments, it is good practice to provide shallow haunches, typically 50 mm deep and 800 mm long, Figure 9.12 (a). These are helpful in striking the side shutters of precast beams, and in addition stiffen the slab and reduce the longitudinal shear stresses.

9.3.3 Top slab of box girder

The slab is effectively part of a portal frame, and is continuous with the webs, which are themselves restrained by the bottom slab. Under these conditions, under UK loading, the slab requires a mid-span thickness of span/30 where the span is measured to the centres of the webs, with a minimum thickness of 200 mm, Figure 9.12 (b). The top slab will normally have haunches at its junctions with the webs to resist the larger support bending moments and to limit the density of reinforcement over the top of the webs, facilitating the casting of the concrete. However, this reinforcement may be controlled by the side cantilevers, in which case it would be logical to size the slab haunches to require the same area of reinforcement. If internal longitudinal prestress is used, the haunches may be sized to house the tendons in hogging moment zones of the deck. The haunches also reduce the longitudinal shear stresses at the root of the slab. These shear stresses will be the sum of the stress due to the vertical shear force and that due to the torsional stresses in the box; at the centre of the slab, there remains only the torsional stress. The haunch thickness is typically between 1/15 and 1/18 of the slab span.

The moment envelope for the slab has the general shape shown in Figure 9.12 (c). There are maxima at the supports and mid-span, and minima in the region of the

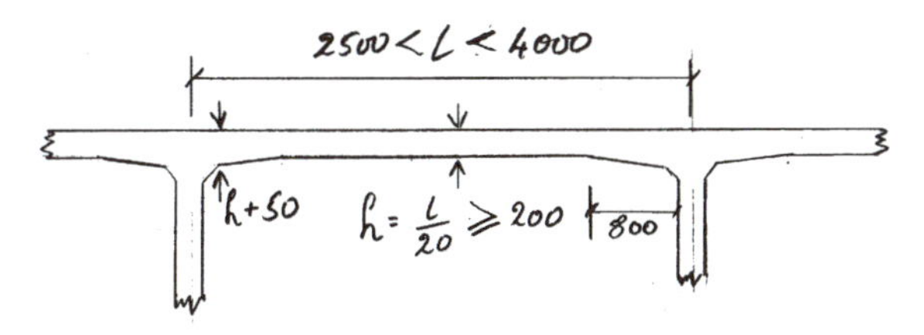

a) Slab on precast beams

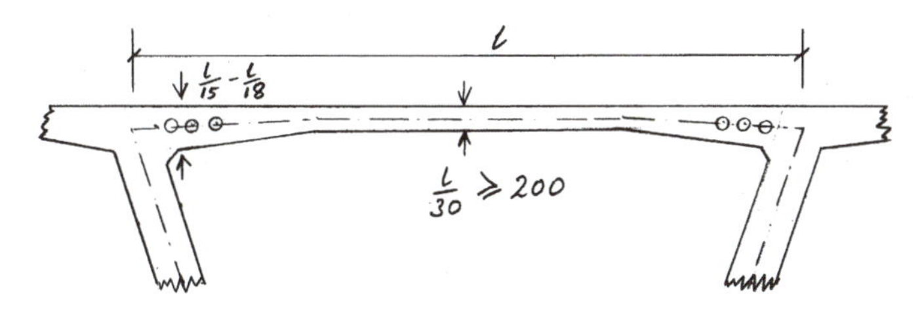

b) Top slab of box girder

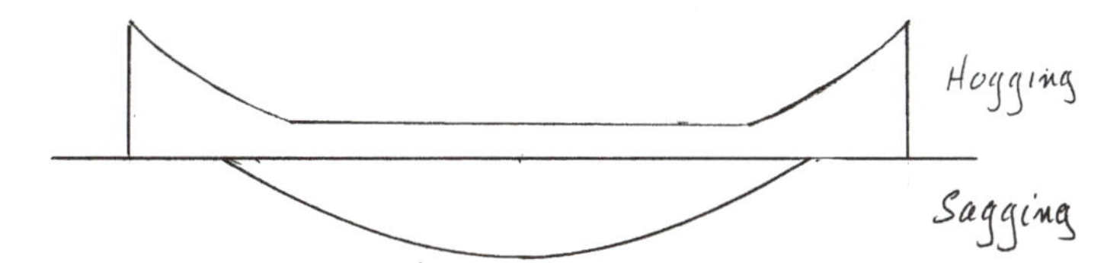

c) Typical moment envelope for top slab of box girder

d) Slab thickness following moment envelope.

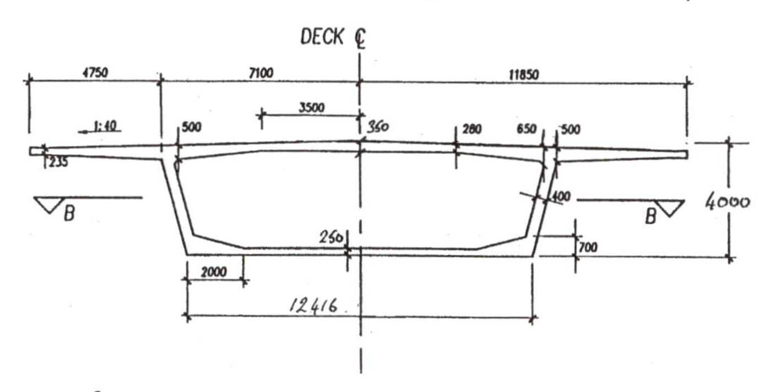

e) Project for Storabaelt approach spans: mid span cross section

Figure 9.12 Top slab configuration

quarter points. Where the road has a crown at the centre line of the box, the thickness of the slab may be tuned to this moment diagram by providing it with a flat soffit between haunches, Figure 9.12 (d).

The influence of the loading code may be seen when considering Benaim's preliminary design for the 4 km long approach viaducts to the Storabaelt Bridge, designed to Danish codes, Figure 9.12 (e). The 164 m long spans had a top slab with a mid-span slab thickness of 350 mm for a 13.6 m wide web spacing, a slenderness ratio of 39. The slab was profiled to follow the crossfall and had a flat soffit, as described above, leaving a minimum slab thickness of 280 mm at the points of least moment.

9.3.4 Ribbed and strutted slabs

Where the box width exceeds about 8–9 m, it becomes relevant to consider lighter alternatives to a plain haunched slab. As for the side cantilevers, the options are to provide a ribbed slab or to provide struts. The arguments for and against ribbed slabs are very similar to those for cantilever slabs (9.2). However, as there is limited room to strike a shutter, a ribbed top slab of a box is more difficult to build.

Struts are normally arranged perpendicular to the deck axis. Although theoretically they could be arranged in a truss pattern, so they would carry some of the shear, this is not usually a practical or cost-effective proposition. The struts normally form a 'K' brace, splaying from the middle of the top slab to the bottom corners of the webs, Figure 9.13. Thus the weight of the slab and of the live loading puts the struts in compression, simplifying the end connections. However, distortion of the box under eccentric live loads on the deck may introduce tensions into the struts that are larger than the dead load compressions. It would then become necessary to provide a tension connection at each end of the struts. Where the struts meet the top slab, the hogging bending moments and shear forces in the slab usually make it necessary to increase the thickness of the slab with a local drop, or to provide a continuous longitudinal beam.

In precast segmental construction the presence of the struts would seriously impede the operation of a mechanised internal shutter. If the slab is sufficiently stiff, it may be possible to strike the internal shutter leaving the slab spanning initially without support. The struts would then be placed once the segment had been removed to storage. Considerable care would be needed to avoid deflecting the slab, which could destroy the accuracy of the match casting. Alternatively, the struts may be wedged temporarily into position in the casting cell once the internal shutter has been removed but before the segments are separated, with permanent connections being made in the storage yard off the critical path of the construction programme. These connections may be effected by bend out bars projecting from the top and bottom slabs, with nodes cast in-situ.

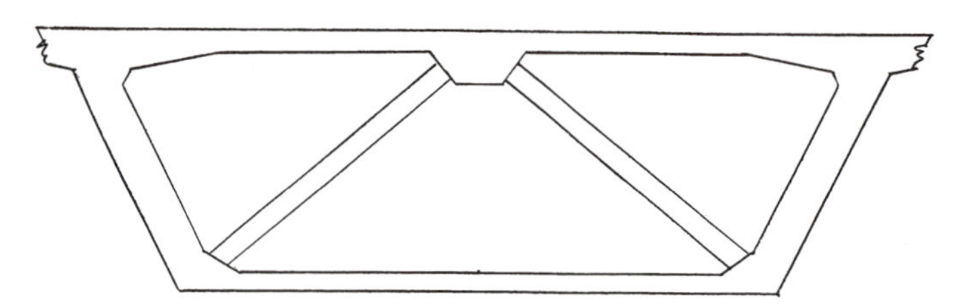

Figure 9.13 Strutted top slab

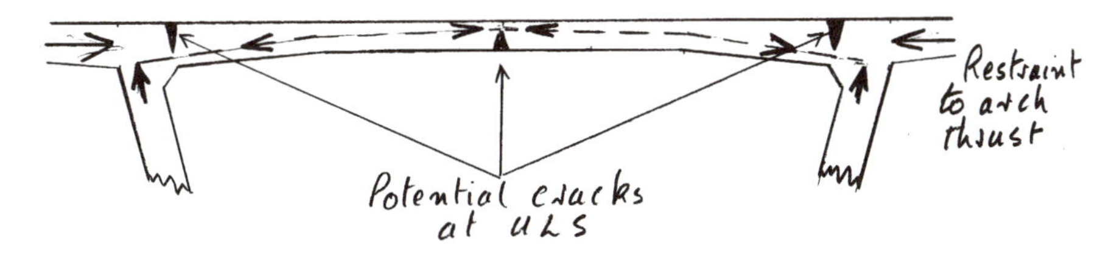

Figure 9.14 Arching action in slabs

9.3.5 Arching action in slabs

It has long been known that conventional reinforced concrete calculations for slabs underestimate their bending strength very substantially. The two principal reasons are the contribution of the tensile strength of concrete in bending, and the presence of arching action.

For slabs that are restrained around their perimeter, the arching action may very significantly reduce the amount of bending reinforcement required, or make it possible to remove it completely, Figure 9.14. This is very desirable, both in the interests of economy and of durability, as the reinforcement in deck slabs is most vulnerable to corrosion induced by de-icing salts.

The restraint that is required is resistance to spreading of the bottom fibre of the support sections of the slab. This may be provided by an effective tie consisting of reinforcing bars or prestressing cables, or by continuity with a slab that completely surrounds the loaded slab and is strong enough to resist the arching forces.

The presence of arching action was recognised by Yves Guyon, who wrote a paper on the subject in 1962 [5]. He carried out load tests on two half-size panels of prestressed concrete bridge deck slabs and found that the collapse load was some 27 times greater than the working load allowed by the contemporary French rules. The paper makes clear the significance of providing haunches to slabs. Designers should consider it good practice to provide such haunches even when the slab is designed by conventional bending theory, both to gain free additional strength and stiffness from the arching action, and to reduce shear stresses in slabs, as has been pointed out in 9.2.

More recently, composite bridge decks have been built where the reinforcement has been omitted from the deck slabs, and restraint to arching forces provided by steel straps placed beneath the slabs [6]. More thought and research is required before this matter becomes fully understood by designers. In particular, the stiffness of the restraint and the behaviour of the slab under different static load patterns and under dynamic loads need to be researched. However, there is no doubt that this is a very important and valuable development in the design of concrete bridge decks.

9.3.6 Slabs between girders

Where the slab is spanning between torsionally stiff girders, as in a twin box or twin rib bridge decks, its thickness may be determined by strength or by stiffness. If the girders are torsionally fixed at the piers, either by twin bearings or by a diaphragm, the slab can generally be sized in the same way as the top slab within a box girder.

If the girders are carried on a single bearing at the piers, care must be taken to ensure that the slab provides sufficient transverse stiffness to the deck. It is common for such

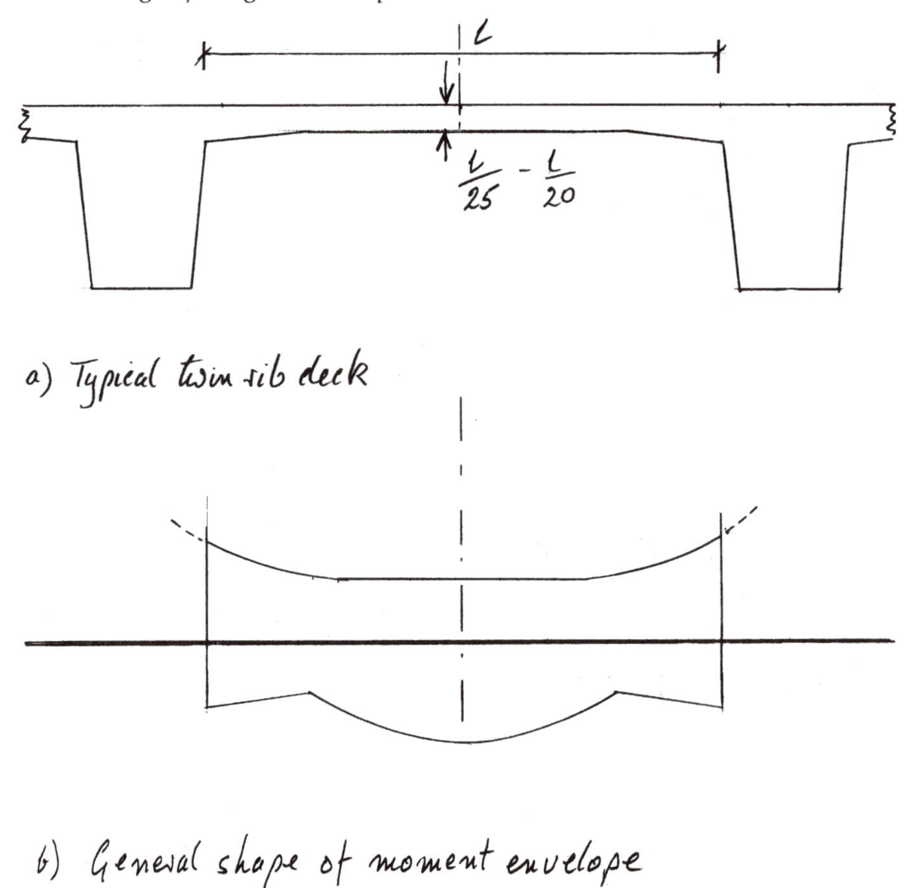

Figure 9.15 Slab spanning between torsionally unrestrained ribs

slabs to have a thickness of the order of (clear span)/25 to (clear span)/20, depending on the intensity of loading, Figure 9.15 (a). Where the deck carries a dual carriageway with a raised central reservation protected by a crash barrier, making it impossible to place full traffic loading at mid-span, the slab may be thinner, Figure 12.17. The moment envelope for such slabs shows significant sagging bending moments at the junction of the slab and beam, due to differential deflection between the beams, also resulting in relatively low hogging moments, Figure 9.15 (b). Consequently, the haunches may be less deep than for the highly restrained slabs described in *9.3.3*.

9.3.7 Longitudinal bending

The top slab carries longitudinal bending moments under concentrated loads as discussed for the cantilever slab (*9.2.7*). In addition, it is common for the top slab to be supported by pier or abutment diaphragms that are usually very rigid, and frequently much thicker than the slab which is effectively built into them. Consequently substantial longitudinal hogging moments exist under the effect of concentrated live loads, and it is not unusual for the slab to need thickening with a longitudinal haunch at these positions. It is good practice and economical to design diaphragms that are cut away from the top slab to eliminate such hard points (9.6). However, at the abutment the slab must be well supported so that the roadway expansion joint does not suffer from relative deflections between the slab and the abutment. The discussion of the dynamic effects on cantilever slabs close to expansion joints applies equally to top slabs.

9.3.8 Heat of hydration cracking

Whenever a bridge deck is cast in-situ with transverse construction joints, there is a risk that as the more recent concrete cools from its peak setting temperature, it will be cracked due to restraint offered by the hardened concrete of the previous pour (3.6). A typical case is when bridge decks are built by the cast-in-situ balanced cantilever method. The bridge segments are usually 3.5 m to 5 m long and the top slabs between 10 m and 20 m wide (including the side cantilevers). Although slabs are usually less than 300 mm thick, limiting the setting temperature, they frequently contain longitudinal prestressing ducts, which create weak planes encouraging the formation of cracks, Figure 9.16. As such deck slabs are always substantially reinforced in the transverse direction, any cracks which form are likely to be within the specified limits, usually less than 0.25 mm. However, as the cracks penetrate through the thickness of the slab, rain water collecting on the deck will pass through, drawing attention to their presence. Sometimes they are only noticed when the tendons are grouted, and water appears at the concrete surface.

If the deck is to be protected by a waterproof membrane, such cracks may be safely ignored. However the designer needs to be forewarned of the possibility of this type of cracking, and he would be wise to alert the client and the contractor in advance.

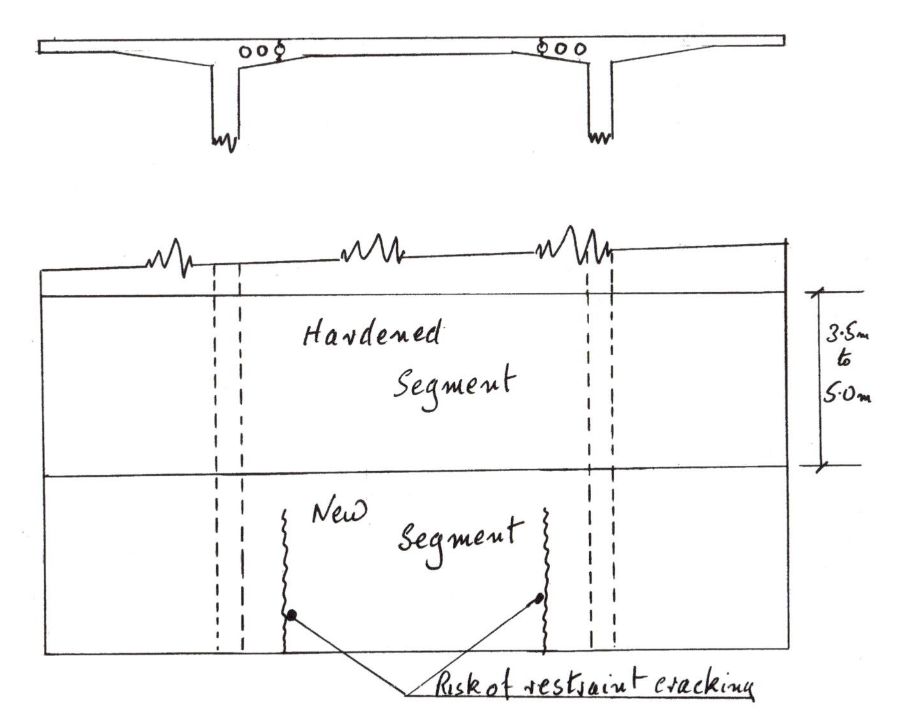

Figure 9.16 Restraint cracking in the top slab of cast-in-situ segments

If no waterproof membrane is specified, it is important to avoid such cracking, particularly if it exposes prestressing tendons to cracks along their length with the consequent risk of corrosion. The concrete mix should be designed to control the initial temperature rise. The main transverse reinforcing bars should be as small as possible, and placed at a pitch of 125 mm or less. Sudden changes in reinforcement density as the bars are curtailed should be avoided, and transverse distribution steel should be designed to restraint cracking criteria. It would also be wise to monitor the first deck sections built to check that the design is performing satisfactorily, or to build a trial panel.

9.4 Bottom slabs

9.4.1 General

The bottom slab of a box girder increases the efficiency of the cross section, reducing bending stresses and the required prestress force, as well as improving the ductility of the section. It also completes the torsion box, re-centring eccentric loads and further reducing the prestress required. When the prestress is internal, it is often used to house tendons near mid-span, Figure 9.17 (a).

The presence of the bottom slab lowers the neutral axis which reduces the eccentricity of the prestress at mid-span and increases it at the support sections. Consequently it is economical to minimise the slab thickness at mid-span with the option of increasing it locally at the support sections where the extra weight has little effect on the overall bending moments. For slender bridge decks carrying exceptionally heavy loads, where the moment variation at mid-span is large, the bottom slab may need to be thicker than the minimum at both mid-span and support sections to control the compressive stresses.

Although the provision of a bottom slab has definite advantages, principally in reducing the prestress force, it very significantly increases the difficulty of casting the deck. As a general rule, box sections are suitable for precast construction, where steel shutters, high intensity vibration and close supervision allow the difficulties in successfully filling the mould to be overcome. Cast-in-situ balanced cantilever construction of boxes is also successful, as the segments are short allowing close supervision and the easy withdrawal of the internal shutter. For the cast-in-situ construction of full spans, the option of using an alternative to a box section should always be considered (*13.2*).

The bridge shown in Figure 12.2 was first built as a twin rib, allowing easy striking of the forms and advancement of the falsework. A bottom slab was then added in a second stage of construction, transforming the deck into a box. This is only likely to be an economically viable technique where the deck is at least 2 m deep, to provide working space, and where labour is relatively cheap.

9.4.2 Shear and shear lag

The bottom slab is stressed in longitudinal shear both due to the bending of the deck, as described for side cantilevers, and in torsion. The attachment of the slab to the webs requires shear reinforcement as described in 9.2. The slab is also affected by shear lag

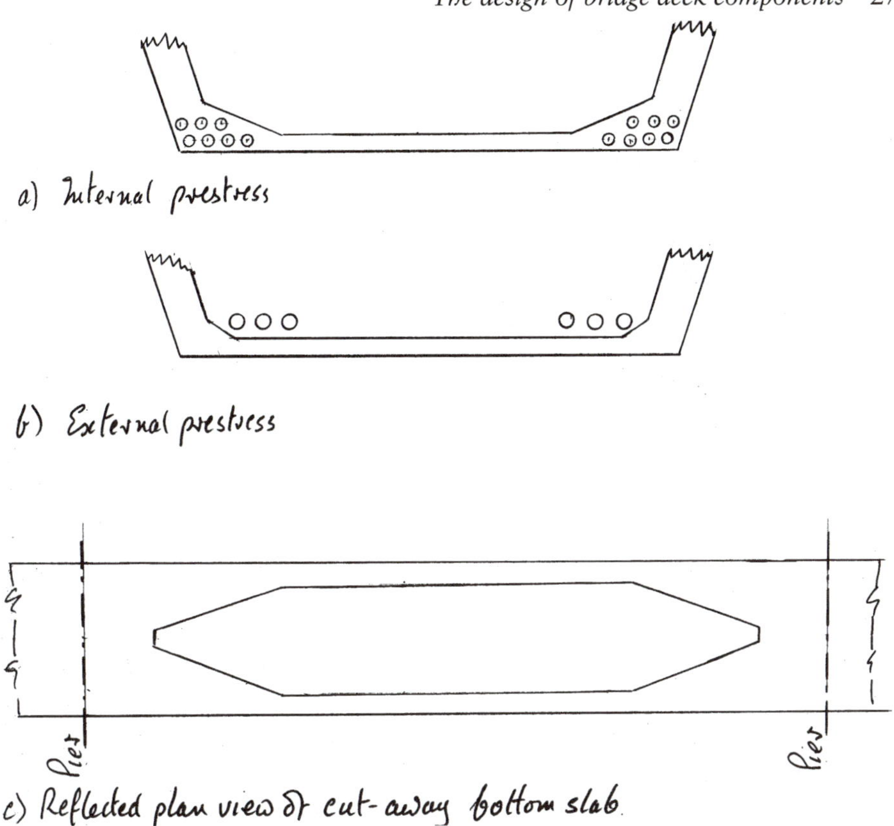

Figure 9.17 Bottom slab configuration

which must be considered in calculating the properties of the cross section and the SLS bending stresses. At the ULS the full width of the slab is mobilised.

9.4.3 Proportions of bottom slabs

Usually the slab carries no live loading, and has to span transversally between webs only under its own weight. As it is normally thinner than the webs, it is virtually built-in, and it is not unusual to find slabs with a span/depth ratio of 1/40. The absolute minimum thickness is recommended to be 180 mm for a cast-in-situ deck and 150 mm for precast. The bottom slabs of the East Moors Viaduct, Figure 8.7, and of the Stanstead Abbotts Bypass Viaduct, Figure 10.4, were both 150 mm thick. These were precast segmental decks and no problems were experienced. However, such thin slabs must only be used in the context of a careful design and a well-managed site; 170 mm for a precast deck and 200 mm for cast in-situ are more usual lower limits.

Bottom slabs are normally haunched at their junction with the webs. This is good practice as it reduces the shear stresses. If the prestress is internal, these haunches may be used for storage of the tendons. When torsional stresses are significant, it is desirable to provide corner fillets inside the box to avoid stress concentrations, and the haunches may be sized to fulfil this role. However, subject to the shear stresses being acceptable, haunches are not strictly essential. In decks with external prestress, the requirement to maximise the eccentricity of the tendons near mid-span generally leads to haunches being omitted, although there is usually room for a corner fillet, Figure 9.17 (b).

9.4.4 *Omission of the slab*

The thickness of the bottom slab for a continuous beam, and indeed the need for it at all as a bending member, depends critically on the span/depth ratio of the beam. Whereas for a span/depth of 1/20 or shallower it is likely that the full width slab will be essential, for a span/depth of 1/15 or deeper, a full-width bottom slab is not necessary.

However, omitting the bottom slab would transform a cross section from a torsion box, which re-centres eccentric loads, to a twin beam system, which would need to be dimensioned quite differently. Consequently, bottom slabs provide benefits even if they are not needed to control bending stresses. As a result, box girders with deep span/depth ratios are normally given a trapezoidal cross section, with the webs sloping inwards to reduce the weight and cost of the bottom slab while retaining its torsional benefits.

For simply supported box section decks, a full-width bottom slab is unlikely to be necessary to control the maximum bottom fibre compressive stresses during construction. However, the comments regarding the benefits of retaining a torsion box are applicable.

In long-span continuous beams, the bottom slab may be cut away locally at mid-span. Although saving little in concrete, the main advantages gained by such an arrangement are the reduction in weight and the raising of the neutral axis at mid-span, economising on prestress, Figure 9.17 (c).

9.4.5 *Trussed bottom slabs*

For box girders that do not need the bottom slab to control bending stresses, it is possible to provide a trussed member that will be much lighter than a solid slab. An example of the possibilities is a truss in the form of precast concrete ‘X’s. The precast units may be placed in the shutter for either cast-in-situ or precast segmental construction, with the web concrete creating the joint. As the torque builds up towards the piers, the ‘X’s may be made thicker. At the piers, the solid slab may be re-introduced if necessary to provide ductility at the ULS. Figure 9.18 shows a preliminary project for a river bridge with a trussed bottom slab designed by the author when at Arup.

Although a good idea in principle, it is relatively rare that such a scheme survives to the detailed design stage. The trussed bottom slab generally is equivalent in weight to a solid slab about 70 mm thick, saving perhaps only 100 mm with respect to a conventional solid slab. It may well be necessary to provide a suspended walkway within the box for maintenance purposes, and a mesh to stop birds entering the box. Clearly there is also the additional cost of moulds, storage and handling for the precast ‘X’s. However, designers should not be unduly discouraged by such considerations. It is a visually interesting and expressive feature which saves materials. It is up to the designer to justify its use.

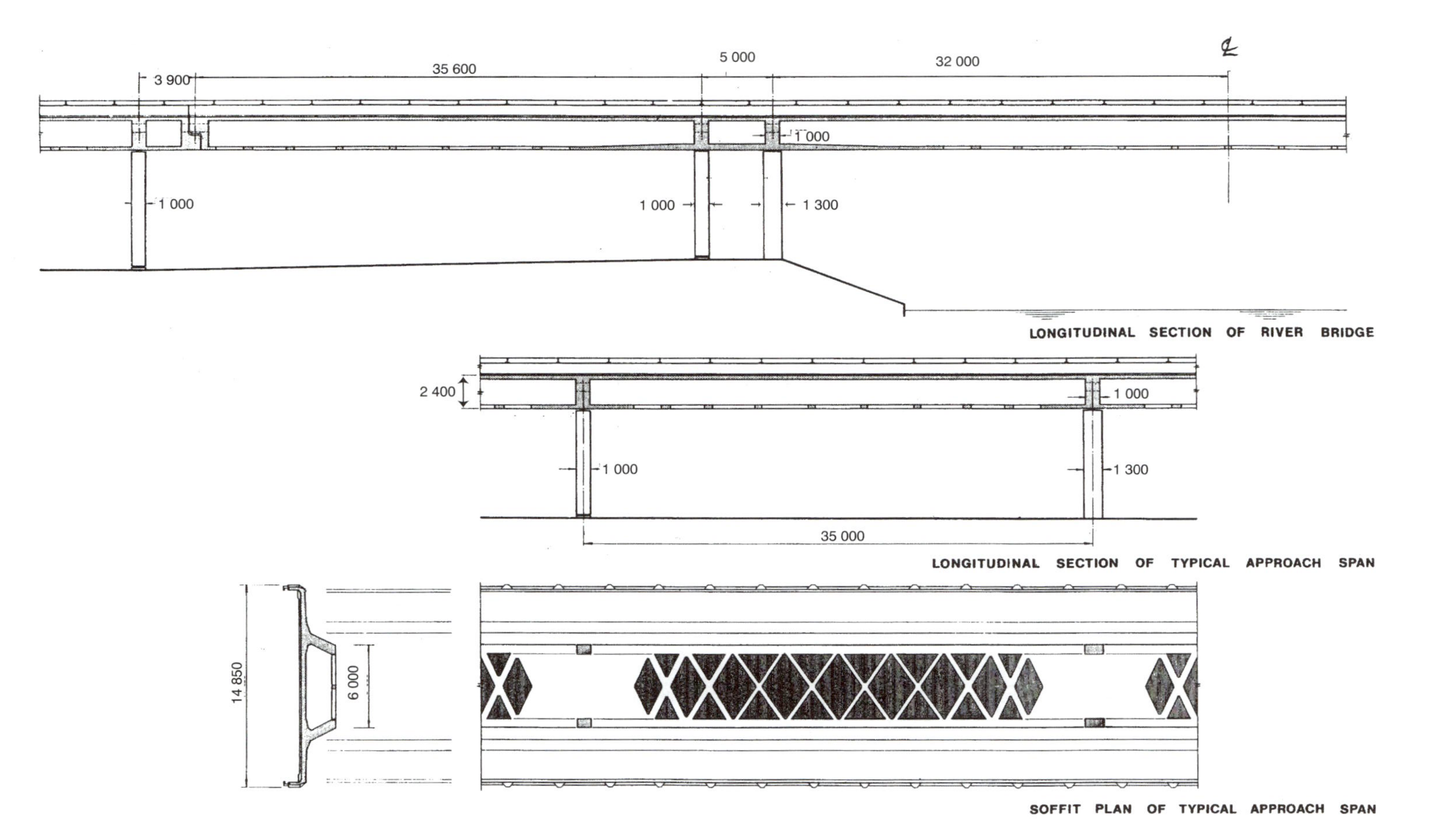

Figure 9.18 Project for bridge with trussed bottom slab (Adapted from Arup drawing)

9.4.6 Out-of-plane forces

When a bridge deck has variable depth, the bottom slab that is compressed longitudinally will be subject to out-of-plane forces. If the variation of depth is circular or parabolic, the compression in the bottom slab creates a distributed force normal to the slab P = F/ r where *F* is the compressive force in the slab due to all loads including prestress, and *r* is the local radius of curvature of the bridge soffit, Figure 9.19 (a). This force must be carried in transverse bending of the slab.

If the bridge deck has a linearly haunched profile, the forces in the bottom slab either side of the angle change give rise to an out-of-plane resultant, Figure 9.19 (b). It is quite likely that the bottom slab will not be strong enough to carry this force in transverse bending, in which case it will be necessary to provide a transverse beam at this location.

Even if the bridge soffit is flat, out-of-plane forces in the bottom slab are created at changes in its thickness, either tapered or stepped, Figure 9.19 (c).

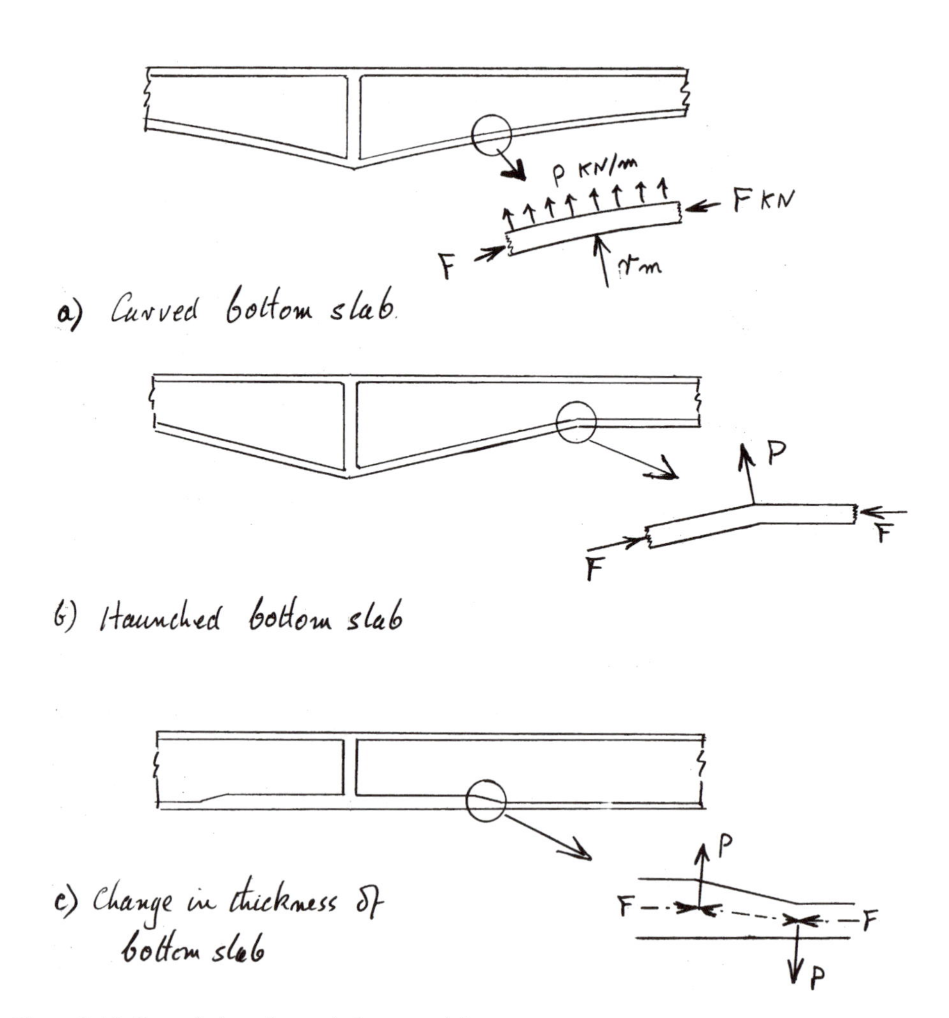

Figure 9.19 Out-of-plane forces in bottom slab

Particular problems arise close to the piers of bridge decks of variable depth. Usually, the deck is of constant depth over the width of the pier. Consequently there is a sharp change of direction of the bottom slab at the face of the pier, with the resulting large vertical force to be catered for. If a diaphragm is placed at this discontinuity, the load can be suspended directly. Usually, the diaphragm is located over the centre of the pier, and is thus some distance from the discontinuity. A thickened slab cantilevering off the diaphragm must then be provided to carry this force, Figure 9.20 (a).

If the deck is carried by piers consisting of twin walls which are continued through the height of the deck as diaphragms, under symmetrical loading the vertical component of the forces in the bottom slab are carried directly down into the walls. However, under asymmetric loading, the force in one of the walls will be reduced or even become tensile, when the vertical component of the bottom slab force must be hung into the diaphragm, Figure 9.20 (b). A similar situation may well occur during construction of such a deck in free cantilever, in an unbalanced phase.

Where the deck profile comes to a cusp over the pier, the vertical component of the bottom slab force may be taken directly into the bearing. However, the bearings are not normally as wide as the flange, Figure 9.21 (a). Thus the force in the unsupported length of slab must be hung from the diaphragm. Note also the unsatisfactory appearance when the cusp of the deck is not visually supported by bearings, Figure 9.22.

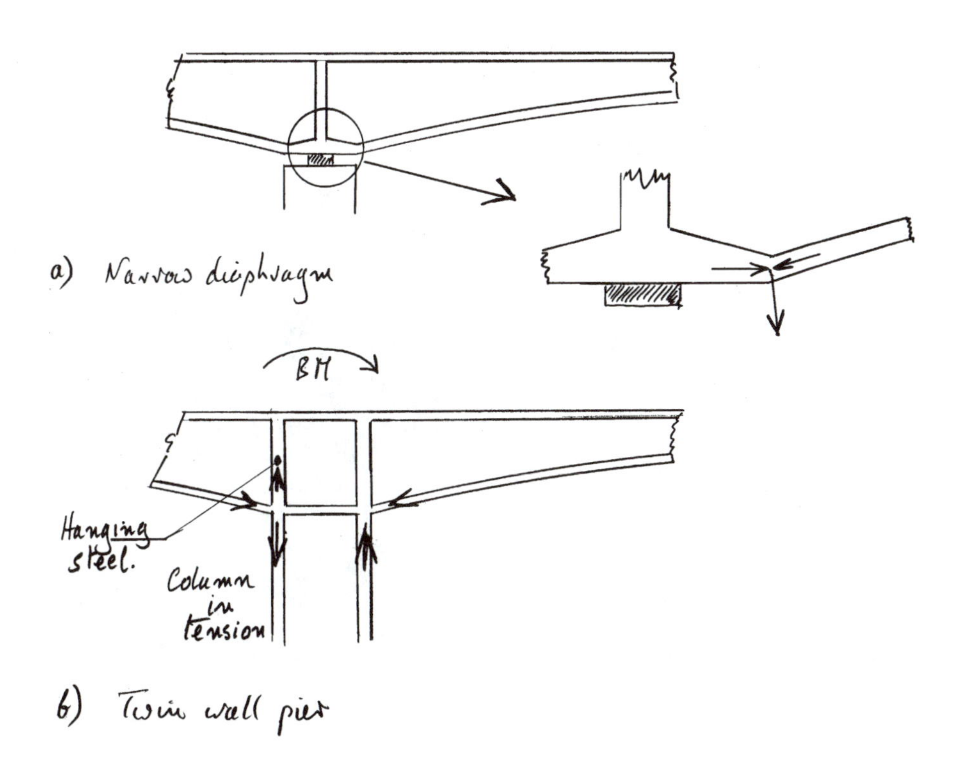

Figure 9.20 Bottom slab at pier

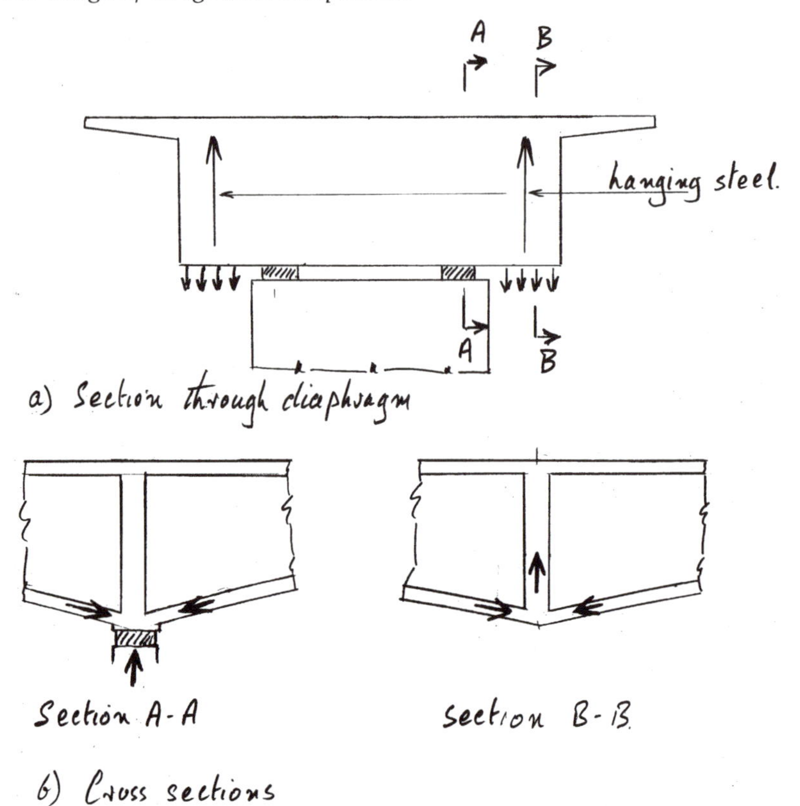

Figure 9.21 Variable-depth deck with inset bearings

Figure 9.22 Variable-depth deck with narrow pier (Photo: Robert Benaim)

9.4.7 Resal effect

The title to this section celebrates the brilliant French designer of bridges (including the Alexandre III Bridge over the Seine in Paris) who, in the nineteenth century, allegedly first documented this important aspect of the behaviour of bridge decks. The compressive force in the inclined bottom slab of a deck of variable depth that is subject to a hogging moment, carries a significant part of the shear force, Figure 9.23 (a). In fact, the inclined fibres of the web below the neutral axis also carry some of the shear force, but this calculation is more subtle and can frequently be conveniently neglected.

Comparison with an equivalent truss, Figure 9.23 (b), makes it clear that there exists an inclination of the bottom slab that would carry the entire shear; the force in the last tension diagonal would be zero. If the applied moment is M, the shear force S, the lever arm l_a and the inclination of the bottom flange with respect to the horizontal is α, the tension in the top member, and the horizontal component of the compression

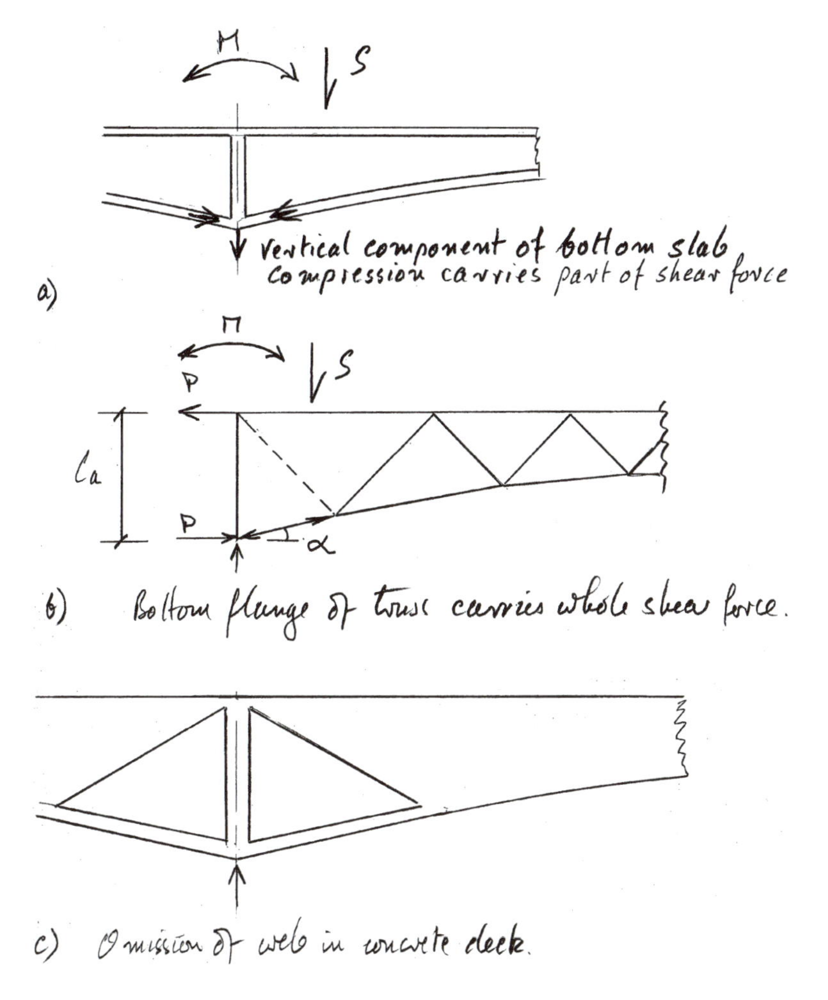
M
S
vertical component of bottom slab compression carries part of shear force
a)
M
S
P
l_a
P
α
b) Bottom flange of truss carries whole shear force.
c) Omission of web in concrete deck.

Figure 9.23 Resal effect

in the bottom member are $P = \pm M/l_a$, and the shear force carried by the bottom flange is:

$$S_R = P\tan\alpha, \text{ or } S_R = (M/l_a)\tan\alpha.$$

For the bottom flange to carry all the shear

$$(M/l_a)\tan\alpha = S, \text{ or } \alpha = \tan^{-1}(Sl_a/M).$$

The web may then be omitted close to the support, Figure 9.23 (c), which is one step in the logical development that leads to trussed webs.

9.4.8 Prestress anchor blisters on the bottom slab

It is often necessary to anchor prestress tendons in blisters in the corner between the bottom slab and the web. The design of these anchors is discussed in Chapter 6. However, it should not be overlooked that these anchors can create large tensile equilibrium forces in the thin slabs, which may cause severe cracking if they are not transversally reinforced or prestressed appropriately.

9.5 Webs

9.5.1 General

The functions of the webs are to carry the shear force in the deck, to support the top slab, and to carry the reactions of curved prestress tendons. For internally prestressed bridges, they also house the prestressing cables as they travel between the extreme fibres of the deck. In a box section deck, the webs are usually subjected to transverse bending moments imposed on them by live and dead loads on the top slab and side cantilevers, and to a lesser extent by the self weight of the bottom slab.

9.5.2 Width of webs

As has been argued in Chapter 8, it is important for the economy of a project that the number of webs is limited, and their thickness minimised. In the central part of the span of most bridges, the web thickness is not controlled by shear forces, but by the minimum that can reliably be built. This minimum depends on whether the bridge deck is precast or cast in-situ, on whether the concrete vibration is internal or external, on the inclination of the webs, on whether the webs carry prestressing ducts, and whether there are prestress anchor blisters at the bottom of the web. Casting and compacting the concrete in the webs and ensuring that it correctly fills the bottom slab or heel is perhaps the most critical activity in the construction of prestressed concrete bridge decks. It is certainly one of the activities most prone to give rise to construction failure and contractual claims. However success or failure is under the control of the designer.

The width of webs in twin rib bridge decks is subject to a different logic to that of other deck types, and is considered in Chapter 12.

a) Width of webs for simply supported precast Tee beams

Precast 'Tee' beams (Chapter 10) have vertical webs with internal prestressing ducts. They have a relatively compact heel, which should be shaped with ease of concrete flow in mind. The quantity of concrete required for each beam is usually about 30 m^3, and rarely exceeds 50 m^3; consequently there are no problems of excessively lengthy pours with the resulting operator and supervisor fatigue. As no significant bending moments are transmitted from the deck into the webs, the beams are lightly reinforced. Furthermore, they are made in the well-ordered environment of a casting yard. As a result of all these factors, the webs may be very slender.

Where steel shutters are used, the concrete may be compacted by external vibrators, and the thickness of the web may be reduced to the physical minimum, that is:

$$\text{Thickness} = 2 \times (\text{cover} + \text{shear links} + \text{lacers}) + D + 5 \text{ mm tolerance},$$

where D = the external diameter of the prestressing duct, Figure 9.24 (a). The nominal diameters of the reinforcing bars should be increased by 10 per cent to include an allowance for their ribs. Generally, this definition of web thickness provides adequate cover to the prestressing duct, but this needs to be checked. Windows in the top surface of the heel allow inspection and the insertion of poker vibrators if necessary.

For the Sungai Kelantan Bridge (*10.3.7* and Figure 10.14), where cover was 30 mm, links 16 mm, lacers 10 mm and the external duct diameter was 87 mm, this approach gave rise to webs that were 220 mm thick. Fifty-six beams 2.7 m high were cast without incident in this low-technology but well-run site in northern Malaysia. For the GSZ project (*10.3.8* and Figure 10.17), where the cover was 35 mm, the links and lacers were 12 mm and the external diameter of the duct was 71 mm, the web thickness was 200 mm.

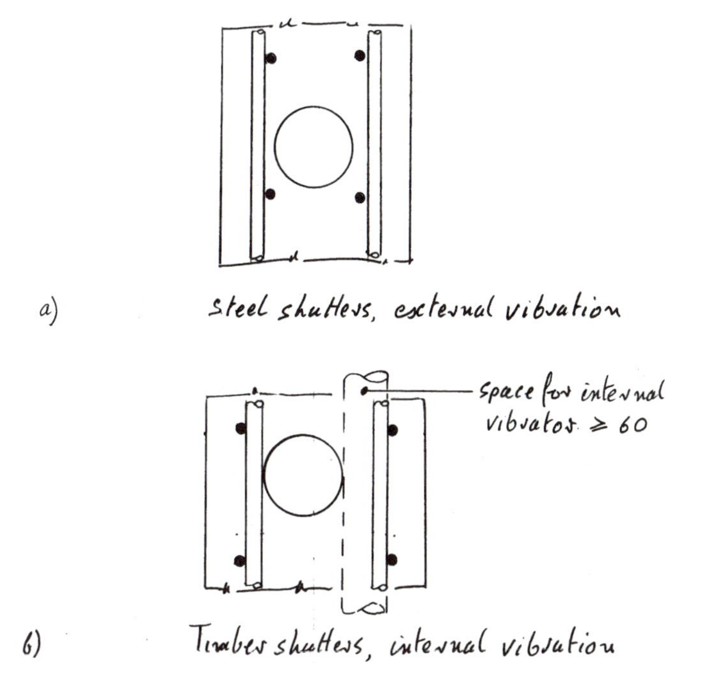

Figure 9.24 Thickness of webs for Tee beams

If the same beams had been cast in timber shutters using internal vibrators, it would have been necessary to increase the thickness of the webs to allow poker vibrators to pass between the prestress ducts and the web lacers. It would under these circumstances be sensible to put the lacers outside the shear steel. With an 87 mm duct, this would give rise to a minimum web thickness of about 270 mm leaving a chimney of at least 60 mm for a 50 mm poker vibrator, Figure 9.24 (b). Clearly this minimum thickness will need to be adapted for different duct and reinforcement diameters and different concrete cover. It should be noted that external vibrators do not work properly on timber shutters; timber shutters shake under the effect of vibrators, which tends to help the concrete flow, but they do not transmit the correct vibration frequency to compact it.

Generally, the parabolic shape of the cable profile for statically determinate beams effectively counteracts the shear force, and it is not necessary to thicken the webs towards the supports. However, at the extreme ends of the beams, for a length usually approximately equal to their height, it is necessary to thicken the webs to house the prestress anchors and contain their bursting forces. For instance, the 2.7 m high Sungai Kelantan beams were thickened to 500 mm, for a distance of 2500 mm, including a 500 mm chamfer.

b) Width of webs near the mid-span of continuous box girders

In continuous beams the tendon profiles do not balance the shear forces as efficiently as in statically determinate beams, leading to heavier shear steel. Also, the webs of box decks are subjected to significant transverse bending moments, further increasing the density of the reinforcement. There is frequently the addition of equilibrium reinforcement related to face or blister prestress anchors. Consequently, the webs of box girder bridges are more highly reinforced than those of precast beams. This more dense reinforcement makes it more difficult to compact the concrete. Furthermore, whereas precast beams have a compact heel that is relatively easy to fill with concrete, box girders have a wide bottom slab that is always troublesome to cast.

The concrete must flow from the web into the bottom slab of the box. Near mid-span the bottom corner of the box usually houses prestressing ducts that are typically parked in the bottom of the web and in the bottom slab haunch. In some sections as the tendons change position to move up the web this bottom corner may be significantly congested with tendons in addition to the reinforcement linking the bottom slab to the web. Prestress anchors may be housed in blisters situated at the junction of the web and the bottom slab requiring a dense grillage of bars crossing the web which is superimposed on the tendons and reinforcement described above, Figure 9.25 (a). The consequence of this inevitable congestion of tendons and reinforcement is that, in some critical positions in the span, it is often difficult to coax the concrete to flow from the web into the blisters and the bottom slab, Figure 9.25 (b). These difficulties in casting and compacting the concrete are made more severe if the webs of the box are sloping.

All these reasons lead to a more conservative approach to sizing the webs of box girders. However, they are all under the control of the designer, who can make a good job of ensuring that his design is buildable, or on the contrary can create the most intractable problems for the contractor. It is reasonable to affirm that the majority of failures of compaction in bridge decks are due to poorly detailed webs, Figure 9.26.

Figure 9.25a STAR Viaduct: typical reinforcement cage for precast segment (Photo: Benaim)

Figure 9.25b East Moors Viaduct: failed pour for trial segment (Photo: Benaim)

Figure 9.26 Failed web pour (Photo: Robert Benaim)

It is principally for this reason that box girders are best built in short lengths, either precast or in situ. See Chapter 13 for a further discussion on the construction of box sections.

When the box girder is being precast in short lengths in steel shutters equipped with external vibrators, in the controlled environment of a purpose-built casting yard, the web thickness may be minimised. However the reinforcement should provide a chimney 60 mm wide to allow the passage of a 50 mm internal poker vibrator to assist in moving and compacting the concrete, as prolonged use of external vibrators can harm the concrete. The normal position for the lacers in the web of a box girder is inside the stirrups, allowing the transverse bending strength of the web to be maximised. Thus the minimum thickness of the web with one row of tendons would be given by:

$$\text{Thickness} = 2 \times (\text{cover} + \text{shear link} + \text{lacer}) + D + 60\ \text{mm} + 10\ \text{mm tolerance}$$

(where D = the external diameter of the prestressing duct), with an absolute minimum of 300 mm, Figure 9.27 (a). The tolerance is necessary as it is impossible to bend and fix the more complex reinforcement of box girders with mechanical precision. Where box girders are cast in-situ in short lengths in timber shutters, such as in balanced cantilever construction, the concrete must be compacted by internal vibrators only, and it is sensible to make certain that there is adequate access by increasing the vibrator chimney to 100 mm.

When there are two rows of prestressing tendons, there should be a 70 mm gap between them, Figure 9.27 (b). If the web is inclined, the vibrator chimney should be at least 100 mm measured normal to the plane of the web, Figure 9.27 (c).

Where box girders are cast in-situ in long lengths, the difficulty of supervising the concreting gang should be taken into account, and a minimum thickness of web

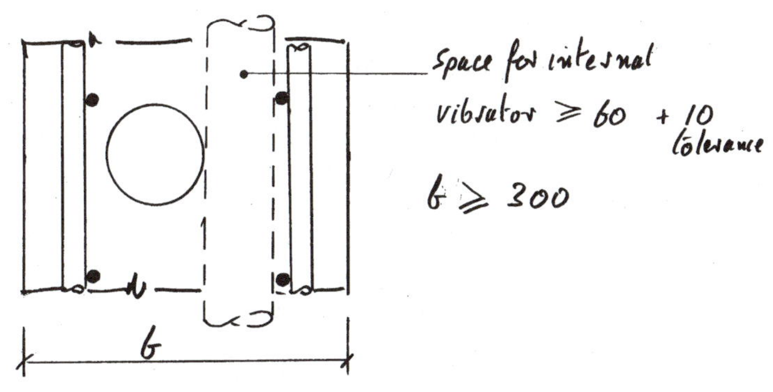

a) Precast segmental girder with vertical webs

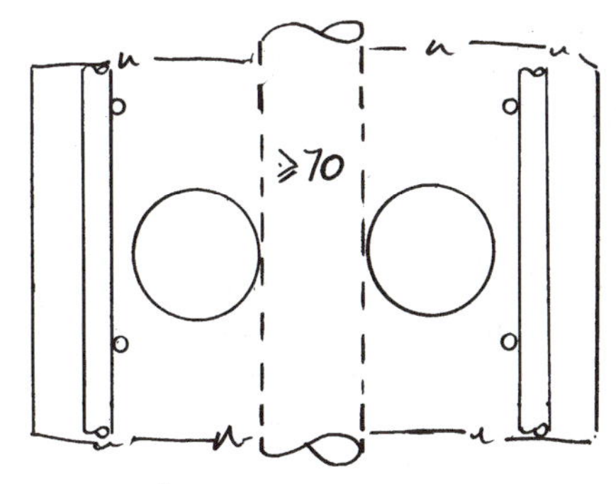

b) Two rows of ducts

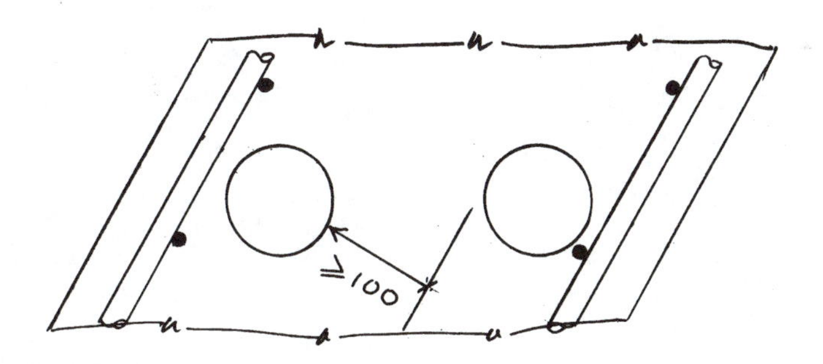

c) Two rows of duct and inclined webs.

Figure 9.27 Thickness of webs for box sections

of 400 mm should be adopted, with a gap between rows of tendons not less than 100 mm.

When small prestressing ducts are used, with an internal diameter of 55 mm or less, it is possible to bundle them together in pairs for the purpose of defining the web width as described above. However, such small tendons are rarely used in bridge deck construction.

When the use of self-compacting concrete is fully established for the construction of bridge decks, it is likely to eliminate the need for internal vibration and the rules on the safe thickness of webs will need to be redefined.

If the prestress is external, the webs will be less congested, and it is suggested that the absolute minimum thickness for webs for box girders should be 250 mm for precast construction, and 300 mm for cast-in-situ decks.

Other factors may determine the web thickness. For instance, in balanced cantilever construction where tendons are anchored on the face of the webs, the minimum concrete thickness may be defined by the requirements of the anchors. However, this constraint can be overcome by locally thickening the web at the anchor positions. On the East Moors Viaduct for instance, the 300 mm thick webs were locally thickened to carry face anchors that required a thickness of 450 mm, and this thickening was integrated with the bottom corner blisters, Figure 9.28. If powerful prestress tendons are anchored in blisters attached to the web, it may need to be sized to safely resist these forces. Finally, if the transverse spans of the side cantilevers or of the slab between the webs are unusually large, the bending moments applied to the top of the webs may determine their thickness.

c) Width of webs near the supports of continuous box girders

The minimum thickness of webs at the support sections is defined by calculation. In most modern codes of practice that consider shear and torsion at the ULS, this minimum is controlled by the principal compression in the concrete.

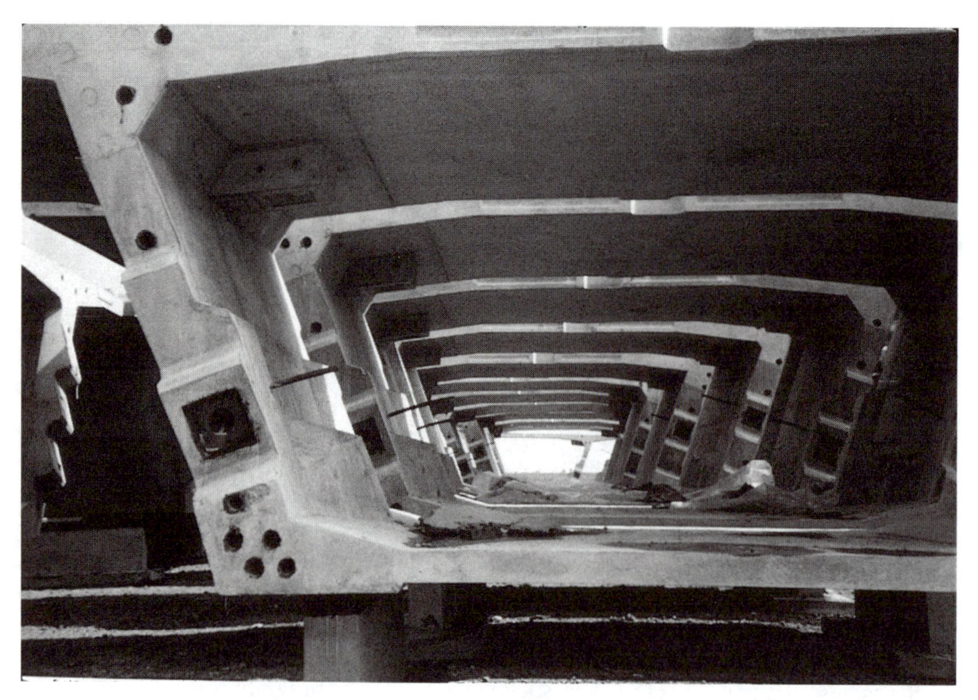

Figure 9.28 East Moors Viaduct: face anchors in locally thickened webs (Photo: Benaim)

A rule of thumb for making a preliminary estimate of the total web thickness required for a continuous bridge when designing to the UK loading code (45 units of HB loading) is $t = LB/200H$, where t is the total web thickness, L is the span, B is the width of the deck and H is the height of the deck at the pier. This assumes that the concrete cube strength is at least 45 MPa and that prestress ducts are in single file in the web. For stronger concrete and external prestress the formula is conservative.

For codes of practice that consider shear at the SLS, the minimum thickness is governed by the principal tensile stress. The rule of thumb is not applicable to such cases.

d) Width of webs near the quarter points of continuous box girders

Between the pier and mid-span, the thickness of the web may vary. The rate at which the thickness can be reduced depends critically on the detailed design of the prestress. At the supports the prestress centroid is high, while at mid-span it is low. This gives the opportunity to incline the tendons in elevation, creating shear of a sign opposite to that due to applied loads. The profile of selected tendons may be designed such that it is possible to shorten the length of web that needs to be thickened, saving weight and materials, Figure 9.29.

The changes in thickness of the webs of box girders along their length add significantly to their cost of construction. One of the main benefits of using very high strength concrete for bridge decks is the possibility of keeping the web thickness constant along the span, without adding too much to the deck weight at mid-span.

9.5.3 Vertical prestressing in webs

When designing to codes of practice that require the principal tensile stress to be limited, it is sometimes cost effective to reduce it by the addition of vertical web prestress. However, prestressing strand extends during stressing by 6 mm/m, and prestressing bars by 3.5 mm/m. Thus for a 2.5 m deep web, prestressing strand will

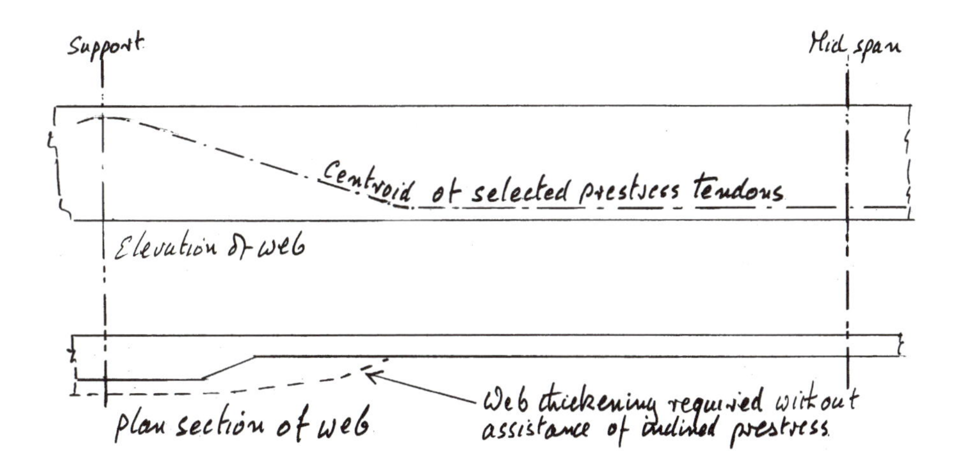

Figure 9.29 Use of inclined prestress to shorten web thickening

extend by 15 mm, and a bar by only 8.75 mm. Consequently, a considerable portion of the jacking force in short tendons will be lost by the pull-in of wedges or the set of nuts. Vertical prestress should not be used on decks less than 2.5 m deep, and close supervision of the stressing operation is necessary, with checks made on the residual force in a sample of the tendons.

9.5.4 Transverse bending moments applied to webs

The webs of box girders are subjected to significant bending moments derived from the actions of the top slab and side cantilevers. These moments are calculated from a frame analysis, and diminish rapidly down the web. The web is also subjected to bending moments due to the distortion of the box under eccentric live loading. The reinforcement for these moments may need to be added to the shear steel, particularly at the top of the web. However, care should be taken that incompatible load cases are not added together; it is unlikely that the maximum bending moments on the web occur concurrently with the maximum shear.

The design of the bending reinforcement in the web should take into account the effect of compression due to the weight of the slab and of any co-existent live loading. For further advice on the design of web reinforcement, subject both to shear and bending, see [1].

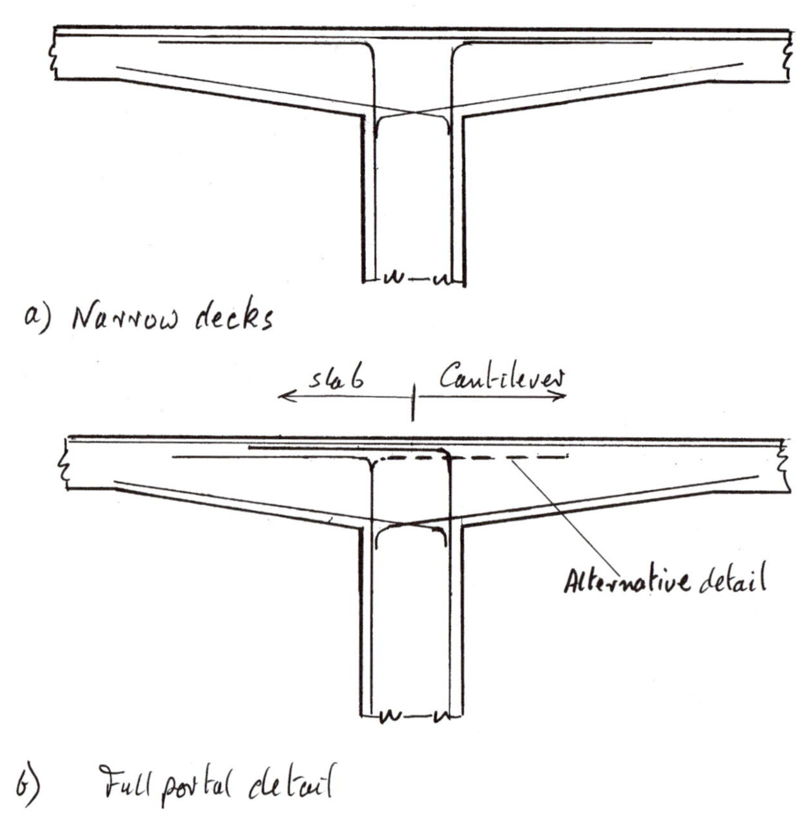

Figure 9.30 Detailing shear links

Reinforcement details of the type shown in Figure 9.30 (a) are suitable for decks up to about 13 m wide, where the bending moments from side cantilever and internal slab are reasonably well balanced. The advantage of this layout is that it reduces the congestion of reinforcement over the web, facilitating the concrete casting. Where the moments turning down into the web from the slab are more substantial, a full portal detail is required, Figure 9.30 (b). Bending moments from the cantilever may be shared between the top slab and the web, while bending moments from the top slab can only turn down into the web. Consequently, portal type corner reinforcement should in general be designed to favour the moment from the slab rather than from the cantilever.

9.5.5 Loads applied at the bottom of webs

For through bridges, and for other special cases, some, or all, the live load may be applied to the web via the bottom slab. The web is thus in tension, and requires hanging steel that is in addition to the shear steel. The bending moments in the web will be combined with a direct tension.

9.5.6 Truss analogy

Truss analogy is described in Chapter 3. However, it is worth mentioning it again in the context of this section on webs, as it is the perfect tool for resolving many difficult design problems; when for instance some loads are applied half way down the web, where holes need to be made in a web or where the web changes in depth close to the support.

9.5.7 Trussed webs

Reducing the weight of the deck and reducing the quantity of expensive prestressing steel are two of the principal aims of the concrete bridge designer. In order to decrease prestress the deck must be made deep, but that in turn increases the weight of the webs. One solution to this dilemma is to use trussed webs.

In the early days of reinforced concrete, when labour was still relatively cheap compared with materials, reinforced concrete trussed bridges were quite frequently built. However, in addition to their high labour content, these bridges suffered from a technical problem that contributed to their eclipse for many decades. As all the nodes of these trusses are inevitably monolithic rather than pinned, the deflection of the truss gives rise to secondary moments in the members. These secondary moments are the greater, the stockier the proportions of the truss. Thus the designer tends to chase after his tail, having to increase the size of members to carry the bending moments, and thereby increasing those moments.

Three developments have combined to make concrete trusses viable structures once again. First, if the deck is prestressed, the deflection of the truss under dead loads is greatly reduced and may even be cancelled. Consequently the secondary moments in truss members are greatly reduced at source. Second, member prestressing may be used effectively to counteract the secondary moments in members. This technology is only viable on deep trusses where the members are long enough to allow prestressing

Figure 9.31 Viaduc des Glacières on the A40 motorway, France (Photo: Robert Benaim)

to be used economically and effectively, requiring a deck depth of some 3.5 m or greater. Third, the development of high strength super-plasticized concrete allows the members to be more slender, further reducing the magnitude of these secondary moments.

Examples of modern trussed web bridges are the Viaduc de Sylans and the Viaduc des Glacières on the A40 motorway near Nantua in France, designed and built by the contractor Bouygues, Figure 9.31.

The Dong Po bridges on the Guangzhou–Shenzhen–Zhuhai Super Highway project in South China will be used to illustrate some of the issues involved in the design of such trussed webs. These bridges were designed by Benaim with the assistance of Leo Leung of the client organisation, Hopec Engineering Design Ltd, a subsidiary of Hopewell Holdings. The design was progressed to full working drawings; construction was cancelled shortly after foundation work had commenced [7, 8]. The cable-stayed Ah Kai Sha Bridge which formed part of the same project, also had trussed webs and is described in more detail in Chapter 18.

The Dong Po Bridges North and South were part of a family of trussed web bridges, with spans ranging from 80 m to 160 m. The two bridges crossed channels of the Pearl River close to Guangzhou. They both were double deck, carrying the dual four-lane motorway of the Guangzhou East South West Ring Road on their upper level and a dual three-lane local road on a lower level. Each pair of carriageways was carried on a separate box girder. Dong Po North had spans of 55 m, 3 × 80 m, 55 m, while Dong Po South had spans of 110 m, 3 × 160 m, 110 m, Figure 9.32. The decision to adopt trussed webs was taken both for economy of materials and for the natural ventilation afforded to the lower level. Highway clearances to the lower level defined the dimensions of the box, which were 14.2 m wide and 7.2 m high. This gave an internal vertical clearance of 6 m, which made provision for overhead road signs and left provision for ventilation fans, Figure 9.33 (a) and (b).

Figure 9.32 Dong Po South Bridge (Image: David Benaim)

The bridges were to be built in balanced cantilever. All the intermediate piers consisted of four columns that were built into the deck, providing a stable base for construction. An 'N' truss configuration was chosen for the webs, which allowed a module of 5 m that was suitable for the method of construction. The thickness of the truss members was fixed at 800 mm. The width of the members varied according to the shear forces acting on the deck, Figure 9.33 (b). For the 160 m long spans, the webs became solid at a point about 30 m from the pier, and then thickened outwards to 1,050 mm for the last 20 m. Figure 9.34 shows the shear force envelope for the family of decks.

The inclined members of the truss were in compression and the vertical members were in tension. The principal difficulty in designing these trusses was the presence of secondary bending moments in the members. These bending moments were both in the plane of the truss, due to its deflections and to the end moments of the top and bottom booms, and out of plane, due to the end moments of the top and bottom slabs. During design, it was discovered that the in-plane secondary moments could be significantly reduced by rotating the diagonal such that it met the vertical slightly inside the point of conjunction of the truss members (below the upper point and above the lower point of conjunction), Figure 9.35 (a).

The tension verticals were prestressed with between one and four tendons, each of 19/15 mm strand. These tendons accomplished three tasks. Firstly they carried the direct tension in the truss member. Secondly, they were displaced about the member centre line in the plane of the truss to help carry the in-plane secondary moments. Thirdly they were displaced out of the plane of the truss to help carry the out-of-plane moments, Figure 9.35 (b).

It was found to be impractical to design the truss members under their combination of direct load and biaxial bending to criteria of zero tensile stresses. Stresses under working dead loads were designed to be entirely compressive, while under the effects of live loads the members were designed at the ULS, with a limitation on crack width at the SLS.

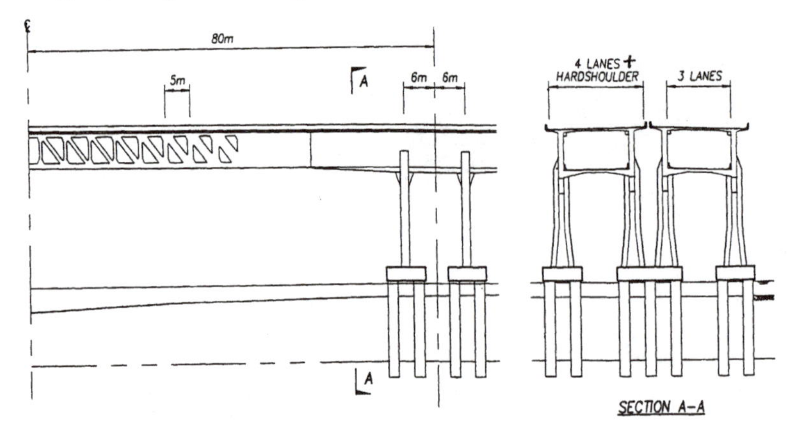

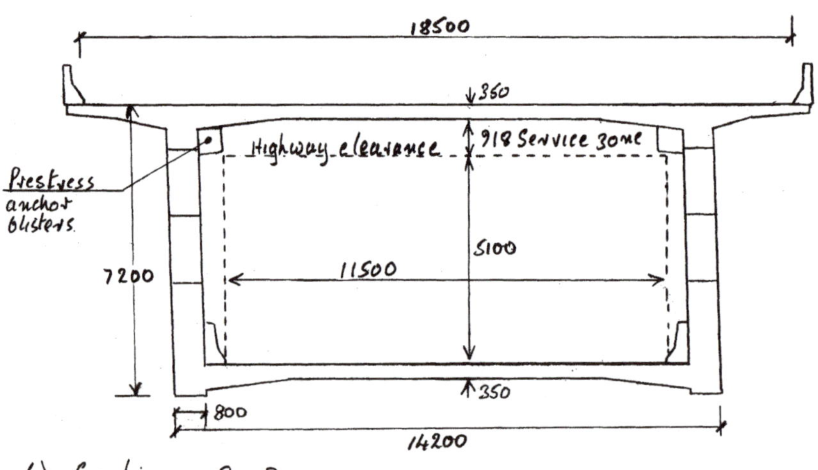

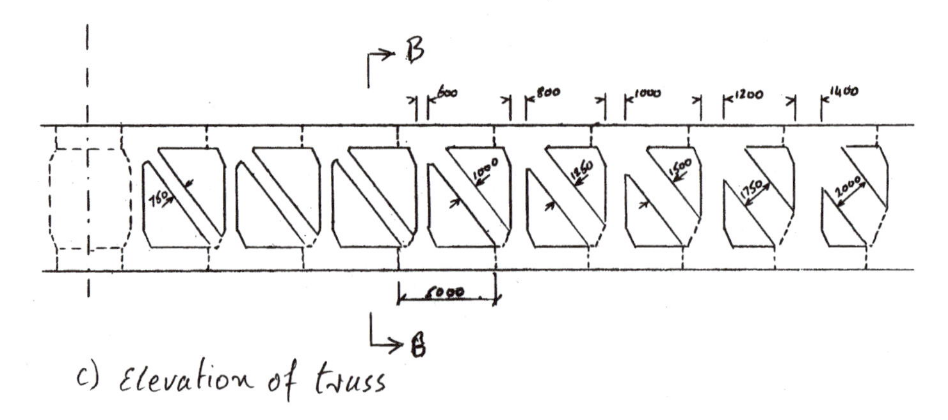

Figure 9.33 Dong Po South Bridge: details

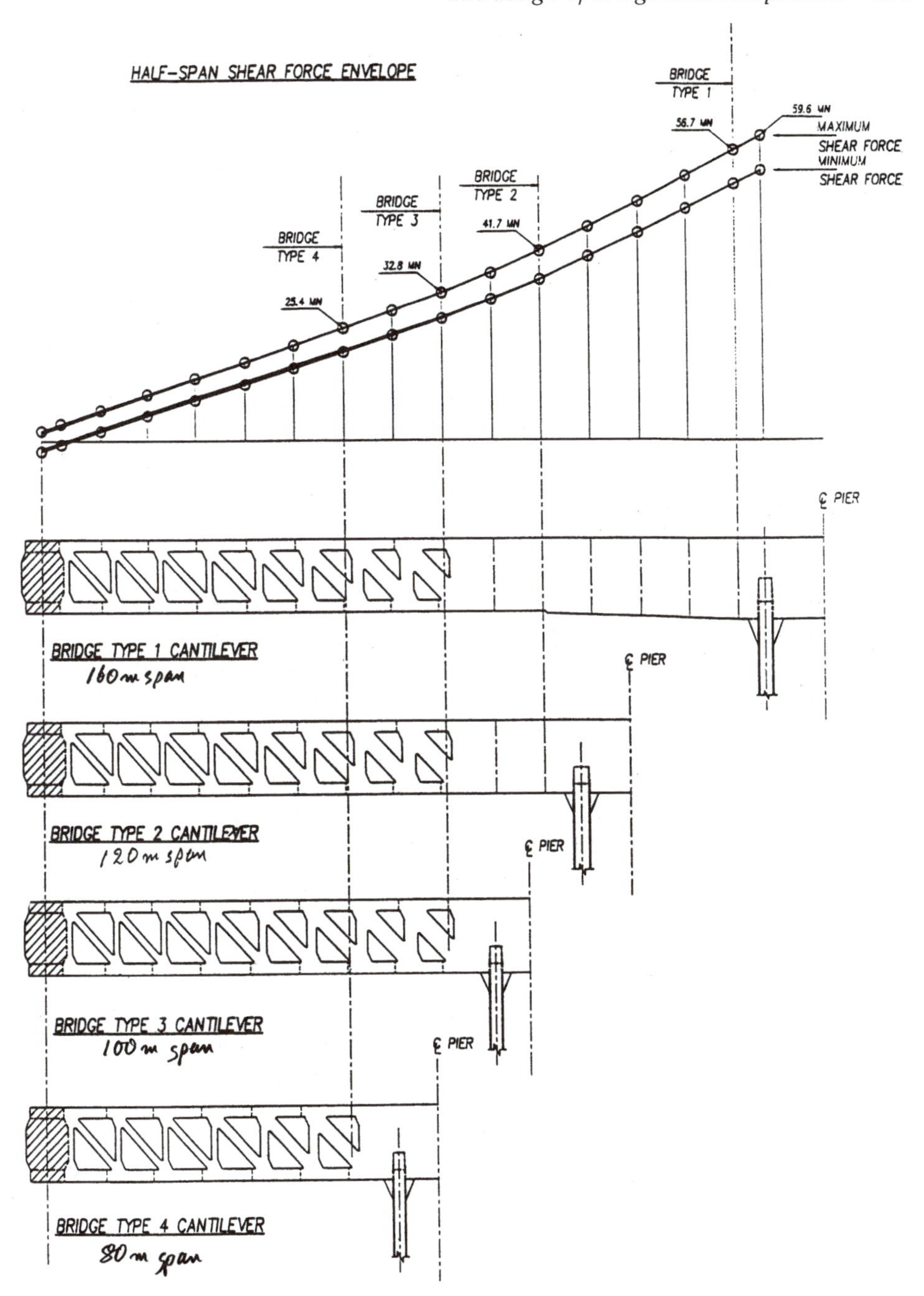

Figure 9.34 Dong Po Bridges: shear force envelope

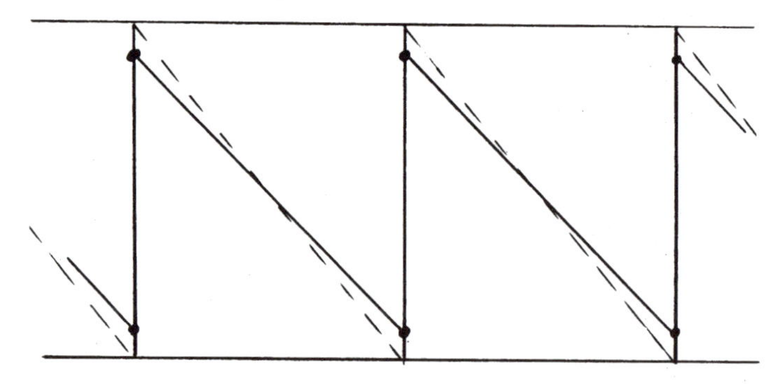

a) Rotation of diagonals

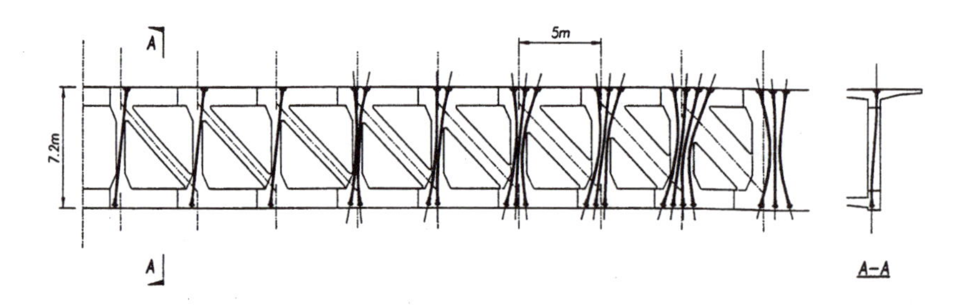

b) Prestress of truss members

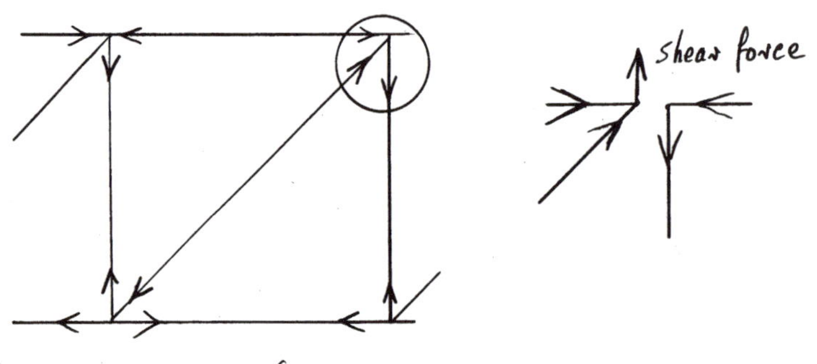

c) Transfer of shear force across a node.

Figure 9.35 Dong Po Bridges: details of trussed webs

The use of precast segmental construction of the bridge deck was considered, but rejected due to the perceived impossibility of making a satisfactory shear connection across a precast node. The full shear force in the deck is transferred across the node of a truss, Figure 9.35 (c). If a vertical joint plane crosses the node, this plane has to transfer the shear, either in friction or through shear keys. However, the node may be put into direct tension by the overall bending of the deck, under normal loading or at the ULS. It did not appear to be possible to design such a shear connection, particularly with the very large shear forces associated with this massive deck.

Clearly, the main issue with casting the trusses in-situ was the possible difficulty of effectively filling the moulds and compacting the concrete through dense reinforcement, particularly for the diagonal and the bottom longitudinal member. This was put to the test by building a full-size test panel, complete with reinforcement, Figure 9.36. It was discovered that the scale of the truss was such that the diagonal could be cast and the concrete vibrated through trapdoors located at several locations in its top surface. Similarly, there was no problem of access to the bottom member. However, for shallower trusses with more slender members, these problems would be likely to dominate the design. The use of self-compacting concrete would make the use of smaller trusses far more attractive.

Figure 9.36 Trial panel of truss (Photo: Engineer Leo K.K. Leung, Executive Director of Hopewell Highway Infrastructure Ltd (Hong Kong Stock Code: 737))

Much of the effort and ingenuity in the design of bridge decks is devoted to minimising the under-employed, expensive and heavy concrete in solid webs. Thus the spans of side cantilevers and deck slabs are made ever longer in an attempt to minimise the number of webs. However, once designers have learnt to use light trussed webs made of very high strength concrete, and once the use of self-compacting concrete becomes fully acceptable to bridge designers, it may well be more cost effective to cut down the span and hence the thickness of the top slab by using more closely spaced trussed webs.

The use of trussed webs for bridge decks with a depth in excess of 3.5 m is a very satisfying solution. However, it requires considerably more design expertise and effort than a conventional solid web.

9.5.8 Steel webs

The logic of attempting to eliminate the under-employed concrete of webs leads the designer beyond trussed concrete webs to the use of steel webs for prestressed concrete decks. These may be in the form of stiffened or corrugated plates or of trusses. The author has no experience of the use of such webs.

It is very well worthwhile carrying out full-scale research by designing and building such decks, as long as all parties to the design and construction contracts are aware that there may be no short-term economic benefits. It is by experimenting in this way that progress is made.

9.6 Diaphragms

9.6.1 General

Diaphragms are transverse beams spanning between main beams or box girders, or within box girders to stiffen the cross section. Diaphragms significantly complicate the construction of any concrete bridge deck. For instance, in cast-in-situ twin rib bridges they impede the launching of the falsework from one span to the next. In precast 'T' beam decks they typically add 4–5 days to the construction of each span, while in cast-in-situ box girders, they make it impossible to launch an internal shutter forwards, imposing a piecemeal dismantling and reassembly. They create hard points for the loaded slabs, making it necessary to thicken them locally and increase their reinforcement and they impede maintenance access from span to span. In precast segmental construction their weight often governs the lifting capacity needed for erection, even when the pier segments are shortened, and they need special moulds which reduce the productivity of the process.

Diaphragms are classed either as 'intermediate', occurring within spans, or as 'support', situated at piers or abutments. Modern methods of bridge deck analysis have demonstrated that intermediate diaphragms are generally unnecessary, and the tendency in design has been to omit them. Their use within box girders has completely ceased, as it has been realised that it is better to carry the bending moments caused by distortion in frame action over a considerable length of box rather than concentrating them on discrete diaphragms. For decks consisting of multiple box girders, support diaphragms between boxes are generally only used at abutments. Consequently, the following discussion concerns only support diaphragms within box girders. (See also the discussion on the use of diaphragms in 'Tee' beam decks in Chapter 10, and on support diaphragms in twin beam bridges in Chapter 12.)

Support diaphragms have three principal uses:

- they transfer the load from the webs of box girders to inset bearings;
- they stabilise box girders against the effects of horizontal loads;
- they resist the sway effect of torsion at piers and abutments.

They may also be used to anchor prestressing tendons, or to deviate external tendons. They may house drainage pipes and provide the anchorage for roadway expansion joints.

The rational and economical design of diaphragms is one of the more difficult tasks in the design of bridge decks, particularly if acute reinforcement congestion is to be avoided. Truss analogy is the best tool for this task. Once a coherent truss has been defined, this technique allows the diaphragm to be sized rationally. However, it is always wise to check the resulting sizing by simple bending theory. Finite element analysis may also be used in conjunction with truss analogy, as described in *3.10*. However, normally a check using bending theory is quite adequate to give the designer confidence in his sizing.

9.6.2 Transfer of load to the bearings

If the bearings are placed directly beneath the web, no diaphragm is required for the transfer of load. The other extreme is when a box girder is carried by a single central column requiring a diaphragm that acts as a stocky double cantilever, carrying the full deck reaction at its ends. Between these two extremes there is the general case of a box girder where the bearings are inset from the webs.

A typical box girder with a single central bearing is shown in Figure 9.37 (a). At the ULS the shear in each web of the box is applied to the support by a 45° shear field, as shown diagrammatically in Figure 9.37 (b) (*3.10.8*). As there is no bearing beneath the webs, the reaction must be lifted by hanging steel up to the top of the web so it may be applied to the diaphragm truss shown in Figure 9.37 (c).

The angle of the compression struts of the diaphragm truss should not be chosen flatter than 45°, although it may be steeper. The thickness of the diaphragm may be assessed by estimating the size of the compression struts, and checking the compressive stresses. The tension forces T_1 and T_2 may be carried by reinforcement or by prestress. Reinforcement must be correctly anchored beyond the nodes of the truss. The tension forces S represent the shear in the diaphragm, and require reinforcement with stirrups. The truss analogy also allows an access hole in the diaphragm to be located and sized.

The hanging steel is additional to the shear steel in the web and in the diaphragm, and consequently the web/diaphragm node may become congested. In this case the hanging steel may be spread over the area shown shaded in Figure 9.37 (d). However, as this requires some of the web compression struts to be deviated out of the plane of the web axis, it is necessary to provide fillets between diaphragm and web as shown. When struts or ties are deviated out of plane, secondary forces are set up which must be taken into account and reinforced for.

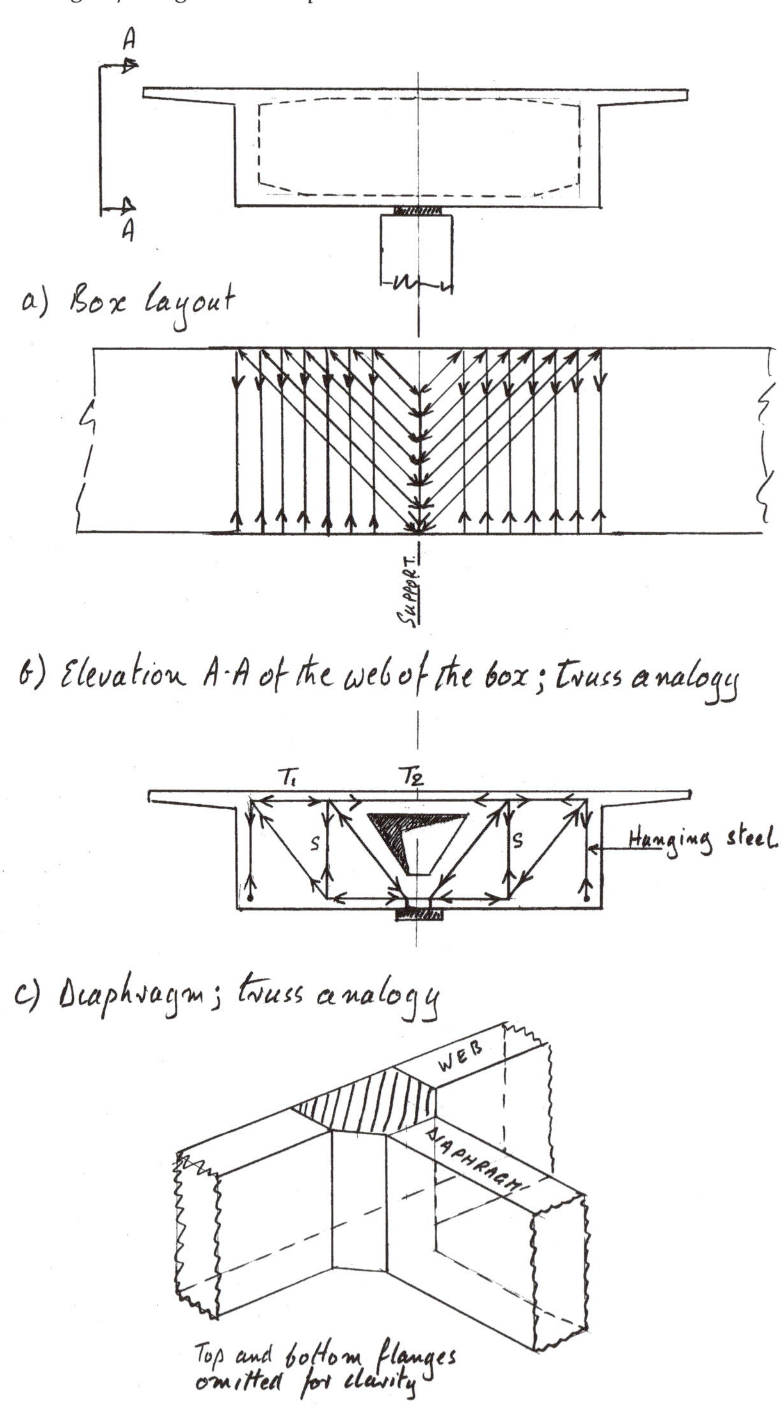

Figure 9.37 Diaphragm for box on single bearing

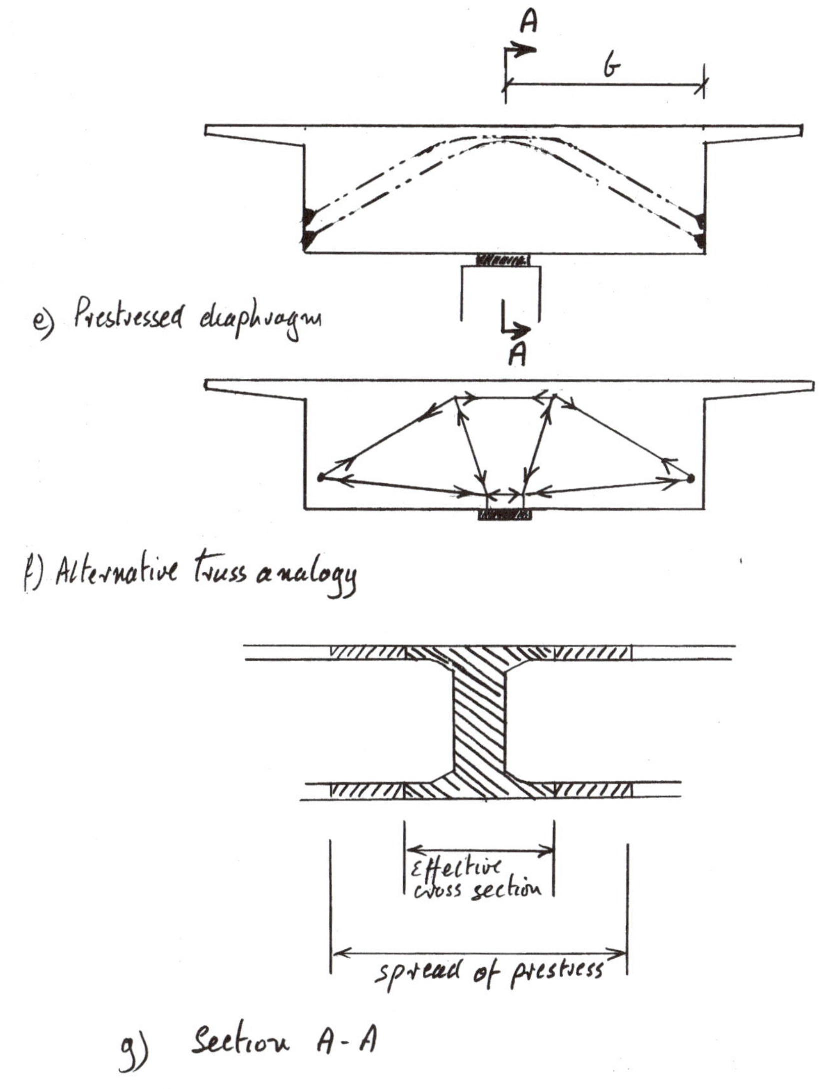

Figure 9.37 continued...

Whenever truss analogy is used in this way, the dimensioning of the reinforcement should be checked against bending theory, and the shear stress in the diaphragm calculated by bending theory should be kept to the threshold defined by the code of practice.

The diaphragm may be prestressed using an alternative truss as shown in Figure 9.37 (e) and (f). The prestress may be dimensioned to control the bending stresses at the SLS (as described in Chapters 5 and 6). The effective cross section of the prestressed diaphragm for calculation of the stresses should include widths of top and bottom slab equal to one tenth of b in Figure 9.37 (e) either side of the diaphragm web, although the prestress compression will spread at an angle of 35° into the top and bottom slabs, beyond the effective section, Figure 9.37 (g). The inclined tendons may be dimensioned to carry the ULS shear in the diaphragm, allowing a narrower diaphragm web. However, it may not be economical or convenient to carry the whole shear in prestress, and it is possible to superimpose the two truss models, each carrying a part of the load. With the prestress arrangement shown in Figure 9.37 (e) the prestress tendons pick up the web shear at the bottom of the web, eliminating the need for hanging steel.

The best design for a heavily loaded diaphragm is offered by partial prestressing, where the prestress maintains the top fibre of the diaphragm in compression under permanent loads, and reduces the requirement for hanging steel and for shear reinforcement. Passive reinforcement is provided to control the width of bending cracks at the SLS, and to make up the shortfall in bending and shear at the ULS. This approach reduces the congestion which is the principal problem of reinforced diaphragms, and avoids the very intense prestress required for an entirely prestressed solution. Unfortunately, some codes of practice do not allow partial prestressing.

The prestressed diaphragm of the River Lea Viaduct on the Stanstead Abbotts Bypass (*15.3.4*) is shown in Figure 9.38. The arrangement of prestressing shown only lifts part of the shear, the remainder being carried by hanging steel. The prestress doubles up over the top tension tie; the diaphragm is fully prestressed for bending. The equivalent truss shows how the prestress and reinforcement work together, and how the access hole in the diaphragm may be rationally sized.

Figure 9.39 shows the situation where a box is carried by twin bearings that are not placed directly beneath the webs. Here, a proportion of the shear which is applied by the shear field from the two spans either side of the support shown in Figure 9.37 (b) may be transmitted directly to the bearing. Only that portion falling below point A on

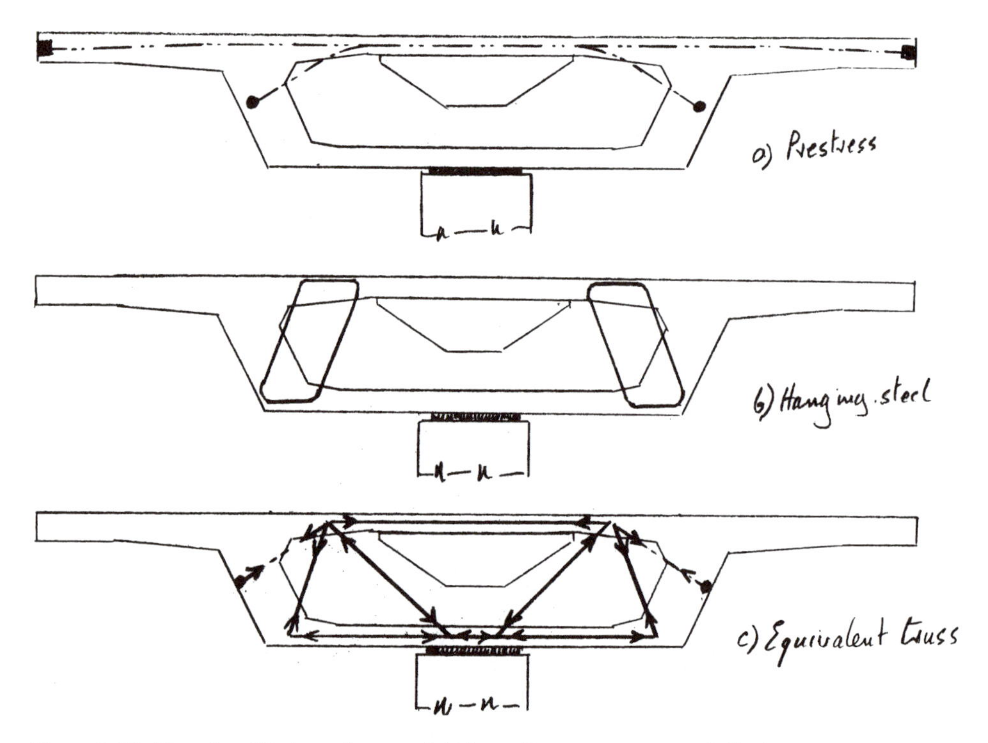

Figure 9.38 River Lea Viaduct, Stanstead Abbotts Bypass: prestressed diaphragm

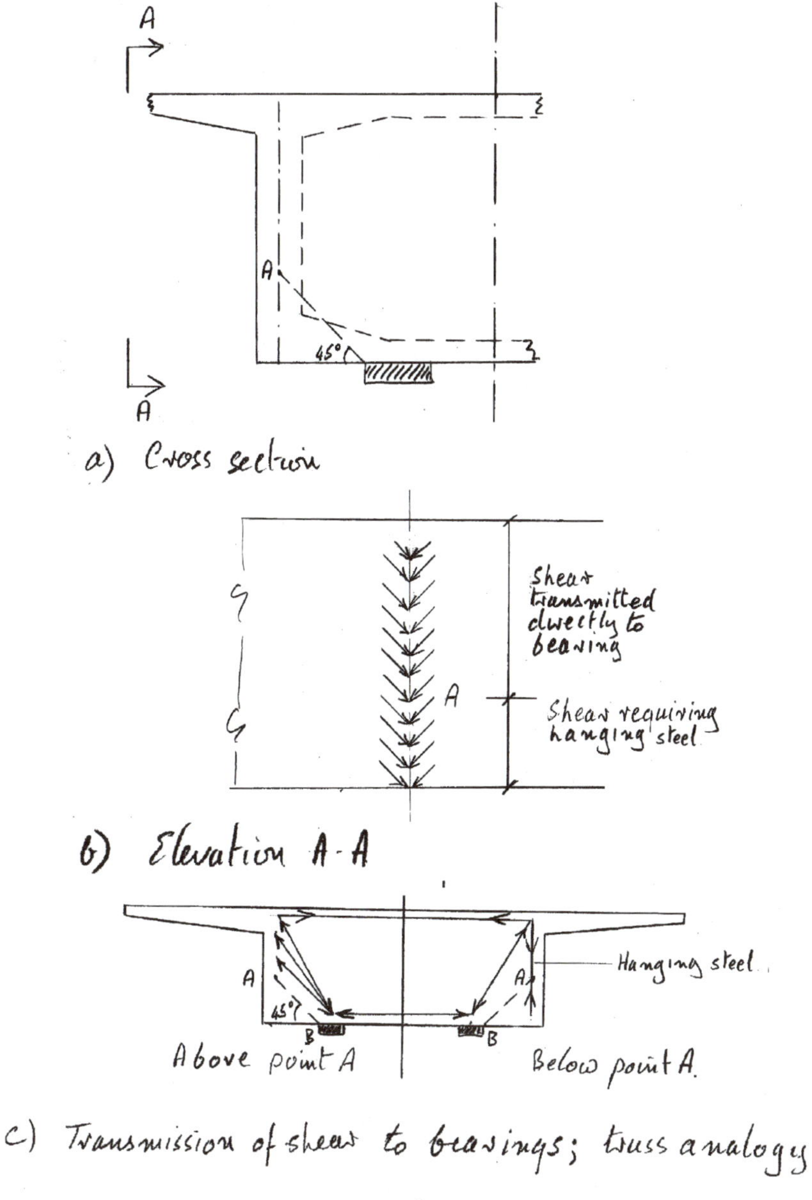

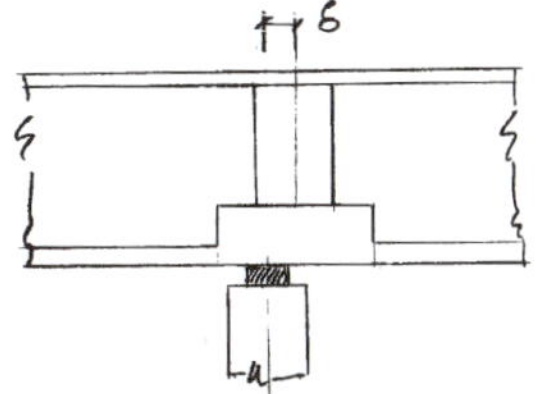

Figure 9.39 Diaphragm action for inset bearings

the web axis, defined by a 45° line from the edge of the bearing, needs to be carried up to the top of the web by hanging steel. Figure 9.39 (c) shows the forces created by the transmission of the two portions of the shear to the bearings. The hung portion of the shear, on the right of the diagram, is transferred to the bearing by an inclined strut and a horizontal tie within the top slab. The directly transmitted portion requires an array of inclined struts that creates a horizontal pressure on the web. The diaphragm needs to consist of a rib on each web which encloses the inclined compression strut, and which acts as a vertical beam spanning between the top and bottom slabs to carry the horizontal pressure. The thickness of these ribs will be determined by the bending, shear and direct compression forces applied to them.

Reference should be made to *3.10.8*, where it is explained that the shear field which distributes the shear in the web uniformly over its height only applies to shear forces that are symmetrical about the diaphragm. The unsymmetrical component of the shear forces will be applied to the bottom of the diaphragm, and must be hung, as for the zone below point A in Figure 9.39 (c).

It is often necessary to provide an upstand rib on the bottom slab to provide sufficient concrete area to carry the horizontal compression strut.

This rib is usually then put into service to allow jacks to lift the unloaded deck for maintenance of the bearings. References [1] and [3] are useful in the design of diaphragms.

An alternative form of bearing diaphragm is a simple beam spanning across the bottom of the box. However, as this has a much smaller lever arm than a full-depth diaphragm, it will be considerably thicker and more heavily reinforced.

When designing a bearing diaphragm for a long continuous viaduct, it is essential to take into consideration the movement of the bridge deck due to length changes caused by temperature, creep and shrinkage. These displace the diaphragm with respect to the bearings, and require either that the diaphragm is thickened so that a sufficient width always remains engaged, Figure 9.39 (d), or that it is designed for torsion.

9.6.3 Diaphragm to resist side-sway

Torque in a box section consists of shear stresses flowing round the perimeter, Figure 9.40 (a). The vertical components in the webs are reacted by the bearings, while the horizontal components in the slabs tend to cause the box to sway. Lateral forces such as wind and earthquake acting on the deck and on traffic, and traffic-induced loading such as centrifugal force also cause sway forces that may cumulate with the torque, Figure 9.40 (b).

The sway force may be taken out by a variety of means. The simplest way, which requires no diaphragm, is to design the box itself to resist the sway in portal action. If the box webs and slabs have not been thickened at the piers, virtually the half span either side of the pier will participate in this action. If the box members have been thickened at the support, the greater stiffness will concentrate the effects close to the pier. If there is a bearing diaphragm, clearly this will have a far greater stiffness than the portal action, and all the sway force will then be concentrated in the diaphragm.

If a diaphragm is provided to resist this sway force, it should consist of a system of cross-bracing that carries the forces to the bearings, converting the sway into vertical push-pull, Figure 9.40 (c). Alternatively, the diaphragm may consist of a portal structure. The external horizontal force (wind etc) must be carried to the guided bearing.

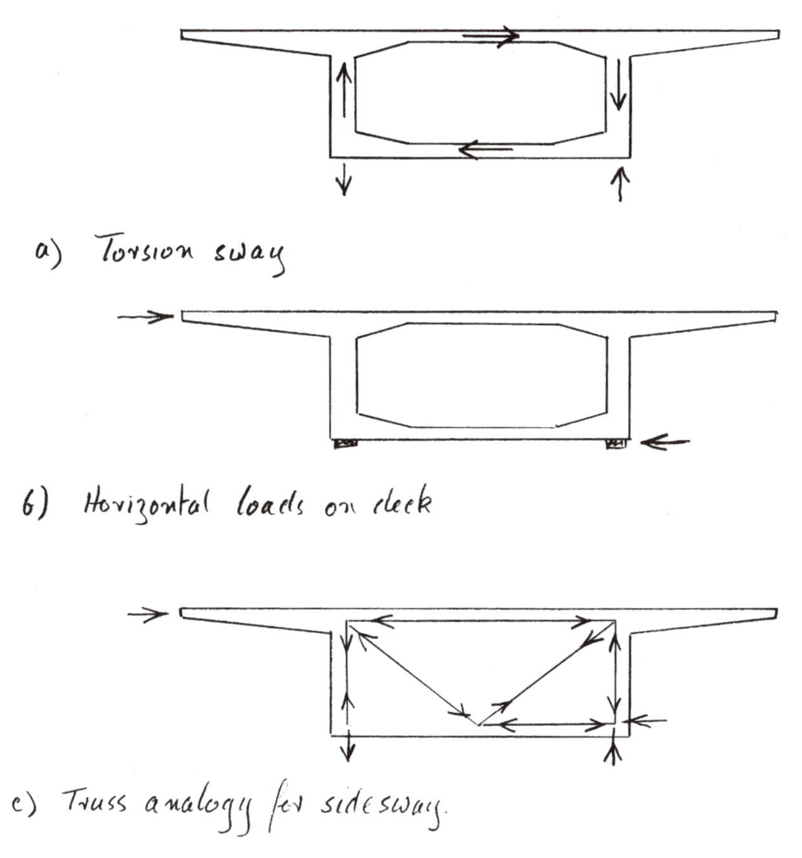

Figure 9.40 Diaphragm action for side-sway

Bridge decks other than boxes must of course also resist external lateral forces. Stocky sections, such as twin rib decks, can carry the lateral loads in portal action, and in general do not need diaphragms (see Chapter 12), while slender structures such as precast beam decks (see Chapter 10) do need diaphragms at the piers.

9.6.4 Trapezoidal boxes

Boxes with inclined webs apply a concentrated transverse compression force to the deck soffit at each pier. This may best be understood by considering each web of the box as an inclined truss, Figure 9.41. A uniformly distributed load on the deck is applied to the truss as a point load on the nodes of the top boom. Thus at each such node the vertical component of the force in the diagonal compression strut is greater than the vertical component of the force in the upright tension member. As the web is inclined, this out of balance creates a transverse tension in the top flange of the deck at each node, h. This is a relatively small effect, as it is caused by the difference in force in the uprights and the diagonals, but it may require additional reinforcement in the top slab. However, at a support, the last diagonal compression strut, which is carrying the whole of the shear in the web, applies a substantial concentrated transverse force, H, which is the sum of all the forces h in the half span, and which could overstress the

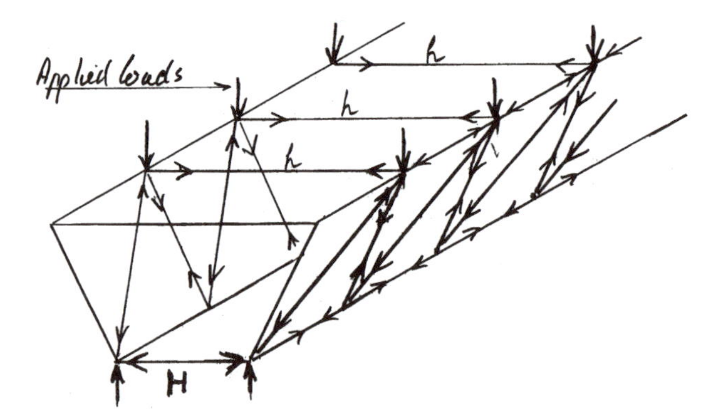

Figure 9.41 Forces applied by inclined webs

bottom slab in the absence of a diaphragm. If a diaphragm is present, this transverse force must be added to the other diaphragm effects, described above.

9.6.5 *Diaphragm synthesis*

One frequently sees diaphragms that consist of a 2 m thick heavily reinforced wall, with a small hole for maintenance access. Clearly no great science is required to justify such a structure both as a bearing and side-sway diaphragm. Such crude diaphragms are frequently very heavily reinforced, and are expensive in materials and troublesome to build. On the other hand, it is possible to design major bridge decks which do not have pier diaphragms. It is always desirable for reasons of economy and for the designer's self-esteem, to design diaphragms rationally.

There is no one design of diaphragm that solves all problems, as there are so many shapes of box and arrangement of bearings. The shape favoured by the author is that shown in Figure 9.42. This arrangement has many advantages, as it encloses the struts that carry the loads to the bearings and those that resist sway. It caters for jacking points located in-board of the bearings and the lower cross-beam can be widened to cater for deck length changes. The shape also allows access between spans and eliminates the hard point over most of the width of the top slab.

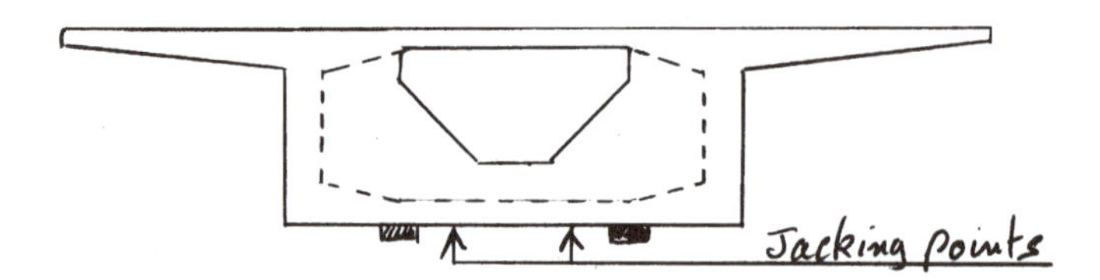

Figure 9.42 Typical diaphragm for box girder

9.7 Deck drainage

9.7.1 General principles

The purpose of draining the deck of a highway bridge is to keep the road safe for traffic. For railway bridges the main purpose is to keep water away from the rails, where it would interfere with signalling currents, and from other electrical equipment, and to avoid standing water that could freeze and disrupt the ballast. These aims may appear self-evident, but are often forgotten.

The drainage of railway bridges is generally simple, as they have no crossfall and drainage gulleys may be placed on the axis of the deck. This section will concentrate on the drainage of highway decks.

The simplest way of draining a deck is to allow the water to drop through holes, and this is sometimes possible. However, the rainwater will be contaminated with petrochemicals, rubber dust and other pollutants. There is also the possibility of the accidental spillage of liquids. Consequently it is usually necessary to design piped deck drainage that carries the water away into a sewerage system.

It is frequently difficult to hide the drainage pipes, and they consequently can have a significant impact on the appearance of a bridge. Also, a piped system will require regular maintenance, and access must be considered in the deck design. It is also necessary to consider what will be the effect on traffic safety if maintenance is neglected.

On high-speed roads, the greatest risk to traffic is unexpectedly to meet standing or flowing water that could lead to aquaplaning. Where a bridge is on a sag curve, or when the crossfall on a deck reverses, the drainage should be designed very conservatively, with redundancy in the number of gullies. The consequence of neglected maintenance should be taken seriously, and, for instance, overflows provided that, in the event of gullies being blocked, allow the water to flow away over or through the parapets before it encroaches on the carriageway. On the other hand a carriageway with a significant vertical gradient with a crossfall towards the hard shoulder runs no risk of flooding, and savings may be made on the provision of gullies.

Drainage design in the UK is based on a rainfall intensity of 50 mm/hour. In such heavy rain traffic could not travel fast and some water on the carriageway would not pose a risk. Although it is quite logical to design the capacity of the drainage system to carry away that quantity of water, it is wasteful to over-specify the number of gullies in an attempt to keep water off the carriageway during such peaks of rainfall.

Whatever system of drainage is chosen, it must be designed to cope with the silt that accompanies the water. This silt may be collected in traps at each gully location, as would be the case in the drainage of roads at grade. This would require deep, bulky gullies that are difficult to accommodate in many bridge deck structures, particularly in side cantilevers. Alternatively the drainage system may be designed to carry the silt off the bridge deck, where it may be collected in a manhole. In an intermediate system, used principally for long decks with box girder construction, the water may be collected in gullies without silt traps, and transferred to a carrier pipe system within the box girder that has silt traps. However silt is dealt with, the deck drainage pipe-work will need to be fully accessible with rodding eyes to clear leaves or debris thrown from cars that may obstruct the pipes. Any drainage equipment that is not cast into the concrete must be designed to be replaceable.

Expansion joints that are marketed as waterproof are unlikely to remain so for long. If surface water may flow across the expansion joint, either by design or because gullies are blocked, it should be assumed that it will penetrate to the abutment shelf, which should be provided with a drainage system to carry it away without overflowing and staining visible surfaces. It is absolutely pointless to provide an abutment drainage system that is not accessible and maintainable, which requires sizing the abutment to allow man access to the shelf. For shallow bridge decks, the abutment structure has to be made deeper than the deck, attracting larger earth pressure forces, significantly increasing the cost of the bridge. Where decks are about 1.8 m deep or more, the height for maintenance access already exists, and the additional expense is less.

Clearly, the design of the drainage system depends on the length and width of the deck, on the longitudinal and transverse falls, and on the speed and intensity of the traffic. Drainage design is an important part of the designer's task that can affect the choice of bridge deck, and should be considered early in the design process. It should not be forgotten that the greatest risk to traffic is from the neglect of maintenance rather than from under-design for exceptionally intense rainfall.

9.7.2 *Specific cases*

a) Drain the water off the bridge

Wherever possible, the water should be allowed to flow off the bridge in the gutter. A generously dimensioned gully and silt trap in the road embankment will be much cheaper and easier to maintain than bridge deck gullies and pipes. Where the bridge deck is integral with its abutments, there is no difficulty in adopting this option. Where there is a roadway expansion joint at the deck end, the abutment should be designed for drainage, as described above.

b) Single gully placed just upstream of expansion joint

For bridge decks with expansion joints, another option is to provide a single gully, generally without a silt trap, on the deck just upstream of the joint and to pipe the water across the joint with a properly engineered system. Such a gully may be hidden by incorporating it within the structure of the abutment diaphragm. The water would then be collected in a manhole with a silt trap in the embankment off the deck. The pipe expansion joint must be designed to accommodate any vertical movement due to the elastic response of the bridge bearings or deflection of the bridge deck, as well as settlement of the abutment manhole. This latter can be avoided by suspending the manhole from the back of the abutment structure.

c) Trough behind the kerb

For larger bridge decks or those where the falls make it essential to collect the water at numerous points along the kerb, the most easily maintainable option is a trough in the footpath behind the kerb. If the falls are appropriate, this trough can take the water off the deck. Alternatively, the water may be collected in a gully placed just upstream of the joint, as in (b). The trough may also be combined with a piped system, with the water transferring from trough to pipe at regular intervals.

d) Hollow kerbs

A commercial version of the trough is the hollow kerb. This is an effective if expensive system. It may be combined with a carrier pipe as described above. Lengths of hollow kerb may be used to solve difficult problems at low points or where crossfall reverses.

e) Piped system in beam bridges

In bridge decks consisting of multiple precast beams, it may be possible to arrange the beams such that the gutter is located between beams, Figure 9.43 (a), ideally in the cast-in-situ slab. The gullies may be associated with conventional silt traps, which are emptied from the road as if it were at grade. A piped system may then be located between the beams, taking the water off the deck. It may be necessary to provide a suspended walkway giving maintenance access to the silt traps and pipe-work.

If it were not possible to locate the gutter directly over a suitable position, gullies without silt traps would discharge into the carrier pipe with a short access pipe, Figure 9.43 (b). Silt traps could be located opposite each gully, or at the bridge ends. It may be necessary for the precast beams to incorporate some of the drainage hardware, which emphasises how important it is to consider, and even to design in detail, the drainage system early in the design process.

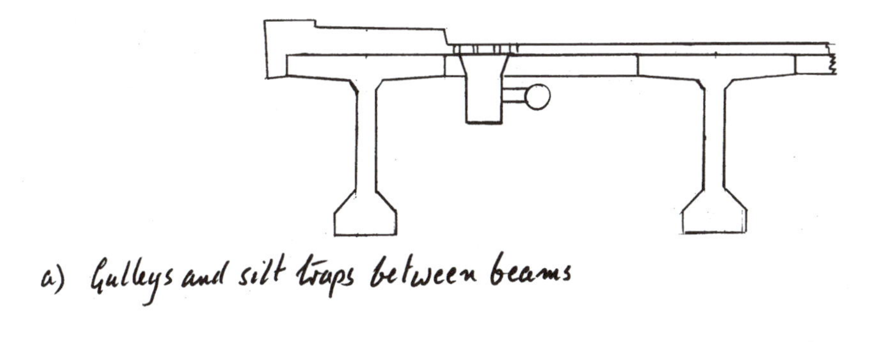

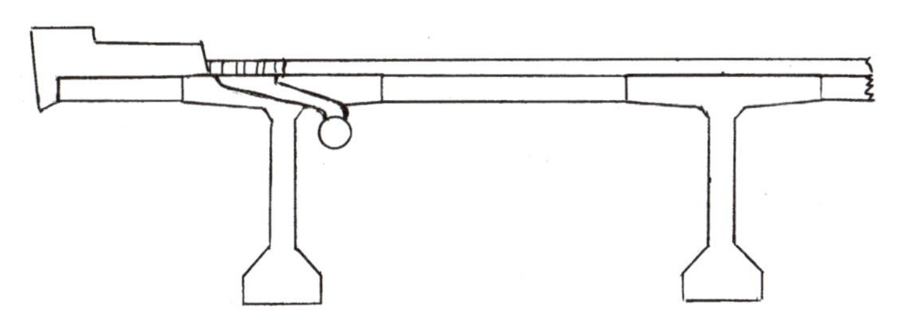

Figure 9.43 Piped drainage in precast beam deck

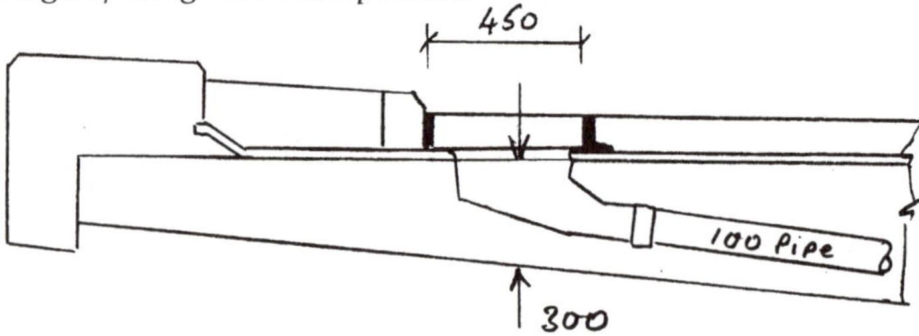

Figure 9.44 Custom-designed gulley in thin cantilever slab

f) Flat gullies in side cantilever

Box section decks with wide side cantilevers regularly pose drainage problems, as the gutters usually are situated towards the end of the cantilever where the slab is thin. Troughs or hollow kerbs may solve the problem, if the gradients permit.

Alternatively, very flat gullies may be custom designed, in glass reinforced plastic (GRP) or, if enough are required, in cast iron. Such custom-made gullies, associated with 100 mm drainage pipes, may be accommodated in slabs as thin as 300 mm, Figure 9.44. It is important in the hydraulic design of such gullies that care is taken to avoid dead areas where rubbish may accumulate. The gullies may need to be located so that the transverse pipes do not clash with the prestressing ducts. These gullies may discharge into a carrier pipe located within the box girder. Alternatively, if the areas and gradients of the road permit, they may be located opposite each pier, and the water piped directly to ground level. If the required gully spacing is less than the span, hollow kerbs or troughs may be used, with discharge pipes at piers.

In countries where rainfall is heavier than in the UK, national regulations may impose silt traps at each gully, as well as minimum pipe diameters, 150 mm for instance. When it is impossible to hide the drainage it is up to the designer to find a system that is neat and does not look like an afterthought. For instance, by adopting hollow kerb drainage if necessary, with discharge pipes placed opposite each pier, the silt trap and the transverse discharge pipe may be housed in a downstand rib that becomes visually associated with the pier.

g) East Moors Viaduct

An original solution was adopted for the East Moors Viaduct. The deck was 13 m wide and 914 m long. The longitudinal alignment had a summit curve, and the S-shaped horizontal alignment required a change in crossfall. Gullies in the kerb discharged into carrier pipes situated beneath a precast footpath, Figure 9.2.

9.8 Waterproofing

Any bridge deck that will be treated with de-icing salts must have a waterproof membrane. Many proprietary products are available with standard details to solve the most common problems.

Water that seeps through the black-top and onto the membrane will be put under high pressure by the weight of heavy wheels above. It will either be forced upwards, disrupting the asphalt, or, finding any flaws in the membrane, downward into the concrete deck. Water is most likely to pond adjacent to kerbs, where the membrane turns up, or adjacent to expansion joints. These areas must be vented, either by placing on top of the membrane small perforated pipes that relieve the pressure and carry the water into the drainage system or to fresh air through the deck, or by providing proprietary vents that penetrate the deck slab. Pipes that penetrate the deck must be detailed so that they do not allow salt-laden water to dribble along the surfaces of the deck, or to be blown onto piers or abutments.

For decks in tropical climates, where salts are not used, it is sometimes considered unnecessary to provide such a membrane. This may well be justified if the deck is unsurfaced. However, if the deck is surfaced it should also have a waterproof membrane, as water seeping through the surfacing cannot evaporate and could provoke corrosion of the deck reinforcement.

9.10 Expansion joints

Most of the unplanned maintenance on bridge decks is related to expansion joints. Furthermore, they allow salt-laden water to seep through and attack piers and abutments. Consequently, the number of expansion joints should be minimised. On short bridges they should be eliminated entirely, *7.13*. On statically determinate bridges, decks may be linked together in groups to space out expansion joints (see Chapters 7 and 10). On longer continuous bridges, wherever possible expansion joints should be provided only at abutments. For instance, the Dagenham Dock Viaduct on the A13 designed by Benaim had a length of 1700 m without intermediate expansion joints.

10
Precast beams

10.1 General

Decks made of precast beams are widely used. Precasting gives the benefit of good control of the quality of construction, while the deck construction is simple and repetitive. Such decks are usually statically determinate, although techniques are available to render them partially or totally continuous. Statically determinate beams are frequently economically competitive with continuous beams, due to the lack of secondary sources of bending moment and locked-in stresses. Modern techniques which allow several spans to be linked to minimise the number of roadway expansion joints have eliminated one of the main obstacles to the use of statically determinate spans.

There are many forms of precast beams, including boxes, troughs and various forms of 'I' beam. One may divide precast beams for bridge decks into two categories, standard precast beams and custom-designed precast beams.

10.2 Standard precast beams

There are many books, pamphlets and design aids describing the use and the analysis of decks using standard beams and troughs, and it is not the intention of this book to cover this ground again [1]. Some general comments on the way that such components may be used most appropriately will be made.

Standard pre-tensioned precast beams such as the 'inverted T', 'I', 'M', 'Y' and 'U' beams originally designed by the Cement & Concrete Association as well as standard AASHTO beams are widely used in the UK, South East Asia and the USA for bridges with spans that are generally below 30 m, but which may attain 40 m for the most powerful units. They are very useful when it is impossible to use falsework, such as for bridging live railways or motorways. They are also useful for small bridges where they may be built by contractors with little bridge expertise.

However they are extensively overused in situations where alternative forms of bridge deck are more appropriate. Every time the author has had the opportunity of analysing the cost of decks designed to use these beams, it has been clear that another form of construction would have been more economical, except in the special cases described above. This is despite the economical advantages that should derive from the standardised forms, pre-existing casting beds and low consultants' fees, as well as the use of pre-tensioning which, once the beds have been amortised, is inherently cheaper than post-tensioning.

One most often sees inverted T and M beams placed at 1 m centres, giving a continuous soffit. Permanent falsework is then placed and a reinforced concrete slab, generally 160 mm thick cast on top, Figure 10.1 (a). This form of construction is not economical unless it is essential to minimise the depth of the deck. It has the further disadvantage that it is not possible to inspect the internal beams during the life of the bridge.

In all cases when it is not necessary to minimise depth, deeper beams, more widely spaced, would result in significant savings. Typically a spacing of 2 m is possible, halving the number of beams, Figure 10.1 (b). Of course each beam will be more expensive, containing more concrete, prestress and reinforcement, but these are marginal costs; so much of their cost, including storage, transport, delivery and placing, is almost independent of their material content. The slab will have to be slightly thicker, say 180–200 mm, but slab concrete is cheap.

The most economical way of supporting the beams is to rest each of them on a laminated rubber bearing carried by a crosshead located beneath the deck, Figure 10.2 (a). This is not very good looking, and a variation is to design a crosshead of inverted T cross section, that is mainly within the thickness of the deck, with a bearing ledge protruding below, Figure 10.2 (b). The bearing ledge may be incorporated into the depth of the deck by using halving joints at the beam ends. However this significantly increases the cost of the beams, which derive their economy from their standardisation. The crossheads form a very important component of the cost of these

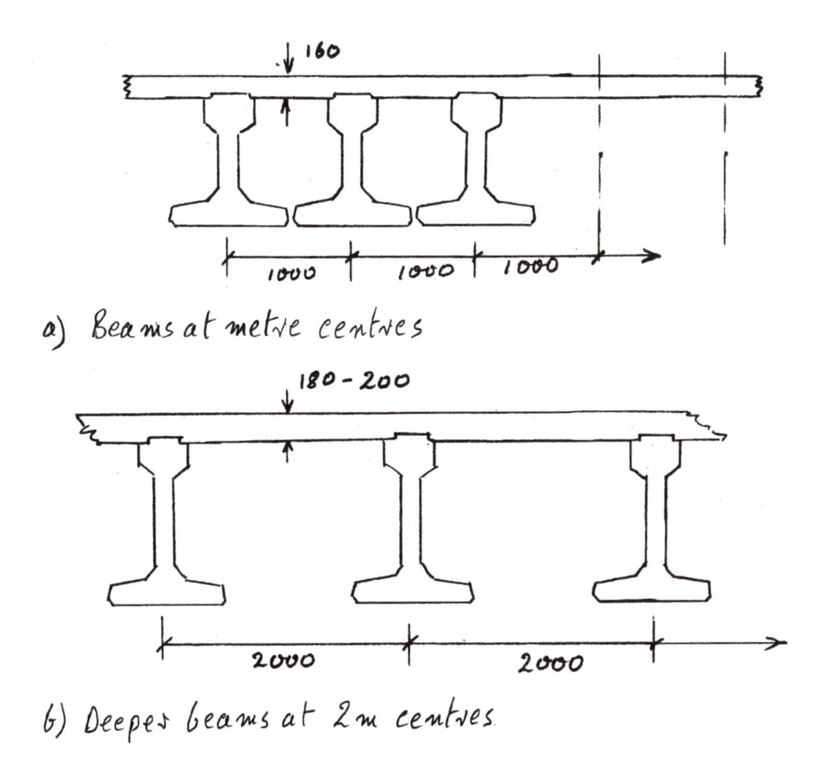

Figure 10.1 Standard precast beams

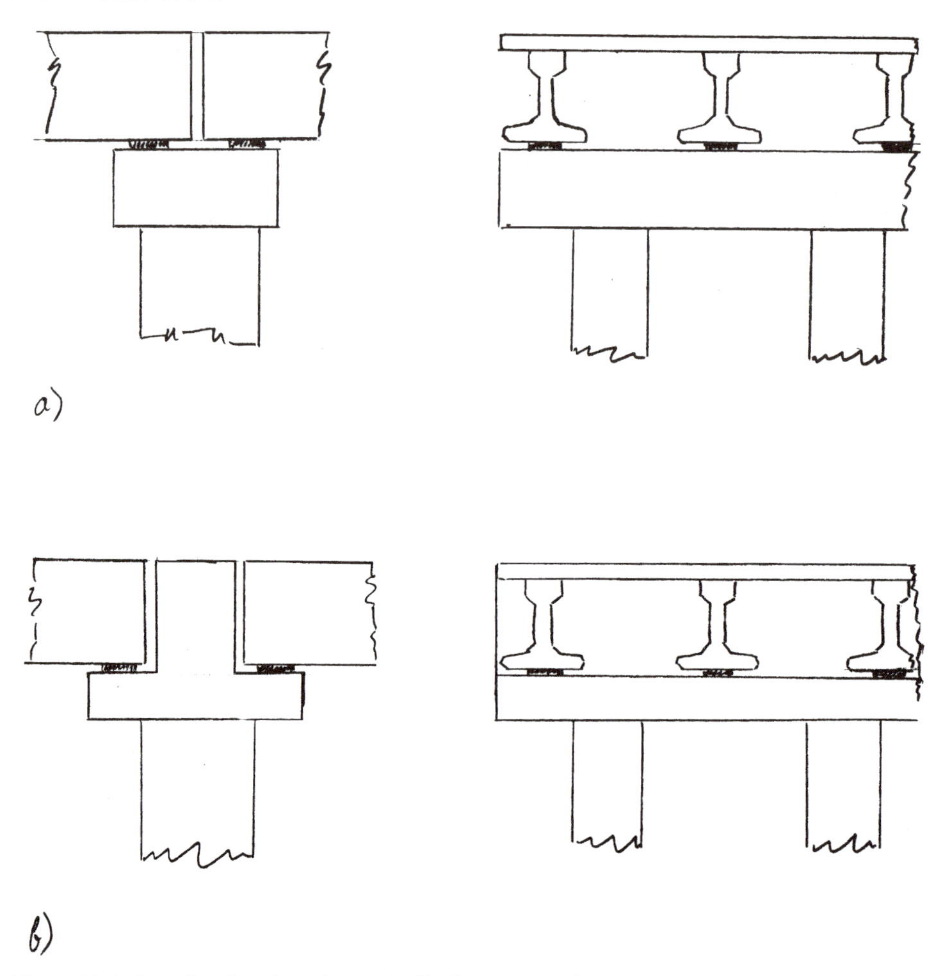

Figure 10.2 Crosshead options for statically determinate beams

types of bridge, and should not be omitted from the cost comparisons of different deck options.

This form of construction leaves the decks statically determinate, with the problem of frequent expansion joints. This may be overcome by linking their top slabs into expansion units several spans long (*10.3.8*). Another disadvantage of statically determinate decks is their camber. Under the long-term effects of prestress they arch upwards significantly, creating an uneven driving surface. Either they must be pre-cambered downwards, which increases their fabrication cost significantly, or the camber must be masked by increasing the thickness of the topping slab near the supports, increasing weight and cost.

It is possible to make such decks fully continuous by supporting the beams on falsework and casting in situ a section of deck several metres long over the piers, Figure 10.3. The continuity is usually created in reinforced concrete. The cast-in-situ section may incorporate a crossbeam, usually within the thickness of the deck, which eliminates the need for a pier crosshead and allows the deck to be supported directly by the pier. Although this solves the problem of joints, reduces the problem of camber

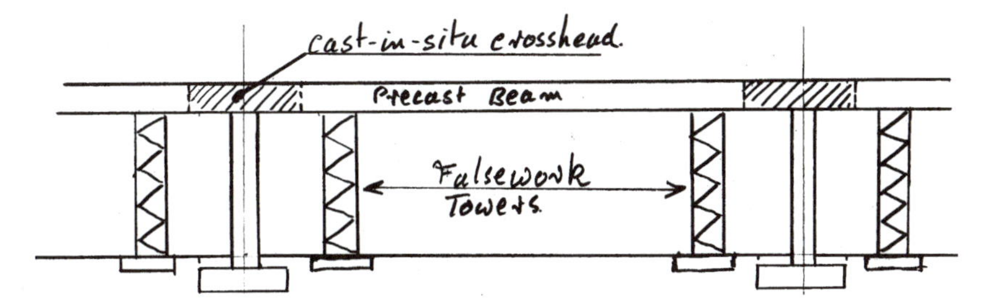

Figure 10.3 Continuity using cast-in-situ cross-heads

and improves the appearance, such decks are not, in the author's experience, ever economical.

A good example was the River Lea Viaduct on the Stanstead Abbotts Bypass (*15.3.4*). The conforming design consisted of an M beam deck with spans of 26 m, made continuous as described above. This deck, which was uneconomical both in the materials used and in the method of construction, was replaced by an alternative precast segmental, twin box girder designed by Benaim and built by Shephard, Hill Construction, Figure 10.4.

It is also possible to create continuity only for live loads. Here the beams rest on conventional crossheads. A diaphragm is cast between the beams of adjacent spans, and reinforcement in the topping slab carries the live load hogging bending moments. Nominal reinforcement is also required to link the heels of the beams, to carry sagging moments at the piers caused by prestress sagging parasitic moments generated by the

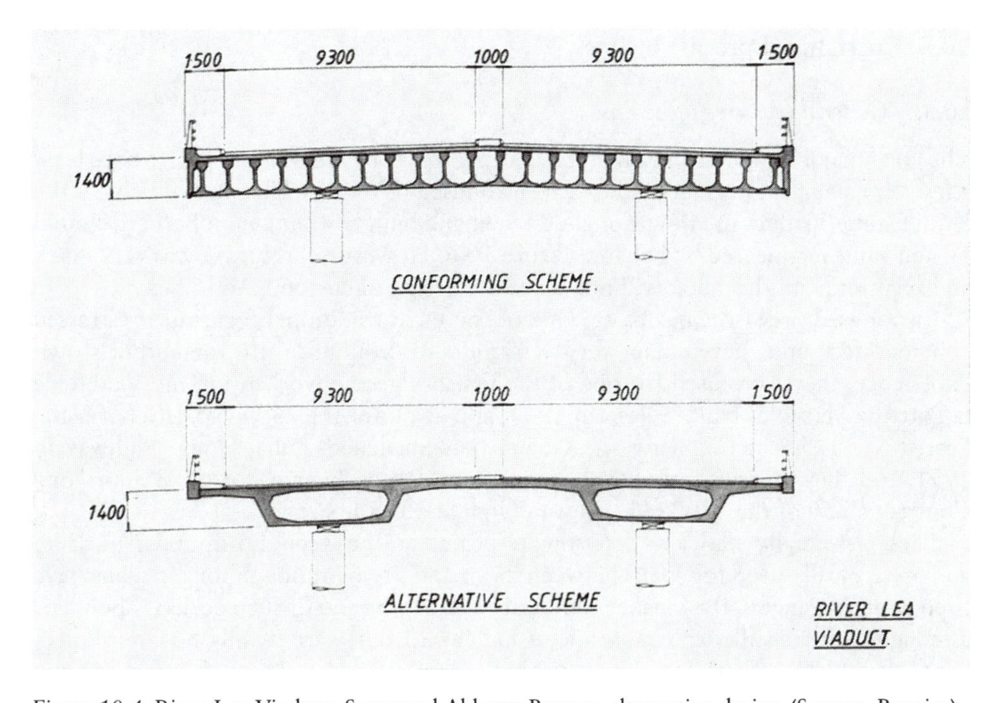

Figure 10.4 River Lea Viaduct, Stanstead Abbotts Bypass: alternative design (Source: Benaim)

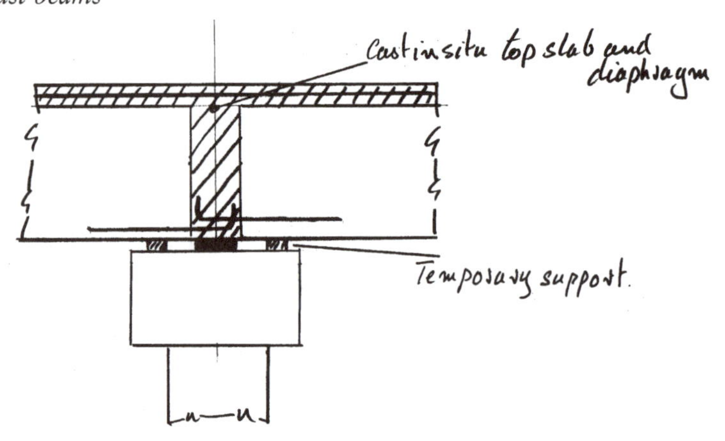

Figure 10.5 Continuity for live loads

upwards creep deflection of the beams, by differential settlement and by temperature gradients in the decks, Figure 10.5. This is probably the most economical form of construction for these standard precast beams.

Other countries have developed different types of standard ranges of precast beams. For instance, in Australia the Supertee beam is used [2], AASHTO beams are widely adopted in the USA, while in Italy very thin webs characterise some of their standard products. Local conditions, including the aggressiveness of the climate, live load intensity, codes of practice and the habits and skills of contractors make it difficult to export such technologies without careful consideration.

10.3 Customised precast beams

10.3.1 General features of beams

The great majority of customised precast beams are 'T' beams although there have been a few very significant bridges built with customised 'U' beams. The most notable is the Kwai Chung Viaduct in Hong Kong, designed and built by Campenon Bernard-Franki JV and value engineered by Benaim, Figure 10.6. However, this bridge was very much an exception, and this book will not cover the design of customised 'U' beams.

Customised precast T beams are one of the most traditional forms of prestressed concrete deck and they remain very versatile and economical. In the author's own experience, they were used for one of the first bridges he worked on, the Viaduc de la Porte de Versailles built in Paris in 1964, and again for the 28 km of decks over the Pearl River Delta in China for the Guangzhou–Shenzhen–Zuhai Super Highway in 1991, and they are currently being used in a design by Benaim for the 3.5 km long Contract C250 of the 350 km/hr Taiwan High Speed Rail Project.

These essentially statically determinate beams are between 1.5 m and 3 m deep, and are typically used for spans between 25 m and 45 m, although longer spans have been built. In general, the most economical span/depth ratio of such decks is about 15, although they are still viable at a ratio as high as 20. They are usually post-tensioned, with a roughly parabolic prestress centroid that balances a large part of the shear force as well as the bending moments.

Figure 10.6 Kwai Chung Viaduct (Photo: Photothèque VINCI et filiales)

A reinforced concrete top slab, typically 200 mm thick, is allowed to span transversally as far as possible and is then supported by the least number of beams. This transverse span depends on the loading code, but is typically between 3.5 m and 4 m. It is sometimes economical to increase slightly the thickness of this slab to accommodate a larger transverse span and reduce by one the number of beams. The side cantilevers are usually short; the cantilever bending moments must remain compatible with the thickness of the slab as no moment turns down into the webs.

As the heels of the beams are not linked, the webs take no transverse bending moment. Consequently, they may be made as thin as possible, subject only to the need to provide corrosion protection to the reinforcement and to compact the concrete in the heel (*9.5.2*). Since the prestress carries a proportion of the shear force, the webs do not usually need thickening, except locally at the ends of the beams to carry the concentrated force of the post-tensioning anchors at their ends. This thickening, to a value dictated by the anchors, is usually extended forwards by approximately a beam depth. The heel is shaped to facilitate the flow of concrete from the web and is dimensioned to avoid over-stressing of the concrete at the first tensioning of the tendons. Its size depends essentially on the span/depth ratio of the deck, all other factors being equal. Modern design practice for highway bridges is to provide diaphragms at the beam-ends only.

Usually the top slab is given a slight haunch, generally 50 mm, with a 100 mm fillet at the junction with the web. As well as reducing shear stresses in the slab and improving the arching action (*9.3.5*), this geometry aids striking of the shutter (*10.3.6*).

One of the reasons for the economy of such beams is that their function is so simple that they may be very confidently analysed, and all the material quantities may be refined down to minimum values. Generally they are prefabricated on a site close to the bridge, eliminating the cost of road transport.

10.3.2 Railway bridges

Precast beam decks are suitable for use as railway bridges, such as the Express Rail Link in Malaysia, designed by Benaim, Figure 10.7. This 50 km long light railway links the new Kuala Lumpur airport with the centre of town, and involved the construction of some 50 structures, including twin beam decks as shown, which typically had a span of 40 m. The live loading was defined as 'RL', to the British Standard BD37/88.

Bending moments on the slab may be minimised if the tracks are placed close to the beams, Figure 10.8, although derailment loads still have to be carried over the full width of the slab. The slab for rail bridges is usually thicker than for highway decks in order to control its flexibility. A bridge deck carrying heavy rail would be stockier with the tracks more closely aligned over the beams, and a thicker top slab.

When the heavy loads typical of railways are applied to a deck, offset from the beam positions, the top slab will deflect and give a small inclination to the webs, Figure 10.9. As a consequence, the heels of the beams will deflect horizontally at mid-span while being restrained by the support diaphragms. They will thus be subjected to a transverse curvature, leading to transverse bending stresses. These stresses may affect the sizing of the heel and that of the longitudinal prestress.

If these transverse stresses are significant, the designer may choose to increase the thickness of the top slab to reduce the deflections of the heels, to change the layout of the deck to place the tracks closer to the axes of the beams, or to increase the longitudinal prestress to provide a residual compression on the bottom fibre that is sufficient to neutralise the transverse bending tensions. This may also require the heel to be increased in size if it has been designed to the limit for vertical actions only.

On some projects, a bottom slab has been introduced between the heels, transforming the deck into a box. However, any kind of bracing between the heels will introduce bending moments into the webs, which is likely to require them to be thickened. The most economical ways forward are likely to be to modify the layout of the deck to minimise the problem, or to leave the heels free to bend laterally while increasing the prestress.

It is not recommended to adopt intermediate diaphragms for rail bridges, as the concentration of forces that occurs at the junctions between diaphragms and webs is likely to give rise to intractable design problems and to concentrations of reinforcement that detract from the economy of the system.

Precast beam bridges should be kept simple; if the design development leads them into increasing complexity, it would be better to select a different type of deck.

10.3.3 Beam layout

The highest compressive bending stresses in precast beams occur in the heel when the prestressing cables are being stressed, and the beam is only carrying its self weight. In fact, the size of the heel is determined by this construction phase. As the beam is

Figure 10.7 Express Rail Link, Malaysia (Photo: Benaim)

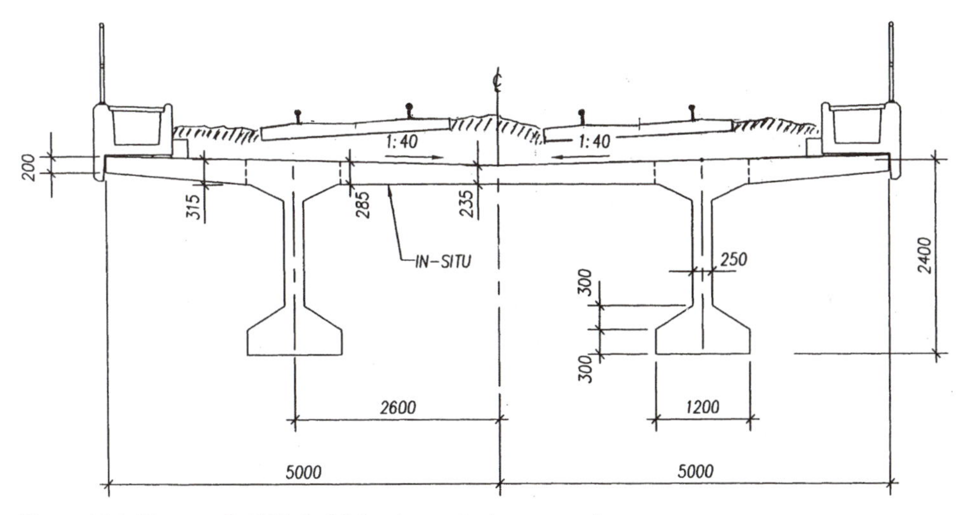

Figure 10.8 Express Rail Link, Malaysia: typical cross section

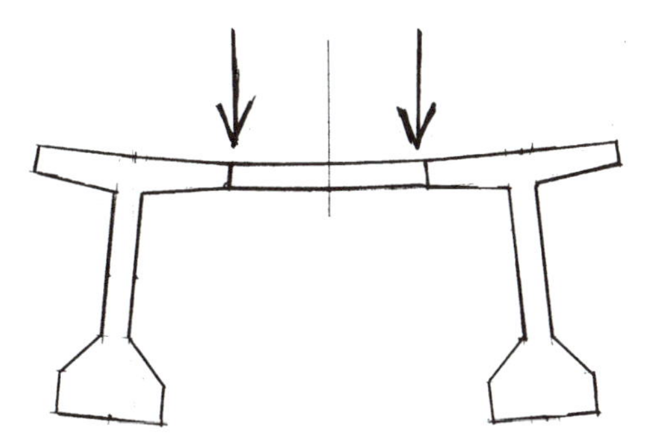

Figure 10.9 Transverse deflection of heavily loaded deck

statically determinate, the external loads only apply positive bending moments, and consequently the heel is progressively unloaded by the application of:

- addition of the cast-in-situ portion of the top slab;
- loss of prestress;
- deck finishes;
- live loads.

Thus, the essence of the design of these beams is firstly to minimise the number of beams in the cross section, and then to size the heel such that its stresses vary from the safe limit under the initial prestress combined with its self weight, to the allowable tensile stress under long-term working loads.

The very rare accidents that have occurred during the construction of these beams have generally been due to the overstressing of poorly compacted concrete in the heel when the prestress is applied.

The strength of the heel is sometimes increased by using compression reinforcement. This is not usually recommended, as it is more economical to increase the size of the heel. In exceptional circumstances, where for instance, a few beams in a series are called on to carry heavier live loads due to a local widening of the deck or to a single longer span, this is an expedient that allows the standard formwork to be maintained.

Three different types of top flange are used. The web may be stopped below the slab with a thickening serving to stabilise the beam during handling, Figure 10.10 (a). This arrangement minimises the weight of the prefabricated beams, making handling in the casting yard and launching more economical. There is a possibility of torsional instability when a beam with a small top flange is lifted.

The connection between beam and slab is simply created by stirrups projecting out of the top of the webs. The shutter for the top flange has to be propped off the heels of the beams, which may be relatively inaccessible. Alternatively, precast formwork such as Omnia planks may be used, which are relatively expensive. This type of beam uses more prestress than those described below, as the neutral axis of the beam alone is lower, giving a smaller prestress eccentricity. A further disadvantage is that the narrow top flange does not give a safe working platform for installing the slab formwork.

The other extreme is the beam shown in Figure 10.10 (b). Here, the bottom lift of the top slab over its full width is incorporated into the precast beams. This 'pre-slab' must be profiled to the crossfall of the road. Once the beams have been erected, there is a full-width working platform on which the reinforcement of the top slab may be assembled and the remainder of the top slab poured, with no need for any other significant falsework. The 'pre-slab' must be designed to carry the weight of the wet concrete and the other construction loads such as men and compressors. This requires some careful definition of the most onerous conditions that are likely to be met. For instance, if the concrete is discharged from hoppers, the possibility of a substantial thickness being deposited locally on the end of a prefabricated cantilever must be considered.

The principal disadvantages of this arrangement are the increased weight and size of the beams to be handled and launched, the additional weight of reinforcement due to the need for top steel in the precast pre-slab, the difficulty in providing economically for sagging reinforcement at the centre of the slabs, and the fact that the beam spacing should be restricted to about 3 m, increasing the number of beams required. For a

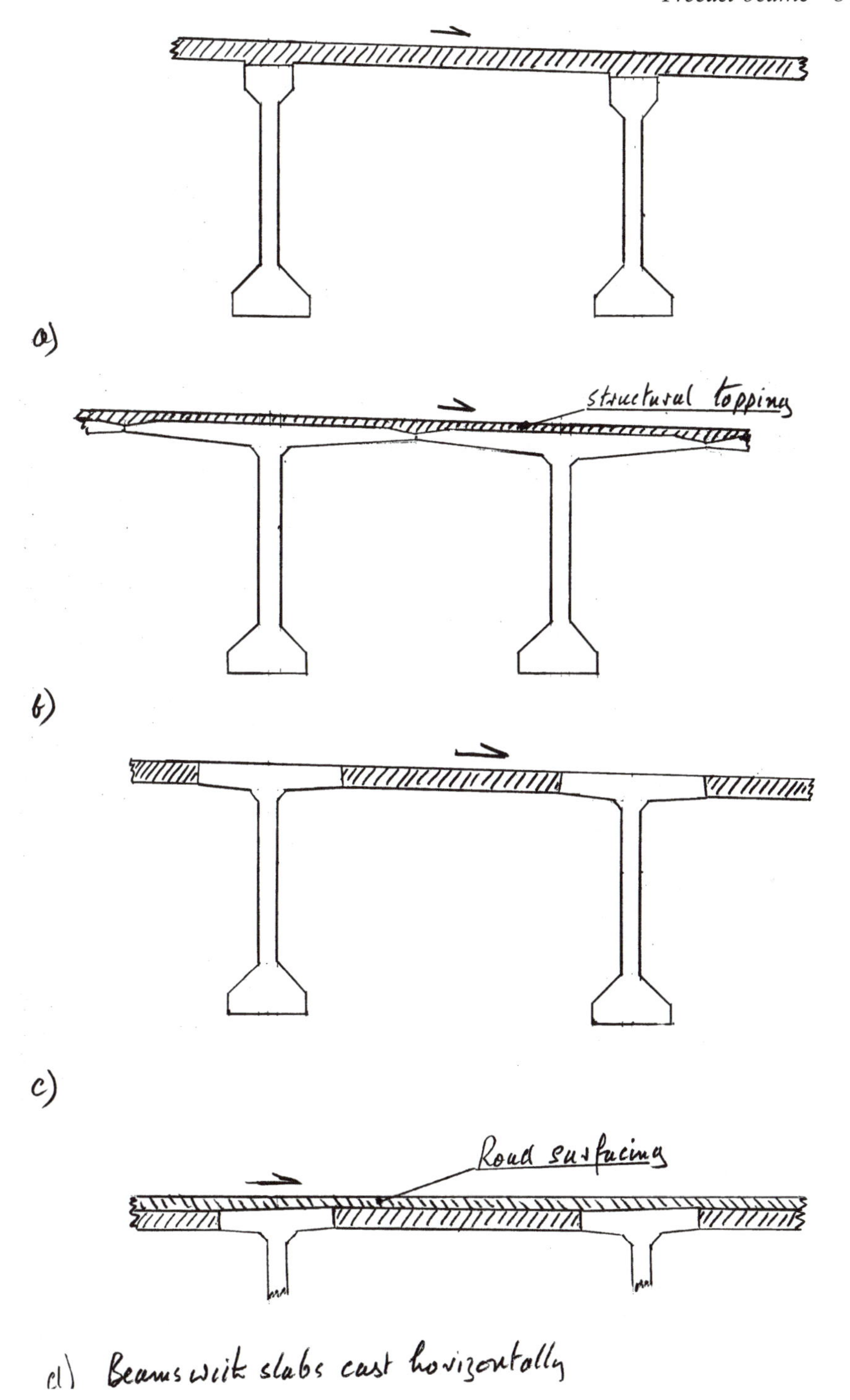

Figure 10.10 Options for customised precast beams

spacing greater than about 2.5 m, it is most likely that the top slab will have to be thickened at the root of the cantilever, increasing the self weight of the deck.

Figure 10.10 (c) shows the third arrangement, generally favoured by the author. Here the top slab is precast over its full depth, but over a reduced width. The remainder of the top slab is then cast in-situ. The width of the precast slab may be chosen to suit the circumstances; the wider the slab the less concrete there is to cast in a second phase, but the heavier the beams. Typically, the top slab of the beam is 1.2 m to 1.7 m wide.

Once the beams have been launched into place, the top flange gives a good working surface. The shutter for the cast-in-situ slab may be simply suspended from the precast stub cantilevers. The cast-in-situ or prefabricated parapet may be attached directly to the precast beam, or alternatively the width of the deck may be increased by casting on a short in-situ side cantilever.

The top slab cast with the beams is usually inclined to the crossfall of the road. However, if the precast slab is narrow, it may be cast horizontal, with the cast-in-situ slab more steeply inclined, Figure 10.10 (d). The road surfacing will then be of variable thickness.

The construction joint between the precast and the in-situ slab needs to be designed in shear friction, requiring projecting reinforcement. Although a vertical joint is perfectly safe if correctly designed, an additional factor of safety may easily be introduced by giving a small inclination to the joint surface, which in any event should be roughened. A variety of reinforcement arrangements for different widths of cast-in-situ slab are shown in Figure 10.11. If the projecting reinforcement of adjacent beams overlaps, more care is required in placing the beams. The top slab may also be transversally prestressed, in which case all projecting steel may be omitted, subject of course to thoughtful design.

If the deck is not to be waterproofed, consideration must be given to the risk of water seepage through the slab construction joints.

10.3.4 Prestress layout

The reasoning governing the layout of the prestressing cables in precast Tee beams has been described in Chapter 5.

10.3.5 Reinforcement layout

As there is no transverse bending moment in the webs, and much of the shear is carried by the prestressing tendons, typical highway bridge decks are generally lightly reinforced, with typically 16 mm shear links near the beam-ends, and 12 mm links elsewhere, with 10 mm distribution steel. Railway bridges with their heavy live loads could well require larger diameter reinforcement. Heavier steel will be required immediately behind the prestress anchors, and in the zone of dispersion of the prestress. Also, as there are no bending stresses in the deck due to differential settlement and temperature gradients, and no significant heat of hydration effects, there is no need to provide longitudinal passive reinforcement for a fully prestressed design, other than corner bars for the links, and a light surface grillage on the external surfaces of the heel, using 10 mm or 12 mm bars. The top slab will of course be more heavily reinforced to carry the traffic loading.

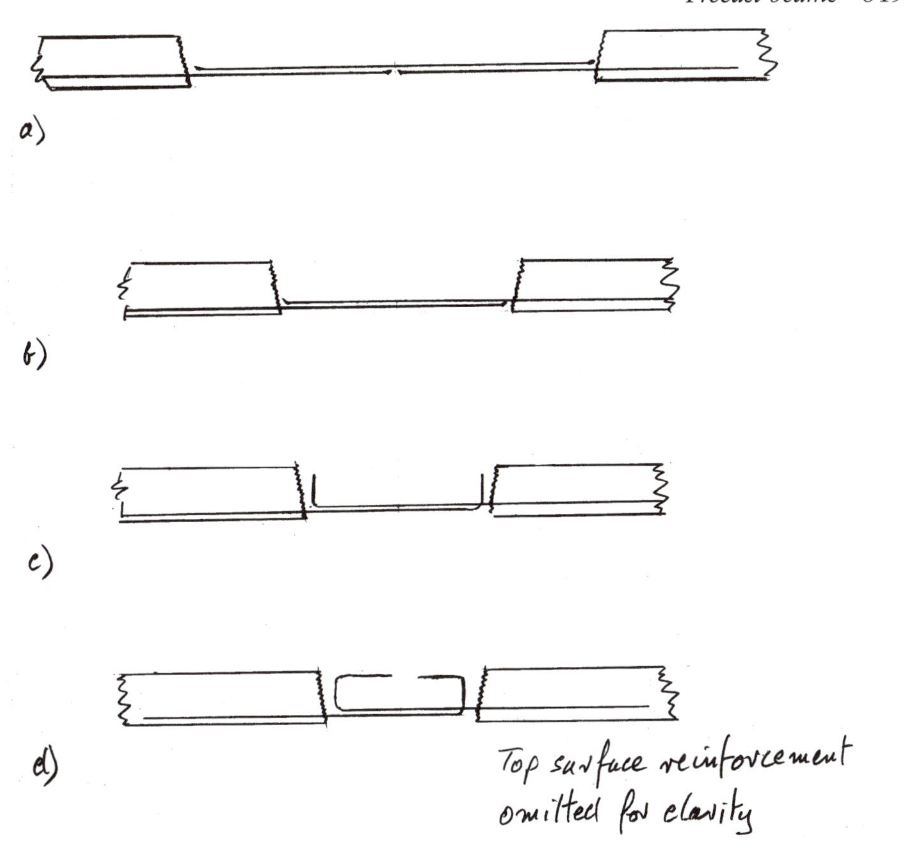

Figure 10.11 Arrangement of slab reinforcement for progressively shorter cast-in-situ section

10.3.6 Construction technology

Precast beams are suitable for very large, highly mechanised sites as well as for more rural projects. Two contrasting sites on Benaim projects were the 630 m long bridge across the Sungai Kelantan at Pasir Mas in northern Malaysia, described in *10.3.7*, and the twin 17 km long GSZ Superhighway viaducts described in *10.3.8*.

If the economy of this form of construction is to be fully realised, it is necessary to minimise the thickness of the webs, which in turn requires the use of steel shutters and external vibrators.

The most common method of striking the side shutters of precast 'T' beams is by rotation about their base. It must be ensured that the steel shutter cannot jam against the concrete; indeed the movement of separation from the concrete should always have a small component normal to the concrete surface, Figure 10.12. The shallow haunch of the slab and its root fillet help this striking action. The crossfall of the deck must not be forgotten in the determination of this geometry.

If steam curing is adopted, it is feasible to achieve a concrete strength of 20–25 MPa the morning after casting. Although this is adequate for the bending stresses in the beam under the effect of its self weight and a first phase of prestress, it would not be enough beneath the prestress anchors. In order to allow early stressing of the

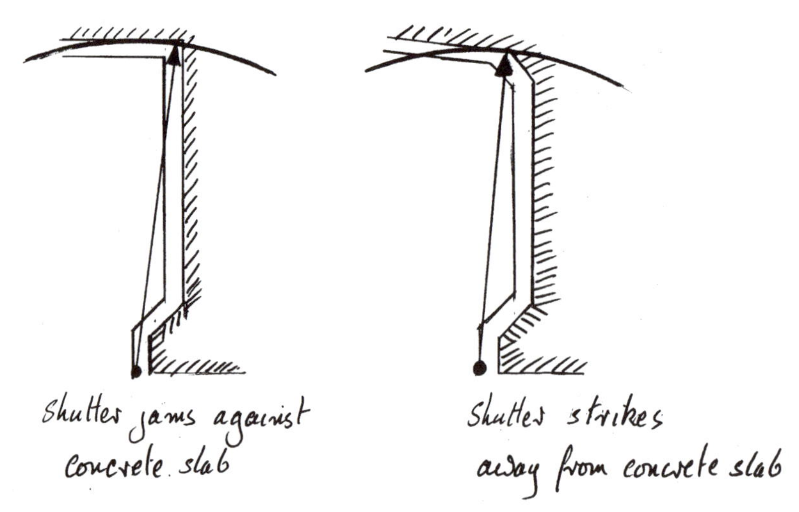

Figure 10.12 Concrete geometry and striking of shutters

tendons, precast end blocks that incorporate the tendon anchors are frequently used. Alternatively, the first tendons to be stressed may be provided with oversize anchor plates.

Beams of this shape have also been cast in-situ. However, it is difficult to achieve the degree of control that is typical of a precasting yard, and the formwork must allow inspection of the heel to confirm its integrity before significant prestress is applied.

There are two basic methods of placing the beams. The launcher may remain on one line, and once the beams have been launched they must be rolled sideways on the crosshead into their final position. Alternatively, the launcher itself may be capable of sliding sideways across the deck, placing the beams in their final position. Lateral sliding of a launcher using long-stroke hydraulic jacks is a very quick operation, taking no more than a few minutes if the falsework has been well designed, and this is likely to be the quickest form of erection for large decks that justify the expense of mechanisation. In countries where precast beams are used regularly, dedicated launchers of various load and span capabilities are often available for hire.

10.3.7 Bridge across the Sungai Kelantan at Pasir Mas, Kelantan Province, Malaysia

This bridge consisted of 14 spans of 45 m, Figures 10.13 and 10.14. It was originally designed with 9 AASHTO type beams in the cross section, and was redesigned by Benaim with four customised T beams, Figure 8.3, requiring a total of 56 precast beams. The consulting engineers responsible for the overall project were HSS Consult Consulting Engineers and it was built by the Kedeco-Kelcon Joint Venture with the help of Pilecon.

Each beam had a height of 2.7 m, a length of 45 m and weighed 130 tons, Figure 10.15. They were spaced at 3.367 m, had a 1.5 m wide top slab incorporated with the beam and 220 mm thick webs. The webs were reinforced with 10, 12 and 16 mm links, with 30 mm side cover and the 10 mm lacers were placed inside the links. Each beam was prestressed with two cables of 12 No 12.9 mm strands in ducts with

Figure 10.13 Sungai Kelantan Bridge (Photo: Robert Benaim)

Figure 10.14 Sungai Kelantan Bridge: casting yard (Photo: Robert Benaim)

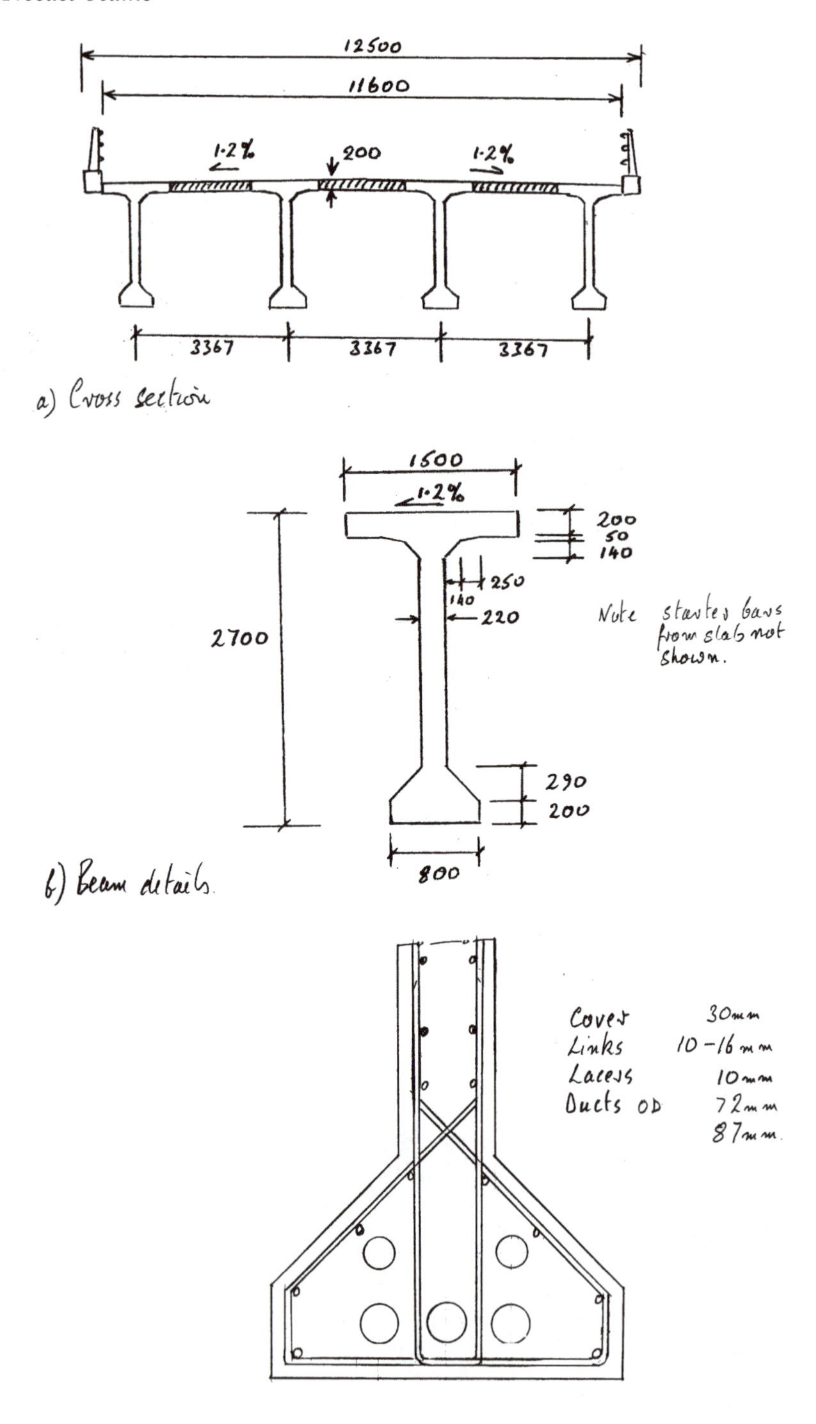

Figure 10.15 Sungai Kelantan Bridge: details

diameters of 66 mm internal, 72 mm external, and three cables of 19 No 12.9 mm strands in ducts with diameters of 81 mm internal, 87 mm external, Figure 5.12. The cast-in-situ reinforced concrete top slab was 200 mm thick.

The limited number of beams did not justify a highly mechanised site. Eight soffit forms were laid out, in line with the bridge, Figure 10.14. The steel shutter, equipped with external vibrators, could be moved from soffit to soffit. Consequently it was possible to work on several beams at the same time, preparing the steel cage, casting the concrete, leaving the beams in place to gain strength, and then tensioning the tendons. The beams were then picked up directly by the launcher which was improvised from Bailey bridge sections, and placed in position. The slabs between the precast beams were then cast using timber shuttering, suspended from the top slabs of the precast beams.

10.3.8 GSZ Superhighway, Pearl River Delta Viaducts, Guandong Province, China

The Pearl River Delta Viaducts extended for 17 km of which over 14 km used precast Tee beams. The viaducts were built by Slipform Engineering Ltd, the contracting arm of Hopewell Holdings [3]. Benaim were engaged to design the decks and foundations by Tileman (S.E.) Ltd, a subsidiary of Slipform Engineering.

The Superhighway consisted of two adjacent carriageways, each carrying three lanes of traffic plus hard shoulders. The 15.4 m wide decks required five beams, at 3.45 m centres, Figure 10.16. The spans were mainly 32.5 m, although there were many special cases, principally adjacent to interchanges; the precast beams were 2.3 m deep, included a top slab 1.6 m wide, and weighed 75 tons. The cast-in-situ deck slab and the webs were 200 mm thick, Figure 10.17. Each 32.5 m long beam was prestressed by five tendons consisting of 6 No strands of 15.7 mm, in ducts with diameters of 61 mm internal, 71 mm external. The side cover to the webs was 35 mm; the links

Figure 10.16 GSZ Superhighway, Pearl River Delta Viaducts (Photo: Engineer Leo K.K. Leung, Executive Director of Hopewell Highway Infrastructure Ltd (Hong Kong Stock Code: 737))

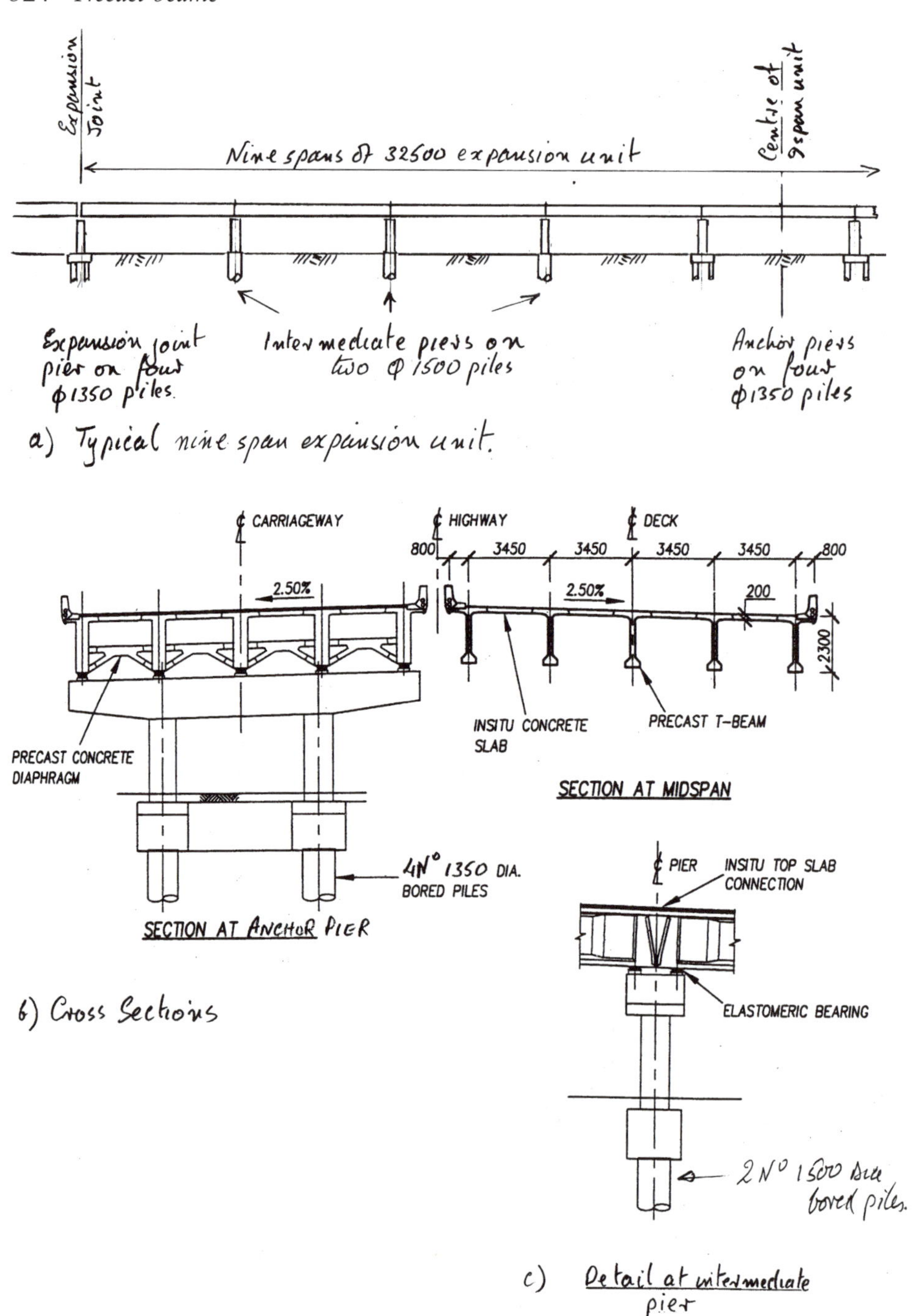

Figure 10.17 GSZ Superhighway, Pearl River Delta Viaducts: deck details

Figure 10.17 continued

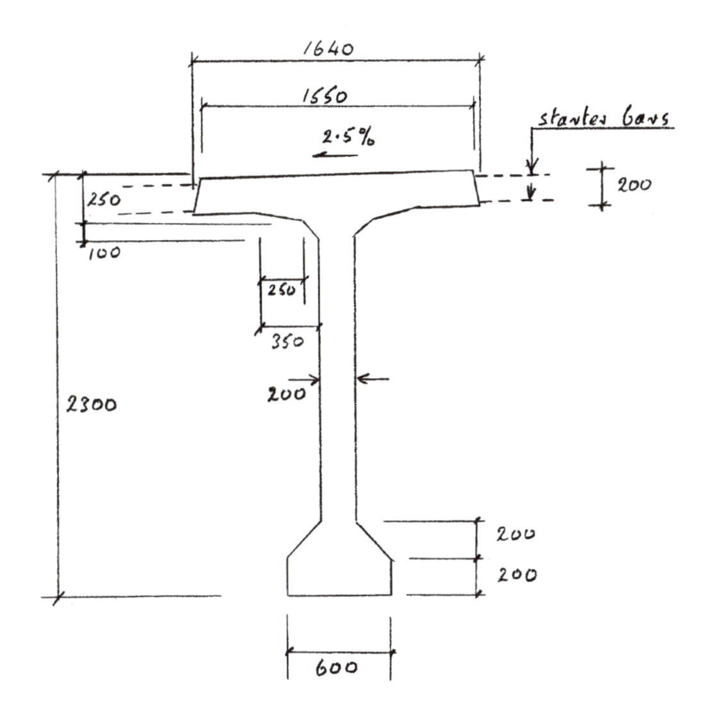

d) Beam details.

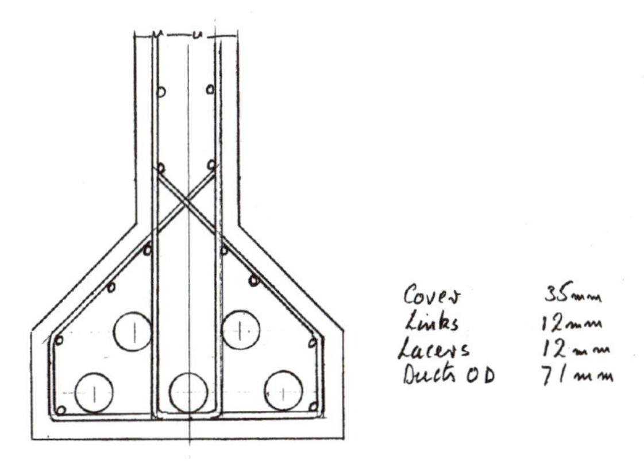

e) Reinforcement details

were 12 mm diameter with 12 mm lacers inside the links. Over 5,000 beams were required.

The beams were made principally in one main precasting yard, Figure 10.18, equipped with 24 steam cured steel moulds, designed to produce one beam per day per mould. The reinforcement was prepared in jigs and placed in the mould. The concrete was cast in the evening and cured overnight. A light first phase of prestress applied next morning allowed the beam to be lifted clear of its mould and placed in a storage area, awaiting the complement of the prestress. As the site was crossed by three large rivers, there were some problems of access, and several smaller strategically placed yards were also employed.

Purpose-built launchers were designed to place all five beams of a deck within two days. The slabs between beams were cast using a launching gantry that carried the formwork for three consecutive spans. This gantry was articulated so that it could adapt to the horizontal and vertical curves of the highway alignment. The precast end diaphragms of the decks were designed to allow the gantry to pass above them.

Most of the piers consisted of a crosshead carried by two columns, each resting on a single bored pile, although at interchanges single columns with crossheads were also used, Figure 10.16. Groups of nine spans were linked together by short lengths of 180 mm thick cast-in-situ slab, Figure 10.17 (c), to improve the ride and reduce maintenance (*7.15.2*). For this reason the ends of the webs were sloped forwards, to provide sufficient length for the link slab to deflect under the effect of the deflection of the beams and of the small vertical compression of the elastomeric bearings.

The 17 km of viaduct were completed in 21 months from the installation of the first pile.

Figure 10.18 GSZ Superhighway, Pearl River Delta Viaducts: main casting yard (Photo: Benaim)

11

Solid slabs, voided slabs and multi-cell box girders

11.1 Slab bridges, general

Slab bridges are under-used principally because of the lack of refinement of the preliminary costings carried out by most contractors' estimators. The unit costs of formwork, concrete, reinforcement and prestress tendons should clearly be lower for a solid slab deck than for more complex cross sections such as voided slab or box girder decks. However, in the early stages of a project, when options are being compared, this is frequently overlooked. Contractors who have experience of building prestressed concrete solid slabs and who have evidence for the low costs and for the reliability and speed of the construction process, tend to use such structures more readily.

Slabs allow the designer to minimise the depth of construction and provide a flat soffit where this is architecturally desirable. Their use is limited principally by their high self weight. Typical medium-span concrete bridge decks with twin rib or box cross sections have an equivalent thickness (cross-section area divided by width) that generally lies between 450 mm and 600 mm. Thus when the thickness of a slab exceeds about 700 mm, the cost of carrying the additional self weight tends to outweigh its virtues of simplicity.

Thicker slabs may be appropriate in constrained urban sites, where it is not possible to place columns in a regular pattern. For such randomly supported decks where beam and slab or box girder decks become excessively complicated and difficult to build, a solid slab with a thickness up to 1000 mm may be more economical than the alternatives.

Slab bridges are best analysed by grillage if the geometry is reasonably regular. For complex geometry, or for supports that are not organised into linear 'piers', finite element analysis is most appropriate.

Solid slab bridges may be of reinforced or prestressed concrete construction.

11.2 Reinforced concrete slab bridges

Reinforced concrete slab bridges are used generally for spans not exceeding 15 m, where a span/depth ratio of 20 gives a thickness of 750 mm. The ratio of the end/internal span for a reinforced concrete bridge is ideally 70 per cent. Such bridges can usually be made integral with their piers and abutments (see *7.13*). Reinforced concrete bridges shorten less than prestressed bridges, which helps an integral design. On the other hand, the rotations at the end of the side spans due to the self-weight deflections can give problems for the design of integral abutments unless the side spans are short.

It is difficult to calculate accurately the self-weight deflection of reinforced concrete bridge decks. Whereas the nominal *E*-value of mature concrete of a specified strength may be known with reasonable accuracy, the *E*-value at seven days, when the span is likely to be struck, is much more variable. Furthermore, the effective moment of inertia at zones of higher stress will be reduced by concrete cracking.

As an example, consider a 750 mm deep bridge deck with spans of 15 m being built span by span. The instantaneous self-weight deflection of the new span calculated with $E = 20{,}000$ MPa will be approximately 100 mm. However, if the *E* were to be 15,000 MPa and the cracked inertia 75 per cent of the uncracked value, the calculated deflection would increase by nearly 80 per cent, to 180 mm. As creep will increase this deflection by a factor of two to three, the difficulty in predicting final deflections and in providing suitable pre-cambers should be clear.

For decks built span-by-span, creep gives rise to uncertainty on the final distribution of bending moments between span and support sections (see *6.21*). The uncertainty is much greater for reinforced than for prestressed sections. Although this uncertainty may not affect the true ultimate strength of the deck it certainly does affect the cracking behaviour at the SLS.

In conclusion, reinforced concrete solid slab bridges are best used in the following conditions:

- one-off short-span cast-in-situ bridges where the contractor does not wish to mobilise prestressing expertise;
- spans below about 15 m, with span/depth ratio not shallower than 20 for continuous structures and 17 for statically determinate structures;
- preferably where the full bridge length may be cast in one operation.

11.3 Prestressed concrete slab bridges

11.3.1 General arrangement

For the designer, the main difference between prestressed and reinforced concrete solid-slab bridge decks is that the self-weight deflections of the former are entirely cancelled by the effect of the prestress; fully prestressed slabs deflect upwards slightly under long-term permanent loads. Thus the only limits on span/depth ratio are those defined by acceptable bending stresses, and of course economy since the cost of the structure rises with excessive slenderness. A span/depth ratio of 30 is probably optimal for a continuous deck (although 33 is quite practical, and 35 feasible), allowing a 600 mm thick slab to span economically up to about 18 m, and if necessary up to 21 m.

Prestressed concrete slabs are very valuable for bridge decks where the supports are skew or for irregularly shaped decks where the columns do not define organised spans. As the prestress opposes the deflections due to permanent loads, the transverse bending moments and torques due to these loads are also eliminated or reversed (*12.5*).

An additional advantage of slender prestressed continuous slabs is their great flexibility. In areas where settlement of supports is to be feared, either through mining subsidence, variable ground or due to settlement of the abutments under the weight of fill, these slabs may be designed to accept large movements without the need to resort to hinged structures.

Slabs may be 'creased' to follow a crossfall that is variable across the width without the unnecessary expense of varying the slab thickness. For normal crossfalls up to at least 5 per cent any 'folded plate' action may safely be ignored.

11.3.2 Transverse bending

Slabs decks are subjected to transverse bending moments and shears under live loading, and under localised dead loads such as footpaths, central reservations and parapets. These moments may be resisted by transverse reinforcement, or by light transverse prestress. The transverse bending under the effects of unequally distributed loads across the slab may also be controlled by varying the intensity of the longitudinal prestress across the width of the slab.

11.3.3 Crossbeams

For regular slabs with piers consisting of rows of supports, reinforced or prestressed crossbeams may be incorporated beams, within the thickness of the slab, or downstand beams. Supports should normally be arranged so that the span/depth ratio of the crossbeam does not exceed about 5.

For a reinforced concrete incorporated beam designed at the ULS, it is conventional to consider it as a Tee beam, with a web width, for calculation of shear strength, equal to the width of the bearing plus the thickness of the deck, and a slab width equal to the web width plus 1/10 of the distance between points of inflection for the span of the crossbeam beam either side of the web, Figure 11.1. This slab is at the top for sagging moments and at the bottom for hogging moments. The depth of this virtual slab for calculation at the ULS should not exceed 40 per cent of the thickness of the slab, to maintain a ductile section.

For prestressed incorporated beams designed at the SLS, the section properties of the crossbeam should be calculated with widths of the web and of the top and bottom slabs defined as for reinforced concrete. The thickness of the top and bottom virtual slabs at the SLS may be assumed to be 30 per cent of the depth of the deck. It must not be forgotten that the prestress force will be dissipated in the deck at an angle of 35°

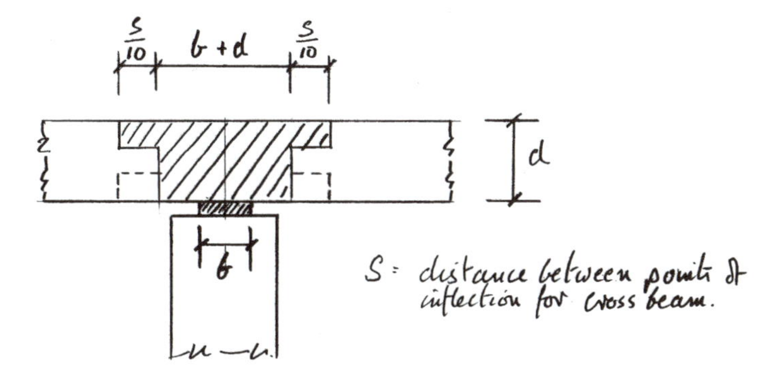

Figure 11.1 Incorporated crossbeams

(*5.26* and *9.6.2*), so only a proportion of the compression due to prestress (P/A) will be useful for the crossbeam, although the moment effect (*Pe*) will be entirely effective.

Generally crossbeams may be designed on the assumption that dead loads and UDL live loads are uniformly distributed along the beam, while concentrated live loads are applied directly to the beam. This is correct for downstand beams but conservative for incorporated beams where the slab itself will span transversally to concentrate the loads over the points of support. If the scale of the project justifies it, or if support points are widely spaced, a more accurate calculation of incorporated beams should be carried out using a carefully defined grillage or a finite element analysis.

Where the columns supporting the slab are spaced widely, drops or mushroom heads to the columns may be used to control the local bending moments and shears, Figure 11.2. The roof slab of Canada Water underground station on the Jubilee Line Extension in London, designed by the Benaim-WORKS JV, carried highway loads, and was supported by mushroom heads on the columns, Figure 11.3.

11.3.4 Prestress arrangements and typical details

a) Prestress layouts

The main cables are draped to carry part of the shear forces, although stirrups are still required for the ULS. These stirrups also act as chairs to carry the tendons. If the slab is designed to Class 1 prestressing (*4.2.3–4.2.5*), the only passive longitudinal reinforcement required, other than a nominal mesh to control heat of hydration effects, is the support bars for the stirrups. The reinforcement content for decks fully

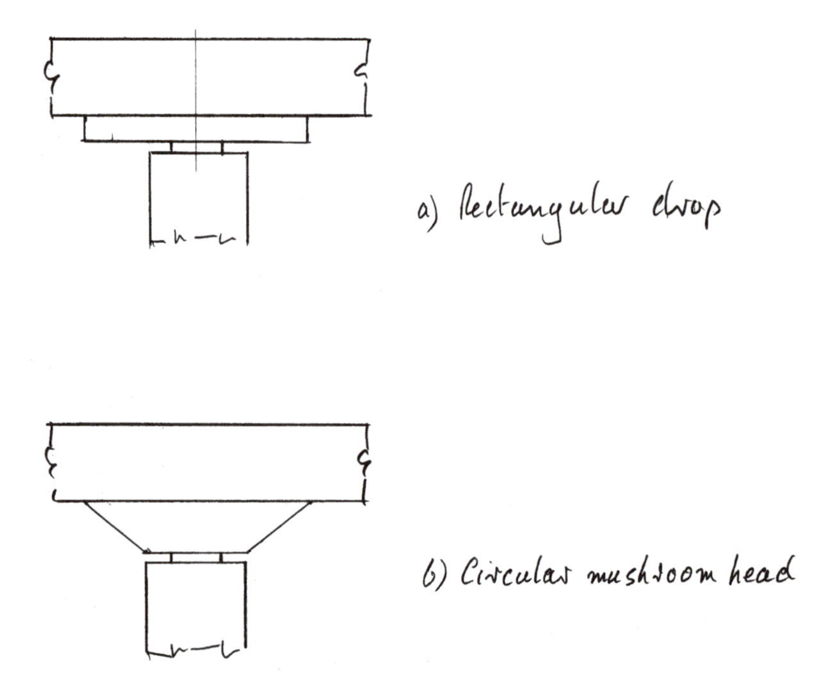

Figure 11.2 Drops and mushroom heads

Figure 11.3 Canada Water underground station: roof slab carrying bus station (Photo: Robert Benaim)

prestressed longitudinally and reinforced transversally is typically 45–60 kg/m^3. Much of this reinforcement will be concentrated in the crossbeams.

Fully prestressed solid slabs are wasteful of prestress, as they have a section efficiency of only 0.33 (*5.4*). The prestress lever arm is thus only $d/2$ – cover + $d/6$, where d is the depth of the slab. Where the code of practice permits, it is sensible to design the slab as Class 2 prestress, or even better as partially prestressed, and to increase the passive longitudinal steel to control crack widths under live loading.

In typical four-span motorway over-bridges, the cables may extend the full length of the deck, or may be crossed over the central pier. The slab decks, typically 400–600 mm thick, minimise the height and volume of fill for the approach embankments. The four spans may be cast in one continuous pour, avoiding all heat of hydration problems. The slabs may also be built span-by-span (*15.2*).

b) Deck end arrangements and bearings

Usually, slab decks are carried by laminated rubber bearings. The spacing of these bearings should preferably not exceed five time the thickness of the slab. The need to anchor the last compression strut of the equivalent truss must not be forgotten (*3.10.2* and *5.21*). Thus a prestress anchor with a factored ultimate strength equal to the ultimate reaction carried by the bearing should be located adjacent to each bearing and as low as possible in the slab. If this is not possible, this anchor force must be provided by passive reinforcement, acting either alone or in combination with the tendon.

The prestressing tendon anchorages should be distributed as evenly as possible along the end of the deck to minimise the equilibrium forces due to the prestress compression as it fans out from the anchorages to uniformity in the slab. Either additional transverse reinforcement or prestress must be provided to resist these forces.

The end zone of a prestressed slab can become congested due to the addition of the reinforcement necessary for the crossbeam actions, equilibrium steel, primary prestress anchorage steel, bars to anchor the expansion joint and longitudinal bars. Careful planning and detailing is necessary.

11.3.5 Construction of slab bridges

Where a multi-span deck is built span-by-span, there is a risk that each new deck cast will suffer heat of hydration cracking adjacent to the construction joint with the previous deck (*3.6.4* and *9.3.8*).

As is usual where such cracking is likely, transverse reinforcement should be provided to limit the opening of such cracks to a nominal 0.2 mm. If it is desired to eliminate such cracking entirely, a light transverse prestress should be applied to the older concrete, adjacent to the construction joint, just as the new concrete starts to cool from its peak heat of hydration temperature. The aim should be to impose a transverse compressive stress at the centre of the deck and at the location of the construction joint of approximately 1 MPa.

Slab bridges are rarely so long that mechanised falsework is justified. Usually, general scaffolding is adopted.

11.3.6 Conclusion

When compared with reinforced concrete slabs, prestressed slabs offer:

- reduced tonnage of materials to handle;
- increased speed of construction due to the early striking of forms, allowing a rapid turnaround of falsework;
- predictable self-weight deflections;
- the most slender of decks where this matters;
- flexibility to cope with relative settlement;
- reduced approach embankment heights due to the slenderness of the deck;
- good durability as the deck will be crack free under permanent loads;
- for skew bridges, or decks carried on irregularly placed columns, the cancelling out of transverse moments and torques under permanent loads.

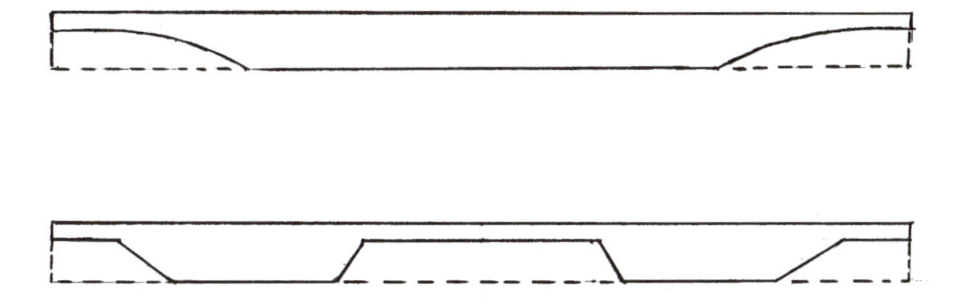

Figure 11.4 Development of solid slabs

11.3.7 The development of solid slabs

Solid slabs may be enhanced by creating side cantilevers. Such a modified slab retains the benefits of simplicity, but sheds some unnecessary material and weight, improving its economy and its appearance. A further step is to remove unnecessary material from the centre of decks, Figure 11.4. However, these modifications have the effect of reducing the bottom modulus of the section and hence increasing the stress range on the bottom fibre. If the bottom fibre stresses are already at their limit, either the concrete strength must be increased or the section must be deepened. As the deck is deepened and the rib width reduced, this leads to the family of prestressed concrete decks known as ribbed slabs (Chapter 12).

Another development of the solid slab is to lighten it by introducing voids, leading on to the concept of voided slabs (*11.5*) and multi-cell box girders (see *11.7*).

11.4 Solid slab portal bridges

11.4.1 General

For single spans, solid slab portals provide a useful option. For instance, where a single-span bridge deck rests on cantilevered retaining walls, a portal is most likely to offer a more economical alternative, Figure 11.5. Such bridges may be entirely reinforced concrete, may have a prestressed concrete deck and reinforced concrete walls, or may be entirely in prestressed concrete. Portals may be built ‘top down’ by making their uprights with diaphragm walls or contiguous bored piles. If the ground is very poor, the portal may be founded on a full-width slab, creating a box structure, Figure 11.5 (c).

The portal action gives rise to an outward reaction at the foundations, while the earth pressure on the side walls gives rise to an inward reaction at the foundation. The direction of the resultant depends on the ratio of span/height, on the intensity of dead and live loads carried by the slab, and on the earth pressure coefficients of the backfill. If the height of the portal and the depth of fill above it are at the designer’s discretion, he can arrange for the resultant load at the foundation to be substantially vertical. For a pad footing, this may not be of great significance. However if the foundation is piled, avoiding raked piles may make very significant savings, Figure 11.6.

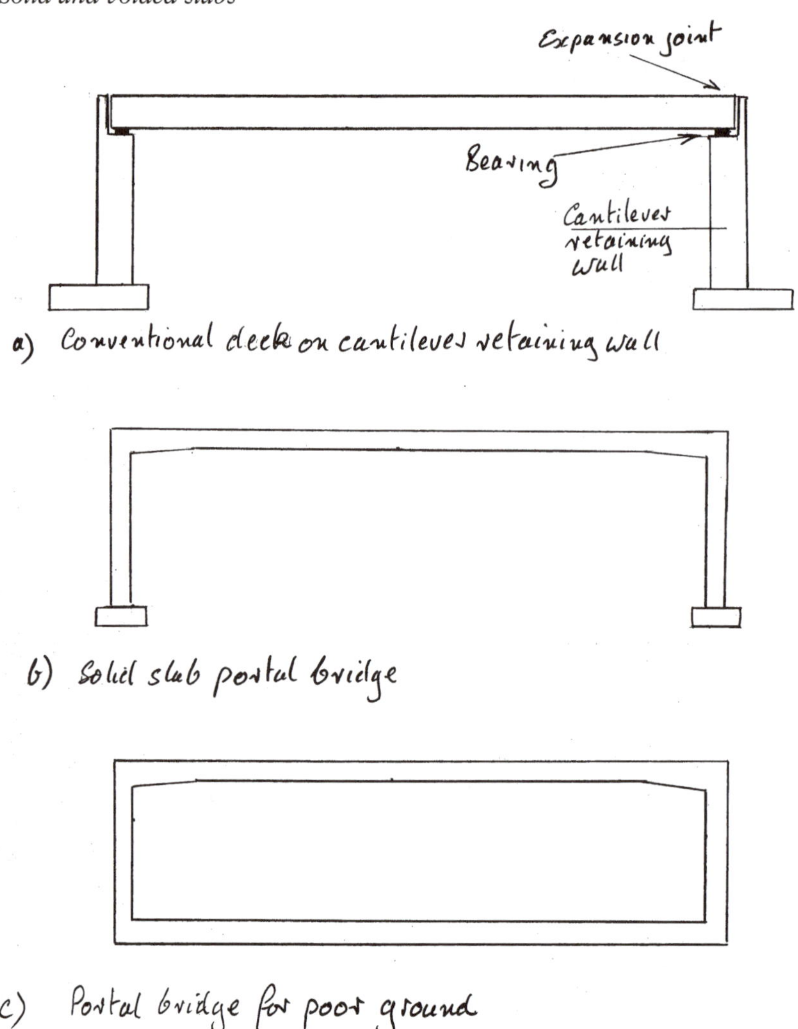

Figure 11.5 Solid slab portal bridges

11.4.2 *Reinforced concrete portals*

A good starting point for sizing a reinforced portal structure carrying highway loading would be to choose a deck slab thickness equal to span/30, with haunches 50 per cent thicker, extending some 7 per cent of the span, if this does not interfere with the clearance diagram. Clearly, if the portal carries a great depth of fill, the slab depth will need to be greater. Haunches are always economical. They provide the twin benefits of attracting moment away from mid-span and then providing a greater lever arm to resist this moment economically. Even very short haunches are valuable in reducing the hogging reinforcement. Figure 11.7 shows a section of the Singapore Central Expressway, designed by Benaim. However, if the haunch is too steep, it will not be entirely effective; the maximum useful slope being about 35°, Figure 11.8 (a).

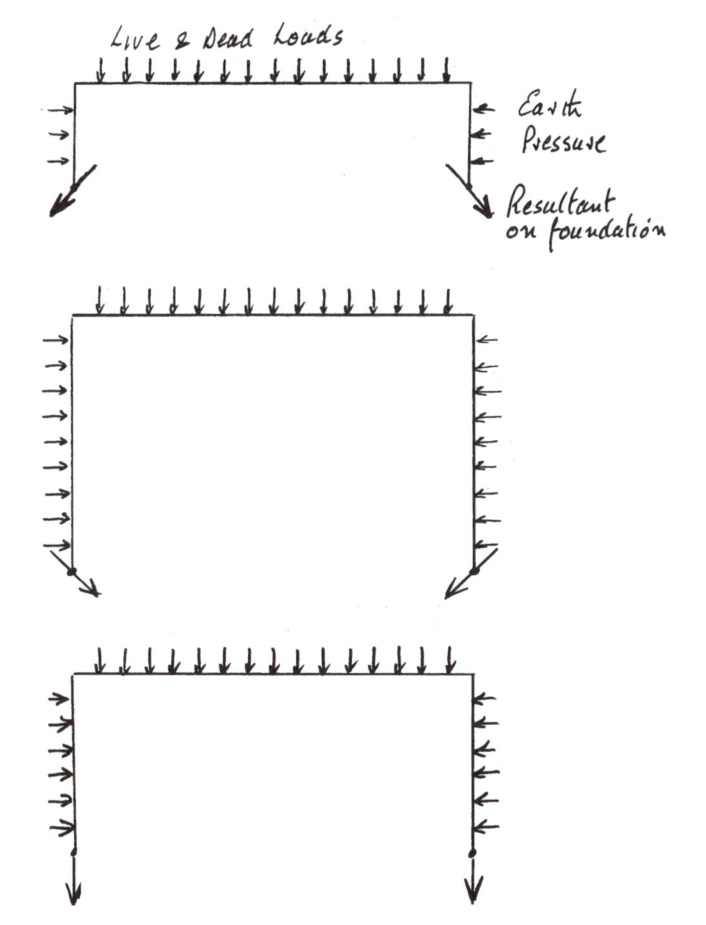

Figure 11.6 Resultant forces on portal foundations

Figure 11.7 Singapore Central Expressway (Photo: Robert Benaim)

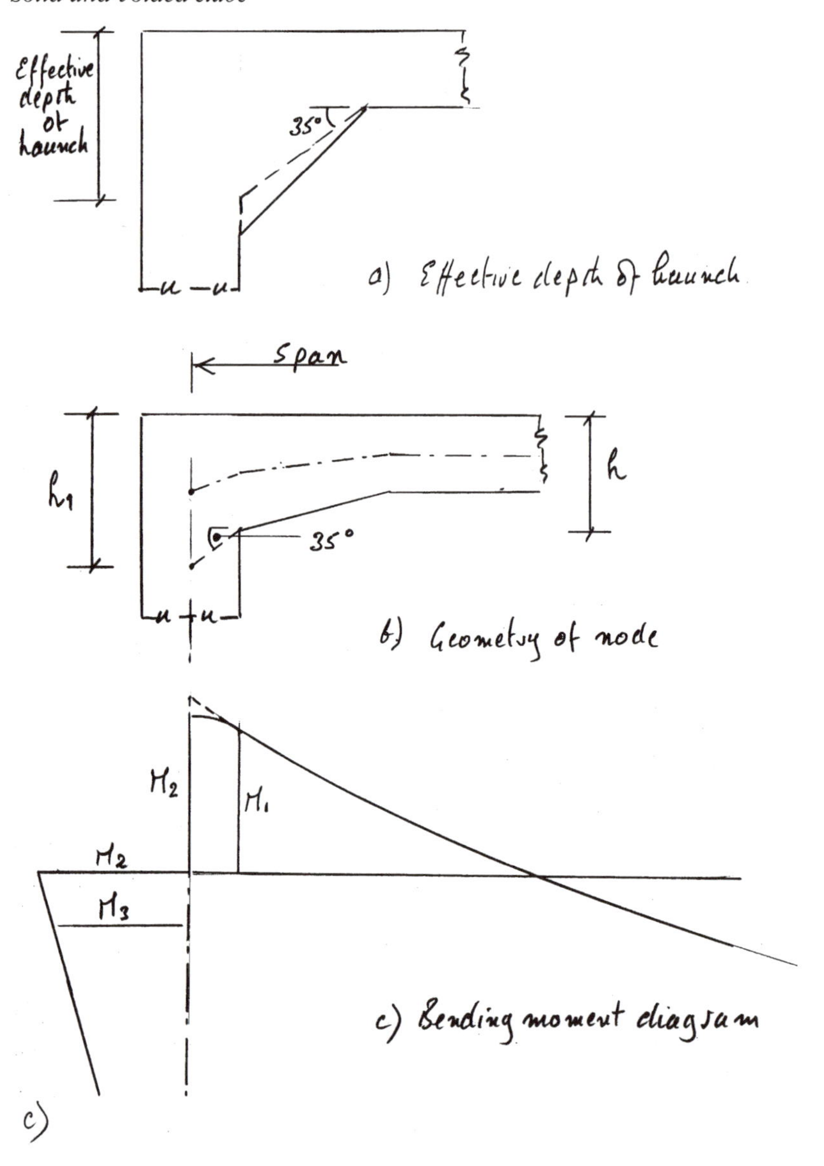

Figure 11.8 Modelling of portal node

The uprights have to carry the same bending moment as the haunch, but with the benefit of a compression force due to the weight of the roof. Thus they may be slightly thinner than the haunches.

The modelling of the node between slab and upright must be carefully considered. The theoretical span for the calculation of the bending moments in the portal should be taken to the point of intersection of the centre lines of members. The peak bending moment beyond the face of the upright will be rounded as described in *6.13.5*, whereas the depth of the slab will be increased at an angle of 35°. The maximum hogging reinforcement will be the greater of that calculated using the moment M_1 at the face of the wall with the depth of the haunch h, or the rounded moment M_2 at the axis of the wall with the enhanced depth h_1, Figure 11.8 (b). Usually the former governs. The moment M_2 turns down into the upright, where the design moment is M_3 at the bottom of the haunch.

A strut-and-tie model should be used to understand the forces within the node, Figure 11.9 (a). This makes it clear that there must be continuity of the tension force all round the outside face of the node. It is not unknown for portal nodes to be detailed with bars given anchorage lengths from the face of the members, without continuing round the corner!

Figs 11.9 (b) and (c) show the wrong and right ways to detail this corner. Arrangement (b) would require the concrete for the upright to be poured through the dense horizontal starter bars, which is not practical. Contractors would be tempted to bend up these bars for the pour, and then bend them back; very poor practice likely to weaken or fracture the bars. Arrangement (c) may influence the level at which the construction joint is made, or the size of bar used (as this determines its lap length).

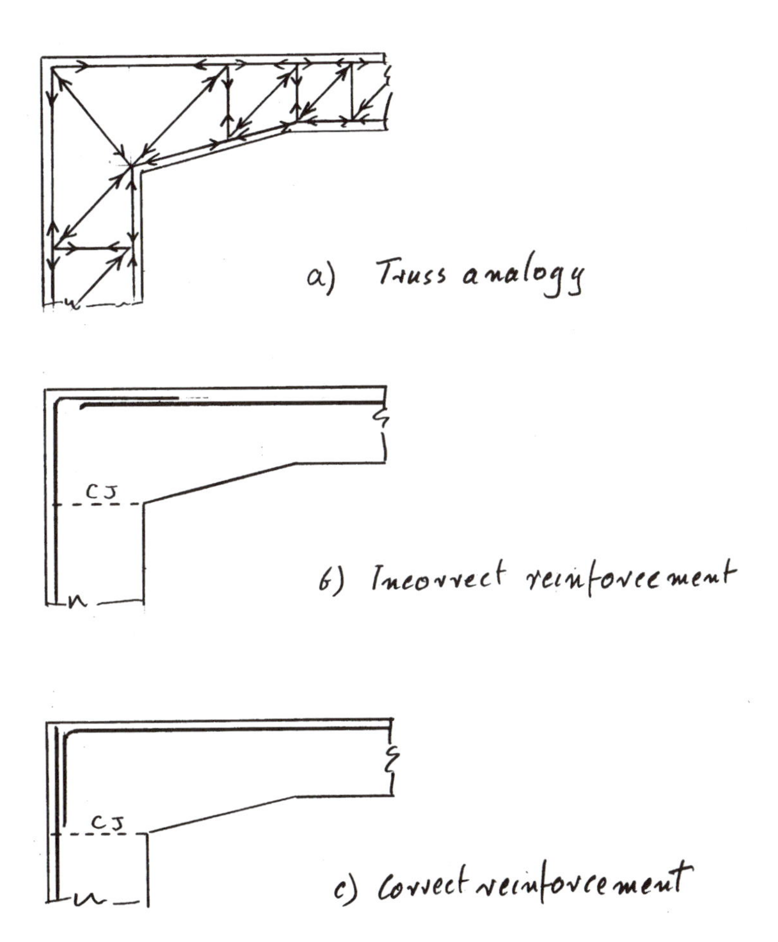

Figure 11.9 Detailing of node

11.4.3 Prestressed concrete portals

Prestressing the slab allows it to be thinner when this is important, and has the other advantages described in *11.3.6*. A span/depth ratio of 35–40 is quite practical for a haunched slab as described above. The detailing of the node is again important, as it is necessary to create effective continuity between the prestress in the slab, and the reinforcement in the uprights. Normally this requires that the prestress anchorages are placed outside the vertical steel, Figure 11.10 (a). It also requires that the reinforcement of the uprights be anchored beyond its intersection with the tendons. If the uprights are also prestressed, the anchors for the vertical tendons must be above the horizontal tendons. This may limit the eccentricity of the horizontal tendons over the supports, Figure 11.10 (b).

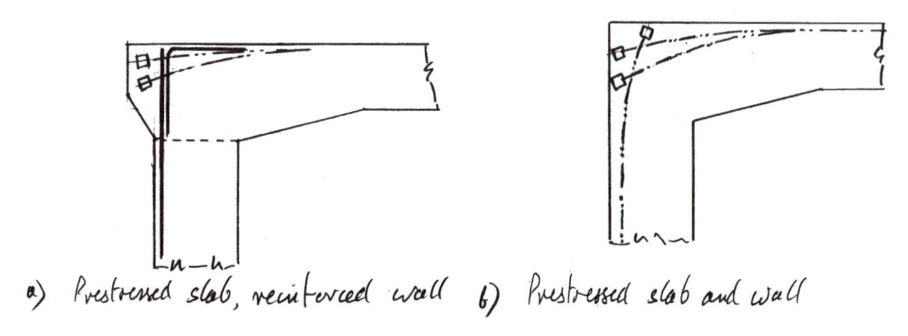

Figure 11.10 Nodes for prestressed portals

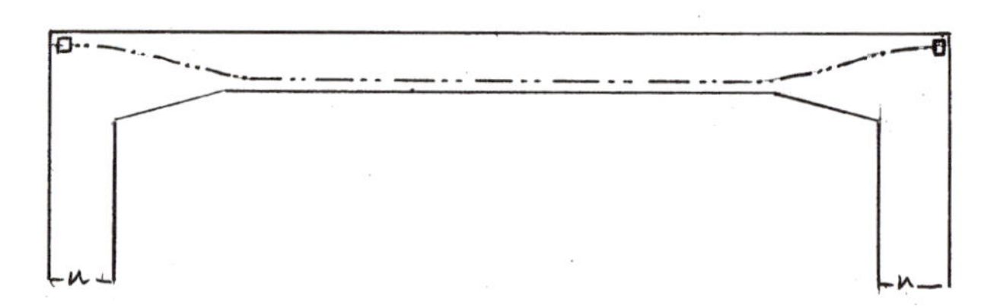

a) Prestress profile designed to maximise sagging parasitic moments.

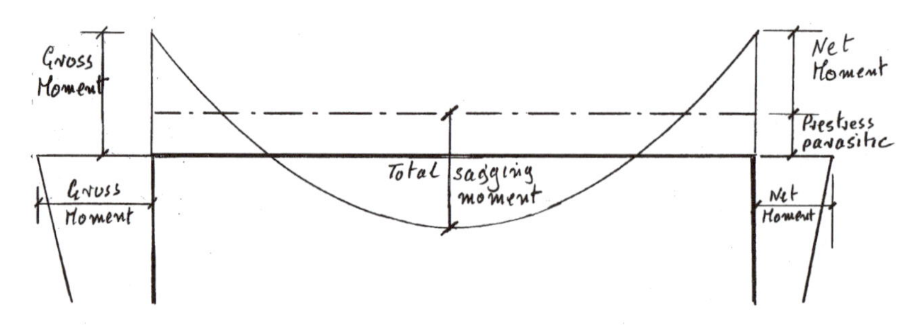

b) Effect of prestress parasitic moments

Figure 11.11 The effect of prestress parasitic moments on portals

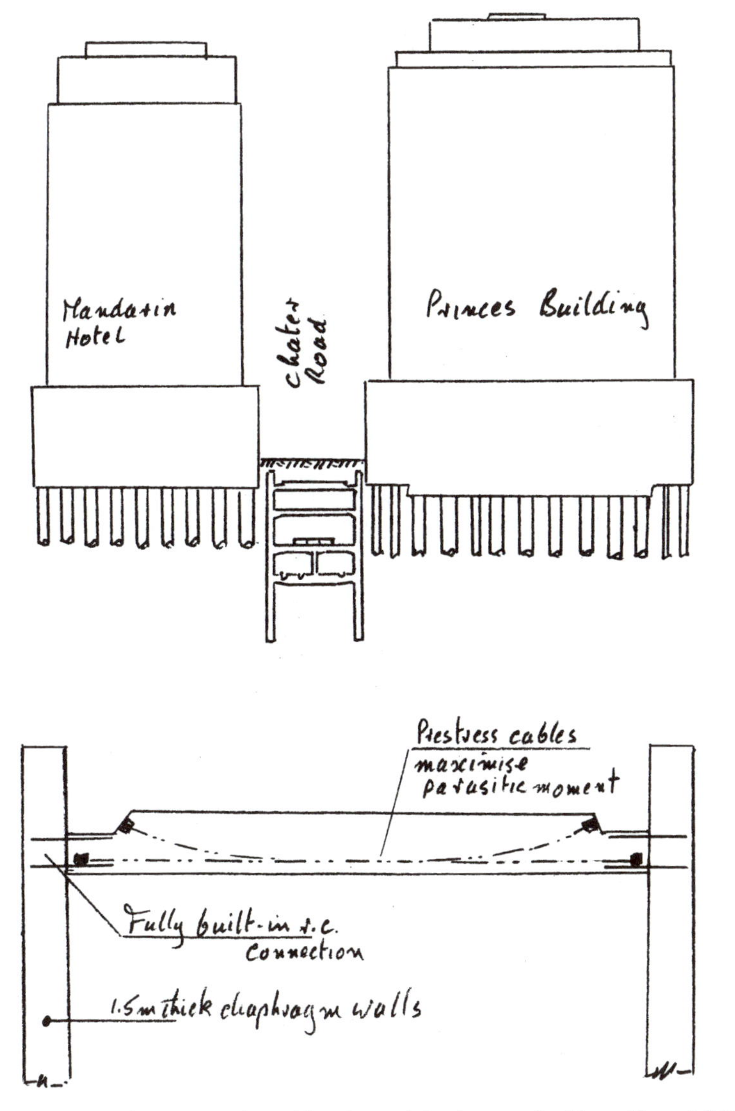

Figure 11.12 Diagram of prestressed roof for Central Station on the Hong Kong MTR

The positive prestress parasitic moment due to the slab tendons will reduce the gross hogging moment at the top of the uprights, Figure 11.11. This may be very valuable where the thickness of the uprights, and thus their moment capacity, is limited such as when they are built as diaphragm or secant pile walls. Under these circumstances, prestressing the slab may be the only way to make the walls work. Figure 11.12 shows diagrammatically the roof of Central Station of the Hong Kong Mass Transit, designed in this way by the author when at Arup. Without the benefit of the prestress secondary moments, the diaphragm walls would have been unable to carry the hogging moments derived from the loads on the slab, and the junction between walls and roof would have had to be pinned, with deleterious consequences for the design.

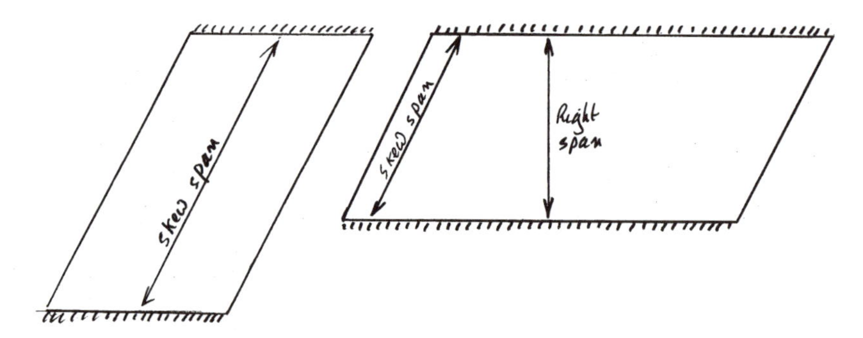

Figure 11.13 Skew portals

11.4.4 Skew portals

Portals are very suitable for skew crossings. Their behaviour depends principally on the ratio between the span and width of the deck. Figure 11.13 shows a narrow bridge where the bridge spans principally on the skew and a wide bridge of the same skew. Over the central part of the wide bridge the deck spans right while at the ends it spans on the skew. Either the thickness of the deck must be adequate for the longer span with locally increased reinforcement, or some kind of edge strengthening must be used, such as structural parapets or a thickened slab.

Where the deck spans on the skew, the support moments are higher than for a right span of the same length due to the stiffening effect of the angled support, and the shape of the sagging moment diagram is also distorted. For the detailed design it is necessary to carry out a carefully defined grillage analysis or a finite element analysis.

11.5 Voided slabs

11.5.1 General

In general terms, prestressed concrete decks become more economical as they become deeper, with span/depth of 15–18 being typical. The solid slab is the only exception to this rule; its economy derives from its extreme simplicity. However, when a prestressed solid slab exceeds a thickness of 700 mm, providing a span of up to 23 m, its weight starts to become excessive, and the designer should consider alternative deck types.

The most logical alternative is to abandon the slab form and adopt a ribbed slab deck (Chapter 12). The depth of such a deck is likely to increase substantially, a 23 m span requiring a depth generally in excess of 1.15 m. However, the depth allowable may be limited, or a flat soffit may be required. For these special cases, a voided slab deck may well be suitable.

It should be clear to the designer that by introducing voids, he has sacrificed the essential simplicity of the slab. The cost of the voids, including the measures taken to hold them steady during concreting and to resist the up-thrust of the wet concrete, is at least as much as that of the concrete saved. The unit costs of the concrete, reinforcement and prestressing will all increase due to the greater intricacy, and hence to the greater labour required. Also, the quantity of passive reinforcement will be significantly increased, probably exceeding 110 kg/m^3 of concrete. The principal benefits are greater deck efficiency and lower weight, leading to economies in the quantity of prestress, and savings in the foundations.

11.5.2 Voids

Voids may be circular, quasi-circular such as octagonal, or rectangular. Rectangular voids are assimilated to multi-cell boxes and are covered in *11.7*. The diameter of circular voids should be a maximum of 240 mm less than the deck thickness, and preferably 300 mm, and not greater than 0.7 of the deck depth. The ribs between voids should generally be 250–300 mm wide. The voids are usually stopped short of the lines of piers, abutments and construction joints to create incorporated crossbeams, Figure 11.14. Intermediate crossbeams are not required in such decks. Skews may be easily accommodated.

Various methods have been used to create voids. The commonest is to use expanded polystyrene, which has the advantage that it is light and easy to cut. In theory, polystyrene voids can be made of any shape, either by building up rectangular sections, or by shaping standard sections. In practice, the labour involved in building up or cutting sections is not economical, and cylindrical voids are usually used. These cylinders may be cut away locally to widen ribs, or to accommodate prestress anchors, drainage gullies etc.

The voids must be held down to resist buoyancy, and this is usually achieved by metal straps, wide enough to avoid cutting into the polystyrene under the very considerable up-thrust. These straps may be anchored to the passive reinforcement cage. However, as in most bridges this cage is not heavy enough to resist the flotation forces and would be lifted with the voids, the cage itself should be anchored through the soffit formwork. The penetrations of the formwork may be masked by siting them in formwork features such as grooves.

Another solution that has been adopted is to anchor the voids to the reinforcement, and then to pour a first layer of concrete enclosing the bottom layer of steel, providing the weight to resist flotation. Before this lower layer has set, but once it has ceased to be fluid, casting is continued. Although effective, this technique requires very good control of the casting process to avoid the twin dangers of creating a horizontal cold joint, or of re-mobilising the up-thrust of the concrete.

Various other types of void have been adopted, such as thin spirally wound steel tubes. These may well need internal stiffening to resist the unbalanced concrete forces as one rib is cast before the other and need to be held down as for the polystyrene void formers. They are much less adaptable to the needs of buried prestress anchors, gullies etc.

Concrete drainage pipes or precast man-hole rings provide cheap and serviceable voids where adaptability is not required. Their high self weight makes it possible to hold them down to the reinforcing cage without the need for further anchorage through the bottom shutter.

11.5.3 Prestress arrangements

Each rib generally accommodates either one or two prestress tendons. The latter is the preferred arrangement as at the end of the deck one tendon may be swept up to carry

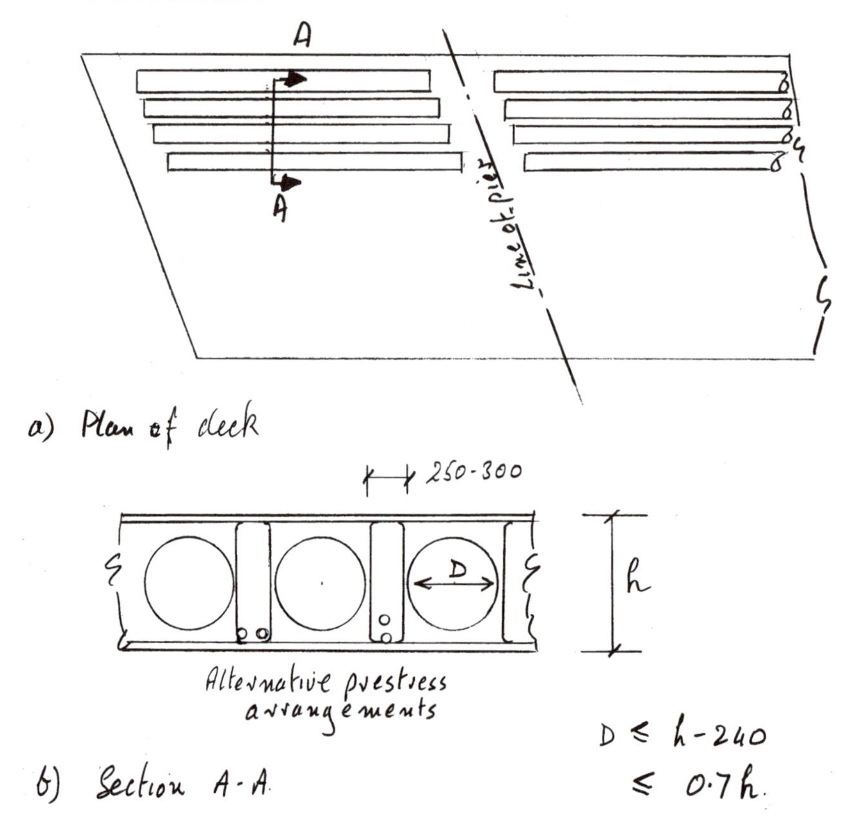

Figure 11.14 Voided slab deck

a proportion of the shear force, while the other may be kept low to tie in the final concrete shear strut. Also, two tendons provide a more even distribution of anchor forces on the end face of the deck, reducing the equilibrium reinforcement required.

11.5.4 Pier and abutment crossbeams

The comments on the design of pier and abutment crossbeams made in *11.3.3* concerning solid slabs apply equally to voided slabs. As the deck is deeper than a solid slab, there is more incentive to use incorporated rather than downstand beams.

11.5.5 Reinforcement arrangement

Voided bridges have shown a tendency to crack along the line of the voids, principally on the bridge soffit. This is due either to a heat of hydration effect, or more probably, to the expansion of the void due to this heat. In order to prevent this, it is necessary to reinforce the concrete above and below the voids with at least the anti-brittle fracture rate in direct tension of 0.6 per cent of the minimum concrete thickness (*3.7.2*).

Transverse bending reinforcement is calculated as for a solid slab. Grillage analysis is marginally different as the transverse members must take into account the reduction of both bending and shear stiffness due to the voids. For normally dimensioned decks it is not necessary to check the transverse bending as a Vierendaal girder, and it is also unnecessary to provide corner bars trimming the voids. For decks with abnormally high transverse bending, due for instance to an irregular pier arrangement, it is prudent to check the Vierendaal action and to provide corner bars.

11.5.6 Casting the concrete

Concreting voided decks requires more care than for solid slabs, as there is the risk of incomplete filling and compaction beneath the voids. Consequently it is a slower and more labour-intensive operation. The normal method for decks with circular voids is to work across the deck, filling one rib and observing the concrete flowing out from beneath the adjacent void.

11.5.7 The development of voided slabs

The development of voided slabs is similar to that of solid slabs. In decks where the maximum stress on the bottom fibre is less than the permissible limit, it is cost effective to create side cantilevers and to remove material from the centre of wide slabs, creating effectively a voided ribbed slab, Figure 11.15. Alternatively, the slab may be provided with haunches at intermediate supports, allowing the slab to span further.

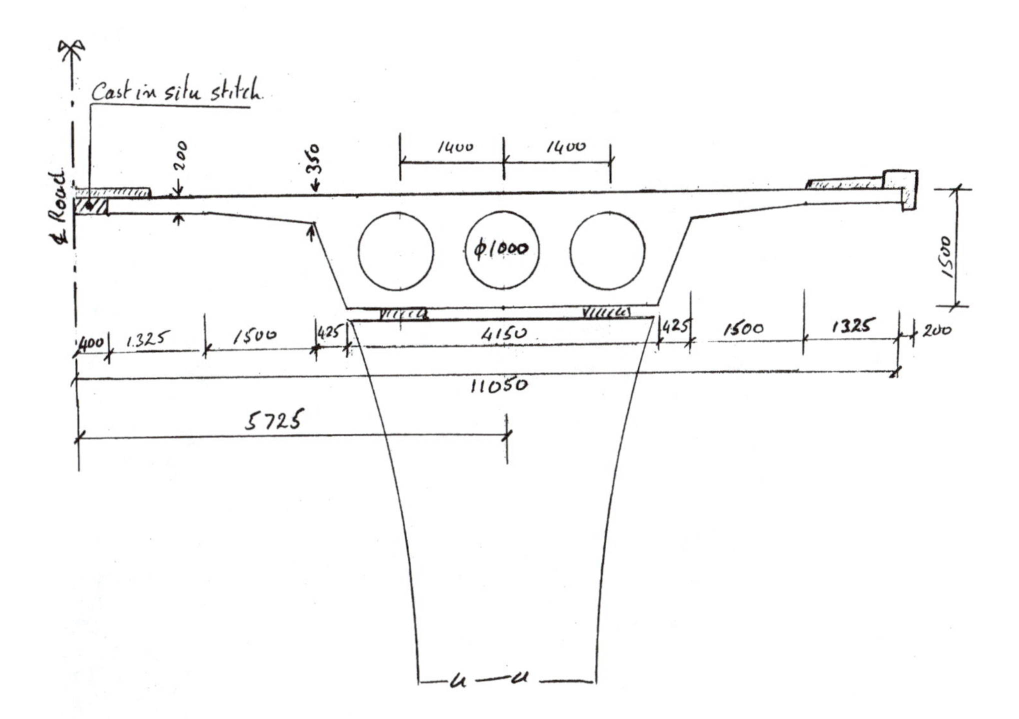

Figure 11.15 Voided ribbed slab

11.6 Case history: River Nene Bridge

The 139.2 m long River Nene Bridge carries the dual carriageway Nene Valley Way over the River Nene and a single-track railway, close to Northampton. This bridge was designed by the author when at Arup, with the participation of Arup Associates. The highway alignment was fixed, as were the clearances required over the railway and the access road on one bank of the river, leaving available only a 700 mm structural depth. Adopting a continuous solid prestressed slab would have required five internal spans of 20.5 m (span/depth of 29.4), and two end spans of 18.4 m. This number of spans was not suitable, functionally or aesthetically.

A four-span solution was adopted (29.7 m, 39.9 m, 39.9 m, 29.7 m), using a 700 mm voided slab which was haunched over the three internal supports, Figure 11.16. The bending moment was attracted from the spans to these haunches by building them into their piers. In order to allow the bridge to change in length under temperature variations and concrete shrinkage and creep, each pier was split into five thin precast concrete planks. The planks of the central support, which were not subject to length changes, were shorter and hence stiffer than those of the intermediate piers so they would attract a larger proportion of the longitudinal external loads applied to the deck. The voids were made of polystyrene, Figure 11.17.

Figure 11.16 River Nene Bridge (Photo: Arup)

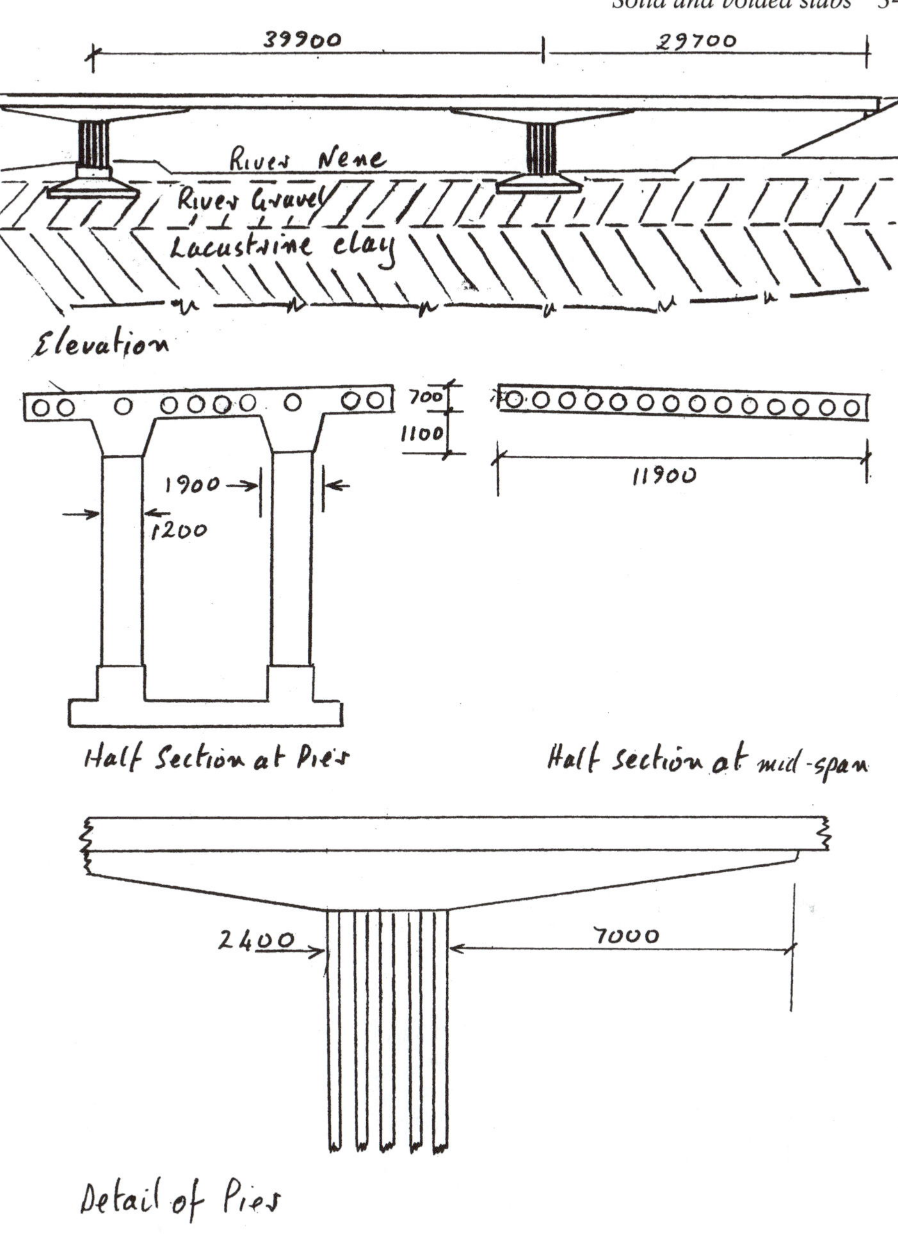

Figure 11.17 River Nene Bridge: details

11.7 Multi-cell box girders

11.7.1 General

The logical conclusion of the development of voided slabs is the multi-cell box. This is understood to mean shallow decks, less than about 1.4 m deep, with contiguous rectangular voids. Many theoretically minded engineers have been mislead by the apparent economy and efficiency of this form of deck. However simple considerations of constructibility and a clearer view of material economy should convince thinking designers that this is a deck form to be avoided. The reasons for this anathema are described below.

The main reason for not using this form of deck is the slow, labour-intensive and risk-prone construction procedure, due principally to the virtual impossibility of casting the cross section in one pour. The bottom slab is cast first with a kicker for the webs from which projects the web reinforcement, Figure 11.18. The top surface of the construction joint in the webs must be tooled to remove laitance, and to provide a rough surface. The void shutters are erected and the deck is completed by casting the upper parts of the web and top slab. Finally the void formwork is withdrawn piecemeal through holes left in the top slab (or is abandoned).

The ducts for the prestress tendons, which are housed in the bottom slab near mid-span, cross the horizontal construction joints in the web at a shallow angle. It is very difficult to protect these flimsy empty ducts during the tooling of the construction joint. Undetected damage to the ducts will lead to leakage of grout into the tendons, which then need major surgery before they can be stressed.

The preparation of this horizontal construction joint, with the interference of the projecting shear reinforcement and the prestress ducts is very labour intensive.

The horizontal construction joint on the visible side face of webs is very difficult to build neatly. The top surface of any pour has a very high water/cement ratio due to bleeding of the mix, and unless the joint preparation is very thorough, gives rise to a porous layer of concrete. Thus as well as being unsightly, this joint also represents a

Figure 11.18 Multi-cell box girder under construction (Photo: Robert Benaim)

weak plane in the corrosion protection of the shear reinforcement. Furthermore the staged construction of the cross section gives rise to heat of hydration stresses and cracks, and increases the amount of reinforcement required.

This is just the type of labour-intensive, delay-prone deck construction that incites contractors to avoid prestressed concrete. The responsibility of the designer is not just to consider economy and beauty, he should be producing designs that can be built reliably, to cost and programme, and which allow contractors to make a profit.

If the deck is not of a regular plan shape, or if the crossfall of the deck varies along its length, the difficulty of construction escalates into a catastrophe as the internal shuttering of the cells has to be tailored to the changing geometry of the deck.

11.7.2 Material economy

The material economy of multi-cell box girder decks is a figment of the designer's imagination. The absolute minimum thickness for the top slab of a highway bridge spanning across 2 m wide voids is 170 mm, subject to checking the requirements of punching shear to the relevant loading and reinforced concrete codes of practice. However with the addition of 30 mm of concrete, a far more robust 200 mm thick slab may span up to 6 m between webs.

There are more webs than are necessary to carry the shear force in the deck. In attempting to make the webs as thin as possible to save weight, the deck becomes much more difficult to build. It is economical to minimise the number of webs and to make them thicker for easier casting.

The absolute minimum thickness of the bottom slab of a bridge deck is 150 mm. With the addition of a further 20–30 mm, this bottom slab may span 6 m or more.

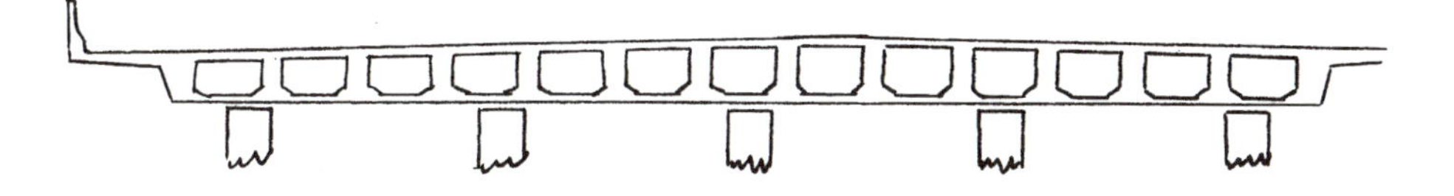

a) Multi cell box varying in width, depth and crossfall

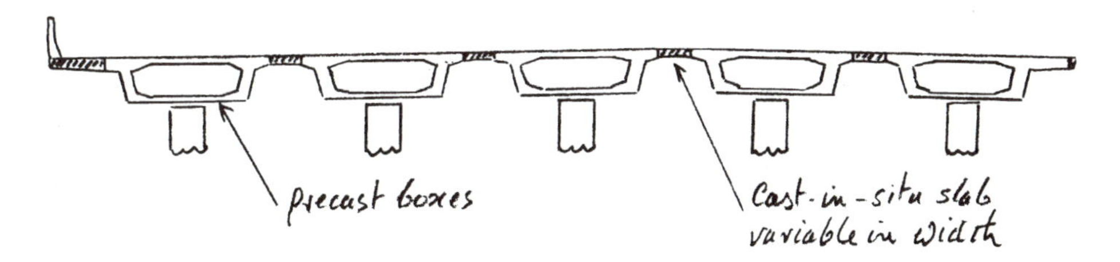

Figure 11.19 Multi-cell box with alternative

The need to clad both faces of each web and each face of the top and bottom slabs with a mesh of reinforcement inevitably leads to the use of expensive small bars, slow fixing rates and a high rate of reinforcement, typically over 230 kg/m^3.

As a consequence, multi-cell box decks are heavier than well-designed ribbed slab or single-cell box section decks, have more reinforcement content, and are far more labour intensive and costly to build.

11.7.3 Development of multi-cell box decks

The designer faced with one of these decks, perhaps specified by a client, should investigate alternatives. These may be voided slabs, ribbed slabs for cast-in-situ construction, or precast segmental single cell boxes attached by their top slabs. The author was faced with such a problem when asked by a contractor to assist in the tender for the deck shown in Figure 11.19. As the deck width and the crossfall varied along the deck, the height and width of each cell was variable and the proposed design was virtually unbuildable. The alternative proposed was a series of identical precast segmental boxes (Chapter 14), linked by cast-in-situ top slabs whose width could be extended as necessary. The variable geometry of the conforming design could thus be matched exactly with a standard precast product. Alternatively, if it had been essential to maintain a flat soffit, the cross section could have been made far easier to build by adopting circular voids.

12
Ribbed slabs

12.1 General

When the maximum bending stress on the bottom fibre of a solid slab deck is below the allowable limit, it is possible to remove material without increasing the depth of the deck, leaving a shallow ribbed slab as described in *11.3.7*. However, if more depth is available, the deck may be made more economical by making the ribs deeper and narrower.

As explained in Chapter 8, it is important for economy to limit the number of webs in a deck. In fact, for most bridge decks up to a width of 20 m and more, two ribs are found to be adequate. Such bridges are usually called 'twin rib bridges' and this chapter will concentrate on this deck arrangement. Figure 12.1 shows one of the Doornhoek bridges in South Africa, a typical twin rib deck designed by the author at Arup with the architectural assistance of Humphrey Wood of Renton Howard Wood Levine for the design of the columns.

Where the depth is not constrained, the width of the ribs is defined by the ULS bending or shear at the supports. This width is maintained constant throughout the span. Economy is achieved by the simplicity of the construction, despite a relatively high consumption of concrete and prestressing steel. The prestressing tendons have simple profiles; the reinforcing cage is uncomplicated and may be prefabricated; casting and compacting the concrete is quick; a complete span may be cast in one continuous pour. Most twin rib bridges do not need diaphragms except at abutments where they support the roadway expansion joint. However, pier diaphragms are necessary in some particular situations, as explained in *12.3* below. Where there are no pier diaphragms, the central tunnel formwork can be launched forwards from span to span without dismantling.

These decks are well received by contractors in both developing and developed economies. In the former, they are suitable for an unskilled work force; in the latter, they allow high productivity from expensive labour. With a semi-mechanised falsework of the type shown in Figure 12.2, used for the railway viaduct forming Contract 304 of the Tsuen Wan Extension of the Hong Kong Mass Transit Railway, and designed by the author at Arup, construction proceeded at the rate of a span in two weeks. With a mechanised self-launching rig and prefabrication of the reinforcing cage, developed from that used for the Viaduc d'Incarville, Figure 12.3, two spans per week is possible (*15.2.5*).

Figure 12.1 Doornhoek Bridge, South Africa: typical twin rib bridge (Photo: Arup)

Figure 12.2 Contract 304 HKMTR: semi-mechanised falsework (Photo: Robert Benaim)

Figure 12.3 Viaduc d'Incarville, Autoroute de Normandie: self-launching rig (Photo Robert Benaim)

Twin rib bridges are essentially a cast-in-situ solution for spans between 20 m and 45 m, and are very well adapted to the span-by-span construction of long continuous viaducts.

12.2 Behaviour of twin rib decks

12.2.1 Length of load and end fixity of the slab

A particular characteristic of twin rib decks without diaphragms is that the end fixity of the slab varies with the loaded length. When a short load (measured along the span) is applied to the centre of the slab, the twisting of the ribs is inhibited by the slab either side of the loaded patch and consequently the slab is substantially fixed ended, with the support moments carried in torsion in the ribs. As the load becomes longer, the ribs are freer to twist and the end restraint of the slab reduces until, when the full length of the deck is loaded, it behaves as simply supported on the rib axes, Figure 12.4. As vehicle loading is made up of both long and short loads, this behaviour must be understood, and reflected in the appropriate calculation model. For instance, under the effect of lane loading applied to complete spans the deck slab is weakly restrained while under localised wheel or truck loading it is virtually fixed ended. Both these loading cases may co-exist.

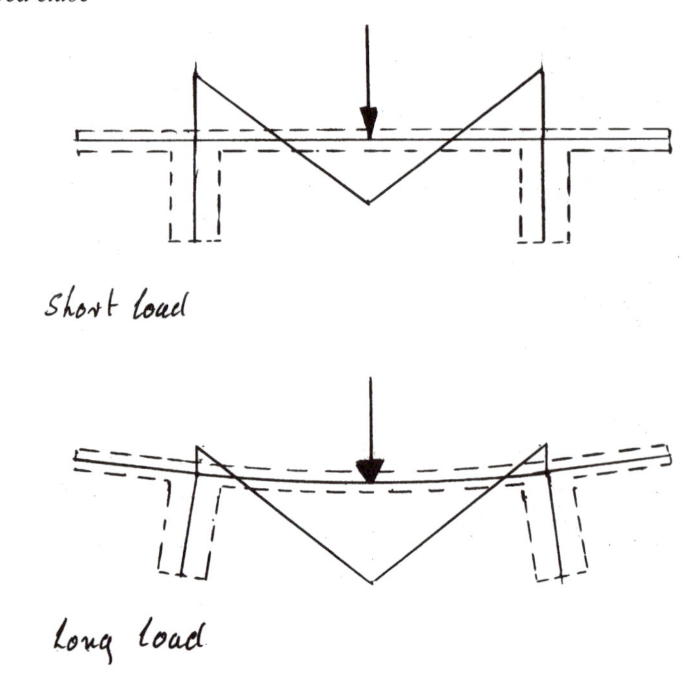

Figure 12.4 Effect of length of load

12.2.2 Restraint to twisting of the ribs

As the torque in a rib depends on its rate of change of twist, the ribs of decks without diaphragms exhibit relatively low levels of torque. As the ribs rotate under the effect of loads applied to the slab, their heels move laterally. If this lateral movement is inhibited by the piers or bearings, the ribs will be less free to rotate and the end fixity of the slab will increase. For instance, if the deck is carried on a crosshead and both bearings are guided longitudinally, under symmetrical loading on the deck, the ribs would be fully restrained against rotation, Figure 12.5 (a). In order to free the ribs, one of the bearings would need to be free to slide laterally, Figure 12.5 (b). A common situation is where each rib is carried by a separate column, and the ribs are either pinned or longitudinally guided on both columns. In this case, the ribs are elastically restrained in torsion, the degree of restraint depending on the transverse bending stiffness of the columns, Figure 12.5 (c). When pier diaphragms are present, they impose zero twist at each pier, and consequently create relatively high rates of change of twist, and high levels of torque in the ribs.

12.2.3 Transverse distribution of loads

Consider a point load applied close to mid-span, and located on the centre line of one rib. The load causes the rib to deflect, creating a differential deflection with respect to the unloaded rib. As the rotation of the torsionally stiff ribs is restrained by the stiffness of the slab either side of the loaded area, bending moments and shears are generated in the slab, transferring some of the load to the unloaded rib. Typically 15 per cent of a point load applied to the mid-span of one rib will be shed to the adjacent rib.

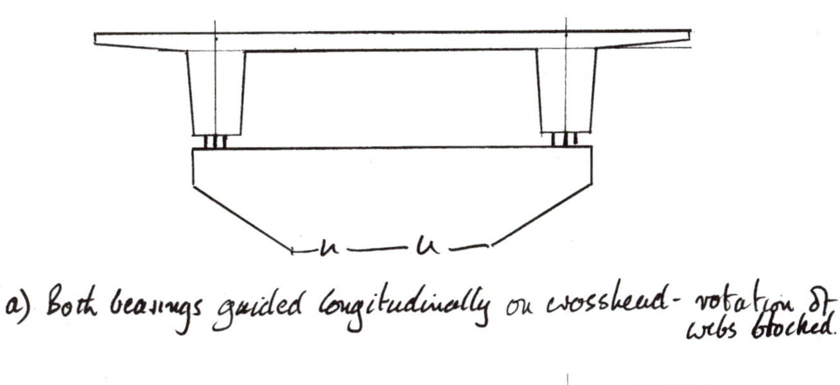

a) Both bearings guided longitudinally on crosshead - rotation of webs blocked.

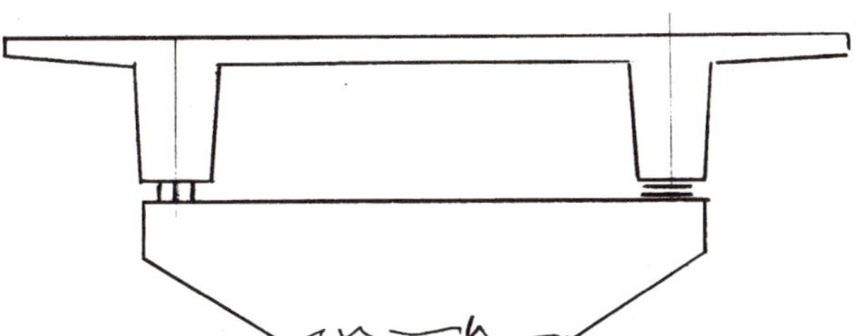

b) One bearing free to slide transversally - webs free to rotate

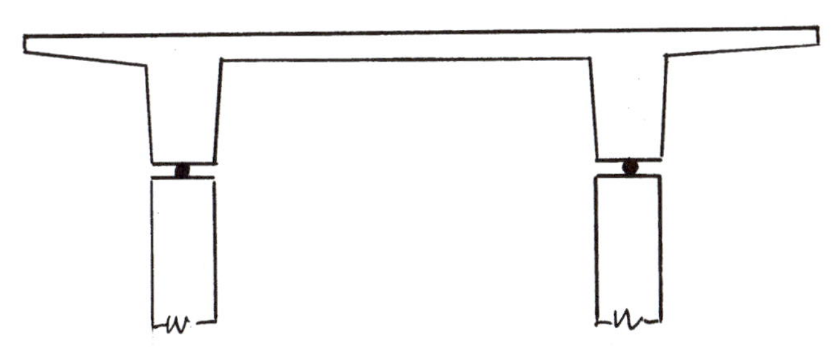

c) Both bearings pinned (or guided longitudinally) on columns - webs elastically restrained in torsion

Figure 12.5 Rotational restraint of webs by substructure

Using simple grillage analysis [1], one can create influence lines for the proportion of longitudinal bending moment carried on each rib as a unit load is moved across the deck, Figure 12.6. This calculation may be repeated at specific design sections along the span and can be used to derive the bending moments in each rib caused by the distributed or concentrated live loads. Similar influence lines can give the designer information on the bending moments in the spans adjacent to the loaded span, as well as on shear and torsion in the ribs.

As the loaded length becomes longer and the ribs are less torsionally restrained, they rotate more, reducing the shears and moments in the slab and hence the transfer of load between ribs. Consequently, in creating these influence lines, the unit load employed should have a finite length, characteristic of the highway loads to be applied.

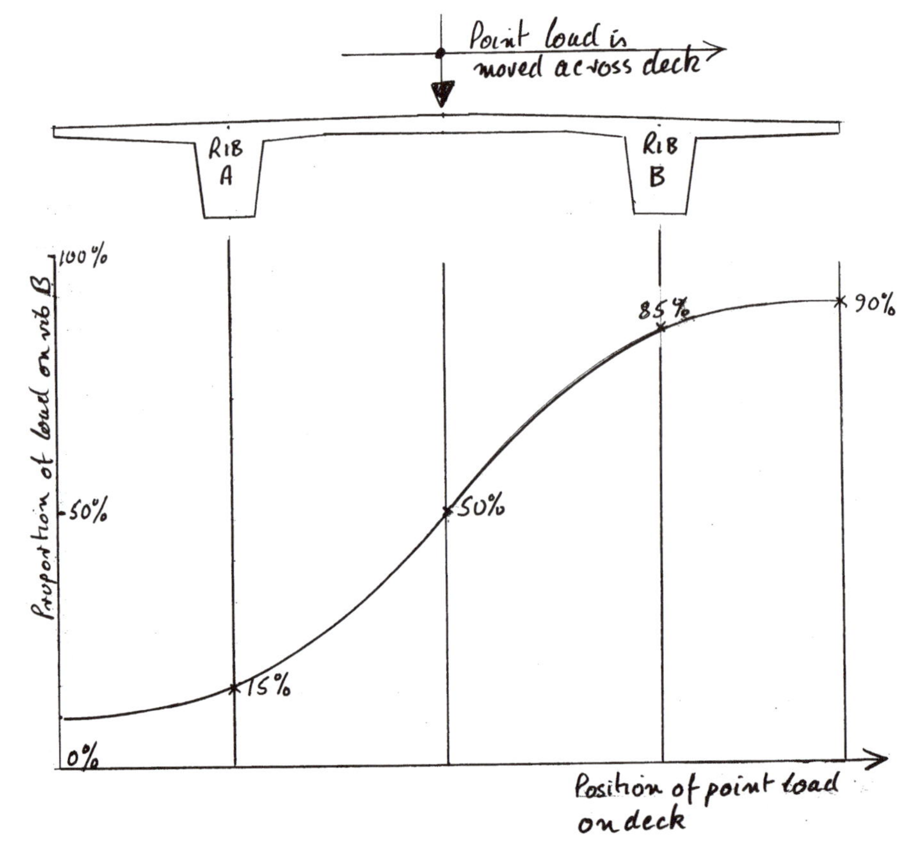

Figure 12.6 Typical influence line for share of mid-span sagging bending moment carried on rib B

For instance, when working to the UK codes of practice where the three critical loads are uniformly distributed HA lane loads, HB vehicle loads and a Knife Edge Load, one would draw three sets of influence lines, one for a full span loaded, one for a 10 m loaded length and one for a point load.

If the load is placed on a rib immediately adjacent to the pier, clearly the loaded rib does not deflect, and no load is transferred to the unloaded rib. For instance, if a multi-axle vehicle is centred on one rib, and placed with the forward axle close to the pier, the load of that axle will be carried entirely by the loaded rib. Axles placed further from the pier will distribute a proportion of their load to the adjacent rib.

In the early stages of a design, great accuracy is not required; it is more important that the basic behaviour of the deck should be understood, and a few well-chosen influence lines will usually suffice. For detailed design, the more extensive use of influence lines can make it clear where the deck should be loaded for maximum effect.

12.2.4 Transverse strength and stiffness of decks without pier diaphragms

It is always necessary to check that the horizontal strength and stiffness of decks without diaphragms are adequate. For highway decks up to about 2 m deep subjected

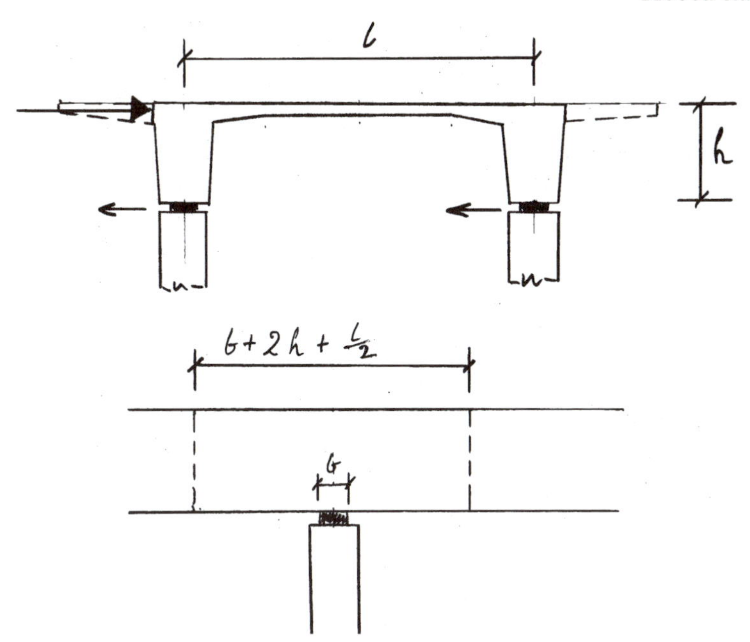

Figure 12.7 Portal resisting horizontal loads

to normal wind loading, the strength and stiffness are generally satisfactory. Horizontal forces on the deck are resisted by a portal at each pier position consisting of the two webs and a length of slab. The length of the portal may be conservatively estimated as $b + 2h + l/2$, where b is the width of the bearing, l is the spacing of the ribs and h is their height, Figure 12.7. For decks subjected to extreme wind loading or to severe seismic loading or for a deck of unusual proportions or subjected to exceptionally high horizontal forces, the designer should satisfy himself that the deck is adequately stiff and strong in the transverse direction.

Railway loading generally includes very long and heavy distributed loads, for which the top slab will be statically determinate, combined with high centrifugal forces for curved decks. The tracks are best placed over the ribs if possible. However, if they are placed on the slab or on the side cantilevers, the slab may need to be thickened either for strength or to control deflection. It may also be necessary to introduce pier diaphragms to provide adequate lateral stiffness.

12.3 The use of diaphragms

12.3.1 Intermediate diaphragms

The purpose of diaphragms within a span is to improve the transverse distribution of load. As explained in Chapter 8, intermediate diaphragms do not improve the transverse distribution of load on decks with two beams or ribs. Thus they are never used on twin rib bridge decks.

12.3.2 Pier diaphragms

The purpose of diaphragms at piers for a ribbed slab deck is to stabilise the deck under the effect of transverse wind and traffic loads on the deck, and of seismic loads applied to the deck through the piers. Pier diaphragms may also be used to offset the bearings from the rib centreline if required for any reason. Figure 12.8 shows such a use of diaphragms in the Tai Ho Viaduct in Hong Kong, designed by the author at Arup with the architectural assistance of Humphrey Wood, where the crosshead was shortened for visual effect.

By impeding any twisting of the ribs about a longitudinal axis, diaphragms act as torsion sumps increasing both the dead (*12.4.2*) and live load torque in the ribs. If the deck is sharply curved, the presence of diaphragms will also increase both the dead and live load torque that is due to the curvature. The torque generated in the ribs adjacent to the pier diaphragms, when combined with the shear force, is likely to define the width of the rib, and in all cases greatly increases the rib reinforcement when compared with a deck without pier diaphragms. Consequently pier diaphragms should only be used when there is no better alternative. The first option for increasing the strength or stiffness of the slab should be to increase its thickness.

Figure 12.8 Contract 308 of the HKMTR, Tai Ho Viaduct: twin rib deck with pier diaphragms (Photo: Robert Benaim)

The diaphragms must of course be designed to carry wheel loads that are applied to them directly, or that are carried to them by longitudinal spanning of the slab. The top slab must also be designed for longitudinal hogging moments where it frames into the diaphragm.

Pier diaphragms hinder the launching of mechanised falsework, which is important in the construction of long viaducts. Where diaphragms are necessary, it is sometimes possible, but not very convenient to leave them out in a first phase of construction, casting them once the falsework rig has moved on. This was the solution adopted for the Viaduc d'Incarville shown in Figure 12.10.

12.3.3 Abutment diaphragms

Abutment diaphragms are required for the reasons described in *9.2.6*. The presence of the diaphragm then attracts other loads, as described above. It is important to keep the end diaphragm as flexible as possible compatible with its tasks as this will minimise the torque in the ribs.

12.4 Proportioning of twin rib decks

12.4.1 Height and width of ribs

Generally, twin rib decks are up to 2 m deep, with 2.5 m being exceptional. Span/depth ratios generally are a minimum of 1/20, with 1/18 being more typical. Consequently suitable spans range from 20 m where this type of deck takes over from solid or voided slabs, to about 45 m. Figure 12.9 shows diagrammatically how the proportions of a twin rib deck of 35 m span evolve towards a voided slab as the depth decreases.

As the design intent for ribbed slab bridges is to keep the dimensions of the rib constant over the full span, economy requires that the width be minimised. The ribs are often slightly tapered towards the soffit. A taper of about 1/12 allows the tunnel shutter between the ribs to be struck without being collapsed or folded. The outside face does not need to be tapered for that reason, as the shutter is not trapped. However, it is often tapered to the same degree to improve the appearance of the bridge. Parallel sided ribs may be designed, as long as the tunnel shutter may be collapsed or struck in stages, Figure 12.10.

The width of the ribs may be defined by one of five criteria:

- SLS compressive bending stresses in the deck;
- ULS bending adjacent to the pier;s
- ULS shear and torsion;
- width necessary to accommodate bridge bearings;
- width necessary to support the top slab.

For very shallow twin rib decks, the width of the bottom flange may be controlled by the need to provide an adequate bottom fibre modulus to reduce the variation of stress under live loads. However, this is unusual, the width generally being controlled by one of the other criteria listed.

Generally, the critical criterion is Ultimate Limit State bending at the piers. There is always a trade-off between section depth and rib width. As the section is made shallower,

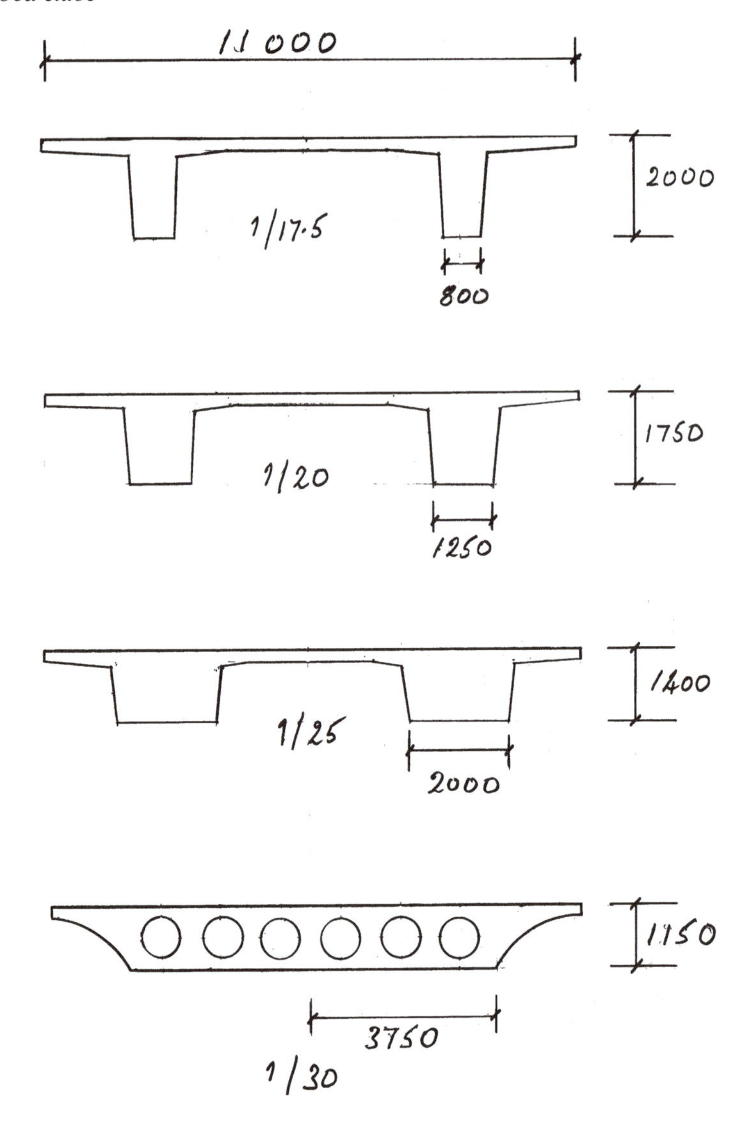

Figure 12.9 Evolution of twin rib deck with decreasing depth

the rib needs to be made wider in order to provide adequate ultimate strength. For the reasons given in *12.6* below, it is often necessary to reinforce the bottom fibre at sections close to the pier with substantial reinforcing bars. These should always be taken into account in the estimation of ultimate strength, and thus must be restrained by adequate links.

The ultimate shear stresses in the ribs due to vertical loads generally do not define the width of the rib. However, in decks with ribs that are restrained against rotation, the combined torsional and shear stresses may define the rib width.

When the deck is deep, the rib width necessary to ensure ultimate bending strength may be inadequate to accommodate bridge bearings. If the rib is widened for this

Figure 12.10 Viaduc d'Incarville: parallel-sided ribs with mechanically operated formwork (Photo: Robert Benaim)

reason, it may be economical to reduce its depth until the ULS bending criterion again becomes critical.

When a deck is wider than about 15 m, it may be economical to widen the ribs to reduce the span and hence the thickness of the slab and of the side cantilevers. Although widening the ribs in this way would appear to be wasteful of materials, this may not be so. If the depth is unchanged the concrete quantity will clearly be increased, and this will increase the weight and hence the cost of the foundations. However, as explained in *6.19*, the high neutral axis typical of twin rib bridge decks leads to an inefficient use of prestress. Thickening the rib lowers the neutral axis, and allows savings to be made in prestressing steel. In general, the reinforcement for the ribs is not increased, while the reinforcement for the slabs and cantilevers is reduced. It may also be possible to reduce the depth of the deck until the ULS bending criterion governs. To the author's knowledge, a careful comparison of the relative costs of widening the ribs compared with thickening the slab has not been carried out.

In order to reduce the weight of the deck and to further increase its efficiency (*5.4*), voids may be introduced into the concrete, see *12.10.2* and Figure 11.15 and Figure 12.17. Although the voids allow savings in the prestressing steel and foundations, they cost at least as much as the concrete they replace, they make casting the concrete more difficult and they increase the amount of reinforcing steel in the deck. The designer must make up his own mind whether to use them or not.

12.4.2 Thickness and span of the slab and side cantilevers

The ratio of the span of side cantilevers to the span of the slab must be chosen so that, under permanent loads, the cantilever bending moment at the rib centre-line approximately equals the fixed end bending moment of the slab. This ensures that under permanent loads the rib will not rotate. If the side cantilevers and centre slabs were of constant thickness and loaded only by their self weight, this would give a slab span that is 2.45 × cantilever length (all spans measured to the rib centreline). However, in reality the cantilever is usually thinner than the slab, but is loaded at its end by a footpath or parapet. The correct ratio will be different for each bridge, but will usually lie between 2.3 and 2.8. If this ratio is not respected, for a deck

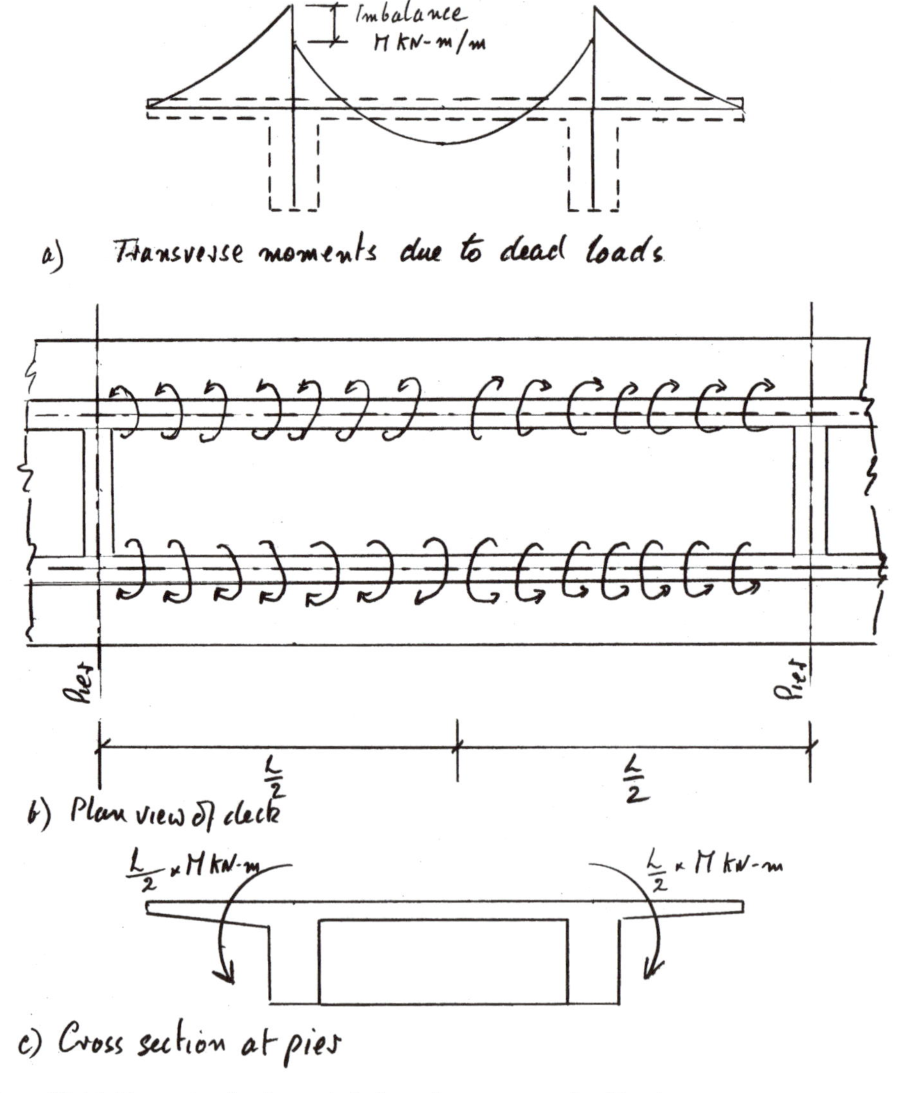

Figure 12.11 Torque in ribs due to imbalance in transverse dead load moments

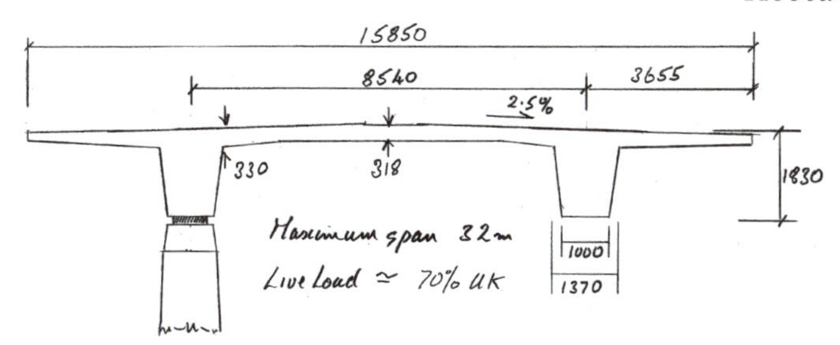

a) Doornhoek Bridge

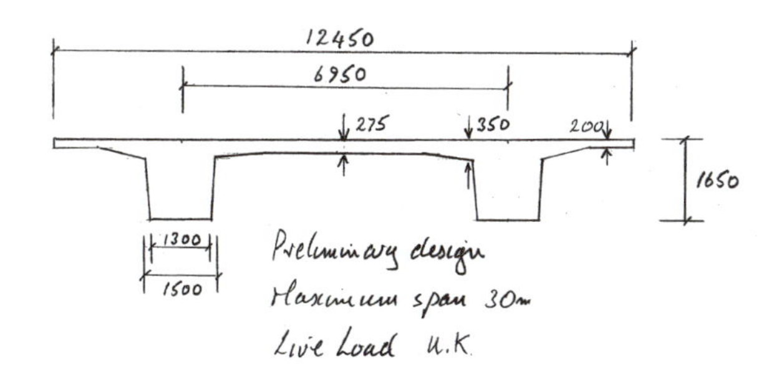

b) Red Jacket Pill Viaduct

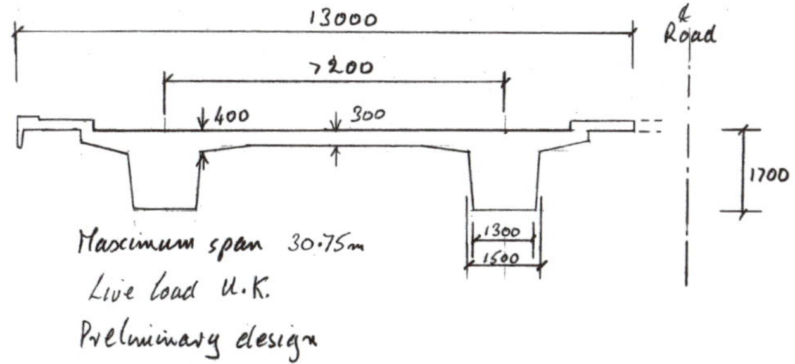

c) Seasalter Viaduct

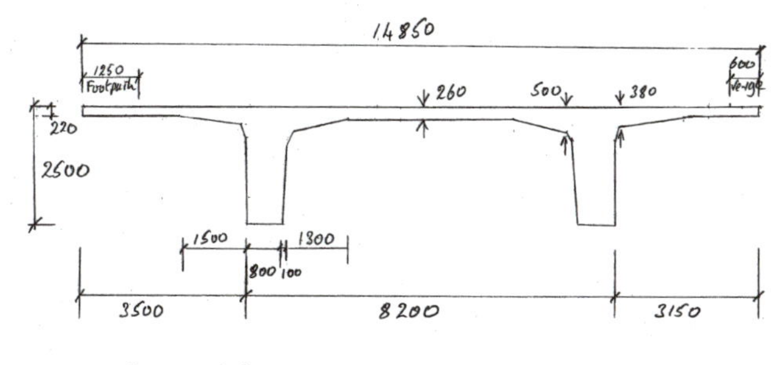

Span 45m
French loading and codes
Pier diaphragms

d) Viaduc d'Incarville; Autoroute de Normandie

Figure 12.12 Typical twin rib decks

without pier diaphragms the ribs will rotate until the moments at the rib centre line have been brought into equilibrium. This rotation will apply bending moments to the abutment diaphragms, and, amplified by creep, will affect the design of the bearings, and in extreme cases may affect the serviceability of the deck. For a deck with pier diaphragms, any imbalance of the transverse moments will cumulate as torque in the ribs at each pier, Figure 12.11.

The issues that control the design of the intermediate slab are closely related to the behaviour of these decks described in *12.2*. The slab is subjected to two superimposed modes of bending; 'field' bending due to the relative deflections and rotations of the ribs, and 'local' bending under the wheel loads themselves. The logic controlling the thickness of the side cantilever and of the intermediate slab is described in *9.2* and *9.3.6* respectively. Typical forms of some twin rib decks are shown in Figure 12.12.

12.5 Ribbed slabs and skew bridges

Ribbed slabs are particularly suited to skew decks. A conventional box section deck with side cantilevers carried on two bearings on each skew pier is heavily stressed in torsion as the twin supports impose a bending axis that is not perpendicular to the bridge centre line, Figure 12.13 (a).

However, each rib of an equivalent twin rib deck may bend about a perpendicular axis and there will be no significant torsion due to the skew geometry. In the heavily skewed deck shown in Figure 12.13 (b), the piers beneath one rib are located approximately opposite the mid-spans of the other. Live loads will deflect the ribs causing transverse bending and shear in the slab, transferring load to the pier section of the other rib. Thus a skew arrangement of piers may even reduce the live load bending moments in a twin rib deck, leading to savings in prestress.

This skew arrangement of piers also highlights an essential difference between prestressed and reinforced decks. In a reinforced deck, the ribs deflects downwards under both permanent loads and live loads, causing such heavy bending of the slab that the arrangement would probably not be feasible. On the other hand, in the prestressed deck described above, the ribs deflect slightly upwards under the combined effects of prestress and dead load, and downwards under live loads, leading to relatively light transverse bending of the slab.

It is frequently the case that the cancelling of self-weight deflections by prestress makes feasible structural arrangements that would not be viable in reinforced concrete. The engineer should always consider the deflection of his structures.

12.6 Heat of hydration effects on twin rib decks

These effects were highlighted by the severe cracking during construction of a deck designed by the author. The deck consisted of two wide, solid prestressed concrete ribs 1.5 m deep, connected by a thin reinforced concrete top slab, with side cantilevers. The design of the prestressing was carried out in the normal way by analysing the deck at the tenth points of each span. In common with most ribbed slab bridge decks, the bottom fibre was very lightly compressed under self weight plus prestress. In fact, at the first tenth point of the main span adjacent to the intermediate piers, the stress

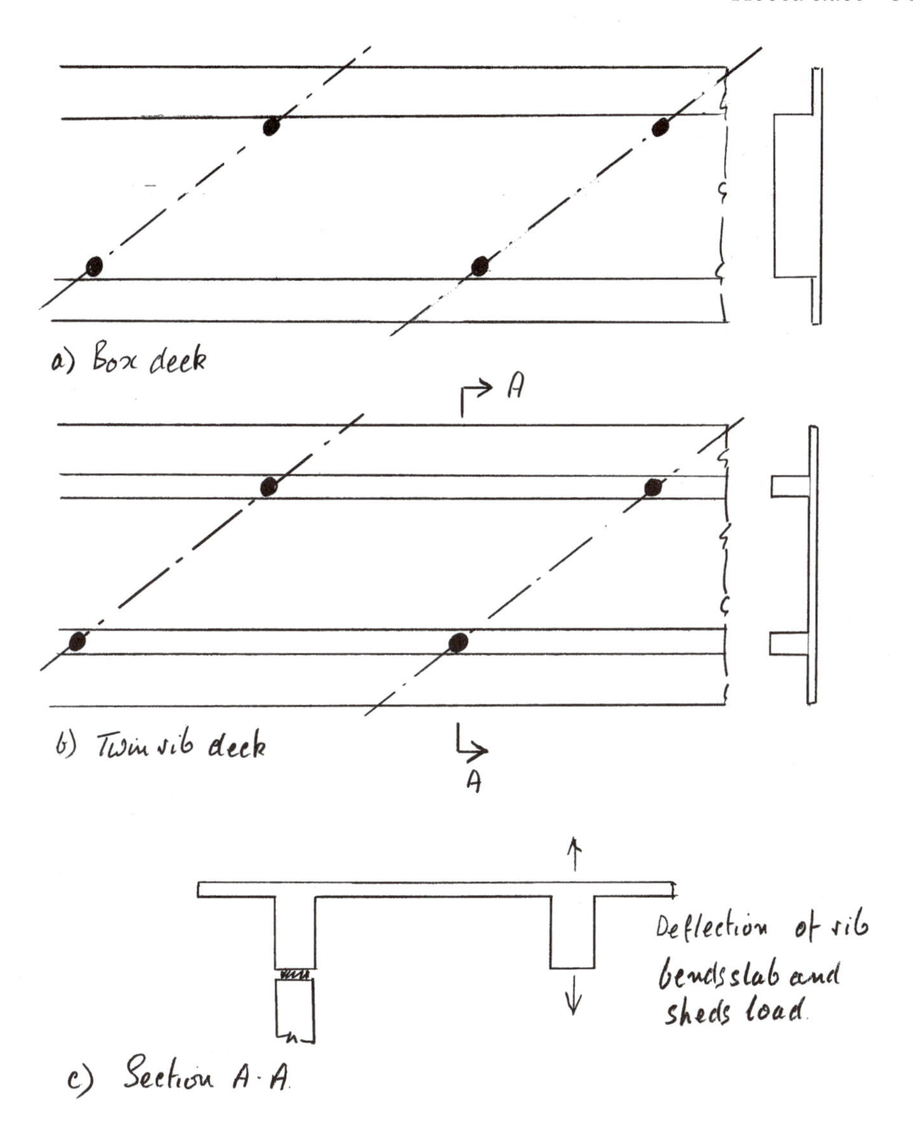

Figure 12.13 Effect of skew on decks

before application of the deck finishes and under initial prestress was –1 MPa, which conformed to the code of practice.

When the formwork to the ribs was struck, cracks over a millimetre wide were found at approximately the first 1/20th point of the main span, in all four locations. The cracks extended throughout the full height of the ribs, and only disappeared in the slab haunches, Figure 12.14.

Back analysis of the 1/20th point showed that the tensile stresses on the bottom fibre were higher than at the 1/10th point, reaching approximately –2.5 MPa for the reasons explained in *6.13.3* (see also Figure 6.14). However, the theoretical tensile zone was still only some 350 mm high and these bending stresses explained neither the width nor the vertical extent of the cracking.

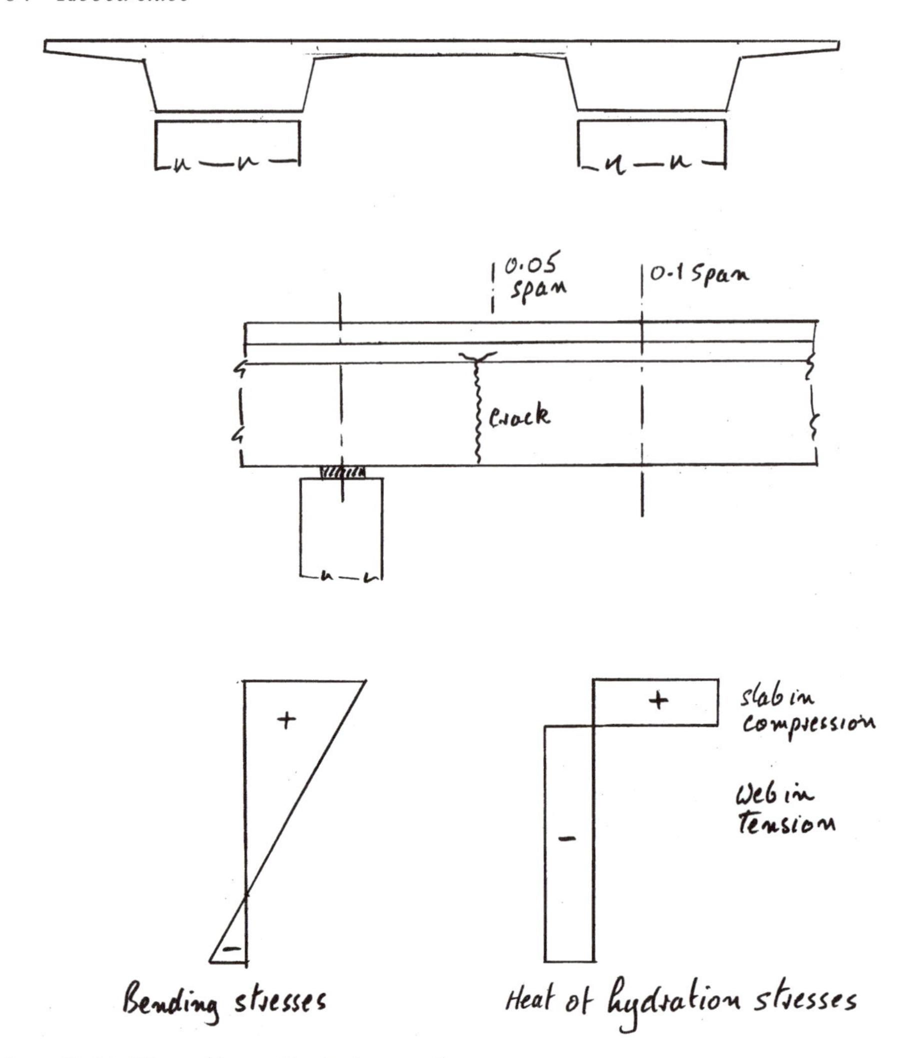

Figure 12.14 Effect of heat of hydration tensile stresses

The principal reason for this cracking was the effect of heat of hydration (3.6). Whereas the thin top slab had a heating and cooling cycle that lasted approximately 36 hours, the massive ribs had a cycle of some three days. By the time the rib cooled, the slab was already stable, and restrained the shortening of the rib causing it to be stressed in tension. The tensile bending stresses acted as the trigger that exhausted the tensile strength of the rib, and defined the location of the cracks. The length of the cracks corresponded to the stress diagram caused by the differential temperature of rib and slab. The cracks were so wide because the reinforcement on the bottom fibre, although respecting the code of practice, was below the threshold required to give ductile behaviour, consisting only of the stirrup corner bars, and small-diameter distribution steel. The cracking was made worse by the fact that in the absence of significant bottom fibre reinforcement, the cracked inertia was locally reduced, raising the neutral axis, and increasing the positive prestress parasitic moment.

All cast-in-situ twin rib decks should be protected from heat of hydration stresses by the addition of substantial bottom fibre passive reinforcement for 15 per cent of the spans either side of the piers, where the bending stresses are lowest under initial prestress. It is suggested that in order to preclude brittle fracture (*3.7*) the reinforcement should consist of the greater of 0.15 per cent of the rib area, or 0.6 per cent of a strip 400 mm deep.

12.7 Prestress layout

As twin rib type decks have a high neutral axis, the prestressing cables lie predominantly below it and give rise to high positive prestress parasitic moments as described in *6.19*, leading to an increase in the prestress force of about 20 per cent. However, twin rib decks are so quick and easy to construct that the savings in manpower and the great benefits of trouble-free construction are not outweighed by the extra materials necessary.

Where partial prestressing is allowed by the code of practice, it should be adopted, yielding decks that would be more economical and more ductile than fully prestressed designs.

12.8 Substructure for twin rib bridges

Where arrangements at ground level and geotechnical conditions are suitable, the deck is most economically supported by a column beneath each rib founded on a single large-diameter bored pile, giving a most economical and satisfactory structure, Figure 12.15. A tie beam between columns below ground level may be necessary, depending on ground conditions. Subject to the other constraints of the bridge articulation (Chapter 7) further savings may be made by building the columns into the deck, eliminating bearings.

For bridge decks without diaphragms, the jacking points for the replacement of bearings must be on the axes of the ribs. If this cannot be accommodated by the piers, temporary support will be required from the foundation which should be designed with this in mind. If necessary, facilities for stressing a temporary bracket to the pier may be included.

12.9 Construction technology

Twin rib bridge decks are most suitable for span-by-span construction which is described in *15.2*. This may use simple falsework, as for the Doornhoek Bridge, Figure 12.16, a semi-mechanised rig as in Contract 304 of the HKMTR, Figure 12.2, or a fully self-launching gantry as in the Viaduc d'Incarville, Figure 12.3.

If the deck is shorter than about 150 m, the option exists to build it on a full-length falsework, with prestressing tendons stressed and anchored only at the abutments. However, the friction losses due to wobble of the ducts (*5.17.2*) may make such long cables uneconomical. They also become progressively more difficult to install. A typical 150 m long bridge, 13 m wide, would require some 1,100 m^3 of concrete. As access for the liquid concrete and for poker vibrators is so good, this volume could be placed in one continuous pour working from several fronts. An additional advantage of such a procedure is aesthetic, in that there would be no construction joints to spoil the finish.

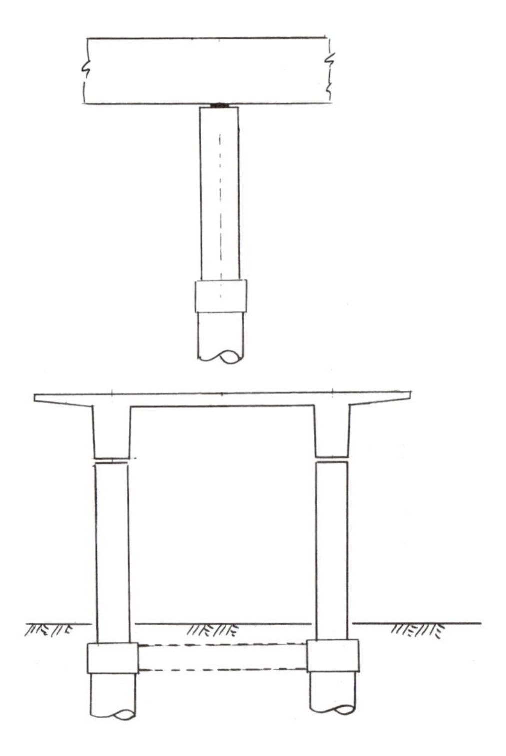

Figure 12.15 Twin pile foundation option

Figure 12.16 Doornhoek Bridge: typical falsework for span-by-span construction of short bridge (Photo: Arup)

12.10 The development of ribbed slabs

12.10.1 General

Twin rib type decks are typically used for bridges with spans up to 35 m and deck widths up to 15 m. Such decks have ribs that may be up to 2 m deep and 1–1.4 m wide, side cantilevers of 3 m and slab spans of up to 7 m. However, they can be stretched to cover greater spans and widths. The 15 m wide Viaduc d'Incarville had spans of 45 m, with a rib depth of 2.5 m. The Super Twin Rib decks described below can take the width above 20 m under UK loading, wider under lighter live loading. In some cases, the extra material implicit in the simplicity of the constant section solid web may not always be justified, and twin rib bridges can be designed with profiled webs for particular situations.

12.10.2 Super twin rib decks

The principal prestressed concrete options for Structure 18 of the Harlequin Interchange, a 375 m long, 20.3 m wide bridge deck with spans of 39 m designed for tender to carry the M4 motorway, were either a precast segmental twin box deck, or a cast-in-situ deck. As the total number of precast segments would have been only about 220, which is considered below the economical threshold, Benaim proposed, as an alternative to the official scheme, the Super Twin Rib deck shown in Figure 12.17. Each rib is carried by a single bearing on each pier, minimising the torque on the ribs. Although the piers are slightly staggered, giving a skew deck, each rib is effectively right. The 2.4 m deep ribs have been lightened by 1.8 m diameter voids which were to consist of concrete drainage pipe sections. Unfortunately the alternative was not accepted.

12.10.3 Profiled ribs

Under some circumstances it may be economical to design twin rib bridges with profiled webs, similar to those of precast Tee beam decks. This is usually only justified if the benefits of simplicity have already been lost by the details of the project, or

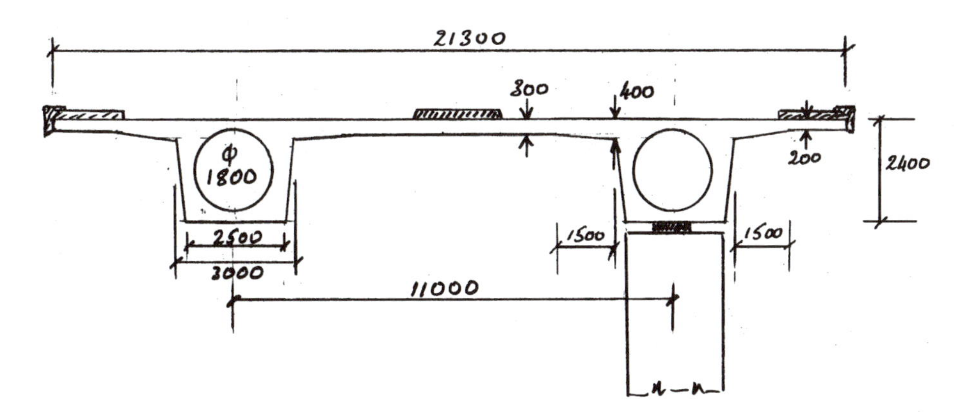

Figure 12.17 Super Twin Rib project

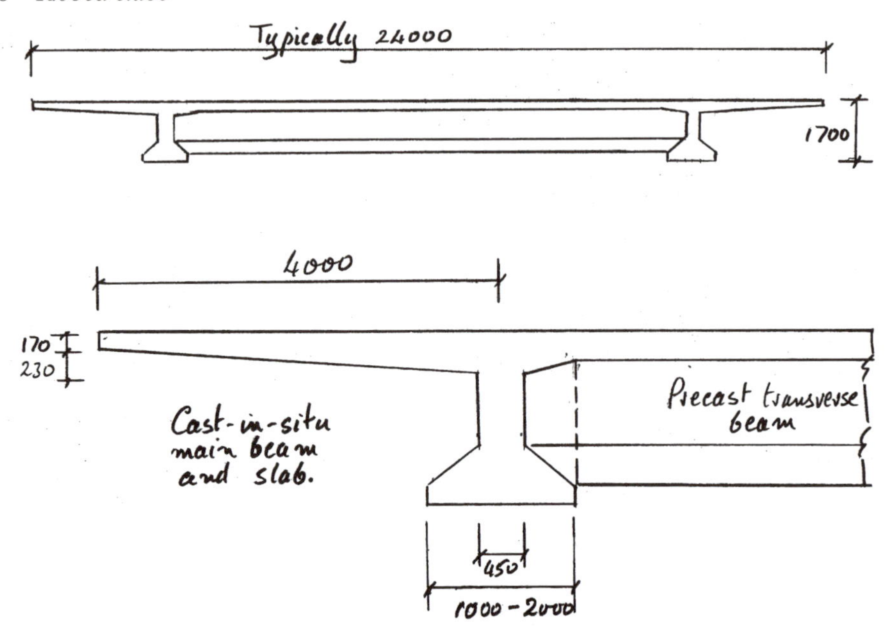

Figure 12.18 RVI Interchange, Bangkok: outline project

if mechanised systems of construction allow the more complex shapes to be cast economically, usually on very large schemes.

This was the situation for the RVI project, consisting of a major interchange on a highway outside Bangkok, designed in outline by Benaim. It required several kilometres of deck to be built in a short time. However, the decks were constantly changing in width with many bifurcations, making it very difficult to find a suitable system of mechanised construction. The option chosen consisted of a development of the twin rib deck. Each rib together with its side cantilever would be cast in-situ by a self-launching falsework rig, with two rigs working in parallel. The rigs would thus be free to follow diverging paths. The ribs would be cast between clamshell side shutters that swing clear for launching. Under these conditions, it was economical to refine the ribs into an 'I' section, with a bottom flange that varied in width from 1 m to 2 m, Figure 12.18. The structure between the ribs consisted of precast, post-tensioned Tee beams of length up to 25 m, tailored to their position in the alignment. The Tee beams would be placed by ground-based crane into position and supported by the falsework rigs, before the ribs were cast. Thus tolerances in the length of the Tee beams could be absorbed. The deck would be completed by casting the ribs and a structural topping to the Tee beams. The live loading was a development of the AASHTO code, and was lighter than that used in the UK.

13

Box girders

13.1 General

The box girder is the most flexible bridge deck form. It can cover a range of spans from 25 m up to the largest non-suspended concrete decks built, of the order of 300 m. Single box girders may also carry decks up to 30 m wide. For the longer span beams, beyond about 50 m, they are practically the only feasible deck section. For the shorter spans they are in competition with most of the other deck types discussed in this book.

The advantages of the box form are principally its high structural efficiency (*5.4*), which minimises the prestress force required to resist a given bending moment, and its great torsional strength with the capacity this gives to re-centre eccentric live loads, minimising the prestress required to carry them.

The box form lends itself to many of the highly productive methods of bridge construction that have been progressively refined over the last 50 years, such as precast segmental construction with or without epoxy resin in the joints, balanced cantilever erection either cast in-situ or coupled with precast segmental construction, and incremental launching (Chapter 15).

13.2 Cast-in-situ construction of boxes

13.2.1 General

One of the main disadvantages of box decks is that they are difficult to cast in-situ due to the inaccessibility of the bottom slab and the need to extract the internal shutter. Either the box has to be designed so that the entire cross section may be cast in one continuous pour, or the cross section has to be cast in stages.

13.2.2 Casting the deck cross section in stages

The most common method of building box decks in situ is to cast the cross section in stages. Either, the bottom slab is cast first with the webs and top slab cast in a second phase, or the webs and bottom slab constitute the first phase, completed by the top slab.

When the bottom slab is cast first, the construction joint is usually located just above the slab, giving a kicker for the web formwork, position 1 in Figure 13.1. A joint in this location has several disadvantages which are described in *11.7.1*.

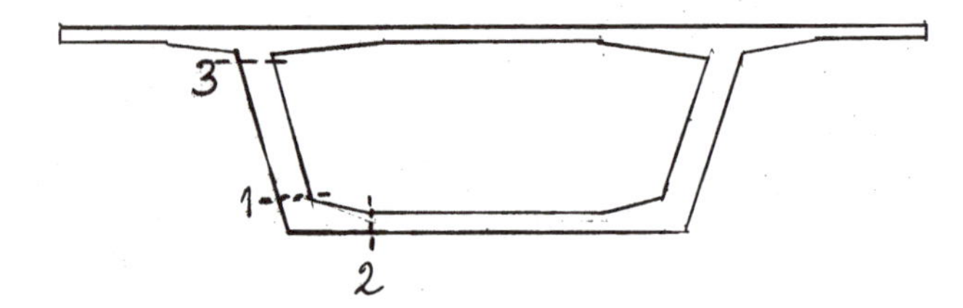

Figure 13.1 Alternative positions of construction joint

Alternatively, the joint may be in the bottom slab close to the webs, or at the beginning of the haunches, position 2. The advantages of locating the joint in the bottom slab are that it does not cross prestressing tendons or heavy reinforcement; it is protected from the weather and is also less prominent visually. The main disadvantage is that the slab only constitutes a small proportion of the total concrete to be cast, leaving a much larger second pour.

The joint may be located at the top of the web, just below the top slab, position 3. This retains many of the disadvantages of position 1, namely that the construction joint is crossed by prestressing ducts at a shallow angle, and it is difficult to prepare for the next pour due to the presence of the web reinforcement. In addition, most of the difficulty of casting the bottom slab has been re-introduced. The advantages are that the joint is less prominent visually and is protected from the weather by the side cantilever, the quantity of concrete in each pour is similar and less of the shutter is trapped inside the box.

Casting a cross section in phases causes the second phase to crack due to restraint by the hardened concrete of the first phase. Although the section may be reinforced to limit the width of the cracks, it is not desirable for a prestressed concrete deck to be cracked under permanent loads. Eliminating cracks altogether would require very expensive measures such as cooling the second phase concrete to limit the rise in temperature during setting or adopting crack sealing admixtures

13.2.3 Casting the cross section in one pour

There are two approaches to casting a box section in one pour. The bottom slab may be cast first with the help of trunking passing through temporary holes left in the soffit form of the top slab. This requires access for labourers to spread and vibrate the concrete, and is only generally possible for decks that are at least 2 m deep. The casting of the webs must follow on closely, so that cold joints are avoided. The fluidity of the concrete needs to be designed such that the concrete will not slump out of the webs. This is assisted if there is a strip of top shutter to the bottom slab about 500 mm wide along each web. This method puts no restriction on the width of the bottom slab, Figure 13.2 (a).

Alternatively the deck cross section may be shaped so that concrete will flow from the webs into the bottom slab, which normally has a complete top shutter, Figure 13.2 (b). This method of construction is most suitable for boxes with relatively narrow bottom flanges. The compaction of the bottom slab concrete needs to be effected by external vibrators, which implies the use of steel shutters. The concrete may be cast down both webs, with inspection holes in the shutter that allow air to be expelled and the complete filling of the bottom slab to be confirmed. Alternatively, concrete may be

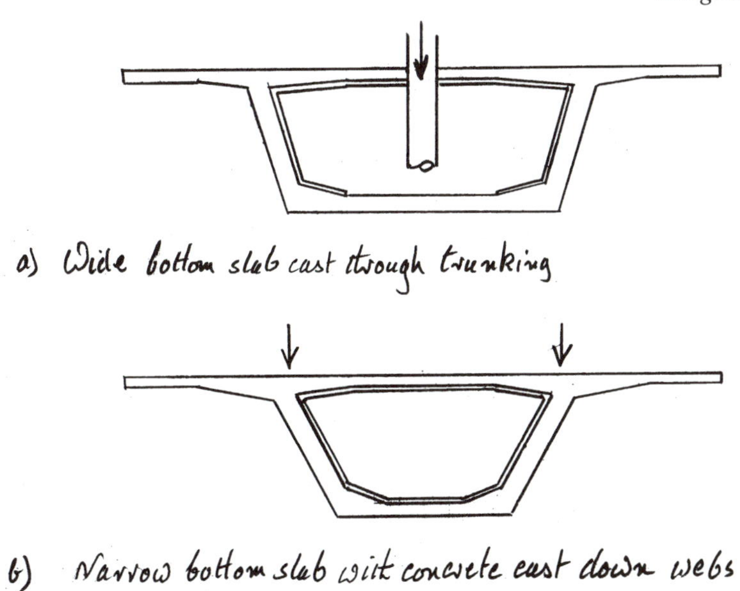

a) Wide bottom slab cast through trunking
b) Narrow bottom slab with concrete cast down webs

Figure 13.2 Casting deck in one pour

cast down one web first with the second web being cast only when concrete appears at its base, demonstrating that the bottom slab is full. The concrete mix design is critical and full-scale trials representing both the geometry of the cross section and density of reinforcement and prestress cables are essential.

However the section is cast, the core shutter must be dismantled and removed through a hole in the top slab, or made collapsible so it may be withdrawn longitudinally through the pier diaphragm.

Despite these difficulties, casting the section in one pour is under-used. The recent development of self-compacting concrete could revolutionise the construction of decks in this manner. This could be particularly important for medium length bridges with spans between 40 m and 55 m. Such spans are too long for twin rib type decks, and too short for cast-in-situ balanced cantilever construction of box girders, while a total length of box section deck of less than about 1,000 m does not justify setting up a precast segmental facility. Currently, it is this type of bridge that is least favourable for concrete and where steel composite construction is found to be competitive.

13.3 Evolution towards the box form

Chapters 11 and 12 described how solid slabs evolve into ribbed slabs in order to allow increased spans with greater economy. The principal advantage of ribbed slabs is their simplicity and speed of construction. However this type of deck suffers from several disadvantages, notably:

- the span is limited to about 45 m;
- live loads are not efficiently centred, resulting in a concentrated load (such as an HB vehicle) being carried approximately 1.7 times for a deck with two ribs, requiring additional prestress force;

- the section has poor efficiency, leading to the requirement for a relatively larger prestress force;
- the deck cannot be made very shallow;
- the piers need either multiple columns to carry each rib, or a crosshead that is expensive and visually very significant.

Box section decks overcome all these disadvantages.

13.4 Shape and appearance of boxes

13.4.1 General

A box section deck consists of side cantilevers, top and bottom slabs of the box itself and the webs. For a good design, there must be a rational balance between the overall width of the deck, and the width of the box. Box sections suffer from a certain blandness of appearance; the observer does not know whether the box is made of an assemblage of thin plates, or is solid concrete. Also, the large flat surfaces of concrete tend to show up any defects in the finish and any changes in colour. The designer should be aware of these problems and do what he can within the constraints of the project budget to alleviate them.

13.4.2 Side cantilevers

Side cantilevers have an important effect on the appearance of the box. The thickness of the cantilever root and the shadow cast on the web mask the true depth of the deck. If the deck is of variable depth, the perceived variation will be accentuated by these two effects, Figure 13.3 (*15.4.2*). In general, the cantilever should be made as wide as possible, that is some seven to eight times the depth of the root (*9.2*).

13.4.3 The box cross section

Boxes may be rectangular or trapezoidal, with the bottom flange narrower than the top. Rectangular box sections are easier to build, and are virtually essential for the longest spans due to the great depth of the girders. However, they have the disadvantages that their appearance is somewhat severe, and that their bottom slabs may be wider than necessary.

The visual impact of the depth of the box is reduced if it has a trapezoidal cross section. This inclination of the web makes it appear darker than a vertical surface, an impression that is heightened if the edge parapet of the deck is vertical.

The trapezoidal cross section is frequently economical as well as good looking. In general, the width of the top of the box is determined by the need to provide points of support to the top slab at suitable intervals. The cross section area of the bottom slab is logically determined at mid-span by the need to provide a bottom modulus sufficient to control the range of bending stresses under the variation of live load bending moments. For a box of rectangular cross section of span/depth ratio deeper than about 1/20, the area of bottom slab is generally greater than necessary, resulting in redundant weight. Choosing a trapezoidal cross section allows the weight of the

Figure 13.3 River Dee Bridge: effect of side cantilever on the appearance of a variable depth deck (Photo: Edmund Nuttall)

bottom slab to be reduced. Close to the piers, the area of bottom slab is determined by the need to limit the maximum bending stress on the bottom fibre and to provide an adequate ultimate moment of resistance. If the narrow bottom slab defined by mid-span criteria is inadequate, it is simple to thicken it locally.

For a very wide deck that has a deep span/depth ratio, this logic may give rise to webs that are inclined at a very flat angle. The designer should be aware of the difficulties in casting such webs, and make suitable allowances in specifying the concrete and in detailing the reinforcement.

Also, an important consideration in the design of box section decks is the distortion of the cross section under the effect of eccentric live loads (*6.13.4*). The effect of this distortion is reduced in a trapezoidal cross section.

Boxes may have a single cell or multiple cells. In Chapter 8 it was explained how important it is for economy to minimise the number of webs. Furthermore, it is more difficult to build multi-cell boxes, and it is worthwhile extending the single-cell box as far as possible before adding internal webs.

13.4.4 Variation of depth

Once the span of a box section deck exceeds about 45 m, it becomes relevant to consider varying the depth of the beam. This is not an automatic decision as it depends on the method of construction. For instance, when the deck is to be precast by the counter-cast method (Chapter 14), if the number of segments is relatively low it is likely to be more economical to keep the depth constant in order to simplify the mould. On the other hand, if the deck is to be built by cast-in-situ balanced cantilevering, it is relatively simple to design the mould to incorporate a variable depth, even for a small number of quite short spans.

Clearly, this decision also has an aesthetic component. The depth may be varied continuously along the length of the beam, adopting a circular, parabolic, elliptical or Islamic profile, Figure 13.4. Alternatively, the deck may be haunched. The decision on the soffit profile closely links aesthetic and technical criteria.

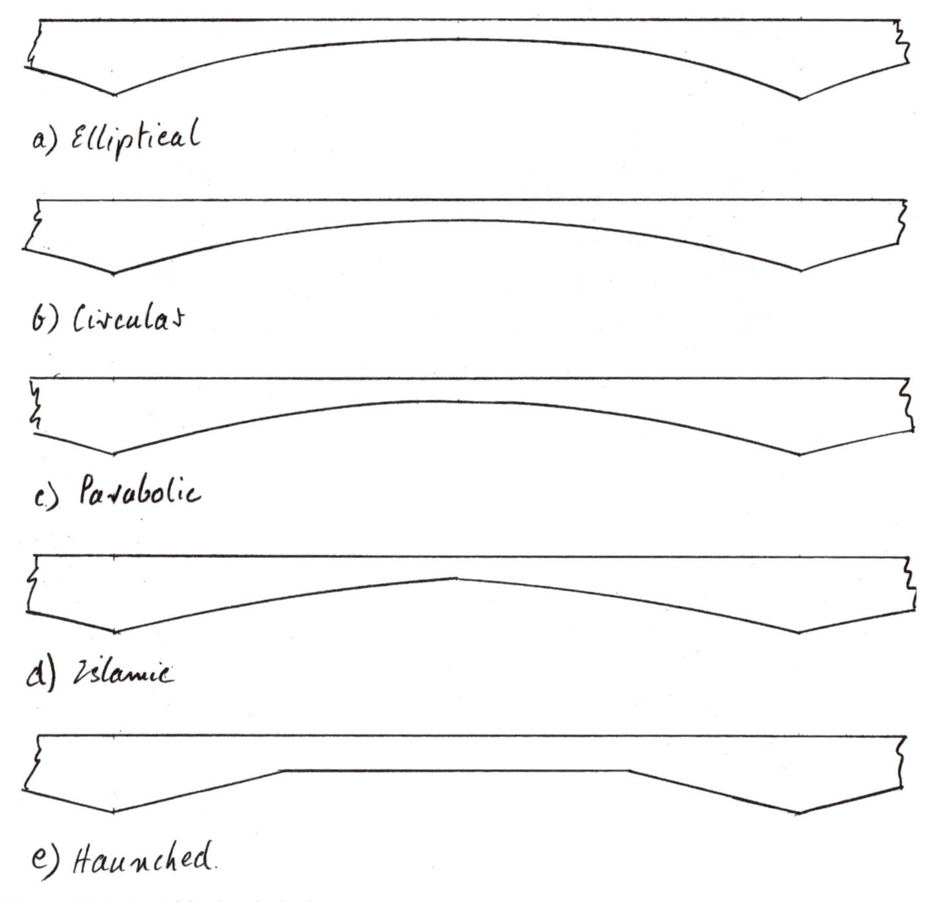

Figure 13.4 Variable depth decks

For instance, when the depth varies continuously it is often judged that an elliptical profile is the most beautiful. However, this will tend to create a design problem towards the quarter points, as at these locations the beam is shallower than optimal, both for shear resistance and for bending strength. As a result, the webs and bottom slab may need to be thickened locally, and the prestress increased. However, the economic penalty may be small enough to accept. The Islamic form is likely to provide the most flexible method of optimising the depth at all points along the girder, but the cusp at mid-span may give a problem for the profile of the continuity tendons while for long spans the greater weight of the deeper webs either side of mid-span implies a significant cost penalty. Also, the appearance may not be suitable for the particular circumstance.

When the change in the depth of the box is not too great, haunched decks are often chosen for the precast segmental form of construction, as they reduce the number of times the formwork must be adjusted, assisting in keeping to the all-important daily cycle of production. However, here again there is a conflict between the technical optimisation of the shape of the beam and aesthetic considerations. The beginning of the haunch is potentially a critical design section, both for shear and bending. This criticality is relieved if the haunch extends to some 25–30 per cent of the span length. However, the appearance of the beam is considerably improved if the haunch length is limited to 20 per cent of the span or less.

When variation of the depth is combined with a trapezoidal cross section, the bottom slab will become narrower as the deck becomes deeper, Figure 13.5. This has an important aesthetic impact, as well as giving rise to complications in the construction. When a deck is built by the cast-in-situ balanced cantilever method, such as the 929 m long Bhairab Bridge [1] in Bangladesh designed by Benaim, Figure 13.6, the formwork

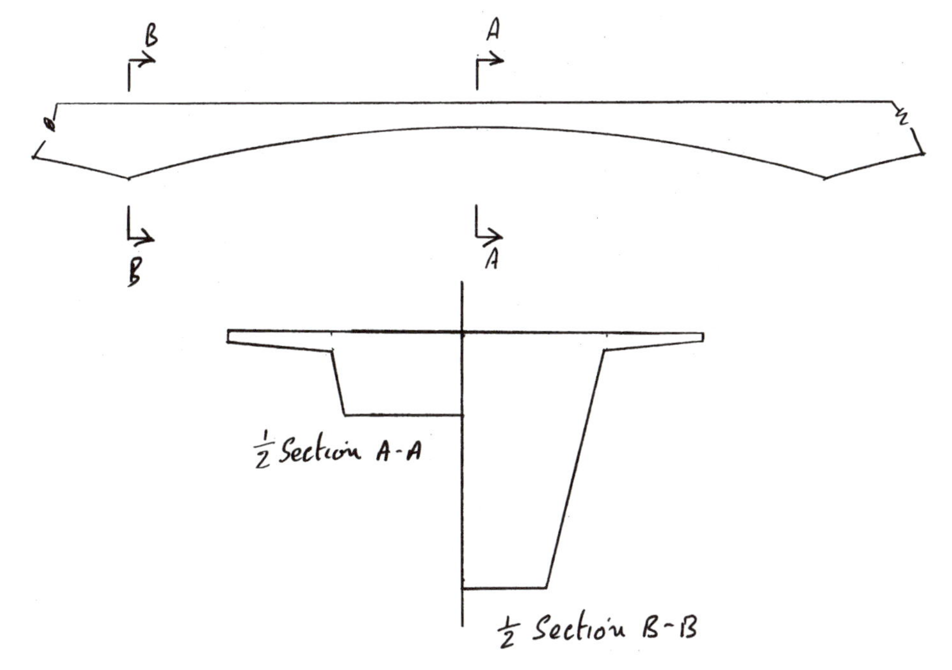

Figure 13.5 Variable depth with trapezoidal cross section

Figure 13.6 Bhairab Bridge, Bangladesh (Photo: Roads and Highways Department, Government of Bangladesh and Edmund Nuttall)

may be designed to accept this arrangement without excessive additional cost. However for a precast deck it is better to avoid this combination, as the modifications to the formwork increase the cost and complexity of the mould and interfere with the casting programme. It is easier to cope with a haunched deck than a continuously varying depth, as in the former case the narrowing of the bottom slab is limited to a relatively small proportion of the segments, and the rate at which it narrows is constant.

If the bottom slab is maintained at a constant width, the web surfaces will be warped. For a deck that has a continuously varying depth, the timber shutters of a cast-in-situ cantilevering falsework can accept this warp, whereas this may not be the case for the steel shutter of a precast segmental casting cell. However, for a haunched deck the warp would be introduced suddenly at the beginning of the haunch, which would probably be impossible to build, and would look terrible.

A successful detail employed on several occasions by the author on precast segmental decks is to adopt a trapezoidal box which runs the full length of the span, and to add a parallel sided haunch. Refinements are to define the haunch by a step in from the trapezoidal section and to finish the haunch with a small step rather than fairing it into the soffit; a detail that was adopted on Benaim's STAR project in Kuala Lumpur, Figure 13.7 and Figure 9.9. If this is not done, the distinction between the inclined web and the vertical haunch will not be clear, leading to visual ambiguity and a lack of crispness in the appearance. This detail has the disadvantage that it complicates the web reinforcement. A further technical refinement is to keep the centre line of the web approximately straight over its full height by thickening the inside face of the web over the depth of the haunch. The web thickening provides additional concrete at the bottom of the section close to the piers, often making it unnecessary to thicken the bottom slab. If the thickening is started from a point just above the step, a local reduction in thickness of the web may be avoided, Figure 13.8. An additional advantage

Figure 13.7 Rectangular haunch on STAR (Photo: Robert Benaim)

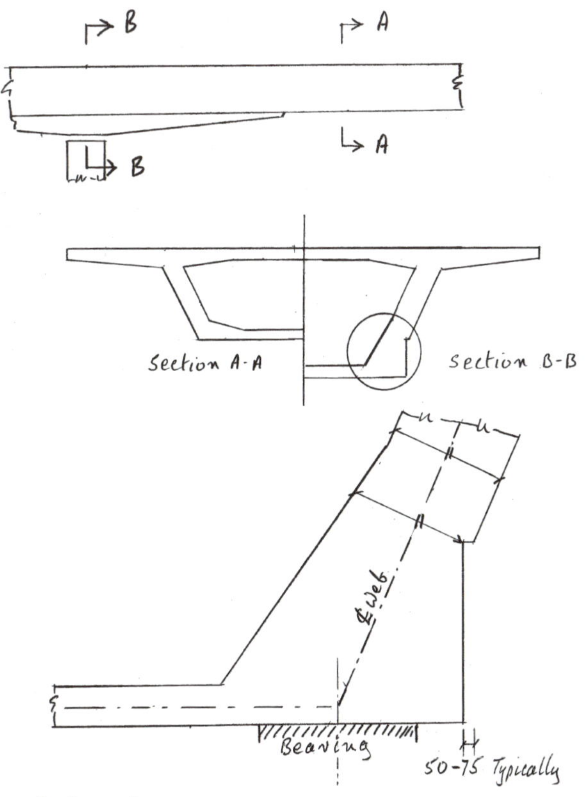

Figure 13.8 Rectangular haunch

of this geometry is that it makes it possible to locate the bearings directly beneath the axis of the webs, greatly simplifying the pier diaphragm.

13.5 The number of webs per box

One of the principal aims of the bridge designer is to minimise redundant material. This is the discipline that is the basis not only for economy, but also for technical innovation and for beauty.

Economy of materials in the design of box sections is achieved principally by minimising the thickness of the deck members. This has the dual benefit of reducing the dead weight bending moments and shear forces, and reducing the cross-sectional area to be prestressed. The benchmarks shown in Chapter 8 indicate the target quantities the designer should be aiming for.

As was explained in 8.2 and 9.5.2, it is much easier to build a few thick webs than several thinner webs; the number of webs should be reduced to a minimum.

A 200 mm thick top slab haunched to 350 mm is adequate for a clear span of about 6 m. 200 mm side cantilevers also haunched to 350 mm are adequate for a span of about 2.8 m. Hence, this minimum thickness slab will allow an overall structural box width of about 12.6 m (excluding parapets), once the thickness of the webs is included, Figure 13.9. As the width of the deck increases, the designer has the option of increasing the span and thickness of the slab, or adding a third web. Up to a clear span of about 9 m, the more economical option is, without exception, to keep to two webs and thicken the slab up to a maximum of about 300 mm, haunched to 500 mm. This haunch thickness will allow side cantilevers of up to 4 m, yielding an overall structural box width of some 18 m. These figures are appropriate for UK loading; most other loading codes are lighter, and consequently slabs of a defined thickness will span further. Some designers add a third web when it is clearly not essential, presumably in the mistaken belief that the additional cost of the web will be balanced by the reduced cost of the top slab. There is no doubt that the economic logic is to maximise the length of the side cantilevers and the span of the top slab in order to keep to two webs for as long as possible. Together with the inclination of the webs to reduce the width of the bottom slab, this also gives a good-looking deck.

For boxes that are still wider, the thickness and weight of a free-spanning solid slab starts to become excessive, and it is necessary either to add a third web, to change to

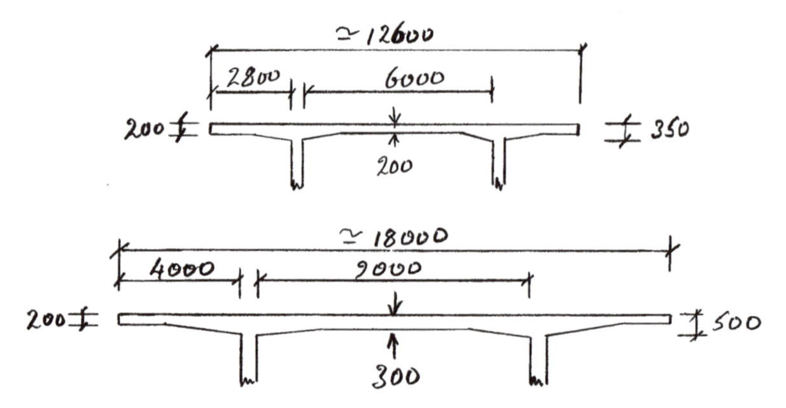

Figure 13.9 Typical width of single cell box designed to UK loading

a ribbed or strutted slab and side cantilever (9.2 and 9.3) or to increase the number of boxes in the cross section. The choice will depend on the methods of construction adopted, and on the preferences of designer and builder. The author's preference is, wherever possible, to build single-cell boxes, accepting the resultant complications to the construction procedure, or to increase the number of boxes.

When three webs are used, it should be noted that the shear force is not evenly distributed, with the centre web taking more than one-third. The prestress should be distributed between the webs in proportion to the shear they carry.

13.6 Number of boxes in the deck cross section

13.6.1 General

Once it has been decided that it is appropriate to adopt a box section for the bridge deck, the designer must decide how many boxes should be used across the width of the deck. This decision governs not only the material content of the deck, but also the arrangement and cost of the substructure, and the type and scale of the temporary works that need to be mobilised. It is intimately linked to the other basic decision that must be made, whether to cast in-situ or to precast.

There are many factors that influence these linked decisions. Some of the most important are the following:

- the scale of the project and subdivision of the scale; many short bridges; several longer bridges; one very long bridge (Chapter 16);
- complication of the project; change of crossfall; variable width of the deck; bifurcations of the bridge deck; presence of a family of bridge decks; presence of slip roads etc;
- criteria for sub-structure and foundations (Chapter 7).

13.6.2 Constant-width decks

Consider a 15 m wide deck carried by a single-cell box. This would typically have 3.25 m long side cantilevers, and an overall box width of 8.5 m. If the span were 50 m and the depth of the box 2.8 m, the thickness of the webs at the piers would be 625 mm (9.5.2), while at mid-span the webs could be reduced to 350 mm or 400 mm, Figure 13.10 (a). The top slab would be 250 mm thick with 450 mm haunches. If two boxes were to be used, the side cantilevers would typically be 1.25 m, the overall width of each box 4.25 m, and the slab between boxes 4 m, Figure 13.10 (b). For the same span and depth, the webs would be 350 mm thick at the piers, and this thickness would be maintained at mid-span. The top slab would be 200 mm thick with 350 mm haunches. This is not a very economical scheme as there is too much web material at mid-span, and the 200 mm top slab will be under-used. Furthermore, the side cantilever length is somewhat short for the appearance of the deck and the 3.55 m internal width of the box is too small for economical striking and handling of the shutter for precast segmental construction. One bonus is that the webs may be of constant thickness, simplifying construction.

The substructure will be more expensive due to the increased dead weight of the deck. Also, if the pier consists of one central column a crosshead will be necessary to carry the two boxes. If the pier consists of one column beneath each box, the cost of the foundation to carrying a live load that consists of a single heavy vehicle (such as the British HB Vehicle) is increased as it is carried twice.

If the deck is made of precast segments a 15 m wide segment, 3.5 m long is likely to weigh of the order of 80 tons, while the narrower segments will weigh about 30 tons, and may more easily be transported by road. The casting cells for the smaller boxes would be smaller and cheaper, the lifting equipment lighter, and the precasting run doubled.

The practical maximum width of deck that may be carried by these two small boxes without thickening the top slab is defined by 2.8 m long side cantilevers and a span between boxes of 6 m, giving an overall width of 20.1 m, Figure 13.10 (c). The webs would need to be increased in width to about 450 mm at the piers. A single box could still be used for this width, with typically a box that is 10 m wide at the top

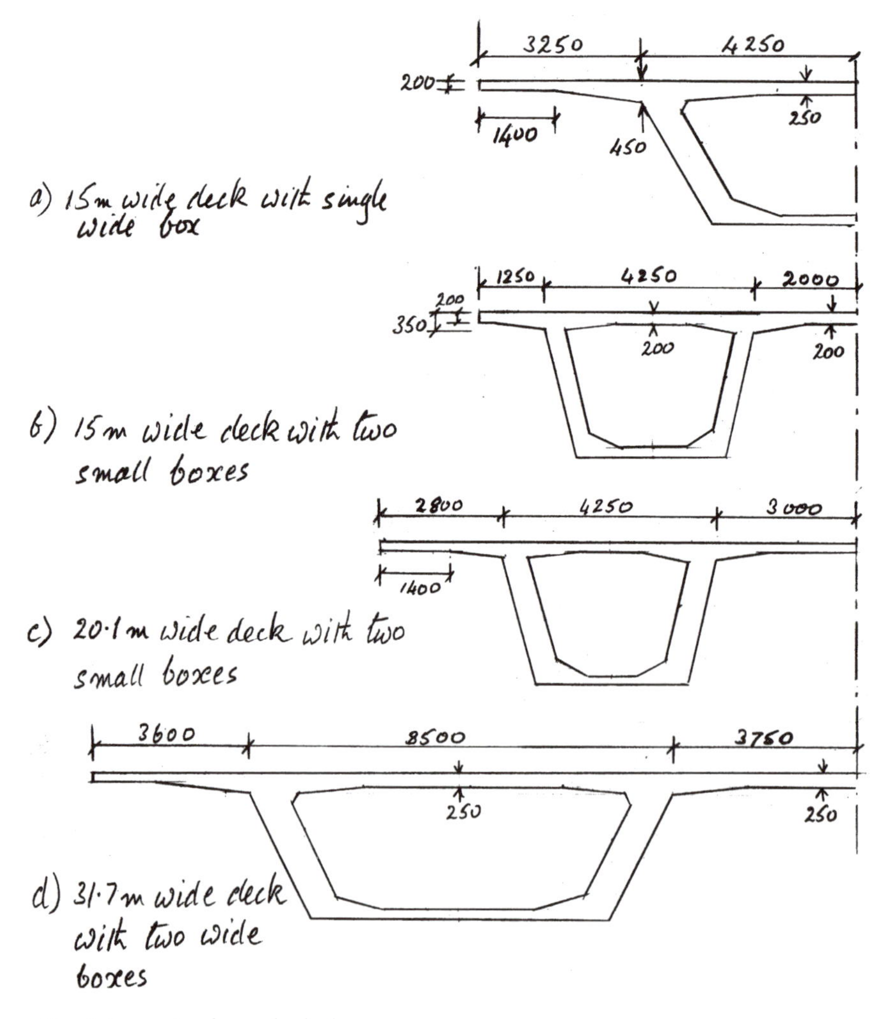

Figure 13.10 Strategies for wide decks

and side cantilevers that are 4.5 m long. However, the thicker top slab required for the British code of practice erodes the weight advantage, and the two box scheme becomes more competitive. A ribbed or strutted cantilever or top slab may regain the weight advantage for the single box, at the expense of a more complicated construction process, Figure 9.6 (c). For lighter loading codes, the single box with solid slabs remains economical over greater widths, as demonstrated by the Benaim project for the approach spans of the Storabaelt crossing, where a 23.7 m wide deck was carried by a single box with solid slabs. The overall box width was 14.2 m, with 4.75 m long cantilevers, 500 mm thick at the root, Figure 9.12 (e).

As the deck becomes wider still, two of the 8.5 m wide boxes, with 3.6 m side cantilevers and a 7.5 m span, 250 mm thick slab between boxes give rise to a 31.7 m wide deck, Figure 13.10 (d). This is clearly not the limiting width for a two-box deck, as the top slab may be thickened to 300 mm with 500 mm haunches, leading to a box that was 10 m wide, 4 m side cantilevers and a 9 m slab between boxes, yielding a 37 m wide deck. Even wider decks may be achieved by adopting a ribbed slab. However, the option remains to adopt more small boxes.

13.6.3 Variable-width decks

When the width of a deck varies along its length, considerable care must be taken to design a scheme that is practical to build. In general, it is not good practice to change the width of the box itself, as this involves substantial modification to complex formwork. The variation in width should be taken out in the side cantilevers and the slab spanning between boxes, initially by reducing the width of the constant-thickness sections, and then if necessary cutting short the haunches.

For instance, the 15 m wide deck shown in Figure 13.10 (a) could be simply reduced to a 12.2 m width by cutting off the 200 mm side cantilever, and then further reduced to 8.5 m, with some complications to the edge parapet that would have to suit the haunch thickness. If the width was to increase locally up to 16.5 m, the side cantilevers could be stretched to 4 m (span/depth of 9) with heavier reinforcement. For still greater width, the top slab could be thickened upwards locally by 50 mm to allow side cantilevers of 4.5 m, or the 450 mm thick side cantilevers could be prestressed to permit the greater span. Although these are expensive solutions, they may be better than adopting a less economical option over the majority of the length of the bridge.

However, if the width were to start at 10 m, and increase linearly to 21 m, the two small boxes could initially be cut down, and then stretched slightly by reinforcing the slab more heavily at the widest point. This maximum width may be further increased by thickening the top slab and increasing the side cantilevers and the span between boxes. If the width varied from 10 m to 30 m, it would be necessary to adopt the small box scheme, and to introduce a third box once the width became greater than 21 m.

The choice of the method of adapting to a variable width of deck can have a very significant impact on the appearance of the deck, particularly if the width of the side cantilever is changed. Where possible, the side cantilever should remain constant, or should vary only slightly, and gradually. However, often this is not possible, and the various methods of widening the deck must be evaluated for both economy and appearance.

This is illustrated by the Dagenham Dock Viaduct on the A13 highway in London, Figures 13.11 and 13.12, designed by Benaim. This bridge was 1,700 m long and,

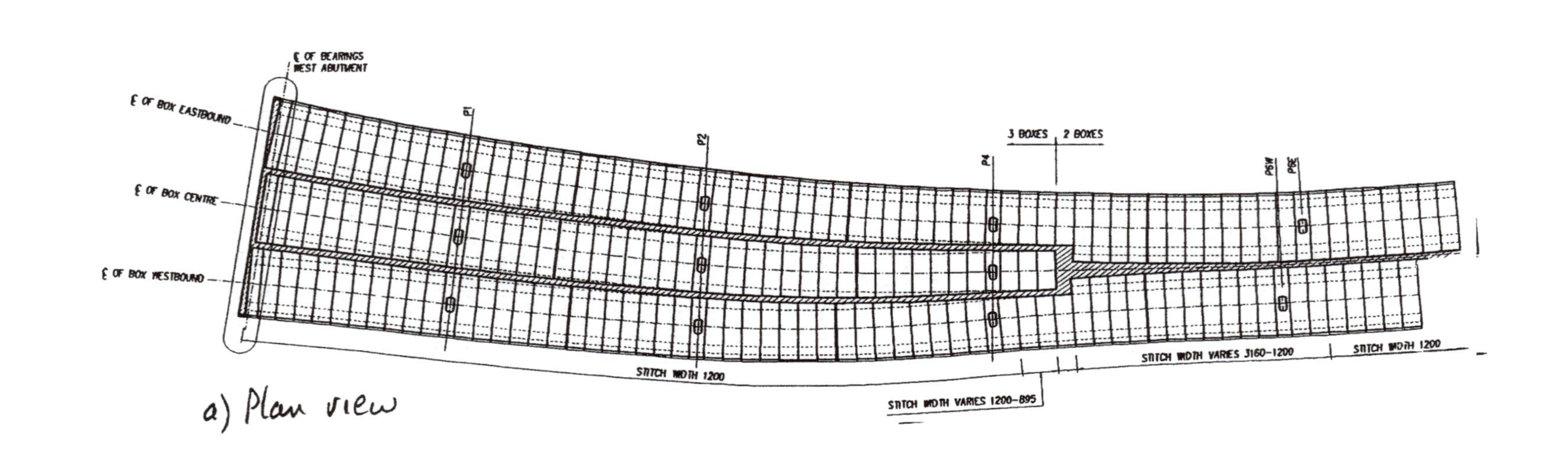
℄ OF BEARINGS
WEST ABUTMENT
℄ OF BOX EASTBOUND
℄ OF BOX CENTRE
℄ OF BOX WESTBOUND
P1
P2
P4
3 BOXES
2 BOXES
P6W
P6E
STITCH WIDTH 1200
STITCH WIDTH VARIES 1200–895
STITCH WIDTH VARIES 3160–1200
STITCH WIDTH 1200

a) Plan view

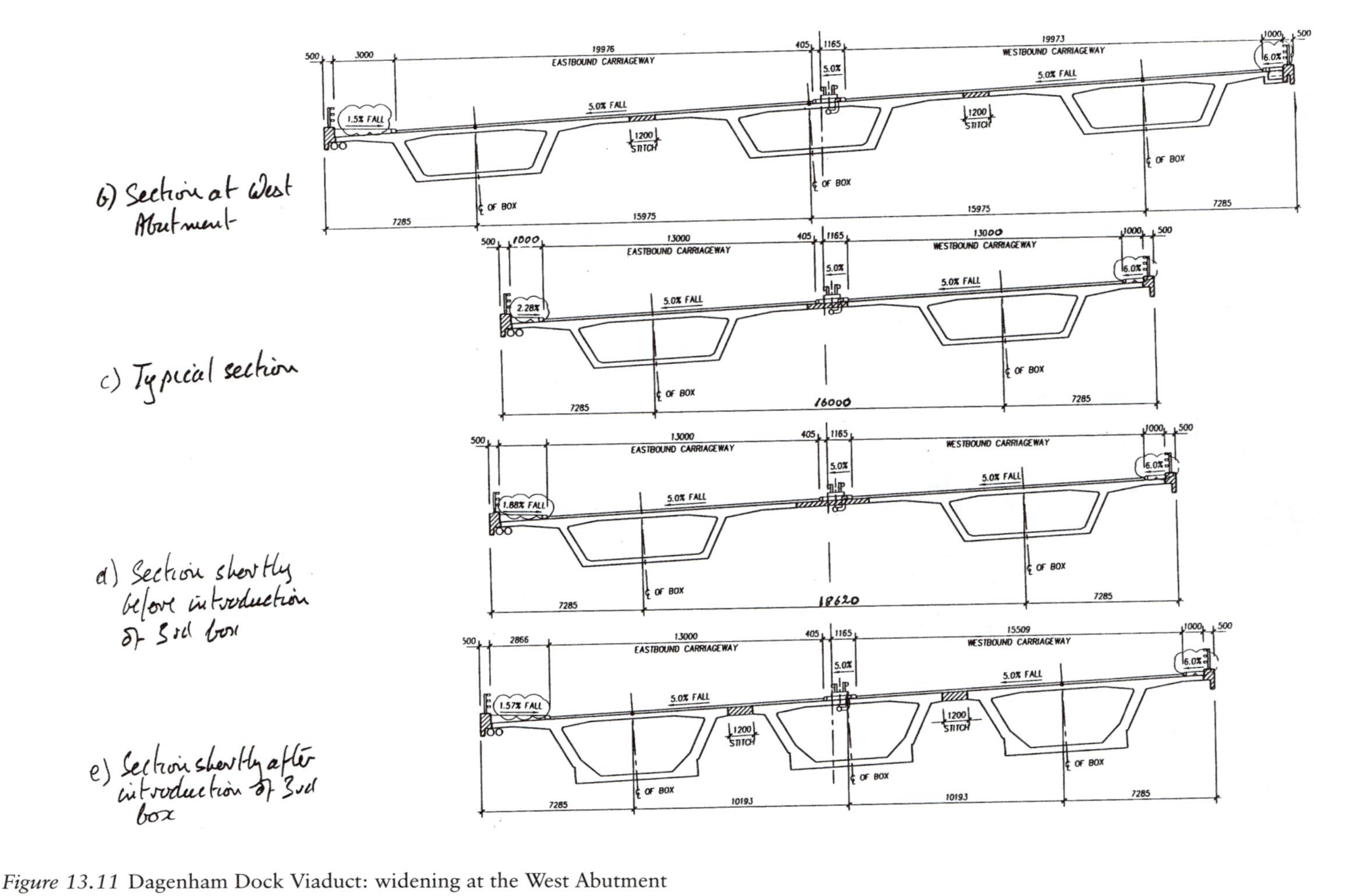

Figure 13.11 Dagenham Dock Viaduct: widening at the West Abutment

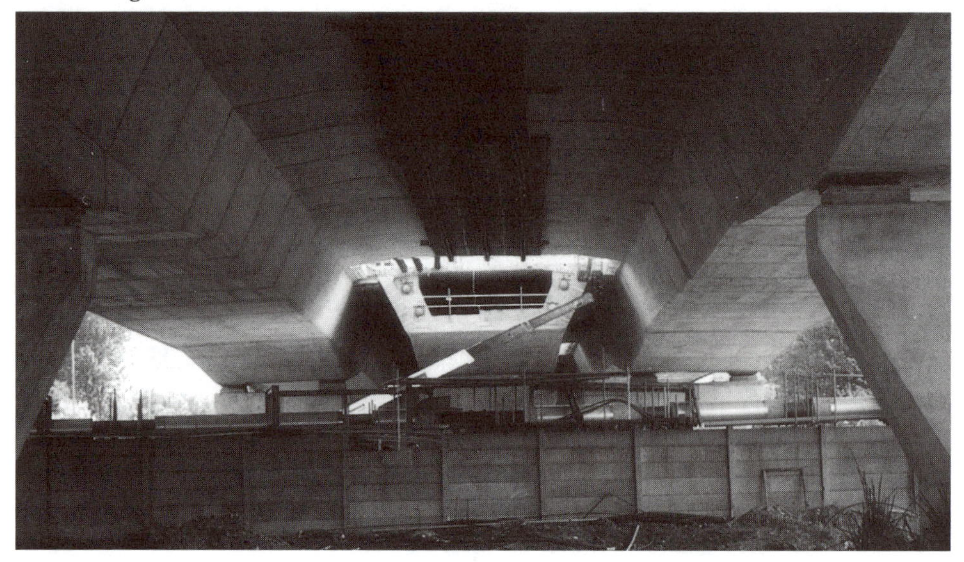

Figure 13.12 Dagenham Dock Viaduct: introduction of third box (Photo: Benaim)

over most of its length, was 30.57 m wide. However at each end it widened, to a maximum of 46.52 m at the West Abutment. It was made up of two trapezoidal precast boxes that were 8 m wide across the box. As the deck widened, the side cantilever was kept constant, while the slab between boxes was widened. When this slab reached its maximum span a third box was introduced, which initially required the inner cantilevers of the external boxes as well as both cantilevers of the new box to be cropped short. On this bridge the additional box was introduced at the point of inflexion of a span.

Alternatively, additional boxes may be introduced at an expansion joint. Figure 13.13 is based loosely on the project for the approach viaducts to the Ting Kau Bridge in Hong Kong, designed for tender by the Benaim–Rendell Joint Venture. The decks were to be made by the precast segmental technique. The slip roads adopted a narrower box than the main line. Where the main line was locally too wide for two boxes, the cast-in-situ intermediate slab was carried by post-tensioned beams, also cast in-situ. The beneficial effect of the sagging parasitic moment due to the beam prestress made it possible to connect these beams directly to the webs of the boxes.

These examples demonstrate the type of options open to the designer. The decisions taken will depend on many factors, among which the most important is how much of the deck is wider or narrower than the average. If most of the deck is of constant width, with a local widening at one end, for instance, the designer should opt for a scheme that gives maximum economy over the greater length, even if the widened portion is locally uneconomical.

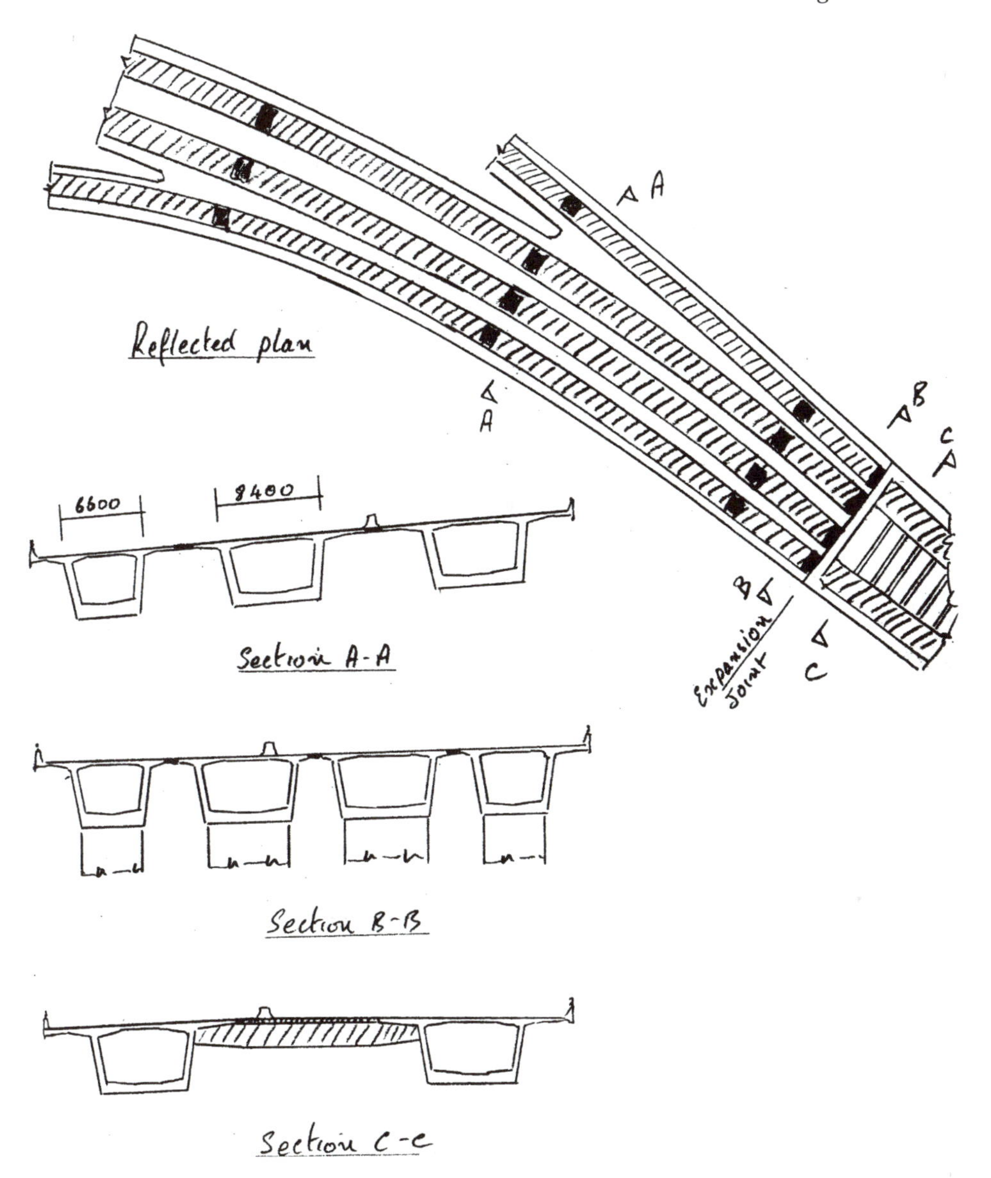

Figure 13.13 Typical detail for bifurcation of multiple box section deck

14
Counter-cast technology for box section decks

14.1 General

As has been stated elsewhere in this book, building box girder bridges in long lengths in-situ is a slow and difficult task, prone to defects and delays, tying up expensive falsework for relatively long periods. These difficulties have been the stimulus for the adoption of segmental methods of construction, such as cast-in-situ balanced cantilever and incremental launching. However, both these methods are only applicable to bridges of suitable layout and geometry. Many bridges, particularly in towns, have span layouts and geometry that are not regular, constrained by the urban fabric.

This difficulty was tackled in bridges built in the 1960s and 1970s, including some major urban viaducts, such as the A40 Westway viaduct in London, and the Mancunian Way viaduct in Manchester, by precasting the deck in short transverse segments, and assembling them on falsework. Narrow, unreinforced joints were cast in mortar or small-aggregate concrete to complete the deck, which was subsequently prestressed. Although this allowed complex box sections to be cast under factory conditions, this method of construction combined some of the disadvantages of both precast and cast-in-situ construction. Falsework was still required, in addition to the typical precasting costs of the multiple handling of heavy segments, their transport and their careful adjustment on the falsework. On curved alignments, the precast deck segments had to be made trapezoidal in plan, as the joints were too narrow to accommodate the taper. In addition, coupling tendon ducts across the joint and pouring the joint itself were difficult and time consuming.

In 1962, the French contractor Campenon Bernard built the Choisy-le-Roi Bridge across the river Seine, adopting a brilliant variation of this form of construction. The bridge deck was cast in short transverse slices as before, but each segment used the previous one as an end shutter. Consequently they were a perfect fit, and could be reassembled on site without the need for a cast-in-situ joint. This greatly speeded up erection, and made it possible to adopt several methods of deck assembly, adapted to a wide variety of alignments. But of course it removed any possibility of adjusting the position of the segments; the precast alignment had to be perfect.

The precast segments may be fabricated in either the long line or the short line method.

14.2 Long line casting

The long line precasting method may be used if the deck is straight or of constant curvature over its entire length. A soffit form is erected in the casting yard either one span or a half span long, depending on the method of erection of the deck, Figure 14.1. This soffit form must represent the finished alignment of the bridge deck, taking into account any pre-cambers that may be necessary to compensate for deflections due to dead loads and prestress.

Deck segments are then cast successively within a travelling shutter. The length of segments is controlled by the lifting capacity available, but is typically 3–5 m. Each segment is cast against the previous one, to create a joint that will be a perfect fit when the segments are erected. A suitable bond breaker is applied to the hardened face of a segment before casting the new segment.

As each segment is cast, the foundation of the soffit form will deflect under its weight, and the pressure of the fresh concrete may rock the adjacent segment backwards. These deflections will distort the vertical alignment of the span. If this distortion is to be kept within tolerable limits, it is important that the soffit form is founded very rigidly.

The segment removal must be done carefully, to avoid damaging the matching faces. Shear keys or other features on the matching faces must be designed to facilitate this separation. As a segment is removed the soffit form will rebound, and consequently a buffer of two or three segments should be left adjacent to the construction head to avoid causing further distortions.

If the deck is to be erected span-by-span with a cast-in-situ stitch in each span, the soffit form would represent a length between construction joints.

If the deck is to be erected in balanced cantilever with a cast-in-situ stitch at mid-span, the simplest arrangement is for the soffit form to represent a double cantilever. However, space in the casting yard and cost may be saved by a soffit form that is only half a span long. With this arrangement the pier segment is cast first. When a complete half span has been cast and the segments have been removed to storage, the pier segment is turned through 180° and used as the origin of the other half span of the balanced pair.

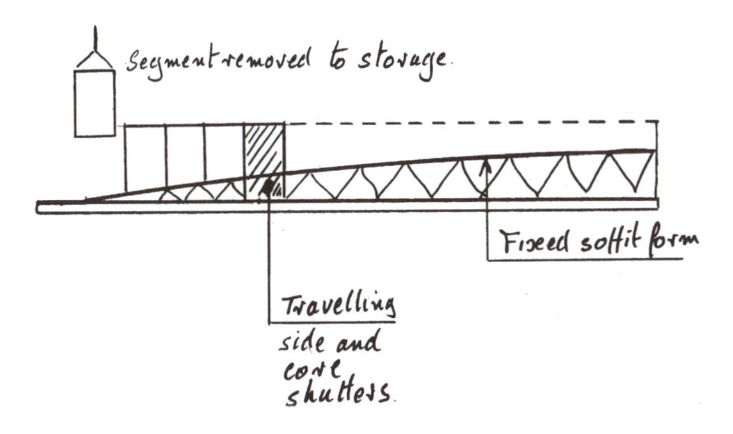

Figure 14.1 Typical long line casting bed

The position of the pier segment when the first half span is cast must be ascertained by very precise surveying, using studs or other markers cast into the top slab. When it is rotated to act as the origin of the second half span, it must be brought back to the same alignment about all three axes. Any error in this orientation will cause a kink in the horizontal or vertical alignment, or introduce a twist into the half span.

An alternative manner of solving this problem is to omit the pier segment when casting the half spans. It is then cast between the two adjacent segments. This arrangement gives a longer base for surveying the position of the segments.

Although the long line method of building match-cast segments is relatively simple to carry out successfully, it is now rarely used. A well-founded soffit form may be costly, particularly if piling is required, and the rate of casting is not high. However it is not to be discarded, and if a good foundation is available cheaply, offers a low-tech version of counter-cast technology that may be used for bridges much smaller than the accepted lower economical threshold for this technique.

14.3 Short line casting

14.3.1 General

Most counter-cast segmental bridges are built by the short line method, which may be used for virtually any shape of deck alignment. One of its principal advantages is that the deck is fabricated in a factory environment, rather than that of a building site. The majority of segments may be built to a daily cycle, with the labour force working to a regular routine. The result is a high quality product, built to close tolerances, with good finish and durability.

However, it depends for its success on a designer who understands the theoretical basis of the alignment control and recognises the contractor's need for productivity. It also requires careful preparation and execution by the contractor, and on the use of a well-designed and engineered casting cell. Once these conditions have been met, the system is remarkably trouble free, and even contractors with little experience of modern bridge-building techniques may be very successful, casting segments at a rate of one per working day, and erecting decks at the rate of several spans per week.

On the other hand, if the design or the construction has been carried out by a team that has not understood the essence of the technology, this can be the most troublesome of construction methods. Errors in the cast alignment are only discovered when the deck comes to be erected, by which time half the segments have usually been cast. Correcting such errors is in every case a very time-consuming process, which may involve recasting segments out of sequence, or introducing additional grouted joints.

Short line precast segmental construction requires at least one complex casting cell, storage capacity for a large number of segments and means of lifting heavy segments in both the casting yard and at the construction front.

If the deck may be erected using cranes or shear legs, this method is generally economical for 250 segments or more, representing a length of box between 750 m and 1,000 m. If a self-launching erection gantry, typically weighing at least 200 tons and costing upwards of £750,000 is required to erect the deck, the economic threshold is more likely to be 500 segments.

It should be noted that if a contractor has the necessary kit from a previous contract, the economic threshold may be much smaller. The Weston-super-Mare railway viaduct designed as an alternative by Benaim for the contractor DMD, consisted of two adjacent boxes each approximately 150 m long, and requiring only some 90 segments in total. It was built using moulds and lifting equipment already amortised on a previous contract.

14.3.2 The basis of the technique

The bridge deck is divided into transverse segments, each generally between 3 m and 5 m long. They are fabricated in a specialised casting cell between a fixed stop-end and the previous segment to have been cast, which is called the 'stop-end segment', Figure 14.2. The relative position of the two adjacent segments in the completed deck is calculated as a function of the deck alignment. This relative position is then reproduced in the casting cell by adjusting the orientation of the stop-end segment. As the new segment is cast, the weight and pressure of the fresh concrete will disturb the alignment of the stop-end segment, and consequently the relative angles between the two segments will be incorrect. This achieved relative orientation of the two segments is then measured with the greatest feasible accuracy. The error that has been cast into the alignment is calculated. The stop-end segment is then taken away to storage, and the new segment moved forwards to become the stop-end segment in its turn, clearing the mould for the next segment to be built. The position of the new stop-end segment is adjusted to correct the errors of the previous cast. Thus the deck alignment is created by successive approximations, each cast correcting the errors of the previous one, but creating new errors that must be corrected in their turn. Any deck built by this method has an alignment that snakes about the theoretical line. However, the angular error between adjacent segments is generally less than 0.003 radians, and is quite invisible, even on close inspection.

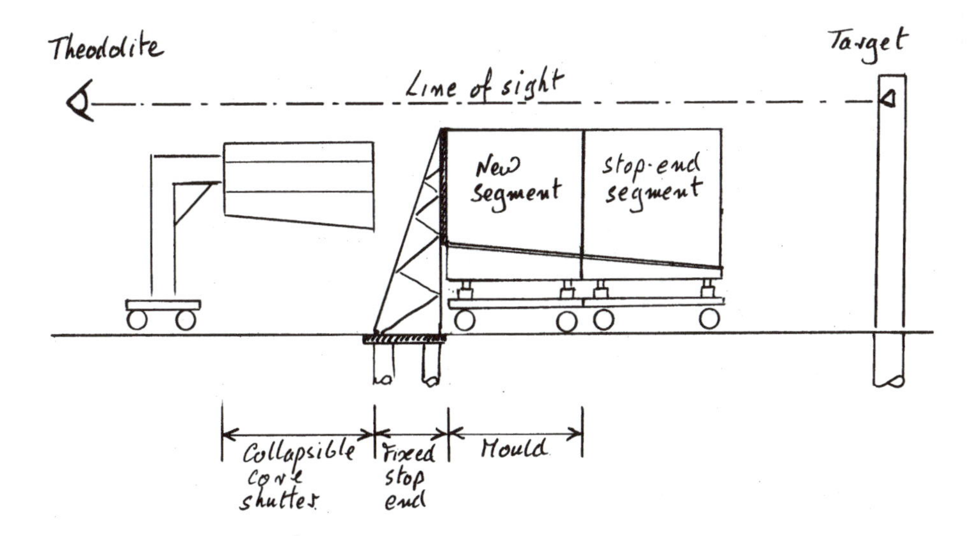

Figure 14.2 Typical short line casting cell

14.3.3 The casting cell

A typical casting cell, Figure 14.3, consists of a fixed stop-end, steel side shutters for the outside face of the webs and the side cantilevers, a core shutter, Figure 14.4, two mechanised trolleys, Figure 14.5, which carry the soffit shutter, and a surveying set-up.

Figure 14.3 STAR: casting cell ready to receive reinforcement cage (Photo: Benaim)

Figure 14.4 STAR: side shutters, stop-end and core shutter beyond (Photo: Benaim)

Figure 14.5 STAR: trolley for casting cell (Photo: Benaim)

The stop-end must be of robust steel construction, and is fixed to the floor. In some instances, where there is more than one segment length to cast, its position may be adjustable.

The side shutters are fixed in position longitudinally, and strike by rotating about their feet or by sliding sideways, usually operated by hydraulics. They are equipped with external vibrators. They must be robust, and should be made of steel sheet generally 8 mm thick or more, suitably stiffened. The side shutters must seal for grout leakage against the stop-end and the stop-end segment, and if they are too lightly built, will give endless trouble.

The core shutter is usually introduced through the stop-end. It is one of the more critical components of the cell, as it must be capable of being adjusted to accommodate the variable internal features of the deck design within the constraints of the daily cycle.

The soffit shutter for the mould is carried on a rail-mounted trolley. Once the segment has been cast, and takes up the stop-end segment role, it remains on its trolley which is equipped with hydraulic jacks that allow it to be rotated about the three principal axes to set up the geometric adjustments that are the essence of the method.

The survey of the segment geometry is carried out by a precise theodolite capable of measuring to an accuracy of 1/100 mm. The line of sight is completed by a target situated beyond the casting cell, usually outside the casting shed. The target support must be founded at a depth which insulates it from seasonal ground movements or settlement due to the fill that may have been used to create the platform for the casting yard. Usually this requires a foundation pile. It must also be shielded from the sun which would cause it to deflect laterally.

The successful outcome of this method of construction depends on using a casting cell of high quality. Top class design, materials and fabrication are worth paying for, and the cost difference with a cut price product will be recovered many times over for a long bridge deck. The author has had experience of a site in the UK with several casting cells whose construction was too light, and strengthening beams had to be attached at each cast to limit the deflections of the side shutters under the pressure of the concrete and to resist the shaking caused by the external vibrators, and then detached to allow the shutters to be struck. The cells were eventually abandoned and a better product purchased. On the site of a very long viaduct in South East Asia, the local contractor needed a large number of cells. He had some made by Ninive Casseforme [1] in Italy, and then had a local firm fabricate others. When more cells were needed, the contractor opted for the better quality Italian product despite the greater cost and the delay of shipping.

14.3.4 The calculation of the casting geometry

In order to develop the geometric instructions for precasting the segments, their relative position in the bridge deck must be calculated with precision. The so-called 'second order' effects must be included in this calculation. Any combination of longitudinal gradient, crossfall, horizontal or vertical curvature creates secondary curvatures or twists that must be included in the casting geometry. For instance, if a horizontally curved deck is given a crossfall by rotating it about a longitudinal axis at the pier it will progressively develop a vertical gradient, Figure 14.6, which must be corrected in the casting cell. Similarly if a deck curved in the horizontal plane is tilted onto a vertical gradient, the carriageway will progressively develop a crossfall unless a corrective twist is built in. It needs a clear head to understand these geometric interactions and to transform the mathematical definition of the alignment into a set of casting instructions. Most of the failures of erected alignment when using this technique of bridge deck construction can be traced to mathematical errors in defining the casting geometry. Software is available to ease the task of the designer, but these programs should not be used unless someone within the team understands the mathematics of the transfer of the deck alignment to the casting yard; when mistakes or accidents occur during the casting of a segment, it is essential to understand the basic mathematics in order to develop corrective action.

Prestressed concrete bridge decks need to be pre-cambered to cancel out the deck deflections under long-term dead loads and prestress. Once the geometry of the deck has been converted into a set of casting instructions, the pre-cambers must be added.

Errors independent of the casting geometry may be introduced into the alignment during the construction of the segments, some of which are unpredictable. Consequently, the calculation of the casting geometry should be as rigorous as possible to avoid adding avoidable errors to those that are unplanned. When the deck is erected, even relatively small errors in alignment can be troublesome and require costly remedial works. For instance, the mid-span cast-in-situ stitch for a balanced cantilever deck is usually between 150 mm and 250 mm wide. Consequently, even a positional error of 25 mm between the ends of the two cantilevers will give difficulties in linking the prestressing ducts and in aligning the parapet.

Every effect that may influence the erected alignment should be taken into account. For instance, on the STAR railway viaduct in Kuala Lumpur that was erected in

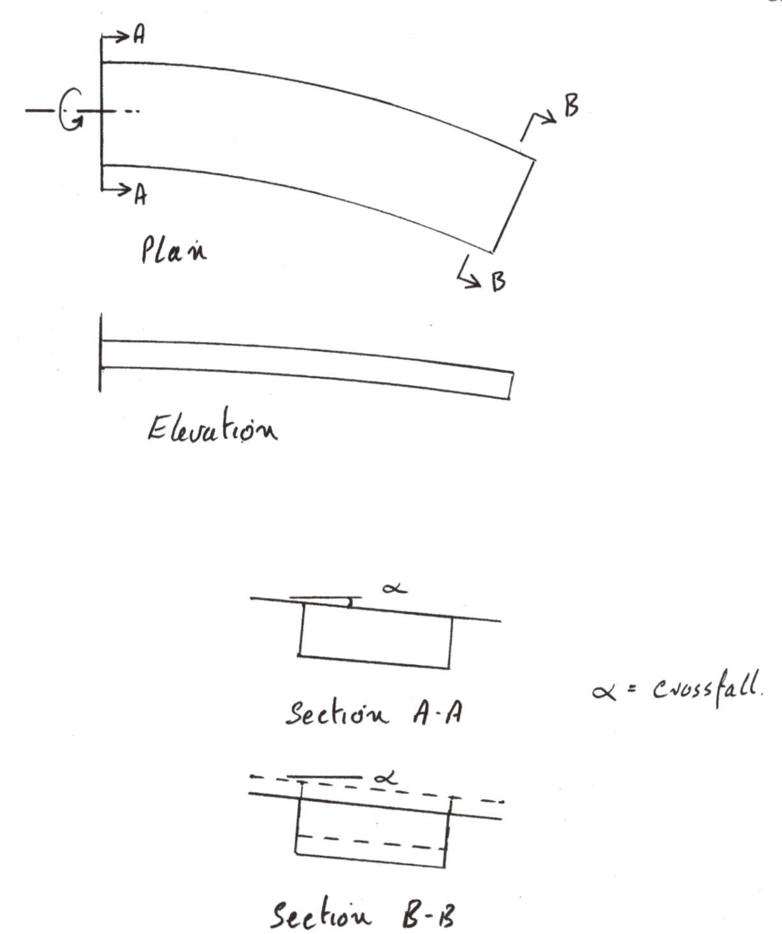

Figure 14.6 Combination of plan curvature and crossfall

balanced cantilever, there were some spans with a length of 52.5 m and a radius of curvature of 130 m. It was realised that this sharply curved double cantilever would apply an eccentric load to the slender columns, which would rotate and displace the deck. These deflections were included in the calculation of the alignment. Although the effect would have been small, this was an avoidable inaccuracy, which if not corrected would have added to the other, unavoidable, tolerances.

14.3.5 Alignment control

There are different methods of achieving the geometric control of the segment casting. The one known best to the author will be described briefly.

The stop-end must be vertical and at a right angle to the horizontal line of sight defined by the theodolite while the casting cell for the new segment must be horizontal. Consequently, the joints in the deck will be normal to the top surface of the deck.

Once the reinforcing cage for the new segment has been lifted into the casting cell, the stop-end segment, resting on its trolley, is rolled on the rails to the correct distance from the fixed stop-end, defining the length of the new segment to be cast. This stop-end segment projects a short distance (≈ 100 mm) into the side shutters and over the new segment trolley. This allows the side, soffit and core shutters to clamp onto the stop-end segment, providing a grout seal.

The stop-end segment is then put approximately (to the nearest millimetre) into the correct relative position to the fixed mould to define the deck alignment. This is done by acting on the adjustments of the trolley in all three dimensions.

The new segment is cast, generally in the late afternoon. Four studs are inserted into the top surface of the segment, over the webs near each end, and a small steel bar is cast into the top surface of the segment on its centre line at the end remote from the fixed stop-end. The concrete is allowed to harden overnight.

The critical segment survey is then carried out, generally before the site has fully opened, so that vibrations due to moving plant do not disturb the measurements. The survey is carried out by the theodolite aiming at the target which defines a centre line and a horizontal plane. The levels of the four studs on the new segment are recorded using a staff, and the axis of the line of sight scratched onto the steel bar. The elevation of the studs on the stop-end segment and the offset of its centre-line scratch are also measured.

The accuracy of these measurements is critical to the method and usually the theodolite is read twice by different operators to avoid the risk of booking or reading errors.

The achieved geometry of the stop-end segment may then be compared with the theoretically correct geometry necessary to create the deck alignment. The error with respect to the theoretical alignment is calculated, and the set-up position of the new stop-end segment for the next cast is chosen such that this error is corrected.

After the survey, the side shutters are removed and the core shutter extracted. The stop-end segment is then taken away to storage, and the new segment rolled forwards on its trolley, becoming the next stop-end segment.

Mistakes or bad practice in the casting yard may create errors in the alignment. Other than gross mistakes such as erroneously booking the results of the survey, there are more subtle errors that may derive from poor procedure or defects in the equipment. As a result, there are in effect two levels of this technology. Where the deck includes cast-in-situ joints at no more than say 50 m intervals when built span-by-span, or 100 m intervals when built in balanced cantilever, a faultless mathematical analysis of the alignment allied to a good understanding of the theory and practice of the method is sufficient to ensure adequate dimensional tolerance of the completed alignment. It is however possible to build decks many hundreds of metres long without cast-in-situ joints, or to cast complicated sections of deck such as bifurcations. These sophistications require a full in-depth understanding of all the possible sources of error in the casting yard and their effects on the erected alignment, allied to very careful procedure.

14.3.6 Length of segments

As the basis of the method requires the precise survey of the relative position of two segments in the casting cell, this becomes less accurate as the segments become shorter. The minimum practical length of match cast segments is about 2.5 m, except for one-off pier segments. The maximum length of segments is generally controlled by the cost of lifting and moving them, which depends on the scale of the site.

As most box section bridges of medium span have an average thickness of the order of 0.5 m, a 12 m wide by 3.5 m long segment would weigh some 53 tons. However, the segments at mid-span are likely to be lighter than those near the piers, where webs are usually thickened. Thus the weight of typical span segments will vary between 45 and 70 tons. In general, a weight of segment that does not exceed 60–70 tons allows the use of readily available cranes and low-loaders.

When the deck is wide, segments will inevitably be heavier. For instance, the segments for a 20 m wide single box will weigh typically between 25 and 35 tons per metre. Thus a 3.5 m long segment will weigh of the order of 90–120 tons.

Clearly, longer span bridges have a greater average thickness of concrete (*8.4*), and consequently the segments will be heavier.

In most box section bridge decks there is a pier diaphragm. In order to limit the weight of the pier segment to that of standard segments, it is common practice to reduce its length, typically to 2–2.5 m. It is also often necessary to reduce the weight of the diaphragm by thinking clearly about its various functions (*9.6*), and removing all redundant concrete. Generally, the pier segment is used as the origin of the counter-cast run, and is sometimes cast in advance in a simple fixed mould.

For long decks, the length of segment adopted may critically determine the number of casting cells that have to be mobilised; an increase in length may reduce the number of cells, at the cost of increasing the weight of the segments. The use of lightweight concrete has proved to be economical when it has allowed the length of the segments to be increased and the number of casting cells to be reduced without increasing the segment weight beyond some critical limit.

By advancing or withdrawing the stop-end segment so it penetrates more or less into the steel shutter of the casting cell, the length of segments may be adjusted slightly during casting, in order to suit slightly varying span lengths. However, this may well be impossible if the bridge deck is sharply curved, as the sealing of the shutter onto the stop-end segment would be compromised, causing grout leaks and untidy joints.

The designer also has the option of adopting two or more lengths of segment, either to adapt them to varying spans, or to limit the weight of the heavier segments for a variable-depth deck. However, if the segments within one casting run have different lengths, the casting programme will be disrupted to some degree, as the mould has to be adjusted. It is good practice to minimise such changes, and where possible keep all the segments of a casting run at the same length.

In general, the maximum span that may be built economically by the precast segmental method is influenced by the height and cost of the mould, and by the weight of segments to be lifted. For instance, a 160 m span would need a mould some 8 m high, and the deepest span segments for a 12 m wide deck are likely to weigh some 50 tons per metre, or 175 tons for a 3.5 m long segment. Clearly for very long viaducts, where the moulds and lifting gear will be well amortised, the size and weight of the segments may not be an obstacle to choosing this method of construction.

In some special cases, very long, heavy segments may be justified. In Benaim's proposed alternative design for the Storabaelt Approach Viaducts, consisting of a 4 km long series of 164 m spans, match-cast segments 23.7 m wide and up to 8 m long were planned, weighing up to 600 tons. This was possible because powerful floating plant with the required lifting capacity was required on site for the construction of the foundations.

14.3.7 Variable spans

The precasting of segments imposes a stricter discipline on the span lengths of viaducts than cast-in-situ construction. However, it is possible to accommodate a considerable variation by using the following measures:

- The length of segments may be modified by up to ±25 mm by adjusting the position of the stop-end segment in the casting yard.
- The length of cast-in-situ stitches may be varied from 150 mm to 1 m.
- The length of the pier segment that may be cast outside the casting cell, may be varied.
- For decks erected in cantilever, unbalanced cantilevers with an additional segment on one side may be adopted.
- As a last resort more than one segment length may be used.

All these measures were required for the Belfast Cross Harbour Bridges, where the tender design and the methods of construction were carried out by Benaim, Figure 15.29 for the Graham-Farrans JV.

14.3.8 Detailed design of the segment

This method of construction relies for its economy on the high productivity of a factory environment. The design of the deck must recognise this, and provide segments which are as repetitive as possible.

In general, the webs and/or bottom slab will change in thickness along the span. There are also likely to be internal blisters for the anchorage of permanent or temporary tendons, generally in the top and bottom corners of the box, and strong points for lifting the segments, Figure 14.7. These features need careful design and should occur at the same position in segments to simplify the core shutter. Good design of the blisters in particular, minimising their protrusion and fairing them well into the concrete, pays dividends in ease of construction.

The core shutter must be capable of being withdrawn with a minimum of labour. It is normally hydraulically powered, and must create the clearance necessary for removal either by retracting the side shutters towards the centre of the deck, or by rotating and folding the side shutters, or by a combination of these two movements. Inexperienced designers often choose powerful prestressing units that need large internal anchorage blisters. If the blisters are too wide compared with the width of the box, there may not be enough room between them to house the mechanical engineering necessary for retraction. Decks with external prestressing, where the anchorage blisters are generally larger than for internal tendons, need particular care.

Falsework designers can accommodate virtually any geometry, but for an ill-considered design it will be at the price of increasing time and labour in use, as the mould will need to be partially dismantled for removal. The bridge designer should keep these problems in mind during the preliminary design, and then involve the falsework designer at the earliest stage in the detailed design. For instance, adopting a

Figure 14.7 STAR: permanent and temporary anchorage blisters (Photo: Benaim)

smaller prestressing unit, with correspondingly smaller anchorage blisters, may make the difference between being able to collapse the internal mould and withdraw it in one piece, or having to partially dismantle it for each cast.

Ideally, the box should be at least 2 m deep and 6 m wide in order to provide space for blisters and for the mechanical engineering of the core shutter. Clearly this is not always possible, but as the box becomes smaller so the need for careful design of the internal features becomes greater. The bridge designer must be aware of the realities of segment production.

In some cases, it may be beneficial to adopt two different sizes of tendon, with powerful units travelling from end to end of the deck or anchoring in pier diaphragms, while those tendons that anchor in internal blisters may be smaller. This approach was used by Benaim in the design of the Weston-super-Mare railway viaduct, where the segments were erected in cantilever with 12 No 15 mm strand cables, and subsequently 27 strand cables were threaded from end to end to complete the prestress. This may be seen in the different duct sizes in Figure 14.17.

If blisters for temporary stressing are required on the bottom flange itself, rather than tucked into the corner, it may be advisable to leave bend out bars or couplers in place and to cast these on after match casting. The segments are stored for a few weeks before erection, giving time for such operations, Figure 14.8.

The extra cost of high strength concrete, which minimises or eliminates the thickenings required to webs and bottom slab, may well be justified by the simplification of the core shutter and the reduction in labour required to cast a span.

Any tapers or changes in thickness to the webs or flanges should be designed to start some 200 mm into the segment, so that the core shutter has a continuous surface

Figure 14.8 STAR: temporary blister cast-on after precasting (Photo: Benaim)

on the stop-end segment on which to bear. Similarly, if a variable-depth bridge deck features a linear haunch, the change in angle should occur a short distance into a segment. The internal surfaces of the segment must have suitable draws so that there is never a parallel shearing movement of shutter and concrete.

A good detail for segmental boxes is to recess the bottom slab up from the bottom of the webs by 50 mm. The bottom reinforcement projecting from the slab gives a convenient plane of support to the lowest row of prestressing ducts. Also, this feature recalls aesthetically the nature of the hollow box, making it a little less bland and less solid looking, Figure 9.28. The web downstand can return at pier positions to provide an aesthetic recall of the pier diaphragm.

On a typical daily cycle, concrete is cast in the evening and the side and core shutters are struck next morning, some 12–14 hours later when the concrete may have achieved a strength of only about 12 MPa. However, the actual average strength depends on the ambient temperature during the night and may vary across the segment. If the side cantilevers are too slender, they may deflect significantly after separation of the segments. This could compromise the accuracy of the match casting, and cause serious problems when the segments are erected. Generally, good results have been obtained by haunched cantilevers up to 2.5 m long that have a slenderness not less than 1/7. If the side cantilevers are longer or more slender, it may be necessary to steam cure the segment.

14.3.9 Shear keys

As the resin used in the joints during erection (*14.3.15*) remains soft for several hours, it is essential that the matching faces be provided with shear keys. If the deck is being erected in free cantilever, two or three segments may be erected within a few hours, before the resin of the first joint has cured, and the shear keys need to be designed to carry their weight. Some designers prefer shear keys that are capable of carrying the full working shear force in the deck (dead load + live load – prestress), so that even if the resin is faulty the deck will survive.

There are two basic styles of shear key: prominent keys or a washboard surface. Prominent keys may be fully engineered to carry the imposed shear forces, and may double as locations for the web anchors, Figure 14.8 and Figure 14.9. They also provide a visual record of the mode and direction of erection of the deck, Figure 14.10. The majority of the end face of the segment remains usable for the passage of ducts or for the location of prestress anchors. It is important that for webs sloping more steeply than those illustrated the shear keys are horizontal, not perpendicular to the webs. In

Figure 14.9 STAR: typical segment details (Photo: Benaim)

Figure 14.10 STAR: directional shear keys (Photo: Benaim)

the latter case, they would locate the segment not just vertically but laterally, and the slightest lateral misalignment would cause the concrete to bear hard and to spall.

In segments that are above about 2 m deep, washboard type shear keys spread the shear force evenly over a large proportion of the height of the web, Figure 14.11. However, they create complications for the location of face anchors and for the passage of ducts. On shallow sections, the proportion of the total section height that may be equipped with washboard type shear keys may be too small for safety. If the web is thick enough, the corrugations may be hidden by a lip of concrete, giving the appearance of a straight construction joint.

It is good practice to adopt half-round shear keys in the top slab to transmit the shear force from wheel loads. A horizontal locating shear key is usually placed at the centre of the top slab, with draws to bring the segment into horizontal alignment as it is pulled in by the temporary prestressing bars, Figure 14.9.

Figure 14.11 AREA contract: washboard shear keys (Photo: Robert Benaim)

14.3.10 The reinforcing cage

The reinforcing cage, Figure 14.12, is generally produced in a jig, which ensures that it is geometrically accurate, and may be placed in the casting cell without the need for adjustment. Usually it is necessary for each casting cell to be served by two jigs to ensure that a cage is produced on a daily cycle.

The production of segments is more an industrial activity than typical cast-in-situ construction, and the preparation of the reinforcing cage needs to reflect this changed emphasis. For instance there are reinforcing bars that must respect cover at each end, such as the transverse bars in the bottom slab, and shear links in the webs; in a long bridge deck there will be many thousand such bars. The conventional bending tolerance on the overall length of the bars would make it impossible to respect the correct cover at each of their ends and in cast-in-situ construction it is normal practice to design them with a lap. In the industrial process, there is no reason why such 'dead length'

Figure 14.12 East Moors Viaduct: reinforcing cage (Photo: Benaim)

bars cannot be made in one piece, with a smaller tolerance on overall length saving steel and time. Such exceptional tolerances must be called for on the drawings.

Prestressing ducts usually consist of spirally wound steel sheet. These ducts may be sprung into curves to suit the prestress profile, and held to shape by wiring them to the reinforcing cage. However, it is obvious that it is not possible to maintain the curvature at the end of a duct, at the segment joint. If the design assumes that the ducts adopt a continuous curve across the joint, the reality will be a series of cusps leading to a higher loss of prestress force due to friction greater than assumed in the calculations. Moreover, most ducts are cut to the segment length, so that a duct inclined at say 10° to the horizontal will be 54 mm short in a 3,500 mm long segment, Figure 14.13. The gap at each end will be filled by the inflatable formers that are used to stiffen the ducts during casting. Consequently there will be a short length at segment joints where the duct is absent and the tendon will be in direct contact with the concrete, further increasing the friction loss.

A good prestress alignment allows for a short straight length of duct across the joint, although there are cases in which the constraints of the prestress tendon alignment make this impossible. When there is a high degree of repetition of the prestress geometry, the ducts may be cut to the exact length and may also be pre-bent.

Where external prestressing is being used, it is not satisfactory to hold the deviator tubes that locate the tendons in diaphragms and blisters by wiring them to the reinforcement, as is habitually done for internal ducts. The orientation and position of these short lengths of tube are critical to the installation of the external tendons and it is important to bolt them to the steel shutter. If a multiplicity of bolt positions is to be avoided, considerable discipline is required in the design of the prestress alignment.

Figure 14.13 STAR: detail of prestressing duct (Photo: Benaim)

14.3.11 The daily cycle

A typical casting cell can be expected to produce one segment per day, which, when taking into account the learning curve, maintenance of the mould and statutory holidays, yields on average about 5.3 segments per six-day week, with both casting and maintenance on the sixth day, about 270 segments per year, or 4.7 segments a week for five days casting and maintenance on the sixth day.

Achieving this daily cycle ensures that the process is truly industrial, with an organised and repetitive daily and weekly routine. Experience shows that where this daily cycle has been achieved, production costs are lower and quality higher.

For normal segments, weighing up to 100 tons, there is no reason not to achieve the daily routine. For exceptionally large or complex segments, it may prove more difficult. If such segments make up the majority of the length of the deck, additional measures, such as the secondary precasting of the bottom slab or webs should be considered. If they are only occasional, then it is necessary to find an organised way of introducing this disruption into the industrial process. For instance, where complex and heavy pier units are used as the origin of casting runs, they may be cast in a separate, non-mechanical mould, off the critical path.

It is normal for there to be a learning curve while the construction team debugs the equipment, learns the best methods of assembling the reinforcing cage and finds how best to fill the mould. The length of this learning curve generally reflects the degree of preparation and preplanning by the contractor. The daily cycle should be attained after about 15–20 segments. A typical learning curve for a six-day casting week would be as follows:

Week 1	1 segment
Week 2	2 segments
Week 3	3 segments
Week 4	4 segments
Week 5	5 segments
Week 6	5 segments
Week 7 onwards	6 segments.

14.3.12 Difficulties in casting the segment

The principal difficulties in casting segments are in ensuring that concrete has flowed around the bottom corner of the web into the bottom flange and that the concrete is well compacted at the bottom of the webs, in particular when there are heavily reinforced anchorage blisters at the junction of web and bottom flange, Figure 9.25 and Figure 14.12. The fluidity of the concrete and the management of the shutter vibrators are both critical to success. Generally the slump of concrete poured into the webs needs to be of the order of 150 mm, while the concrete for the top and bottom slabs typically has a slump of around 100 mm.

The concrete is usually first poured onto the bottom slab soffit shutter through trunking that passes through a trapdoor in the top slab shutter. The concrete is spread by hand, and kept short of the bottom of the web. Concrete is then poured down the webs, aided by poker vibrators and short bursts from the shutter vibrators, until it can be seen flowing out of the bottom of the webs and joining up with the bottom slab concrete, demonstrating that the webs are full. If the bottom slab has haunches, these have a top shutter, which is sufficient to ensure that concrete will not well up in the bottom slab. However, if the bottom slab has no haunches, it may be necessary to place temporary top shutters over part of its width. If the shutter vibrators are used too intensively, they may also cause the concrete to flow uncontrollably out of the bottom of the webs. The learning curve serves in part to tune the details of the construction process, and in particular to learn how to use the shutter vibrators.

The top shutters of the bottom slab haunches and of bottom corner anchorage blisters should have removable inspection panels. Once the correct casting procedures and concrete consistency have been established these panels may never need to be opened again, but in the learning stages they can avoid much remedial work and even discarded segments.

14.3.13 Lifting the segment

Typical segments need to be lifted three or four times before they are assembled with the bridge deck. As segments are unique in their position in the bridge deck, as well as for reasons of safety, the lifting system has to be absolutely reliable.

Lifting systems may be active or passive. Active systems involve stressing a lifting beam to the top slab of the segment with short prestressing bars. The residual force in the bars should be equal to the weight of the segment plus 20 per cent, to cover any dynamic effects during lifting. When the crane applies tension to the lifting beam in order to lift the segment, the contact pressure between beam and segment is reduced, but the tension in the bars does not change. The bars have been fully tested when the lifting beam was stressed on, and any failure due to damaged bars or anchor nuts will have taken place before the segment is lifted. Consequently it is quite safe for the bars to work at 60 per cent of their ultimate load.

As these bars are necessarily short, it is very important to ensure that their residual force is adequate. It is common practice to lengthen the bars by putting spacers either beneath the slab or above the lifting beam, so increasing their extension on stressing and making it easier to achieve the desired residual force. Generally the bars pass through holes in the top slab, and are thus recoverable, although it is also possible to cast bars with couplers into the concrete. With careful design, top temporary anchorage blisters may double up as strong points for lifting, simplifying the core shutter, Figure 14.14. With a prestressed lifting system, only one or two bars per lifting point are necessary.

Generally there are two grades of prestressing bars, the lower grade having significantly greater ductility. It is important that ductile bars are used, as this provides a margin of safety if they should be inadvertently bent, or if stressing plates are not quite normal to the bar.

In passive systems the lifting beam is not stressed down to the segment; the link between the beam and the segment is only loaded as the segment is lifted. Clearly the factors of safety must be much greater, with bars and their anchors typically working at between 15 per cent and 25 per cent of their ultimate load. The lifting device may consist of reinforcing or prestressing grade bars or proprietary products, passing through the top slab anchored by plates and nuts, or cast into the segment. The factor of safety chosen must depend on the vulnerability of the system. Ductile steel may have a smaller factor of safety than more brittle steel, while, as with all tension systems, the possibility of progressive collapse must be considered. For instance, when the failure of one bar would leave the remainder with an adequate factor of

Figure 14.14 East Moors Viaduct: temporary stressing blister doubling as lifting point (Photo: Benaim)

safety, the overall factor may be lower than when no component can be allowed to fail. With passive lifting devices, it is desirable to create redundancy by using multiple bars per lifting point.

Where the satisfactory anchorage of lifting devices cast into the concrete depends on careful supervision during casting, in some circumstances it may be necessary to load test each anchorage before suspending the segment from it. Despite all design precautions, passive lifting systems are more subject to human error and material failure than active ones, and are consequently less safe. Where they involve cast-in products, they are also more expensive. However there are many instances where they are the most suitable for the project.

The lifting beams can be designed with a hierarchy of sophistication. If the deck to be erected is on a vertical gradient or crossfall, the lifting beam may be equipped to incline the segment in both directions, using either a series of alternative lifting positions or hydraulics. Similarly, the segment may be rotated about the axis of the suspension system by hydraulic jacks, or this may be achieved by pulling on cables from the erected deck.

14.3.14 Storing the segments

Depending on the size of the site and the philosophy of the contractor, the storage may be highly organised and mechanised, or segments may be stored more informally. In the most organised sites, the segments are handled by purpose-built transporters, Figure 14.15, and are stored in regular rows, rather like a container yard, Figure 14.16. Segments may be stacked two or three high, Figure 14.17, as long as no significant bending moments are imposed on the webs and slabs. In such a storage yard, it will usually be necessary to build foundations for the rows of segments.

Figure 14.15 STAR: segment handler (Photo: Benaim)

Figure 14.16 STAR: casting factory and storage (Photo: Benaim)

Figure 14.17 Weston-super-Mare railway viaduct: segment stacking (Photo: Benaim)

If a mobile crane is used to handle the segments, the storage yard may be organised more flexibly as the crane can negotiate less than absolutely flat ground. Another flexible arrangement is to use a self-loading low loader, which can put down and recover segments from any spare land around the site, Figure 14.18.

However the storage is organised, it is essential to avoid warping of the segments which must each rest on a three-point support.

14.3.15 Joints with resin filler

When counter-casting was first introduced, all prestressing tendons were housed in ducts within the concrete section. To avoid the risk of corrosion to the tendons, it was essential for the joints between segments to be made waterproof. This was achieved by filling them with a suitably specified epoxy resin. This resin offered the additional advantages of lubricating the joint during the assembly of the segments, avoiding the risk of points of hard bearing between segments in the event of less than perfect match casting and eliminating the inconvenience of rainwater dripping through the top slab or side cantilevers in service. However, in parallel to its advantages, this resin introduced an additional element of cost and risk into the construction; the integrity of the deck depends on the correct specification and use of the resin.

The end faces of the segments are lightly sand-blasted to remove surface laitance to optimise the bond of the resin. However this sand-blasting must be carefully controlled, as if it is overdone, the perfect match of the two surfaces will be lost.

The two-part resin requires mixing immediately prior to its application. Generally, the two parts have different colours so that it is clear when mixing is complete. It

Figure 14.18 East Moors Viaduct: segment transport by low loader (Photo: Benaim)

has a limited open time within which the two segments must be brought together, typically of the order of 40 minutes from start of mixing. This open time depends on the ambient temperature, and when a large variation in temperature is likely to be experienced over the deck-erection programme, it may be necessary to specify different formulations for each season. When the temperature falls below some critical value, the resin will 'freeze' and although it may appear to be hard, will not be cured. When the temperature rises once again the resin recovers its fluidity and continues to cure. However, the segment shear keys may not have adequate shear strength to carry the bridge deck that has been erected in the meantime.

Resins vary in their sensitivity to moisture, and it must be made clear in the specification whether precautions are to be taken to avoid contamination in the event of rain falling directly on the end faces of the segments or on the freshly applied resin. If the bridge is being erected on a gradient, it will be necessary to stop rainwater flowing from the deck over the joint surface.

The designer must ensure that the resin is a high quality and proven product, that mixing and application procedures are clear, and that its quality and use is monitored on site. A model specification for epoxy resin used for the erection of bridge decks has been prepared by the FIB [2].

Some formulations of resin require it to be applied to both surfaces, while others should be applied to one surface only. The resin is normally applied to the concrete face by gloved hand, Figure 14.19. Serrated combs are sometimes used to attempt to control the quantity of resin applied. However, generally the erection team learns quickly to judge this quantity by eye. 'O' rings are placed around the ducts to avoid resin squeezing into them. These rings must either be made of very soft foam which

Figure 14.19 STAR: application of resin (Photo: Benaim)

will compress into the thickness of the epoxy joint, about 0.5 mm, or must be placed in circular recesses.

In order to create intimate contact between the resin and the concrete, the two surfaces must be compressed within the resin's open time. This is achieved by applying a temporary prestress that may be mobilised quickly (*15.5.3*). The minimum compressive stress to be applied is normally specified as 0.15 MPa. The viscosity of the resin and the amount and nature of any filler must be such that the resin will flow under this pressure to achieve an even thickness of approximately 0.5 mm. However, when erecting segments in cantilever, it is usually impossible to arrange for an even stress over the height of the segment during the whole of the resin-curing period.

It should be noted that the compressive stress on the resin must be calculated by simple engineer's bending theory. The author has seen engineers check the stresses during erection using finite element analysis, which demonstrated that there were zones of tension close to the cantilever tips. They then concluded that some temporary stressing must be applied at the extremity of the cantilevers to compress the resin. This is of course nonsense, as it is not elastic concrete stresses that are being checked but pressure on the viscous epoxy resin.

If a long cantilever is being erected, a variation in the thickness of the glue line of say 0.2 mm over the height of each segment would cause a significant error to the vertical alignment of the cantilever tip. If the vertical error at the forward end of each segment due to an uneven glue line was δ, the error Δ at the cantilever tip would be:

$$\Delta = n(n-1)\delta/2,$$

where n is the number of segments in the cantilever. Some simple tests should be carried out on the fresh resin to determine the effect of different intensities of pressure on the thickness of the glue line. These tests should use concrete slabs that represent the thickness of the members of the box section, and which have a surface that has been sand-blasted to the same degree as the segment faces. Once a curve has been drawn of the thickness of the glue line against pressure, the casting geometry may be corrected for this effect. In the author's experience of such tests, they demonstrate that the thickness of the line reduces with pressure up to some limiting value, and that from then on further increase in pressure has no effect.

When the temporary prestress is applied, any excess resin will squeeze out of the joint. It is important that it does squeeze out, as this is the only evidence that the joint is full. However it creates problems of appearance, as it forms a ragged fringe beneath horizontal joints, and runs on vertical joints. After a few hours curing, the resin achieves a Cheddar-cheese like consistency when it is easy to remove with a blade; too soon and it smears on the concrete surfaces; too late and it must be chiselled or ground off.

On one major contract known to the author, the designer thought he could overcome this problem by specifying that the resin should not be applied to a band a few centimetres wide along the perimeter of the section, and by limiting the thickness of resin applied. The result was a series of partially filled joints though which one could see daylight, and an expensive contract for remedial grouting.

It is essential to include in the cost estimate for the works the measures necessary to remove excess resin that has squeezed out of the joints. This may be achieved from scaffolding, or may require a travelling gantry that follows the segment erection.

Resin is also likely to squeeze into the prestress ducts, and if allowed to harden would impede the threading of the tendons. A suitable tool should be pushed down each duct to flatten the liquid resin against the walls of the duct.

If the match-cast faces are damaged prior to the erection of the segments, concrete or mortar repairs must remain two or three millimetres low so that they do not compromise the accuracy of the match casting. When the segments are erected, an extra thickness of resin may be applied to this area, in the hope that the deficiency in concrete will be made up by the resin. However, the cured resin has a lower Young's modulus than concrete, and also creeps under sustained load. As a result, the damaged area is likely to have a lower compressive stress in service, raising the average stress on the surrounding concrete.

14.3.16 Dry joints

When the deck is prestressed with external tendons, it is no longer necessary to waterproof the joints between segments with resin, and they may be erected dry. The lack of lubrication of the joint during erection requires that the plant used for bringing the segments into contact must be capable of fine adjustment of the segment position if damage to the matching faces is to be avoided. It has been reported that some minor spalling due to hard contact has taken place on such dry joints.

The system is less tolerant of damage to the matching faces. Damage that does not reach the edges of the concrete cannot be repaired, and will result in a loss of effective section, unless the complete joint is grouted up after erection. Damage that appears at the edge may be repaired by dry packing or similar techniques after erection.

It is necessary to create a waterproof seal across the width of the top slab to avoid rainwater dripping through. This may consist of a resin filled groove, a grouted duct within the thickness of the slab or other methods that may be devised.

One of the principal benefits of omitting resin is that temporary prestress is not necessary, simplifying and speeding up the erection process. Dry joints are normally used on relatively short segment runs, such as statically determinate spans. However, some authorities refuse the omission of the resin, particularly in areas subject to freeze–thaw cycles.

14.3.17 Steering during erection

If the calculation of the alignment and the casting of the segments have been correctly carried out, it should not be necessary to steer the segments. However when things go wrong, such steering may avoid the need for more expensive and time-consuming remedial works. It consists of making small adjustments to the thickness of the epoxy joint, and of course is only possible on resin-filled joints.

The first requirement is to survey the deck during erection to a high degree of accuracy. If errors in the alignment are to be remedied, they have to be detected very early in the erection of a span so that the small angular corrections possible have a significant effect.

The most trouble-free method is to vary the viscosity of the resin across the joint so that the same assembly pressure gives rise to a different thickness of glue line. However, only very small adjustments may be made by this method. For instance, one deck known to the author was being erected in balanced cantilever. It was discovered that the cantilevers were drooping below the predicted alignment, possibly due to a mathematical error. It was calculated that if all the epoxy resin joints were 0.5 mm thicker at the intrados than at the extrados the droop would be cured. Experiments were carried out on fillers to the epoxy resin that increased its viscosity by the requisite amount. This modified resin was applied over the lower half of the segments, curing the droop.

It is also possible to pack part of the joint with layers of fibreglass to achieve an increase in thickness from the normal 0.5 mm to a maximum of 3–4 mm. The fibreglass layers are tapered out to avoid sudden steps in the resin thickness, Figure 14.20. Although feasible, this method seriously slows erection, and greatly increases the risks of voids in the resin, allowing grout transmission from one duct to another as the tendons are pressure grouted.

Varying the joint thickness has been managed by using shims sized to transmit an acceptable stress to the concrete during erection. These shims may be made of hardened resin, which will not affect the final distribution of forces in the section. However, such shims stop the fluid resin from being compressed by the temporary prestress, which may compromise good adhesion to the concrete. The author has no experience of this method of steering, but correctly designed it may be useful. Shims made of steel must not be used as they have a much higher stiffness than resin, and in service the prestress compression in the deck will transfer to the shims, causing local high compressive stresses with risks of spalling or splitting of the concrete.

Figure 14.20 Byker Viaduct: preparing fibreglass steering packs (Photo: Robert Benaim)

14.3.18 Remedial works to the alignment

Where alignment errors cannot be corrected by steering, remedial works are necessary. It is sometimes possible to create one or more cast-in-situ joints in a span to correct its alignment. The joint is usually 100 mm thick, and is made with small aggregate concrete. It is of course necessary to mobilise some special falsework to hold the adjacent segments rigidly in place. If the prestressing cables are internal, it is also necessary to find a way of connecting the ducts across the joint. These joints will increase the length of the precast span, and it may be possible to take this out in planned cast-in-situ stitches, such as at mid-span for a deck erected in balanced cantilever.

In some decks that consist of a long run of counter-cast segments, such as long-span cable-stayed decks, it has been considered prudent by some designers to include the provision for one or two such cast-in-situ joints which allow errors to the alignment to be corrected. If all goes well, they may be eliminated, widening slightly the cast-in-situ mid-span closure.

One remedial measure sometimes proposed is to replace the 100 mm concrete joint by a poured or injected resin joint, centimetres rather than millimetres, thick. This is not recommended, as the resin has a much lower Young's modulus than that of concrete, and creeps significantly when subject to sustained compression or shear.

When it is not possible to adopt cast-in-situ stitches to correct the alignment, it is necessary to recast segments. This generally involves setting up the two outer segments of a run of three, and then casting a new segment between them. This is a difficult procedure, needing very careful surveying and execution, and is very much a last resort.

15

The construction of girder bridges

15.1 General

There is a close link between the designer's choice of bridge deck type and the most appropriate method of construction. In general, solid and voided slabs and twin rib decks are appropriate for the cast-in-situ construction of complete spans on falsework, while box sections are most easily cast in-situ in short sections, or precast. One reasonable rule of thumb is that if a deck is to be cast in-situ on falsework it should be possible to pour a complete span in one continuous operation.

There is also a close link between the method of construction of a concrete deck and the detailed design of the prestress. For instance, if a deck is designed to be built in balanced cantilever, a change in the method of construction would require a redesign of the deck. For this reason, the descriptions of the prestress layouts and of the various methods of building decks are combined in this chapter.

The two principal components in the cost of building a bridge deck (apart from the material content of the deck) are labour and plant. They are closely interdependent as the cost of labour may be reduced by investing in plant to automate the construction process. The balance between the investment in plant and the reliance on labour depends principally on the scale of the project. The influence of scale on the choice of the type of bridge deck and the method of construction is covered in Chapter 16.

15.2 Cast-in-situ span-by-span construction of continuous beams

15.2.1 General

Span-by-span construction is applicable to decks with spans that lie generally between 20 m and 45 m. Due to the difficulties of casting boxes in-situ, decks are most commonly solid or voided slab or twin rib, and are usually continuous.

However, some contractors have special skills in casting boxes in situ. The sharply curved 50 m spans of a series of viaducts on A8 motorway in southern France, were built at a rate of a span every three weeks on an articulated gantry, and were steam cured in-situ, Figure 15.1. German contractors have great experience in building box section viaducts span-by-span using self-launching gantries. For instance, the bridge carrying the A14 autobahn over the Ahr Valley, built in 1975, used a 2,000 t gantry to cast in-situ spans up to 106 m long, at a rate of a span in 14 days [1].

Figure 15.1 Autoroute A8, France: span-by-span construction of box decks (Photo: Robert Benaim)

Bridges built by this method may be only three spans long, or may extend for many kilometres. Consequently, the falsework adopted varies from the simplest to the most highly mechanised self-launching rigs. A disadvantage of the system is that it is essentially linear; it is difficult to miss out spans or move to a second construction front.

15.2.2 Prestress layout

The construction joint for span-by-span construction is normally situated at about 0.25 of the span. Whereas the optimum for minimising creep and reducing the uncertainty on the final self weight bending moments (*6.21*) is 0.22, the tendons are still too high at this position, and a short distance further into the span makes the arrangement of the prestress anchors easier.

The cables shown in Figure 15.2 (a) are each one span long, and they are all coupled at the construction joint. The disadvantages of this arrangement are as follows:

- The coupler sleeves must swaged on and the tendons coupled before the concrete of the next span may be cast, taking time on the critical path.
- The concrete must gain substantial strength before all the tendons in a span may be stressed, delaying the construction programme.
- Stressing all the tendons for each span is time consuming.
- The space necessary to house couplers tends to force the cable profile lower than desirable.

- Couplers are effectively buried dead anchors where the tendon cannot be changed in the event of a fault. Consequently the stressing force of the cables must be reduced in accordance with good practice, increasing the weight of the prestress steel.
- Couplers have a higher risk of failure than normal anchorages. Thus coupling all the prestress of the deck at the same point, and relying totally on the mechanical engineering of the couplers and the workmanship of their installation may be considered somewhat imprudent.
- A fault in a coupler leads to long delays to the construction programme, as repair requires major concrete surgery.
- The couplers create large voids in the webs at a section of relatively high shear force. Although they will be filled with grout, this will be weaker than the concrete. In general, the webs must be locally thickened to maintain sufficient parent concrete, wasting material and complicating the shutter.
- The couplers are more expensive than two standard anchorages.
- Finally, the cables will be only one span long, which is uneconomically short.

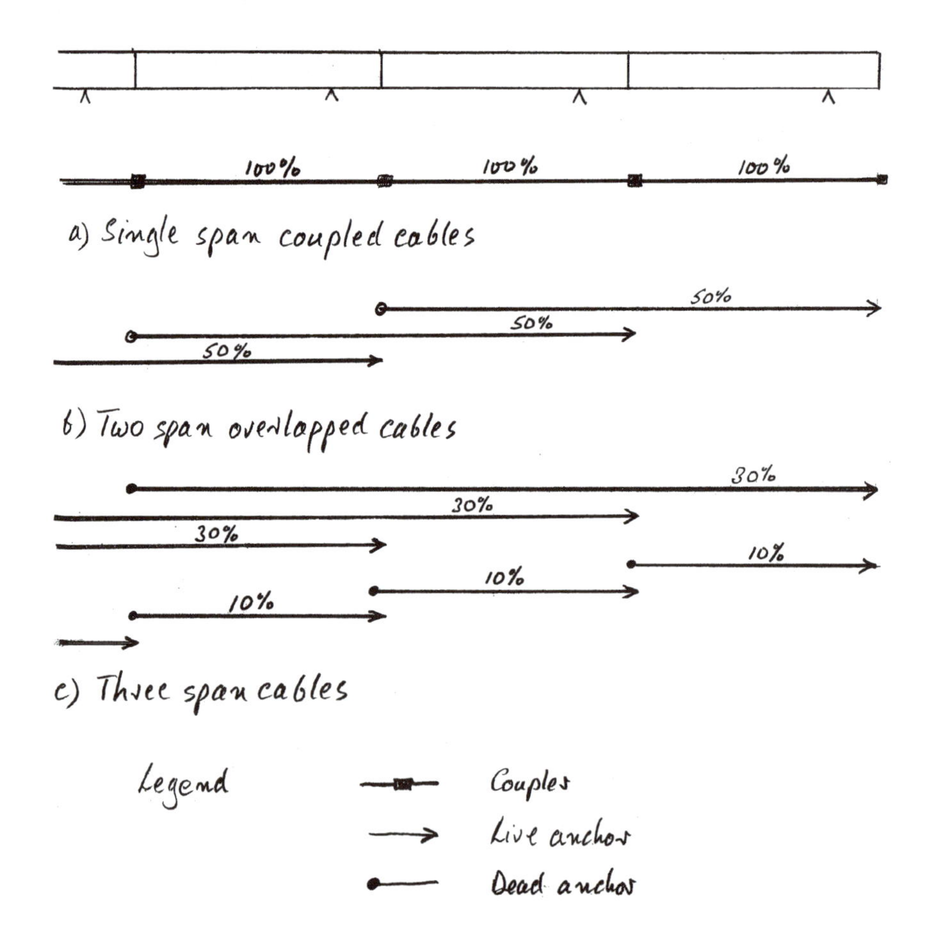

Figure 15.2 Cable arrangements for span-by-span construction

The principal advantage of this arrangement is that it saves the consultant time in the drawing office.

In Figure 15.2 (b) the tendons are two spans long. Thus, at each construction stage, just half the tendons are stressed. This is in most cases enough to carry the self weight of the beam and allow the falsework to be struck. It also allows a quicker construction sequence, and longer, more economical cables. When the self weight may be carried by less than 50 per cent of the tendons, it is possible to extend this concept to cables that are three spans long. In general, stressing only one third of the tendons will be insufficient to carry the self weight of the deck, and it will be necessary to supplement the first stage prestress by additional single-span cables, Figure 15.2 (c).

At the construction joint, the stressed tendons may be coupled, which involves some of the disadvantages quoted above, or may be overlapped. Generally, overlapping tendons are stressed from the front end (the construction joint) with dead anchors at their trailing end as dead anchor access pockets are smaller and less disruptive than live anchor pockets. If the deck has pier diaphragms, live or dead anchors may be located at the junction of the diaphragm and the webs. Figure 15.3 shows alternative ways of arranging the trailing tendon anchor without using couplers. Special precautions must be taken to waterproof top surface anchor pockets, as described in *5.16.2*.

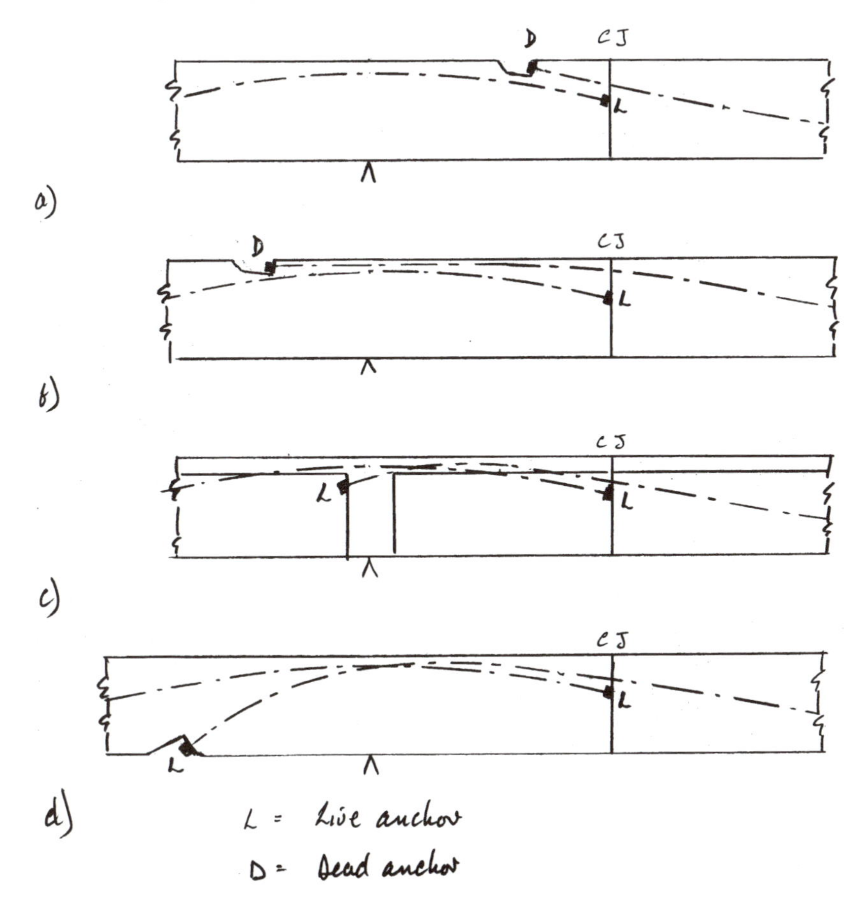

Figure 15.3 Options for the trailing prestress anchors

Finally, for bridges that are less than about 150 m long, the entire deck may be built span-by-span using only half the tendons, the remainder being threaded from end to end.

15.2.3 Typical construction details

When a falsework truss is used for construction, its rear end may be supported on the pier foundation, or it may be suspended from the end of the bridge deck, Figure 15.4. Clearly if the latter arrangement is used, the load applied to the cantilever will affect the self-weight bending moments of the deck. This will be more significant if the falsework has no intermediate supports, but spans freely to the next pier foundation.

If the falsework spans from pier to pier, the weight of the fresh concrete will cause it to deflect at the location of the construction joint. Care has to be taken to avoid a step in the concrete or leakage of the shutter at the joint.

When only half the tendons are stressed, the deck will deflect downwards under its combined self weight and prestress, and the falsework must be designed to be struck when under load. If all the tendons are stressed the deck will deflect upwards, and so will tend to lift itself off the falsework. However, adjacent to the pier there may be a short length where the deflection under prestress is downwards, even with all the cables stressed.

If the falsework is flexible and has no intermediate supports, the striking must be planned with care. Before being prestressed, the weight of the deck is carried by the falsework which deflects downwards. As the cables are stressed, the deck and falsework together will deflect upwards; the weight of the deck is now being shared between the deck itself and the falsework. If the downwards deflection of the falsework under the self weight of the deck is large compared with the upwards deflection of the deck under prestress, the majority of the weight of the concrete will still be carried by

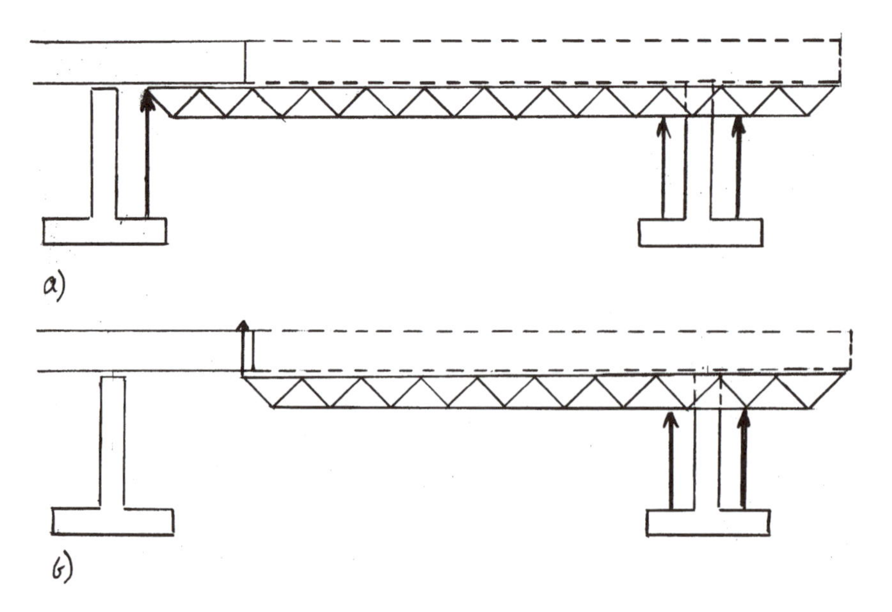

Figure 15.4 Support of falsework truss for span-by-span construction

the falsework. Consequently the bending stresses in the deck due to the prestress are combined with only a fraction of the deck self weight, and dangerous tensile stresses may exist on the top fibre at mid-span or on the bottom fibre at the pier. In some cases it is necessary to stress the cables and lower the falsework in a carefully co-ordinated sequence.

If the striking of the falsework is delayed beyond 36 hours after completion of casting, it is good practice to partially stress some of the tendons to provide some tensile strength to the deck, which is vulnerable to cracking. Usually, an average compressive stress of 1 MPa is adequate.

15.2.4 Stressing cables early

A reinforced concrete beam must be supported by falsework for some time before it may be allowed to carry its own weight, typically 14 days. Adequate concrete strength is required principally to generate bond with the reinforcement and to limit the deflections of the beam. One of the advantages of prestressing is that beams may be struck much earlier as their strength does not rely on bond and self-weight deflections are cancelled out. In the construction of a long viaduct, for instance, early striking of the falsework shortens the time required to build each span with benefits to the construction programme. Consequently it is not unusual for prestressed cast-in-situ decks to be stressed and struck from their falsework 36 hours after casting the concrete, or even earlier in special circumstances.

When only a half or fewer of the tendons are stressed at each construction stage, the maximum compressive bending stress in the concrete under self weight plus prestress is likely to be less than 4–5 MPa. This requires a concrete strength measured on cubes of 20–25 MPa, which, in a temperate climate, can usually be achieved with a 50 MPa concrete within 36 hours.

However, the concrete behind the prestress anchors generally needs a cube strength of 35 MPa before the cables can be fully stressed. To allow early striking of the falsework, the steel base plates of the anchors may be oversized, and the edge distances increased from the standard. Alternatively, the stressing anchors may be encased in a concrete end block that is precast well in advance of the deck construction, Figure 15.5. This also allows the area of the deck that is most congested with reinforcement to be precast under optimum conditions. The end block, which has a high elastic modulus, bears on immature concrete with a lower modulus, and should be designed as a beam on an elastic foundation. The stresses in the parent concrete behind the end block must be checked, taking account of any eccentricity of the stressed anchors.

A bridge deck may take several hours to pour. Consequently, the designer should specify that pouring should commence at the location where he requires the highest strength, to give as much time as possible for the concrete to gain strength, and test cubes must be taken from this early concrete.

The designer should be cautious about loading immature concrete as the strength will be variable across the deck and the test cubes may not be entirely representative of the parent concrete. When the cube strength is less than 25 MPa and the concrete less than 36 hours old, the working bending stresses should be kept to less than 20 per cent of the cube strength.

Figure 15.5 Hong Kong MTR Contract 304: precast end block (photo: Robert Benaim)

15.2.5 Rate of construction

The falsework may be a simple scaffold, a self-launching falsework rig, or something in between. Figure 12.16 shows a simple falsework for the construction of the three Doornhoek bridges each of which had six spans. A more sophisticated falsework is shown in Figure 12.2, where the 1,200 m long twin rib, twin-track railway bridge of Contract 304 of the Tsuen Wan Extension of the HKMTR was built at a rate of one 32 m span every two weeks. It consisted of a platform carried by girders which were supported on the end of the completed deck, an intermediate support, and the forward pier. The intermediate support was founded on a precast concrete pad foundation, despite the poor ground. This was possible as the weight of concrete was only applied for a short time before the span was prestressed and became self-supporting. In similar circumstances the pad foundation may be preloaded to further reduce the risk of settlement. The shuttering for the deck consisted of a series of boxes assembled on the platform. This type of falsework can adapt easily to vertical and horizontal curves. In order to achieve a rate of two weeks per span, it was necessary to stress enough cables to render the span self-supporting some 36 hours after casting the concrete. An initial prestress was applied 24 hours after casting to render the deck resistant to cracking. The prestress anchors were encased in the precast end blocks shown in Figure 15.5.

The Viaduc d'Incarville with 15 m wide, 2.5 m deep twin rib decks of 45 m span, Figure 12.3, was built span by span as part of the Autoroute de Normandie. The author, when with Europe Etudes, designed and patented the self-launching falsework rig that carried the shuttering for the deck. The rig consisted of an overhead truss weighing some 250 tons and an under-slung steel girder weighing approximately 50 tons. The shutters for the side cantilevers and for the outside face of the ribs were suspended directly from the truss, while the shutter for the slab between ribs and for the inside face of the ribs was carried by the steel girder that was suspended from the truss during concreting, Figure 12.10. The reinforcing cage for each rib, complete with prestressing tendons, was prefabricated on the bridge deck behind the construction head, delivered to the rig by a purpose-designed trolley and lowered into the steel shutters. The reinforcement of the slab was fixed in-situ. The pier diaphragms were omitted from the first stage of construction and built subsequently, which allowed the launching of the internal formwork. The lower girder was launched first, and then acted as a bridge to carry the main falsework, Figure 15.6. The rig built the bridge at an average rate of a span every two weeks. As the bridge consisted of only 16 spans, there was no pressure to increase the rate of construction. However, this type of mechanised falsework has the potential to build much more quickly.

It is undoubtedly possible to construct twin rib decks in-situ at a rate of two spans per week, rivalling precast segmental for the longest viaducts. In order to achieve such a rhythm, the reinforcing cage for an entire span complete with post-tensioning tendons, weighing some 50 tons for a typical 12 m × 35 m span, would be prepared behind one abutment, transported over the completed deck, and lowered into the shutter. The casting and compaction of the 250 m^3 of concrete would take about 7 hours using two pumps, starting at the forward end. This would allow time for the prestressing ducts to be coupled to the previous span, and the reinforcement at the construction joint to be arranged. It would also give a few additional hours of maturity to the concrete beneath the prestress anchors. Thus in one twelve-hour shift the reinforcing cage could be placed and the concrete cast. The next evening, when the concrete beneath the anchors had attained a strength of 25 MPa, half the

Figure 15.6 Viaduc d'Incarville: falsework truss being launched (Photo: Robert Benaim)

prestress tendons would be stressed, completing the stressing for the previous span and rendering the new span self-supporting. On the third day, the formwork would be struck, the falsework launched forwards to the next span, the steel shutters prepared to receive the new reinforcing cage and the stressed tendons would be grouted.

Such a technique would be transitional between precasting and casting in-situ, and would use some of the techniques and site organisation used on the Poggio Iberna Viaduct described in *15.9.1*.

15.3 Precast segmental span-by-span erection

15.3.1 General

This method of construction is well adapted to long viaducts with spans that generally do not exceed 50 m. The decks may be statically determinate or continuous. The segment joints may be glued with internal prestress or dry with external tendons.

Generally the spans are erected on under-slung falsework consisting of relatively simple girders placed beneath the side cantilevers of the box section deck, such as in the East Rail Viaduct in Hong Kong, Figure 15.7. This minimises the projection of the girders beneath the bridge soffit. If the bridge deck rests on cross-heads or portals that would impede the underslung girders, if there are deck bifurcations or if highway clearance diagrams make it impossible for falsework to protrude beneath the deck soffit, overhead gantries may be used that suspend the segments during their assembly, such as that designed by Benaim for bridges on the Country Park Section of Route 3 in Hong Kong, Figure 15.8.

Figure 15.7 East Rail Gantry (Photo: VSL, www.vsl.com)

Figure 15.8 Route 3 Gantry (Photo: Benaim)

15.3.2 Statically determinate spans with internal tendons and glued joints

One of the most economical options for a long viaduct is to design it as a series of statically determinate spans. For highway bridges, groups of up to 10 spans may be linked together at the top slab to minimise the number of expansion joints. Although such decks need up to 20 per cent more prestress than continuous decks, their simplicity and speed of erection overcome this disadvantage. Once a span is complete the construction of the following span may start immediately; creating continuity takes up at least one additional day per span. The span length for such decks should in general not exceed 45 m, with a span/depth ratio preferably of about 14–16 for UK loading.

On the 10 km long West Rail Viaducts which were designed by Benaim and are typical of the method, underslung girders, very similar to that shown in Figure 15.7, carry on their top boom a slide track that may be polished stainless steel or rails. The point of support of the deck segments may be some distance from the root of the cantilever, particularly if the deck is curved, and the side cantilevers must be reinforced for this sagging moment, Figure 15.9. The segments are placed on sledges that slide or roll on the track, Figure 15.10. The sledges are each equipped with lifting jacks and small lateral jacks that allow precise positioning of the segments once they are brought into contact. The lifting jacks must have sufficient travel so that they may provide a true alignment to the deck despite the deflection of the gantry.

All the segments for a span are loaded onto the gantry, so that it takes up its deflection. A gap of about 200–300 mm is left between the first segment and the remainder for the application of the epoxy. Once the resin has been applied, the two first segments are brought together and clamped using temporary prestress, the procedure being

Figure 15.9 West Rail, Hong Kong: detail of gantry (Photo: VSL, www.vsl.com)

Figure 15.10 West Rail, Hong Kong: detail of sledges (Photo: VSL, www.vsl.com)

repeated for all segments. As the precise position of each segment is defined by two rigid systems, the sledges and the face of the preceding segment, it is essential to ensure that the force in the sledge jacks has been released, and that the joint is not held open.

When all the segments have been temporarily stressed together, the bearings on the piers are grouted into place, the permanent prestress cables are threaded into their ducts and stressed and the falsework struck, observing the precautions described in *15.2.3*.

West Rail is not entirely typical of such bridges as each span was built into very flexible split piers. Strictly each span was a portal, although the moment restraint offered by the pier leaves was only a small proportion of the free bending moment, and the methods of construction were virtually identical with statically determinate spans.

15.3.3 Statically determinate spans with external tendons and dry joints

Erection of such decks is exceptionally quick and trouble free. The tendons are principally anchored at the end diaphragms, and sometimes at intermediate ribs which double as deviators.

Once all the segments have been loaded onto the gantry and brought into close contact, the ducting for the external tendons must be installed and the tendons threaded. The segments may then be pulled together by the application of a small percentage of the prestressing force, and their successful fit up confirmed. As this force is applied, the segments need to be free to slide minutely sideways on their sledges to take up their final position. Once all the joints are closing successfully, the bridge bearings are grouted into place and the prestress may be completed.

It is essential to leave room at the end of each span for the replacement of the external prestressing tendons. This makes it necessary to adopt a relatively wide column head, with the penalty of increased cost of the piers and a less elegant appearance.

Such decks may be built at a peak rate approaching one span per day although average rates of erection rarely exceed three spans per week.

15.3.4 Continuous bridges with glued joints

The 360 m long, continuous River Lea Viaduct on the A414 Stanstead Abbotts Bypass near London was built span-by-span with glued joints, although it is not entirely typical due to its short spans and shallow deck. The deck area was 9,000 m^2, which is at the lower end of the viability of precast segmental construction. Consequently, the methods of erection were kept as simple as possible.

The design was an alternative to a precast beam deck, and its dimensions were constrained by the original design, Figure 10.4. The 22.5 m wide deck consisted of two adjacent boxes 1.4 m deep, linked by a 300 mm slab. Each 25.7 m long span consisted of six span segments each 3.927 m long and a 2 m long pier segment. The deck was designed with a 150 mm cast-in-situ stitch at the end of the first span segment in each span. The boxes were each carried by a column with a single bearing. The deck was proportioned so that there would be no significant transverse rotation of the boxes under dead load.

The deck segments were cast using the pier segment as the origin, casting the cantilevering segment in one direction, then reversing the pier segment and casting the other five span segments.

The segments were placed onto a self-launching gantry by mobile crane. More typically, the segments for viaducts built by this method are carried to the construction head over the completed deck.

With such a shallow deck, work inside the box was minimised, and the temporary prestressing was located above the top slab, anchored on precast concrete blocks stressed down to the deck, Figure 15.11. Thin plywood was used as a bedding medium between the blocks and the deck, allowing re-use of the blocks. The bars clamping the blocks to the deck were 750 mm long, and as they extend by only 2 mm when stressed, had to be carefully monitored to ensure that the stressing force was not lost as the load was transferred from jack to anchor.

The segments were handled on the gantry by four sledges equipped with vertical jacks running on greased slide tracks. Once a segment had been attached by the temporary bars, it was allowed to cantilever off the previous segment, maintaining a hogging moment in the deck. As a new segment was attached, the previous segment was packed off the girder and the sledges recovered. Although economical in sledges, this method made it essential to align the first pair of segments very accurately, as their orientation could only be changed within very narrow limits subsequently. On larger, more highly automated sites, each segment is carried by three or four sledges equipped with jacks that may be adjusted at any time until the span is struck.

Once all the segments had been assembled, fine adjustments to the elevation of the span were made by jacking the entire girder at the pier falsework, and the bearings on the pier were grouted in. The tendon ducts were coupled across the 150 mm stitch, and the joint cast. When the joint had achieved adequate strength, typically 15–20 MPa, the prestressing tendons could be stressed, the temporary stressing removed and the girder lowered clear.

As the girder was very flexible compared with the deck (*15.2.3*), it was necessary to execute a staged programme of application of the prestress and lowering of the gantry. When the prestress was applied, the contact pressure between deck and gantry reduced firstly at mid-span, throwing more reaction onto the side cantilevers either side of this mid-span zone.

The average speed of construction was of the order of a span per week, with a peak rate of a span in four days. However, this relatively small site was not highly mechanised, and achieving faster construction was not a priority.

The two parallel boxes that constituted the deck were built successively. Once both were complete, they were joined at their top slabs by a 2.5 m wide cast-in-situ stitch. At the time the two boxes were linked, the deferred shortening of the first completed box was proceeding more slowly than that of the second box, Figure 15.12. Consequently the completed deck behaved like a bi-metallic strip, bending sideways, damaging the bearing plinths at the abutments. This effect may be mitigated by equipping the penultimate piers with free sliding bearings. Alternatively, the forces may be calculated and the substructure designed in consequence.

Figure 15.11a Temporary prestressing anchors: River Lea Viaduct, Stanstead Abbotts Bypass, concrete blocks (Photo: Benaim)

Figure 15.11b Temporary prestressing anchors: STAR Viaduct, steel shoes (Photo: Benaim)

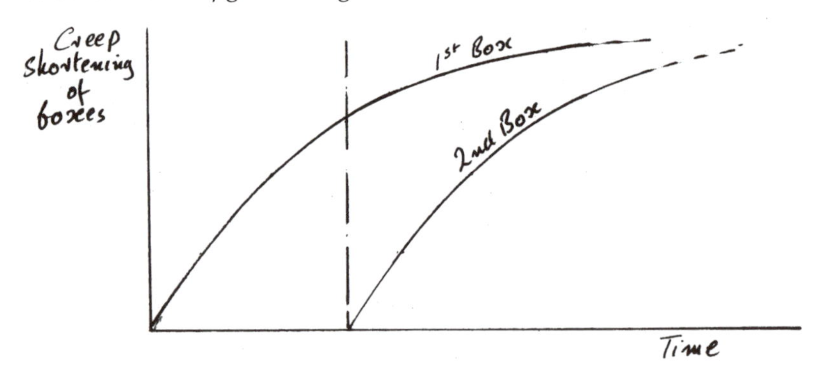

Figure 15.12 Differential creep shortening of boxes

15.4 Cast-in-situ balanced cantilever construction

15.4.1 General

Cast-in-situ balanced cantilever construction is ideally suited to box section bridges of medium or long span, where there is insufficient repetition to justify precasting. The method becomes economical for bridges with a main span of 60 m and above, and remains viable up to the largest span that may be built, currently about 300 m.

Early cantilever built bridges were pinned at mid-span. However, these bridges have in several instances exhibited unexpectedly large creep deflections at mid-span (*7.2.4*). Furthermore, the presence of an expansion joint at mid-span affects the ride of the bridge, and is a maintenance liability. Most modern cantilever built decks are made continuous by casting a mid-span stitch.

15.4.2 Proportions of balanced cantilever built bridges

The method of construction leads to a self-weight moment diagram that is predominantly hogging, and it is clearly economical to provide a greater structural depth over the supports and to minimise the weight of the deck towards mid-span. Furthermore, the construction of the deck in short lengths on a weekly cycle makes it relatively easy to change the geometry of the shutter for each cast. Consequently, most bridges built by this method have variable depth.

The depth of the deck for large bridges is inevitably a compromise between economy of materials, appearance and ease of construction. For spans less than about 100 m, for economy the depth of the deck at the support should ideally be approximately span/14. At this depth, the prestress is economical, the webs do not usually need to be thickened to carry shear, and the bottom slab can also stay at its mid-span thickness. However, the depth of the deck may be forcing the road alignment higher than is desirable, and there may also be consequences for the appearance of the bridge. The support depth may be reduced to span/20, with the consequent greater consumption of prestress, and the greater complication of the construction as the webs and bottom slab will need to be thickened at the supports.

Large bridges generally adopt a support depth of about span/20 to overcome the disadvantages of the great structural depths otherwise attained. For a 300 m span for instance, span/15 would yield a depth of 20 m.

The economical mid-span depth depends principally on the loading code used. With the British code, where the dominant load is a 180 ton vehicle, for bridges up to about 80 m span the mid-span depth should not be less than about span/35. If the deck is more slender, the amount of prestress in the mid-span section becomes excessive, with difficulties in housing and anchoring the tendons. For instance, Benaim's Taf Fawr Bridge has a span of 66 m and depths at pier and mid-span of 4.5 m and 1.9 m; the River Dee Bridge, designed initially by Travers Morgan and re-engineered for tender by Benaim, has a main span of 83 m and depths at the pier and mid-span of 6.1 m and 3.4 m. With most other loading codes it is economical to adopt depths of span/45 or less. As the span becomes longer and the live load less significant, the mid-span section may be thinner. The thinner the mid-span, the more moment is shed towards the supports, which improves the economy of the deck.

If the deck is built into its piers, carried by double piers that provide a high degree of fixity or has short stiff end spans restraining a main span, the mid-span section may be thinner still. For example, the 929 m long Bhairab Bridge in Bangladesh was designed by Benaim to British codes and rests on double piers, Figure 13.6. It has spans of 110 m, and depths of 6 m and 2.7 m at pier and mid-span. Benaim's Kwai Tsing Bridge in Hong Kong, built-in to double leaf piers, Figure 15.13, has three main spans of 122 m, and pier and mid-span depths of 7.5 m and 3 m. One end span was only 60 m long, and was ballasted by an adjacent 36 m span with which it was continuous. The world's longest cantilever span, the 301 m Stolma Bridge in Norway, has end spans of 94 m and 72 m, is 15 m deep at the supports and only 3.5 m deep at mid-span (span/86). This mid-span depth would not be feasible under British live loading.

Constant depth cantilever built bridges rarely exceed a span of 100 m, due to the inherent lack of economy. Generally such bridges have span/depth ratio not shallower than 1/18.

Figure 15.13 Kwai Tsing Bridge, Hong Kong (Photo: Benaim)

The length of the side spans of cantilever built bridges are usually little more than half the main span. This suits the construction technology as explained in *15.4.5*. When the side span is this short, it must be checked that the abutment bearings are not decompressed by factored live loads in the main span. A counter-weight consisting of mass concrete in the last metres of the box may be used to increase the bearing reaction. Alternatively, the deck may be held down by ties which permit the length changes of the deck, but this constitutes an additional maintenance and durability liability. If the deck continues beyond the balanced cantilever spans, the reaction of the adjacent span may be used to hold down the end of the cantilevered span. However, in bridges with very long main spans it may be cost effective to limit the overall length of the deck by curtailing the side spans, which will then need to be anchored down with very substantial forces. This is clearly easier to achieve in Norwegian rock than in British clay!

As described in *7.11* and *7.12*, decks should be built into their piers wherever possible. The principal penalty for building the deck into the piers is the bending moment that is applied to the foundations when the bridge is in service. For most internal piers, where the adjacent spans are equal, this moment is almost entirely due to live loads, the dead loads being virtually in balance. However, short end spans lead to significant unbalanced dead as well as live load moments being applied to the foundations, principally of the first and last pier, Figure 15.14. For the dead load bending moments of the end span and of the first internal span to be equal over the pier, the end span length should be approximately 80 per cent of the internal span. This causes problems in building the unbalanced portion of the end span, and the sagging moment in the end span will be greater than in an internal span. A compromise needs to be found.

With spans above about 150 m, it becomes worthwhile to use different qualities of concrete for different areas of the deck. For instance, the support section may use very high strength concrete, minimising the thickness of webs and bottom slab, although very thin deep webs may start to meet problems of elastic stability. Further out in the span, the concrete will be less highly stressed and the use of more expensive lightweight concrete may be cost effective. Several large bridges, including the Stolma Bridge, were designed in this way.

The Resal effect is described in *9.4.7*. For spans over about 150 m, where the support depth is typically 8–10 m, it is worthwhile considering a trussed start to the span, Figure 9.23.

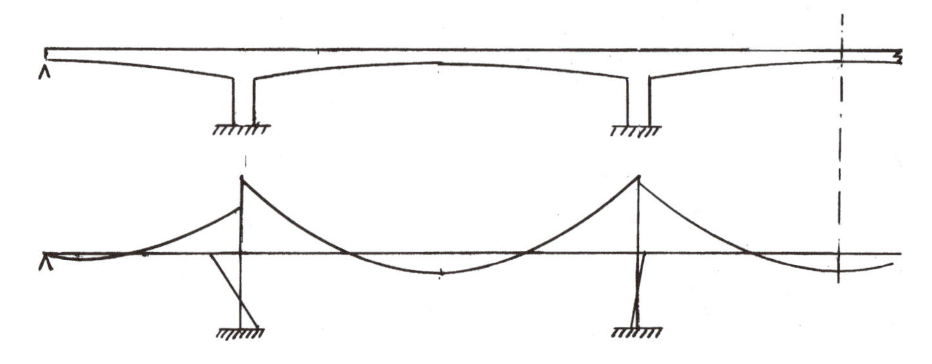

Figure 15.14 Typical dead load moments in deck fixed to piers

15.4.3 Construction of the deck

Typically, a hammerhead at least 6 m long is cast off falsework on each pier. A traveller is then lifted onto the hammerhead and builds a first deck segment. This traveller is then launched forwards, clearing the hammerhead for the second traveller that will build the balancing half span. The pair of travellers then proceeds to build the two adjacent half spans in balanced pairs of segments, Figure 15.15.

As each pair of segments is completed, prestressing cables are threaded and stressed. When the balanced cantilevers are complete, a mid-span stitch, usually 2–3 m long, is cast, using one of the travellers as falsework. Continuity cables are then threaded through the stitch and stressed.

The travellers used to build cantilever bridges are usually simple steel trusses that suspend a platform on which is placed the shutter. They are counterbalanced so that they are stable under their own self weight, and are tied down to the deck to resist the weight of concrete. The main design criterion for the truss is to limit its deflection as the concrete is cast. The bottom slab which is cast first may be cracked by this deflection at its junction with the previous segment as the main weight of concrete is poured. It is common practice to leave a strip some 300 mm wide between the new bottom slab and the previous segment, which is then cast once the webs and top slab are complete.

Each deck segment is normally between 3.5 m and 5 m long. After a short learning curve, it is traditional to cast a pair of segments each week. A reasonable approximation to a construction programme is to assume that the hammerhead takes 6–8 weeks to build, the first pair of segments four weeks, the second pair three weeks, the third and fourth pair two weeks, and one week per pair thereafter.

Figure 15.15 Travellers for River Dee Bridge, Newport, UK (Photo: Edmund Nuttall/Molyneux Photography)

A weekly rhythm has the advantage that concrete curing may take place over a weekend. However this rate of construction is not the best that can be achieved. The economy of this form of construction is dependent on the use of the minimum number of travellers to complete a bridge within a given programme. In some circumstances, speeding up the rhythm to, say, a pair of segments every 4 or 5 days may allow the use of one less pair of travellers.

Construction may be accelerated by prefabricating the reinforcement cage in whole or in part. When this option is adopted, it is advantageous to use under-slung travellers, so that the cage may be lifted in freely, Figure 15.16.

The prestress scheme is usually arranged so that cables are stressed for each pair of segments cast. When respecting a weekly rhythm the cables must be stressed as soon as the concrete has achieved a cube strength of 25 MPa. This is less than the design strength required for standard anchors, but can be achieved by using oversized bearing plates.

The finished cantilever is likely to be deflected downwards, as the self weight downwards deflection is usually greater than the upwards prestress deflection, and always so for spans in excess of 60 m. To obtain a correct alignment pre-cambers are built in during construction. The pre-camber calculations are complicated by the variation of the Young's modulus of the concrete with age. In reality, this causes few problems for spans less than about 100 m, but requires a sophisticated calculation for longer bridges.

Figure 15.16 Bhairab Bridge, Bangladesh: underslung falsework travellers (Photo: Roads and Highways Department, Government of Bangladesh and Edmund Nuttall)

15.4.4 Stability during construction

The stability of the balanced cantilever may be provided by a prop which should be founded on the pier foundation, Figure 15.17, which must be capable of carrying the overturning moments imposed during construction. The props usually consist of two steel stanchions braced together transversally, and, when they are tall, braced off the pier. As the load comes on the prop due to the unbalanced weight of one segment, it will shorten and cause the cantilever to rotate, giving difficulties in the setting out of subsequent segments. It may well be necessary to limit its working stress to control these deflections. The prop is equipped with hydraulic jacks that allow the attitude of the double cantilever to be adjusted for construction of the mid-span stitch and that assist in its unloading and removal.

Concrete props may be used and are usually made of precast circular segments that may be match cast to ease fit-up and may be lightly stressed together. Clearly they would be much stiffer than steel, but are also more difficult to handle.

When a prop is used the deck is normally built directly onto its permanent bearings. The deck support, consisting of the prop and the bearings, must be designed to resist lateral wind loads, including slewing forces when differential wind pressure is applied to the two arms of the balanced cantilever. This slewing resistance may be provided by the two bearings on the pier, although if they are sliding bearings, they will need

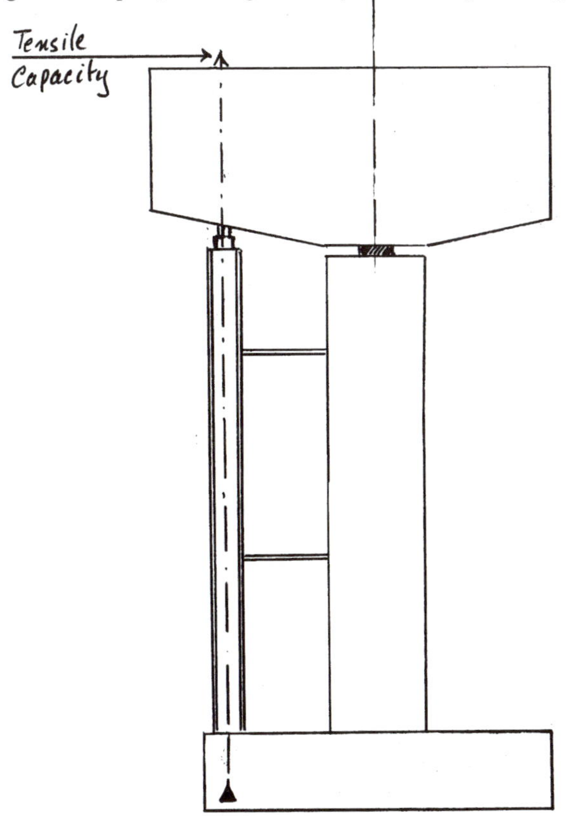

Figure 15.17 Typical eccentric hammer head and prop

to be temporarily locked. Alternatively, the slewing resistance may be provided by one guided bearing and the prop, which would need to have adequate transverse strength and stiffness.

If the pier itself is adequate, the deck may be temporarily stressed down to the pier head, when it is usually carried on four temporary packs and placed on its permanent bearings after completion, Figure 15.15. These packs may consist of reinforced concrete that has to be destroyed once the span is complete, which is time consuming, or of sand jacks. These are short steel cylinders filled with dry sand, with a bearing plate resting on the sand. To lower the deck onto the permanent bearings, the sand is encouraged to flow out of a small hole at the bottom of the jack. This gives a very safe and controlled lowering operation without the need for hydraulics. At one time there existed proprietary resin packs which could be softened for removal by passing an electric current through a built-in electrical resistance. They may still be manufactured.

Stability falsework may be avoided altogether by resting the deck on double piers, Figure 15.16, by building it into a sufficiently robust pier or onto piers consisting of two leaves, Figure 15.13 and Figure 15.31, all of which provide a stable base for cantilever construction, needing no primary falsework.

When stability falsework is required, it must be designed to accommodate an out-of-balance moment due to the successive construction of complete segments. It is not practical to co-ordinate the casting of the two segments of a pair, and this should not be attempted. It must also be checked for reverse moments when the cantilever is nominally in balance.

The stability falsework will be more economical if the construction sequence maintains the out-of-balance moment always in the same direction, requiring a prop on only one side of the pier rather than on both sides. However, the reverse moments, when the cantilever is nominally in balance, may put small tensile stresses into the prop, which will need to be designed in consequence. In order to fine tune the stability falsework, the hammerhead may be made slightly eccentric, so that when both segments of a pair have been cast, there is still a small out-of-balance moment in the direction of the leading segment, compressing the prop, Figure 15.17. If stability is provided by prestressing the deck down to the pier head, the stressing may be eccentric, with resulting savings.

The falsework needs to be designed to ensure the stability of the cantilever during construction for three principal load cases. (Codes of practice may have different conditions that must be met.)

a) In normal construction, out of balance:
   - one segment out of balance;
   - the traveller advanced onto the newly cast leading segment. It is possible to specify that the leading traveller will not be moved forwards until the trailing segment has been cast. However this is a constraint on the contractor's working method and should only be specified if there are significant benefits, such as making it possible to anchor the deck to the column head, omitting the temporary props;
   - construction load on the deck on the side of the leading segment. This should include at least 40 kN at the cantilever end for stressing equipment, jacks, compressors and men, plus a small line load on the deck, say 2.5 kN/m. The

drawings must make clear the areas of the deck where materials may be stored, and the construction loads assumed;
- a tolerance on the weight of the deck, leading to an additional load of 2.5 per cent of the self weight on the side of the leading segment;
- the overturning effect of the vertical component of normal construction wind load, including any differential effects on the two arms of the cantilever to allow for gusting.

b) In normal construction, nominally in balance:
- construction load as above on the side of the trailing segment;
- a tolerance of 2.5 per cent on the weight of the deck on the side of the trailing segment;
- the traveller on the trailing side moved forwards;
- the overturning effect of normal construction wind as above.

c) Disaster scenario:
- the complete loss of a segment on either side;
- the dynamic effects of this loss. The loss of a segment by failure of the supporting shutter during casting is unlikely to be an impulse and an upwards dynamic force of 25 per cent of the segment weight is suggested. The holding-down bars of the traveller should be designed with sufficient redundancy to make it inconceivable that it should fall with the segment;
- the 2.5 per cent tolerance on weight as above;
- construction load as above;
- the overturning effect of the maximum credible wind.

For cases (a) and (b), the falsework providing stability should be working at normal stresses or load factors, with up to 25 per cent overstress allowed on foundations (but not on falsework) for the effect of wind. Any tension on bored foundation piles should not exceed their weight, suitably corrected for the presence of water, and suitably reinforced.

For a deck prestressed down to the pier, all temporary bearings should remain compressed. However, it must also be checked that there is no danger of the deck slewing under the effect of wind, which could be a risk if the two temporary bearings on one side are close together and the temporary bearings on the other side are barely compressed.

For load case (c), the falsework, column and foundations must be checked at the ULS for overturning and for slewing, with load factors of 1.4 on wind, and 1.5 on the weight of the traveller and on construction loads.

When the deck is built into the piers, the load cases (a) and (c) must be considered; load case (b) is generally not relevant.

The weight of the travellers should be estimated with caution as the majority consists of formwork, access platforms and other 'secondary' items which are often inadequately allowed for early in the project.

15.4.5 Building the end spans

End spans should ideally be little more than half the length of the main span. Thus when the balanced cantilever is complete, the end span traveller is at the face of the abutment. The abutment segment can then be cast on shuttering carried by the abutment itself, Figure 15.18 (a).

There are several strategies that allow longer end spans to be built. As the stability falsework is designed to resist the effects of one segment out of balance, it is usually possible to complete the cantilever construction with one additional segment in the end span. If the end span needs to be still longer, falsework may be cantilevered forwards from the abutment foundation, Figure 15.18 (b).

It is also possible to build the unbalanced end of the end span using the end span traveller. Once the point of balance with the main span has been reached, a prop is installed and the traveller continues building, Figure 15.18 (c). Generally, two or three segments may be built before another prop is needed. If the ground is subject to settlement, the props should be equipped with levelling jacks. As permanent top fibre tendons are not required at this section, the construction bending moments can be carried in reinforced concrete action, or by using temporary prestressing bars. Although this is probably an economical way to build a long end span, frequently the programme precludes it, as the traveller is needed elsewhere on the bridge.

The last resort is to build a substantial length of the end span on independently founded falsework, Figure 15.18 (d). In order to control its settlement relative to the abutment, it should consist of beams resting on a few towers equipped with levelling jacks. The one advantage of this method, when the construction programme is of the

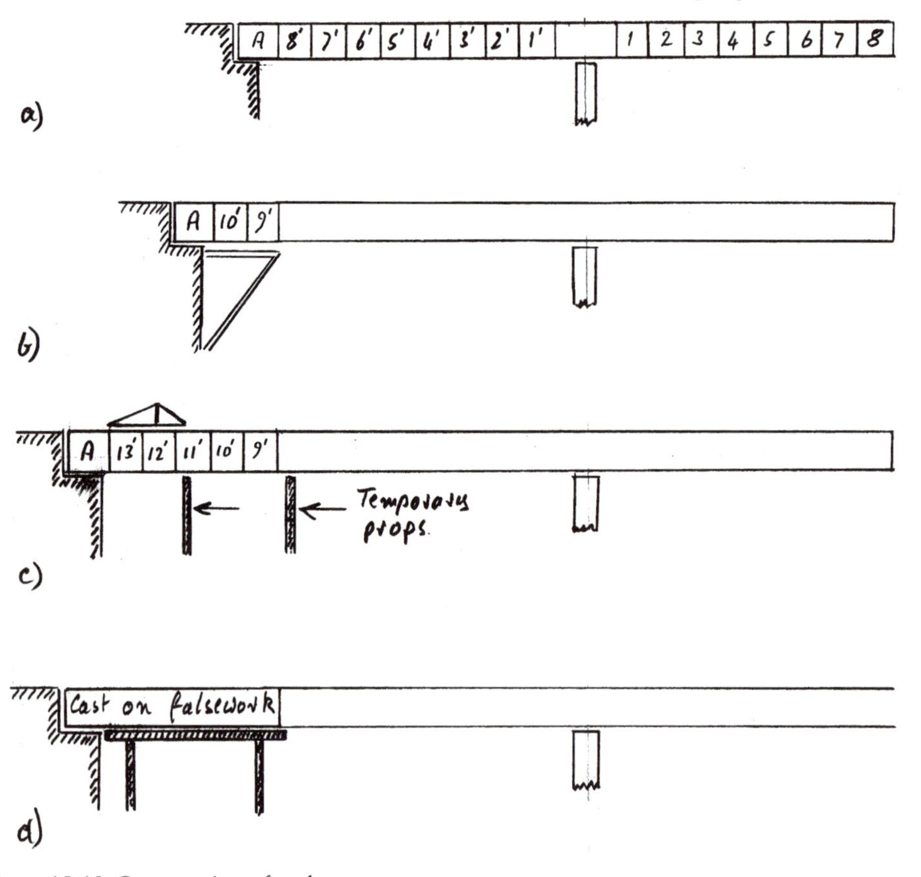

Figure 15.18 Construction of end spans

essence, is that a significant section of the deck may be built off the critical path with resources additional to the balanced cantilevering teams.

15.4.6 Bending moments in cantilever built bridges

The self-weight bending moments in a cantilever built bridge are entirely hogging at the end of cantilever construction, and are balanced by the sagging moments of the construction prestress cables. The weight of the mid-span stitch concrete is carried by the statically determinate cantilevers, and thus produces only hogging moments. When the stitch has hardened sufficiently, the deck is no longer statically determinate; it has become a continuous structure. As the statical definition of the structure has been changed, creep will gradually modify the self weight and construction prestress moments, as explained in *6.21.2*. As creep is a slow process the final bending moment diagram will only be attained after many years.

The removal of the stitch falsework is equivalent to applying an upward force equal to its weight producing hogging moments at mid-span and sagging moments at the supports. Similarly, if a prop has been used to provide stability, its removal equates to applying a force equal and opposite to the prop reaction.

As bottom fibre continuity prestressing cables are stressed, they will produce sagging parasitic moments. Subsequently, deck finishes and live loads are applied to the continuous structure, producing further hogging moments at supports and sagging moments at mid-span.

15.4.7 Prestress layout

The cantilevering prestress cables, called Stage 1 cables, are usually installed in each pair of segments as they are built. As the bending moments close to the pier are changing rapidly, it is frequently necessary to anchor several tendons in each of the early segments. Closer to mid-span it is possible to install them in every other segment, with the un-prestressed segments being carried in reinforced concrete action. The tendons may be anchored in the end face of the webs, in the end face of the thicker parts of the top slab, or in blisters inside the box. Tendons anchored in the face of the webs have the advantage that they may be deflected downwards to help carry the shear, Figure 15.19 (a). This allows the web to be made thinner, saving precious weight. However, their size may be limited by the width of the web, and they have to be stressed before the next segment may be cast. Tendons anchored in blisters may be stressed at any stage in the construction sequence and are not limited in size, although it is good practice to use only one size of tendon in a deck, except for large projects and special circumstances.

Stage 1 prestress is sometimes provided by bars, which are coupled at the face of each segment. Although simple to install, in the author's experience such a form of prestress is considerably more expensive than tendons; not only do the bars work at a lower stress than strand, leading to a greater tonnage of steel, but their cost per ton of steel is higher.

During construction, the Stage 1 tendons need to be near the top fibre to carry the self-weight hogging moments. However, in service the moment diagram will have dropped, and, in the central half of the span, Stage 1 prestress that is too high in the section has to be counteracted by additional bottom fibre, Stage 2 prestress, wasting

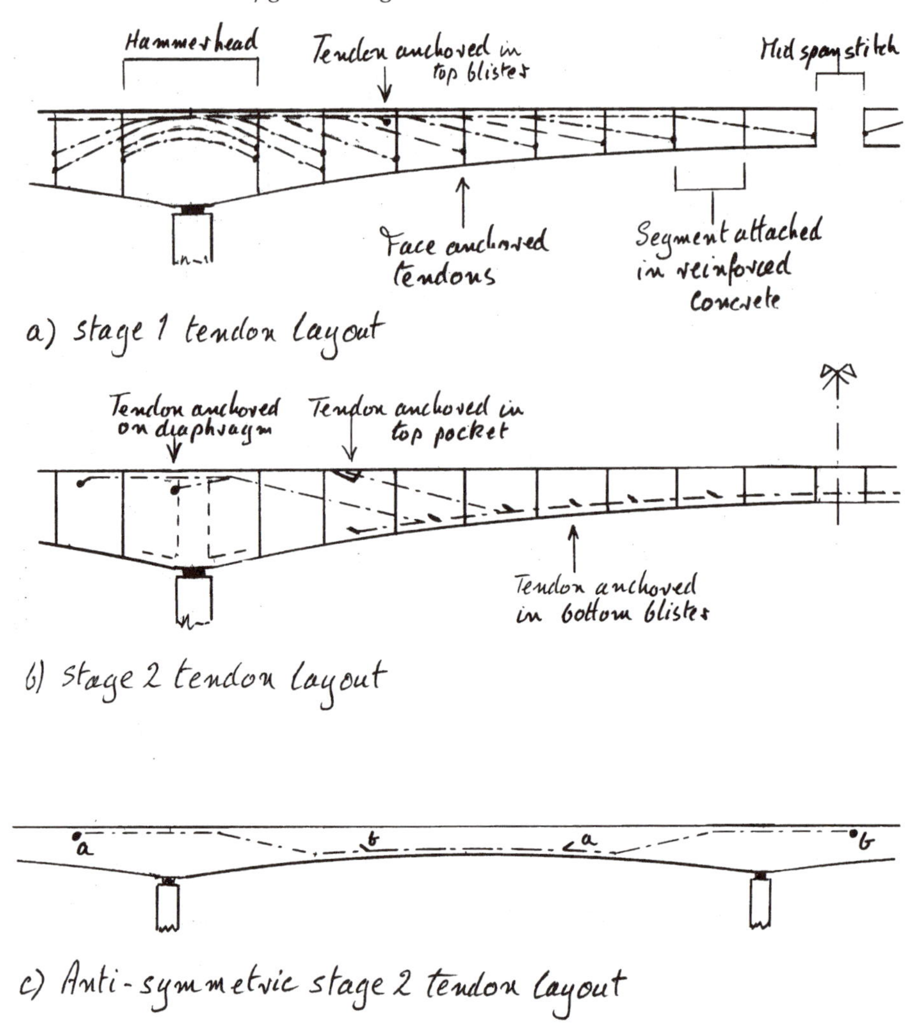

Figure 15.19 Typical layout of prestress tendons for cast-in-situ balanced cantilever decks

materials as explained below. The designer should lower the Stage 1 tendons from about the quarter point of the span as much as the construction moments allow, and use reinforced concrete action or partial prestressing to resist these moments to the limit allowed by the code of practice.

Generally, the top fibre cables required at the support for cantilever erection are not enough to carry the finishes and live load moments and additional top fibre cables are required. These may be straight capping cables stressed between top fibre blisters, or the lengthened Stage 2 tendons described below.

Stage 2 cables provide the sagging moment capacity in the span, Figure 15.19 (b), and they also create substantial sagging parasitic moments, which for preliminary sizing may be assumed to be equal to the mid-span live loads moments. Initially there are no other significant bending moments at mid-span, and the prestress is likely to cause tension stresses on the top fibre. These stresses will be small, and can be carried in reinforced concrete action if the code of practice allows. The parasitic moments increase the length and magnitude of the sagging moment envelope, creating a vicious circle as yet more tendons are required to carry the increased moments, increasing the parasitic moments in their turn. Consequently, the designer should make every effort to reduce the number and length of these Stage 2 tendons. The sagging parasitic moments may be slightly reduced by stressing some top fibre Stage 1 tendons after completing the stitch. Sagging parasitic moments diminish the total hogging moments at the supports, reducing the number of Stage 1 cables. However, the saving in weight of prestress at the supports is always less than the increase at mid-span, particularly when the deck is of variable depth.

Stage 2 tendons may either anchor in bottom flange blisters, or rise up through the web, providing shear relief around the quarter points of the span, a section that is often the most critical for shear. This may allow any web thickening to be moved back several segments. They anchor either in top flange cut-outs or in top flange blisters either side of the pier diaphragm. When they are anchored beyond the pier, they supplement the Stage 1 tendons for service loads.

In some schemes there are anti-symmetric Stage 2 tendons which anchor in a bottom blister, cross the mid-span, and then rise up through the web and anchor in blisters beneath the top flange beyond the pier, Figure 15.19 (c). In heavily prestressed bridge decks, such tendons help overcome a shortage of anchorage points for Stage 2 tendons, as they provide two mid-span continuity tendons for each symmetrical pair of bottom blisters. However, such tendons are difficult to design as their ducts have inevitably to be rotated about each other in the limited space of the bottom slab haunches.

15.5 Precast segmental balanced cantilever construction

15.5.1 General

The most widely used method of erection of precast segmental bridges is balanced cantilever. It is adaptable to spans from 25 m up to about 150 m, and can cope with virtually any succession of span lengths and deck alignments. The upper limit on span is generally imposed by the weight of the deeper segments and the cost of the casting cells, although if there is enough repetition, longer spans are viable. A typical deck consists of pier segments, and a number of span segments that are usually symmetrically placed about each pier, in balanced cantilever. The span is closed by a mid-span stitch, cast in-situ. The joints are usually glued, although this is only essential when internal tendons are adopted.

Cast-in-situ bridges built in balanced cantilever have been discussed in *15.4*. Only the characteristics that are specific to precast segmental construction will be described here.

One significant difference from cast-in-situ construction is that the segments are several weeks old when they are erected, reducing the changes in bending moment due to creep. Whereas typically the creep coefficient for the concrete of a cast-in-situ deck is between 2 and 3, for a precast deck it is of the order of unity. (Clearly these are broad generalisations; the creep coefficient needs to be calculated for each specific case.) Except for very long spans, the effects of creep on the erection geometry are reduced to virtual insignificance by the speed of erection for precast decks; six segments in a day is not unusual.

A considerable proportion of the shrinkage of the concrete will also have been completed before erection. As a result of the reduced shortening of the deck, expansion joints may be further apart, and longer lengths of deck may be pinned to, or built into the columns. Also, early thermal stresses are virtually eliminated by the single-phase precasting, giving rise to crack-free units.

Balanced cantilever bridges may be erected by crane, by shear legs (or beam and winch), or by overhead gantry. The choice of method depends on the scale of the bridge, on the weight of the segments, on the height of the deck above ground level and on the nature of the terrain crossed.

The erection and attachment of the pier segment, the stability of the balanced cantilever, the erection of the end spans and the construction of the mid-span stitch are common to more than one method of erection, and they will be covered first.

15.5.2 Erection of the pier segment

The erection cycle starts with the placing of the pier segment that must be cast with precisely dimensioned downstand bearing plinths that accommodate the transverse and longitudinal falls of the deck. Due to the presence of the diaphragm, this is usually the heaviest segment and is frequently made shorter to limit its weight. For very deep decks the pier segment may be subdivided, either vertically into two or more very short segments, or horizontally, with match cast joints allowing its assembly without casting concrete. The pier diaphragm needs to be very carefully designed to minimise its weight, which involves thinking carefully about the purpose of each component. When designing a major segmental bridge one cannot lazily adopt a diaphragm consisting of a 2 m thick solid wall! It is also possible to cast the diaphragm in-situ once the pier segment has been erected although this slows down the construction programme and is rarely adopted.

As the crane or gantry cannot place the segment accurately enough, it is lowered onto falsework which is equipped with jacks that correct its position before it is fixed to its bearings. The segment must be accurately positioned, in level and in crossfall as these parameters cannot easily be corrected subsequently. In general, the fine adjustment of the longitudinal gradient may be achieved by rotating the complete double cantilever on its bearings, as long as this remains within their rotational capacity. If the deck rests on sliding bearings, or if one is fixed and one sliding, the orientation in plan of the completed cantilever may be corrected by rotating it about a vertical axis. However, if both the bearings are fixed, either the pier segment has to be placed with near perfect orientation, a difficult operation on a short base length, or one of the bearings must be modified to allow a small amount of adjustment before it is locked in position.

Mechanical bearings are conventionally attached both to the pier and to the deck with cast-in anchors or dowels. The connection detail must allow the bearings to be replaced. For precast concrete construction this causes problems of tolerance, as the items that are cast into the pier segment must match exactly those cast into the pier head. However, it is frequently unnecessary to connect the bearing to the segment. Clearly, for all free-sliding bearings the friction between the steel top plate of the bearing and the concrete segment will always be greater than that of the PTFE sliding layer. Consequently no mechanical connection is needed.

For guided and fixed bearings, it is necessary to investigate the critical combinations of vertical and horizontal load, and compare them with the lower bound estimates of friction coefficients between the surfaces in contact. On the East Moors Viaduct for instance, it was found that the coefficient of friction between the top plate of a few of the bearings and the mortar skim was inadequate. Consequently, a pattern of grooves was machined into the plate to enhance this friction, avoiding the need for a mechanical connection to the segment.

When a mechanical link between segment and bearing is required, the connecting bolts must be sized to carry only the residual horizontal force to be transmitted between the deck and the bearing, once friction has been deducted. The simplest method of making the connection is to equip the bearing with an additional top plate to which it is bolted. The segment is placed onto the accurately levelled bearing with an epoxy resin bonding it to this top plate. However, this is an expensive way of solving the problem due to the presence of the additional plate which must be thick enough to accommodate the bolts and must be well protected against corrosion as it cannot be changed.

Alternatively, a light steel frame to which are welded threaded anchorage sockets for the bearing bolts is placed on the soffit form for each of the bearing downstands in the casting cell, and cast into the segment. As the pier segment is removed to storage, a template is made up recording accurately the relative position of the anchorage sockets in both plinths. This has been done by placing a plank of timber across the two downstand plinths and hammering it so that the position of the bolt-holes is imprinted, but other methods may be devised.

This template is used to locate the bearings on the pier head; the dowels anchoring the bearing to the pier head may then be cast in. When the segment is presented, it can be bolted directly to the bearings. This procedure is very quick as the precise orientation of the pier segment has been predetermined, but requires careful working and supervision.

A further option is to cast sockets into the bottom of the segments as described above. The dowels anchoring the bearings to the pier are then left in empty pockets in the pier head, Figure 15.20. When the pier segment is presented, it is placed on its temporary supports and the bearings are bolted up to it. It is then orientated precisely, and the pockets on the pier head containing the bearing dowels are grouted up. This method is less demanding but slower.

15.5.3 Temporary prestress for cantilever construction

The resin used for jointing segments has an open time typically of 45 minutes. As it is not possible to install the permanent Stage 1 prestress within this time, each new segment must be attached to the cantilever by temporary prestress bars that can be installed in a matter of minutes.

A minimum of four bars should be used to provide the redundancy that allows one bar to fail without loss of the segment. The bars used for temporary stressing should be reasonably ductile. Generally the grade adopted should not be stronger than 835/1,030, these being respectively the 0.1 per cent proof stress and characteristic ultimate strengths in MPa. The Macalloy bar of this grade has a minimum elongation at rupture of 6 per cent.

Temporary prestress may be anchored on internal blisters beneath the top slab, cast together with the segment, Figure 15.21. This is certainly the cheapest form of anchorage, but may become complicated if there are also top anchor blisters for the

Figure 15.20 STAR Viaduct: installation of bearings (Photo: Benaim)

Figure 15.21 East Moors Viaduct: interior of box showing temporary stressing bars, web thickenings for face anchors and blisters for Stage 2 cables (Photo: Robert Benaim)

permanent prestress, Figure 14.7. It is necessary to plan the temporary anchorage blisters carefully if the webs of the box change in thickness, or if the bridge deck is on a tight curve. Temporary bars may also be anchored on blisters that have been cast onto the concrete webs or slabs after match casting, using bend-out bars or couplers, Figure 14.8. In addition to the internal blisters, temporary steel or concrete anchor blocks may be used to anchor bars onto the webs or slabs of the box, Figure 15.11 and Figure 15.25. Such anchorages are usually required where the temporary stressing is intense, such as when several segments are erected on bars before installing the permanent tendons.

For all blisters and external anchorages carrying temporary stressing, the bending moments and direct tensile stresses introduced into the slabs or webs must be considered, and usually additional reinforcement will need to be added, Figure 15.22. The blisters or temporary anchor blocks should be placed on the transverse axis of the segment or further forward. If they have to be placed towards the back of the segment, it is essential that the force in the prestressed bars is carried forwards into the segment by reinforcing bars in the top slab; otherwise the top slab may split in front of the anchor, Figure 15.23. The rear anchorage point of the temporary prestress is usually tied in by the permanent prestress. If not, the segment must be appropriately reinforced.

It may also be necessary to mobilise a small temporary prestress force near the bottom of the segment to stop bottom fibre tensions occurring when the permanent prestress tendons are installed.

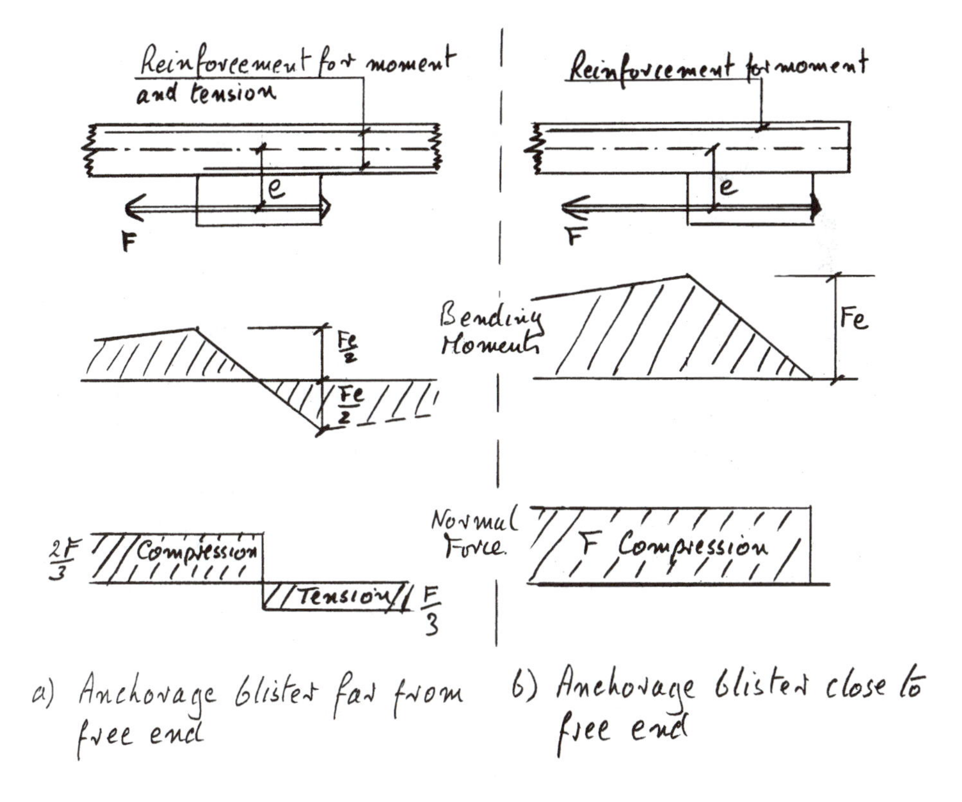

Figure 15.22 Typical moments and forces due to temporary prestress

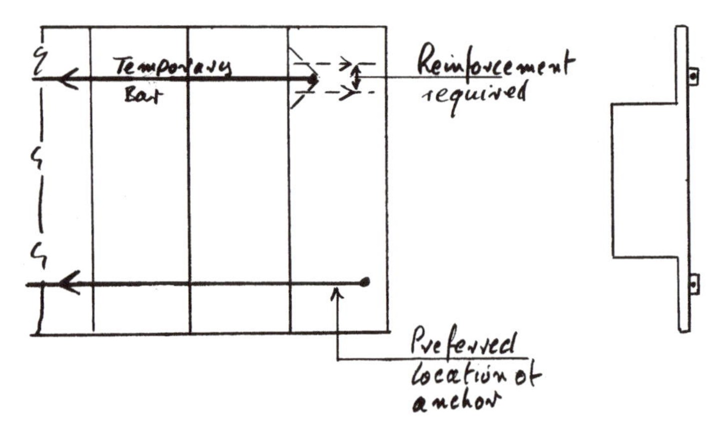

Figure 15.23 Location of temporary bar anchors

Finally, temporary stressing may be incorporated in ducts within the top slab of the box. Although apparently a neat option, this method requires pockets to be left to allow the bars to be de-stressed and dismantled, which is a laborious task. This temporary prestress is sometimes designed to be incorporated into the final prestressing scheme. Although superficially attractive, the bars tend to account for only a small percentage of the total prestress required, at a very high unit price. In general, it is better to design to recover and re-use the temporary prestressing bars.

15.5.4 Prestress layout

The prestress layout for precast decks is broadly similar to that described for cast-in-situ decks in *15.4.7* and only matters specific to precast construction will be discussed.

Precasting to a daily routine requires greater discipline than cast-in-situ construction, and repetition of prestress details should be sought. Thinner webs may be adopted, which can limit the size of face anchored tendons. Alternatively, the web may be increased in thickness to accommodate the anchor chosen, or thickened locally. As the concrete is mature at the time of stressing, a tendon consisting of 12 No 13 mm strand requires a 320 mm thick web, 19 No 13 mm strand or 12 No 15 mm strand require a 450 mm thick web, while 19 No 15 mm strand requires a 500 mm web. For an economically designed bridge, the web in the central half of the span rarely needs to be thicker than 400 mm, and 300 mm is usually adequate.

For example, in the East Moors Viaduct the webs were 500 mm thick close to the pier, thinning to 300 mm in the second segment. The Stage 1 prestress units were 19 No 13 mm strand, face anchored towards the bottom of the web. This size of tendon was chosen as it allowed one or two pairs of tendons to be anchored in each segment except the last which was attached by temporary stressing alone. A local thickening of the 300 mm web to 450 mm was therefore required. This thickening was faired into the bottom corner blisters used to anchor the Stage 2 tendons, Figure 15.21 and Figure 9.28. Intelligent design of the temporary stressing allows the Stage 1 tendons to be lowered from about the quarter point of the span onwards, to optimise their in-service performance, as described in *15.4.7*.

Conventionally, bridges were designed with at least one pair of permanent tendons per typical segment. This allowed the temporary stressing to be removed early, Figure 15.24 (a). However, there are many advantages in anchoring Stage 1 tendons only in every second or third segment, Figure 15.24 (b). Lifting a segment into place, applying the resin to the face, attaching and stressing the temporary bars and then releasing the lifting tackle takes between 45 and 90 minutes. Thus if the erection team does not have to stop after each pair of segments is erected to install the permanent prestress, 2 × 3 segments may be erected in a normal shift. The prestress may then be installed either at the end of the day, or in a night shift. This was first explored by Benaim on the Weston-super-Mare railway viaduct where two pairs of segments were

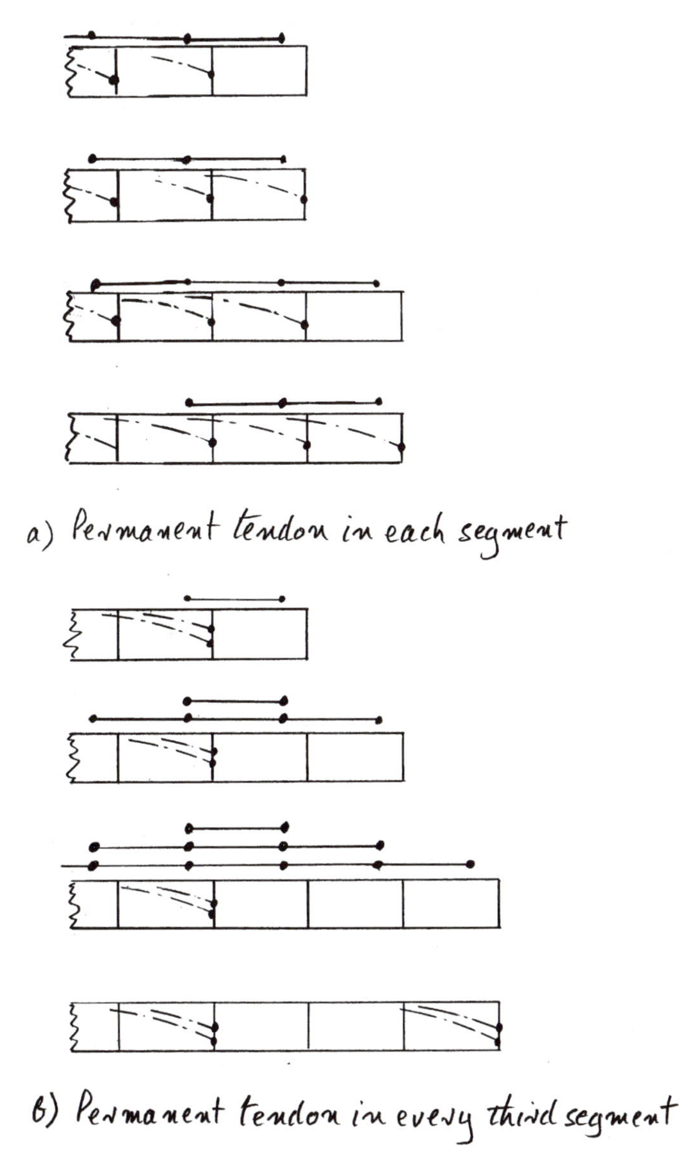

a) Permanent tendon in each segment
b) Permanent tendon in every third segment

Figure 15.24 Typical arrangement of temporary prestress

erected on bars, and then on the 6 km long STAR Viaduct in Kuala Lumpur where up to three segments were erected before the permanent prestress was installed, allowing up to four 37 m long spans to be erected per week. The technique was taken to its logical conclusion on the approach spans to the Belfast Cross Harbour Bridges where complete 28 m spans, each consisting of seven 4 m long segments, were erected in balanced cantilever on temporary prestressing bars, the permanent prestress tendons being threaded through several spans once the mid-span stitches had been cast, Figure 15.25. This last experience demonstrated the limits of the technique, as the temporary prestress was becoming very complex.

If external Stage 1 tendons are used the plastic ducts must be fixed in place after erecting the segments, before the tendons may be installed. As this operation is on the critical path for deck erection, it is necessary to minimise the number of cables by adopting large prestress units, with typically 27–31 No 15 mm strands. Furthermore it is very difficult to manage the duct geometry of a large number of smaller tendons, giving a further reason to adopt large units for external prestressing. These powerful units have to be anchored on large internal blisters or ribs, Figure 15.26. The jacks for stressing such large units can weigh over a ton, and need specialised plant to handle them.

External Stage 1 tendons cannot be tailored as closely to the bending moment diagram as internal tendons, with the result that a significantly greater weight of prestressing steel will be required. This increase in prestress tonnage for external tendons will be exacerbated by the loss of eccentricity over the pier, as the tendons have to remain beneath the top slab. Furthermore, the internal ribs required to anchor these tendons interrupt the match casting, frequently requiring special internal forms, and their weight cancels out any benefit from the thinner webs made possible by the external tendons.

One of the major design aims is to reduce the number and length of the Stage 2 tendons in order to minimise the prestress parasitic moments, as described in *15.4.7*. When the Stage 2 cables are first stressed, at mid-span there is no externally applied bending moment, the only sagging moment is the parasitic moment generated by these cables themselves. Consequently, prestress that is close to the bottom fibre of the box is likely to cause top fibre tensions that are unacceptable for glued joints. In order to avoid these tensions, the first permanent tendons stressed should be close to the lower limit of the kern, or alternatively temporary top fibre prestressing bars or some short

Figure 15.25 Belfast Cross Harbour Bridges: spans erected on temporary stressing alone (Photo: Graham-Farrans JV/Esler Crawford)

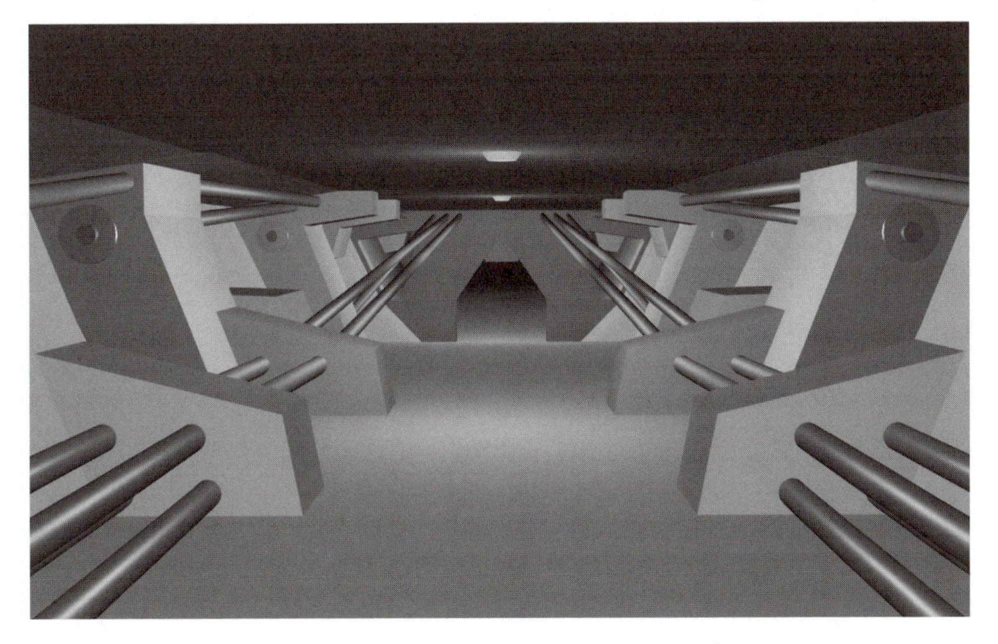

Figure 15.26 Typical arrangement of external stressing (Image: David Benaim)

top fibre permanent tendons may be installed concurrently with the first Stage 2 cables. It is worth stressing the longer Stage 2 cables first, as they generate larger sagging parasitic moments, minimising these top fibre tensions.

It is common to see external prestress used for the Stage 2 tendons. These tendons are anchored at the top of the pier diaphragm and deviated in bottom blisters or transverse beams. This use of external prestress simplifies the casting of the segments. However, the jury is still out on whether this is just a design fashion, or provides real benefits. Certainly, the main benefit of external tendons, that they may be changed during the life of the bridge, is irrelevant if the Stage 1 cables are internal. External cables are usually more expensive per ton than internal cables, they involve a greater tonnage of steel due to the inflexibility of anchoring arrangements, and they introduce another technology onto the site, with the consequent increased risk of things going wrong; simplicity on site is an absolute benefit.

15.5.5 Stability during erection

Much of the section on the stability of cast-in-situ balanced cantilever bridges (*15.4.4*) is equally applicable to precast segmental decks that are erected by crane or shear legs, and will not be repeated. However, there are some matters specific to precast construction.

The segments are attached at the rate of several per day, so the cantilever is never out of balance for more than a few hours at a time. Thus consolidation settlement of a separately founded stability prop is not a concern, frequently allowing a separate pad foundation where the prop for cast-in-situ construction would have had to be piled or supported on the pier foundation. However, if the soil is very poor, it may be unwise to leave the prop loaded by an unbalanced cantilever for weekends, or even in extreme cases, overnight!

The dynamic load case of a segment being dropped due to sudden failure of the lifting mechanism is a realistic scenario when the deck is erected by shear legs. In the absence of a rigorous calculation, the dynamic component of the overturning moment is generally considered to be equal to the weight of the segment × its distance from the axis of the pier or prop. The connection of the shear legs to the deck must be designed with adequate redundancy, to make it inconceivable that the shear legs could fall together with the segment.

If segments are erected by crane, and if the temporary prestress is designed with adequate redundancy so that a bar may fail without loss of the segment, this dynamic load case is probably not relevant.

When a deck is erected by gantry, the sudden failure of the segment-lifting mechanism would apply a shock load to the gantry. The effect on the stability of the deck will depend on the manner in which the gantry is supported. For instance, if it spanned from pier to pier, this accident would have little effect on the deck. However, if the deck carried one of the legs of the gantry, the dynamic effect of the sudden loss of a segment could be transmitted to the deck.

15.5.6 Building the end spans

The discussion on this subject in the section on cast-in-situ construction (*15.4.5*) is in general valid for precast segmental construction. The principal difference between the two technologies is the speed of erection of the latter. It is easier to adopt the progressive cantilevering solution, as the temporary props need only be loaded for a matter of a few days, giving little scope for consolidation settlements. If there are several end span segments beyond the point of symmetry of the balanced cantilever, they may be assembled on falsework using temporary stressing to make them self-supporting. This run of segments may then span freely from the abutment bearings onto one temporary prop, which may be jacked to line up the segments with the balanced cantilever, and then joined with a short stitch, Figure 15.27 (a).

When a long continuous viaduct is divided into expansion lengths, there are inevitably pairs of end spans that meet on a pier. The use of falsework in such a repetitive situation is not economical, and if the viaduct is in town, may not be possible. On the STAR viaduct the expansion joint over the pier was temporarily blocked, and three segments either side of the pier were cantilevered using temporary prestress. These temporary cantilevers were then joined to the balanced cantilevers by a cast-in-situ stitch and the permanent prestress installed. The temporary prestress could then be removed and the expansion joint freed, Figure 15.27 (b).

15.5.7 The mid-span stitch

Once the cantilever erection is complete, it remains to cast the mid-span stitch and connect the new double cantilever to the completed portion of the viaduct. The stitch may be as long as a typical segment giving aesthetic advantages as the segment joints become evenly spaced. This solution was adopted on the East Moors Viaduct. However, it was found to be slow to build and not very economical, as it requires a substantial formwork to be mobilised and a large reinforcing cage to be prepared.

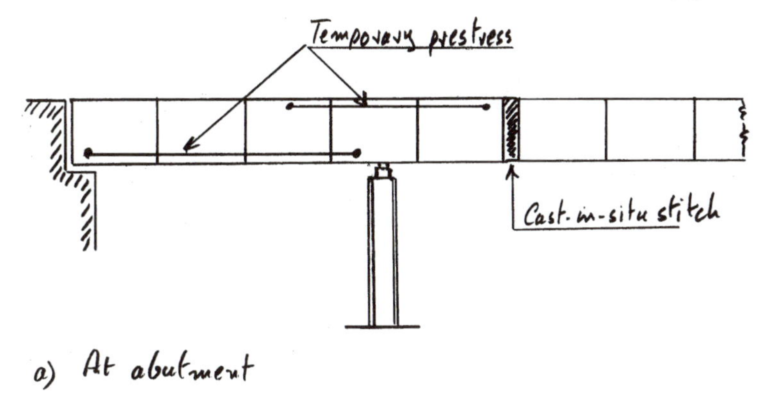

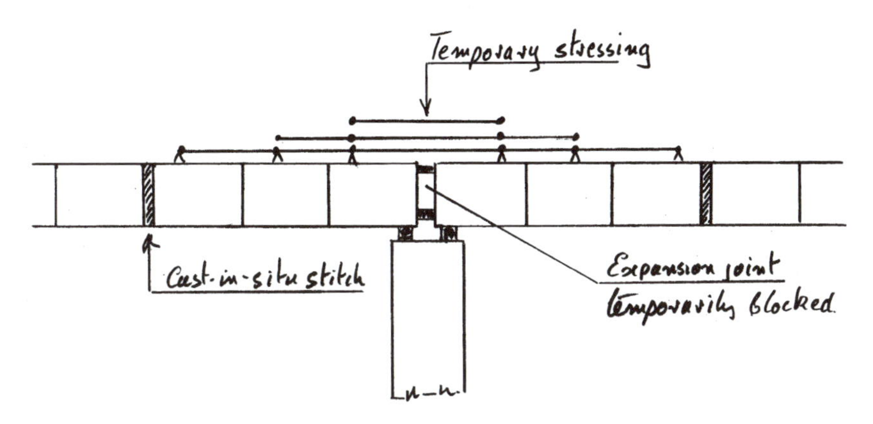

Figure 15.27 Erection of unbalanced end spans

The mid-span stitch is best made thin enough so that it does not need to be reinforced, imposing a maximum width of approximately 250 mm. This gives very little tolerance to correct alignment errors. It also gives little margin to accommodate tolerance on the length of segments. When such a short stitch is adopted, it is important to keep a record of segment lengths during casting, so that action may be taken in good time if, for instance, segments are being cast fractionally long.

The principal difficulty in making this stitch is the fact that the cantilever ends to be joined are constantly moving. The completed viaduct to which the new double cantilever must be attached will be expanding and contracting on a daily cycle. Furthermore, the cantilevers will deflect downward during the day under the action of the sun shining on the top surface, and rise at night. If the stitch were cast without precaution, these temperature effects would crack the fresh concrete before it was possible to stress the continuity prestress.

The simplest case is when the spans do not exceed about 40 m and the joint width is less than about 250 mm. Temporary stressing bars are installed just above or just below the top slab, reacting against hardwood timber packs placed between the top flanges, Figure 15.28.

The force in the bars and the shape and size of the timber packs must be designed to provide stability to the new double cantilever. The disturbing moments that must be resisted are due to the overturning moment of the new double cantilever when nominally in balance (*15.4.4*), to the friction in its bearings as the bridge deck changes in length, and to the weight of the mid-span stitch. The bars must also resist the direct tension due to the bearing friction. Furthermore, it is necessary to check that the

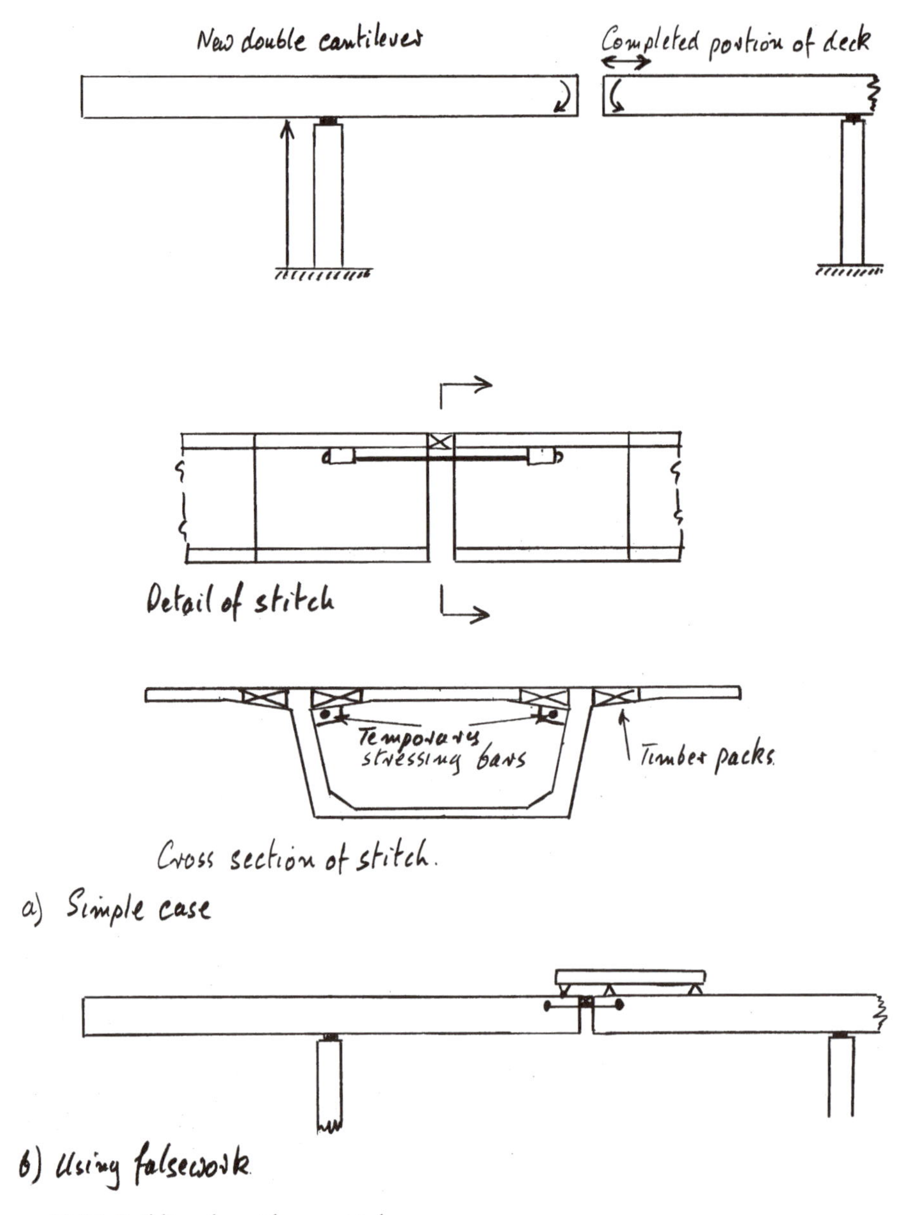

Figure 15.28 Building the mid-span stitch

prestress holding on the end segments of both cantilevers is adequate to carry these forces and moments. If not, additional temporary prestress must be installed.

As soon as these bars and packs are in place, it is necessary to release the locks on the bearings for the new double cantilever so that it can move longitudinally, and to release the stability falsework so it is free to rotate.

When the new double cantilever is carried by fixed bearings, it is necessary to check that the pier is sufficiently flexible to accept the daily temperature movements of the deck. Alternatively the bearings must be provided with a small sliding capacity in the temporary state, to be locked later.

At this point the cantilever tips are effectively hinged at the top flange level. Solar heating during the day will cause the cantilevers to droop, and the joint to open at the soffit. Overnight, and until the sun re-heats the deck next morning, the cantilevers will rise, closing the gap at the soffit. Thus the stitch must be cast in the evening so that the night-time recovery of the droop will compress the fresh concrete. Before the sun's heat next morning, the stitch concrete must have attained a strength of 12–15 MPa allowing either additional temporary bars, or the first pair of bottom fibre tendons, to be stressed, opposing any opening of the joint. As the concrete gains strength, the remainder of the span cables may be stressed, completing the construction of the stitch.

The limits on span length and joint width for this method are conditioned by the stability of the timber packs under the overturning moment on the new cantilever. If the joint is too wide and the overturning moments on the double cantilever too great, it will be impossible to design the packs against overturning and vertical sliding.

When, due to length of the span or the width of the stitch, it is not considered safe to rely on the timber packs to provide stability to the new double cantilever, a relatively light falsework beam, locked to one cantilever and pinned to the other, is designed to carry the overturning moments on the new double cantilever, Figure 15.28 (b). The bars and blocks then become simply a hinge in the top flange that only has to resist the longitudinal tension/compression of bearing drag. The remainder of the procedure is as before.

This procedure has been used successfully on the majority of the precast segmental balanced cantilever bridges designed by Benaim. There are of course variations, due for instance to the stabilising beam having to resist additional moments caused by the launching of a gantry, or the need to close several stitches at once. However the basic principles of locking the top fibres of the cantilevers together longitudinally, not attempting to resist the rotation of the cantilever tips and respecting strict timing for casting the stitch and stressing the first tendons, makes for a reliable and economical procedure.

15.5.8 Erection by crane

The simplest method of erecting bridge decks in balanced cantilever is by using ground-based cranes. If the bridge is over navigable water, barge-mounted cranes may be used.

Crane erection is simple and safe, using equipment with well-established safety factors and operating procedures rather than purpose-made falsework. Also, no temporary loads are imposed on the deck during erection, simplifying the stability falsework. It offers the flexibility of being able to work out of sequence in a long or complex viaduct, Figure 15.29. For a viaduct erected by gantry, crane erection may offer a second front if it is necessary to accelerate the works. Also, a crane may be used to erect pier segments in advance of the arrival of the gantry, saving time on the critical path.

The cranes may be mobile, minimising the lifting capacity needed. If the crane stands to the side of the segment, it must have the capacity to lift the heaviest segment at a reach that is something greater than the half width of the deck. If the crane stands on the axis of the deck being erected, the radius of lifting is little more than half a segment length. However a beam and winch must be mobilised to lift in the last segments of the span.

The crane may also be fixed in position, when it will need much greater capacity to lift a segment at a large radius. The Grangetown Viaducts were designed by Neil Kilburn of the County of South Glamorgan, with the erection technology designed by Benaim. The viaduct consisted of two parallel decks with a series of 72 m spans. A large ringer crane located adjacent to a pier erected the 40 segments of the double cantilevers for both decks, requiring a capacity to lift 50t at a 40m radius, Figure 15.30. The choice of size of crane for any particular site depends on the trade-off between speed of erection and crane hire and operating costs.

Cranes can erect decks very quickly, four to six segments in a day being typical if the segments are assembled on temporary bars without the delay of installing permanent

Figure 15.29 Belfast Cross Harbour Bridges: the flexibility of crane erection (Photo: Graham-Farrans JV/Esler Crawford)

Figure 15.30 Grangetown Viaducts, Cardiff: erection by fixed crane (Photo: Robert Benaim)

prestress. Generally the speed is likely to be controlled by the rate at which segments may be delivered to the crane, and by such details as the rate at which lifting beams and access platforms may be attached to, and detached from, segments. If speed is important, the contractor will need to invest in several sets of lifting beams, so that their attachment is off the critical path.

The main limitation on crane erection is clearly the suitability of the bridge site. On an urban site, the delivery of segments at ground level and the operation of cranes may be too disruptive to traffic. If the bridge deck is very high above the ground, the cost of craneage may escalate rapidly. Furthermore, it is necessary for each span under construction to be stabilised by falsework.

15.5.9 Erection by shear legs (or beam and winch)

Where it is possible to deliver segments at ground level to all points of a viaduct, the segments may be erected by a pair of shear legs (or beam and winch). These are non-slewing derricks equipped with electric winches, resting on the deck, as used for the erection of the Byker Viaduct, Figure 15.31, [2], (*7.15.4*). The shear legs are generally counter-weighted for self weight and stressed down to the deck when lifting a segment. An access platform is carried beyond the segment being lifted, for the prestressing gangs.

The pier segment plus at least one adjacent segment are generally placed by crane. This provides a platform on which the first set of shear legs may be placed. Once an additional segment has been placed, the shear legs are moved forwards and the second set may be lifted into position.

When it is not possible to lift in the pier segment by crane, the pier head falsework may be designed to support the first set of shear legs which places the pier segment itself. The shear legs may be assembled on the falsework.

Generally, shear legs erect only one segment per day at each end of the double cantilever with permanent prestress installed in each pair of segments. After each pair of segments has been erected, the shear legs are moved forwards. However, they may be designed to erect two segments from each position, when the deck should be designed with permanent cables anchored every two segments, and a construction rhythm of four segments per day adopted.

Figure 15.31 Byker Viaduct: erection using shear legs (Photo: J. Mowlem/North East Studios)

15.5.10 Erection by overhead gantry

Gantries are employed to erect viaducts on high piers, in cities or over water. Generally a gantry will first erect the pier segment, launch itself forwards to rest on this segment, receive segments delivered along the deck of the erected viaduct and erect them in balanced cantilever while stabilising the deck. The size, complexity and degree of mechanical sophistication of a gantry depend principally on the maximum span of the deck, on the weight of the segments to be handled, on the radius of curvature to be negotiated, on the rate of construction planned and on the combinations of these factors, Figure 15.32.

A pier head falsework is required to carry the pier segment, and to provide a support to the front leg of the gantry. This falsework may be supported off the pile cap, or may be attached to the pier head. Figure 15.33 shows the pier head falsework being load tested by pulling it upwards; the steel beams above the pier belong to the testing rig and are not part of the pier head falsework.

In the majority of projects, the gantry applies loads to the deck as it launches itself forwards. As gantries weigh between 150 tons and 400 tons, the decks need to be

Figure 15.32 STAR Viaduct: gantry erection (Photo: Benaim)

designed specifically to carry their weight during the launching cycle. Consequently, it is best for the design of the deck and of the method of construction to be carried out by the same team. When the deck has been designed by a consultant in the traditional manner, without considering construction loads, the gantry may be designed to apply loads only above the piers, increasing its length and cost. It may also be possible to strengthen the deck using additional temporary prestress so that it may carry the weight of a shorter, more economical gantry.

The principal advantages of gantry erection are the minimal interference with activities on the ground in urban sites, the ability to cross any terrain, and the speed of erection that is possible. The principal disadvantages are the time required for the design, construction, erection, testing and commissioning, which is generally of the order of 15 months (unless an existing gantry is available for hire), the high first cost which requires a bridge of at least 20,000 m^2 of deck area, and the essentially linear, less flexible construction programme.

There are very many designs of gantry, and it is proposed to illustrate two that built bridges designed by Benaim.

Figure 15.33 STAR Viaduct: load testing the pier head falsework (Photo: Benaim)

15.5.11 STAR Viaduct, Kuala Lumpur

The 6 km long STAR twin track light railway viaduct is described in *7.15.3* and in Figures 7.11, 7.22 and 7.23 as well as in Chapters 9 and 14. The spans were typically 35.1 m, and varied from 22.95 m up to 51.3 m, with radii of curvature down to 130 m. The deck was continuous, with expansion joints at approximately 300 m centres. The deck was made of precast segments, generally 2.7 m long and 8.4 m wide, erected in balanced cantilever. The short segments were chosen to adapt to the tight radii of curvature. Typical segments weighed 29 tons, while pier segments and segments carrying the stations weighed up to 58 tons. Mid-span cast-in-situ stitches were generally 250 mm thick and were unreinforced.

The gantry adopted had been designed by VSL for another deck and needed some modification for use on STAR. It was a cable-stayed design, where the beam was suspended from the main legs extended upwards as a mast, and could be rotated about a vertical axis to cope with the sharp curvature, Figure 15.32. It weighed some 160 tons and imposed loads on the deck during the launching cycle, requiring the deck to be designed accordingly. Consequently, the detailed movements of the gantry had to be established before the prestressing of the deck could be finalised. In some sections of the viaduct, pier segments could be erected by crane in advance of the gantry, saving time on the programme. The deck was erected at rates of up to four spans per week, which is among the fastest ever for glued, balanced cantilever construction. This fast rate of erection was assisted by the routine of erecting up to three segments on temporary bars only (*15.5.4*). The first phase of the work, which consisted of 2.6 km of deck requiring 968 segments, was erected in 18 months after the start of piling, despite several hold-ups due to delays in completing the piers.

As the piers were located in busy streets, the pier head falsework, which was required to receive and position the pier segment, to stabilise the first balanced cantilever segments, and to receive the front leg of the gantry during its launching cycle, was clamped to the pier head by temporary stressing bars, to avoid the need for stanchions resting on the foundations which would have been at risk of damage from traffic. Brake lining pads acted as a slightly flexible friction interface. The consequences of a falsework collapse in the centre of Kuala Lumpur were so severe that Benaim specified that every pier head falsework should be load tested. As this procedure had been defined early in the design process, most of the pile caps were equipped with anchorage points allowing the falsework to be pulled downwards with prestressing bars. In some locations where this was not possible, the falsework was tested by pulling it upwards.

15.5.12 Dagenham Dock Viaduct, A13, London

The viaduct crossed the Ford Motor works in Dagenham, East London. It was 1,700 m long without intermediate expansion joints, and consisted of two parallel box girders, increasing to three boxes at each end where the deck flared to accommodate slip roads, Figure 13.12. The viaduct was built by Tarmac/DMD to an alternative design prepared by Benaim. The spans were mainly 62 m long, with some variations. In accordance with the regulations of the time, all the prestress for the deck was external and the segment joints were resin filled. The Stage 1 construction cables consisted of 31 No 15.7 mm strands made by DEAL [3], anchored every three segments on full-height

Figure 15.34 Dagenham Dock Viaduct: gantry erection (Photo: DEAL S.r.l.; www.deal.it)

ribs; thus three segments were erected on temporary bars. These ribs were located so that the cables could be stressed by a jack suspended from a crane located on the top surface of the deck. The continuity tendons, also made of 31 strands, were anchored on the ribs or on the pier diaphragms and were deviated by concrete blocks. The gantry, which was designed and supplied by DEAL, was a rigid truss weighing some 400 tons, and was capable of handling segments weighing up to 120 tons, Figure 15.34. As it launched, it imposed loads on the cantilevers, for which the deck prestress had to be designed. It built the deck at a rate of 4–6 segments per day.

15.6 Progressive erection of precast segmental decks

In an extension of the method of balanced cantilevering, precast segmental decks may be erected by the method of progressive forward cantilevering. With the segments being handled either by crane or shear legs, the deck is cantilevered forwards on one front only, with each span being carried on a series of temporary props until the next pier is reached, and the installation of permanent prestress renders the span self-supporting. Several spans may be erected without any cast-in-situ stitches, offering the possibility of a fast construction schedule.

Before the permanent prestress is installed, and the deck is only lightly compressed with temporary bars, it is very fragile, and susceptible to being cracked by any movement of the temporary props due to settlement of their foundations, to their expansion and contraction under temperature changes, and by thermal gradients through the deck.

Figure 15.35 Byker Viaduct, Newcastle: progressive erection of decks (Photo: Arup)

It is thus essential that the props be equipped with jacks that are kept at a constant pressure by nitrogen accumulators or other similar devices, monitored by load cells.

The author adopted this method for the construction of the approach spans of the Byker Viaduct, Figure 15.35. However since then, the erection of short spans by balanced cantilevering has proved so quick and trouble free that were the same project to be repeated, progressive cantilevering would not be chosen.

15.7 Construction programme for precast segmental decks

When planning the construction programme for a precast segmental bridge deck, the discrepancy between the rate of casting the segments and the rate of erection, particularly when using a gantry or cranes, must be borne in mind. In general, a casting cell produces approximately five segments per week, while a gantry may erect over twenty segments per week. Also, a casting cell takes about five months to design, build and commission, while a gantry takes typically 15 months. The number of casting cells mobilised should be calculated so that the last segment is produced some four weeks before it is due to be erected.

For example, consider a deck with 600 precast segments. The gantry will erect its first segment in month 15, after which it may erect segments at a rate of 85 per month, erecting the last segment in month 22.

One may expect the first segment to be produced at the end of month 5. The casting yard then has 16 months in which to produce 600 segments, a rate of 37.5 per month, or 8.7 per week. In order to achieve this rate, two casting cells will be required.

If segments are produced at an average rate of 37.5 per month, before erection starts there could be a need to store up to 375 segments. This in-built need for storage in precast segmental construction must be considered, particularly as it is not unusual for erection gantries to be delivered late.

For very large sites, involving the use of thousands of segments, the storage requirements using this method of assessing the number of casting cells would be unmanageable. Consequently, the rate of casting must approximate to the rate of erection, requiring many more casting cells, and increasing the cost of the technique. For instance, for the 22 km PUTRA LRT in Kuala Lumpur, 7,000 segments was cast in 42 moulds and erected by nine gantries (see also *16.8.4*).

15.8 Incremental launching

15.8.1 General

The method of building bridges by incremental launching was first used by Leonhardt. The deck is built in segments behind one of the abutments and pushed or pulled forwards out of the mould by hydraulic jacks. As successive segments are cast the lengthening bridge deck slides over the piers, cantilevering from one to the other. Usually, the deck is equipped with a steel launching nose to control the cantilever bending moments.

It is used almost exclusively to build box girders and is best adapted to that deck form. It is normally used for bridges with spans between 30 m and 55 m, with a plan area in excess of 3,000 m^2. Longer spans may be launched, but they require intermediate falsework towers to cut down the launching span. Bridges over 1,200 m in length have been built by this method. Rates of construction are typically one 15–25 m segment per week. However, there are large variations in the methods and rate of construction.

The major falsework required for the construction of a launched bridge deck consists only of the casting area and the launching nose. Also required are the launching jacks, the launch bearings and the jacking arrangements at the head of each pier.

Conventional launched bridge decks require typically up to 20 per cent more prestressing steel than bridges built by other methods, due to the relative inefficiency of the first stage central prestress. The penalty in weight of prestress is limited if the span/depth ratio of the deck is below 15, with 13 being approximately the optimum. If it is necessary to launch a shallower deck, it may be economical to cut down the launching span by introducing temporary towers. For bridges launched in reinforced concrete (*15.8.5*) there should be no penalty in the weight of prestress.

Due to the limited extent of the falsework, and to the repetitive nature of the process, launching is frequently the most economical method of building a prestressed concrete bridge within the span lengths and overall length defined above and with a suitable alignment, despite the greater depth of the deck and the greater weight of prestress. It may still be competitive for longer spans, but the need for intermediate temporary towers compromises its economy to some degree.

15.8.2 Launchable geometry

The bridge deck segments are built on a fixed soffit form. Consequently the bridge soffit must have a constant geometry over its entire length. Other than a straight line, this may include a vertical or horizontal curve of constant radius, or a combination of the two. However, these constraints are not as limiting as they may appear. If the road or railway carried has a non-circular horizontal alignment, it may be possible to find a constant radius that is a close approximation. The top flange of the box may then be made wide enough to accommodate the alignment, or may be made eccentric to follow its shape.

If the bridge is on a crossfall, it is necessary to build the soffit flat and have webs of different height. It is quite feasible to vary the crossfall along the length of the bridge by adjusting the height of the webs. Similarly, the depth of the girder may be varied along the length of the bridge if necessary.

'S'-shaped bridge decks have been built by this method, launching from each end, with the transition spirals accommodated on a widened deck, and the varying crossfall dealt with as described.

Although launched bridges are usually continuous over their length, it is not uncommon for railway bridges to be divided up into relatively short expansion lengths. The intermediate expansion joints must be temporarily locked by prestress during launching.

15.8.3 Launching bending moments

The self-weight bending moments for the internal spans of a long deck with equal spans are $-0.0833\ pl^2$ at each pier and $+0.0417\ pl^2$ at each mid-span, where p is the unit weight of the deck and l is the span length. When a deck is launched, each section of the bridge deck passes successively from being a support section to a mid-span section.

However, if the front end of the deck had to cantilever from one pier to the next, the bending moment over the leading pier would be $-0.5\ pl^2$, six times higher than over the standard piers. In order to reduce this cantilever bending moment, a steel launching nose is attached to the leading edge of the deck. The launching nose is much lighter than the concrete deck, and its length is typically 65 per cent of a span. Thus it will land on the next pier when the deck is cantilevering just 35 per cent of a span length. As the launching nose is flexible, the bending moment in the deck over the pier will continue to increase as the deck is launched forwards to reach a maximum value of about 1.3 times the typical value; that is approximately $-0.105\ pl^2$, Figure 15.36. When the front of the concrete deck is over the forward pier, the sagging moment in the adjacent span will be approximately $+0.07\ pl^2$, the normal continuous beam moments being relieved only by the cantilevering weight of the nose.

The final launching self-weight bending moment envelope for a deck of equal spans consists of a constant hogging moment of $-0.0833\ pl^2$ and a constant sagging moment of $+0.0417\ \mathrm{pl}^2$, with an increase in both hogging and sagging moments at the front end of the deck. The moments at the tail end of the deck are also atypically smaller, Figure 15.36 (c).

In addition to the bending moments due to self weight, during the launch the deck must resist the effects of a temperature gradient, and of so-called 'differential settlement' of the launch bearings.

For the temperature effects, the most relevant model is the non-linear gradient of the type described in the UK code of practice (*6.13.3*). A simple linear temperature gradient is not adequate to calculate this important effect.

Launched bridges are generally of low span/depth ratio, 14 being typical, reducing to 10 or less when temporary intermediate towers are employed. Consequently, they are stiff in bending and torsion, and are very vulnerable to discrepancies in the vertical

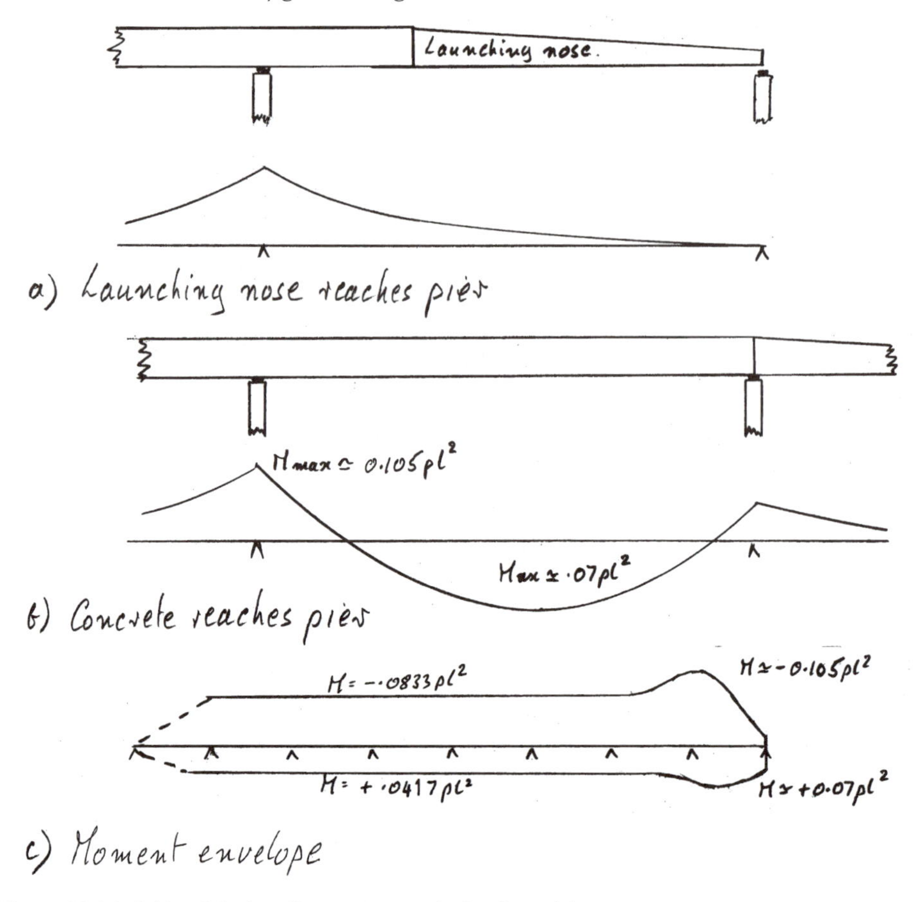

Figure 15.36 Self-weight bending moments during launching

alignment of the supports over which they slide. The 'differential settlement' consists of several components. The actual shortening of piers and foundations as the weight of the deck first comes onto them should be calculated, and introduced into the deck calculations explicitly. Once the first span length of the deck has passed over a pier, the reaction on it stays constant. However, if soil conditions vary along the length of the bridge the foundations may suffer deferred differential settlement, and a best estimate of this effect should be introduced into the calculations explicitly.

Temporary steel falsework towers will shorten elastically more than the concrete piers. Either this should be calculated and included explicitly in the deck calculations, or the launch bearings on the towers should be carried by hydraulic jacks that can adjust their level.

There remain the unquantifiable misalignments of the soffit which may consist of steps at construction joints, movement of the form under the weight of concrete, a slightly incorrect vertical radius given to the whole deck or errors in the levelling of the launch bearings. A typical value adopted for misalignment is ±3 mm, applied on successive piers in the most unfavourable way.

To put this possible misalignment into context, consider a 3 m deep bridge deck being built on a vertical radius of 10,000 m, in segments 20 m long. The height of the chord of the 20 m long segment of the circle is only 5 mm. If the soffit form were 1 mm out of true, giving a 6 mm height to the chord, the bridge would be built to a radius of 8,333 m. As the piers would define the correct vertical radius, the deck would be subjected to a sagging moment as it was launched. On the assumption that the top and bottom fibre distances were respectively 1 m and 2 m, in the short term, the stresses on the top and bottom fibres would be respectively +0.7 MPa and –1.4 MPa (using $E/R = \sigma/y$). In the longer term these stresses would reduce to less than a third of these initial values due to the effect of relaxation, and would become insignificant (*3.9.2*). However, during launching this very small geometric error significantly increases the risk of the section cracking. This example underlines the importance of designing a section that is reinforced to behave in a ductile manner as explained below.

15.8.4 Launching stresses

The choice of the limiting bending stresses to be observed during launching is fundamental to both the satisfactory behaviour of the bridge in the construction phase, and to its economy.

Some authorities insist on there being no bending tensile stresses in the deck during launching under the combination of self weight, temperature gradient and differential settlement. This leads to an excessive amount of first stage prestress, which is both uneconomic and gives a false sense of security. If the deck is designed to be fully prestressed during launching, there is considered to be little justification for providing any more than nominal longitudinal passive reinforcement. This leaves the deck brittle, and highly vulnerable to severe cracking in the event of unexpected bending moments. Such cracking may or may not require repair, but inevitably halts construction while its causes are investigated.

It is more rational to reduce the central prestress and to increase the passive reinforcement to render the deck ductile. For a bridge which is designed for a conservative client and is designated as Class 1 or 2 in service, the launching prestress may be designed to deliver zero tensile stresses under self weight plus predictable settlement effects. The effects of temperature gradients and ±3 mm of misalignment as described above should then be checked at Class 2 (tensile stresses not exceeding –2.5 MPa). The top and bottom slabs should be reinforced longitudinally with a minimum of 0.6 per cent of the slab area, providing for ductile behaviour in direct tension in the event of cracking (*3.7.2*). Such a design will not crack if the construction is carried out to a high standard, but if tolerances are exceeded, cracking will be controlled and within specification. However, it would be better to design the deck as partially prestressed during launching, with crack widths limited to 0.25 mm. This will reduce the central prestress and increase the longitudinal reinforcement, rendering the bridge more ductile. Once the second stage prestress is installed, there is no reason such fine bending cracks should not close, when the deck reverts to Class 1 or 2.

If the bridge deck is designed to be partially prestressed in service, then partial prestressing should also be adopted during launching. It must be checked that there is sufficient reinforcement in the top and bottom flanges to limit crack widths to the specified value and to provide for ductile behaviour at all points along the deck.

Figure 15.37 Blackwater Viaduct, Ireland (Photo: Strabag SE)

15.8.5 Launching in reinforced concrete

A further development is to launch the bridge entirely in reinforced concrete. For launching decks with a span/depth of the order of 8–10, which is typical when intermediate props are employed, this concept does not give rise to excessive quantities of passive reinforcement and makes major savings in prestress, eliminating the 20 per cent deficit noted in *15.8.1*. The permanent cables are installed when the launch is complete, as described in *15.8.16*.

If the final prestress is designed to a code of practice that accepts partial prestressing, much of the launching reinforcement is utilised for the final design. This gives rise to a very economical form of construction that is ductile during the critical launching phase. If the final prestress is external, it is feasible to install tendons that are hundreds of metres long, eliminating, or much reducing, the cost of intermediate anchorages.

This technique was pioneered by Benaim on the 300 m long Broadmeadow Viaduct in Dublin, Ireland, and re-used by Benaim on the design and build project for the 450 m long Blackwater Crossing Viaduct in Cork, Ireland, Figure 15.37, built by Strabag. The spans which ranged from 51 m to 58 m were halved by temporary supports for the launching phase. Once the 2.9 m deep deck had been launched, external prestressing cables were threaded from end to end without intermediate anchors, and the deck designed as partially prestressed.

15.8.6 Design of the webs during launching

The launch bearings should if possible be located beneath the point of conjunction of the axes of the web and of the bottom slab. However with webs which are typically 400 mm thick and launch bearings that are typically 500 mm to 800 mm wide, that is not always possible, Figure 15.38 (a). The bending moments caused by this eccentricity are likely to require heavy reinforcement along the entire length of the deck and to lead to the webs being cracked also along their full length. Furthermore this eccentricity creates rotations of the bottom corner of the box, creating stress concentration at the outside of the launch bearings. In order to minimise these problems, one option is to widen outwards the base of the web. In trapezoidal box girders that results in a short vertical plinth to the web, Figure 15.38 (b), while in rectangular box girders an 'elephant's foot' thickening results, Figure 15.38 (c).

When this is not desirable, the web may be thickened internally with a vertical fillet, Figure 15.38 (d). This reduces the eccentricity at the base of the web and creates a kink in the axis of the web, causing a horizontal resultant under the effect of the launch bearing reaction that induces additional bending moments in the web, Figure 15.38 (e). In general, the concentrated reaction of the launch bearing will have spread over a sufficient length of web to make these moments acceptable. It is essential to carry out this analysis carefully, on the complete box, as significant bending moments are also created in the bottom slab.

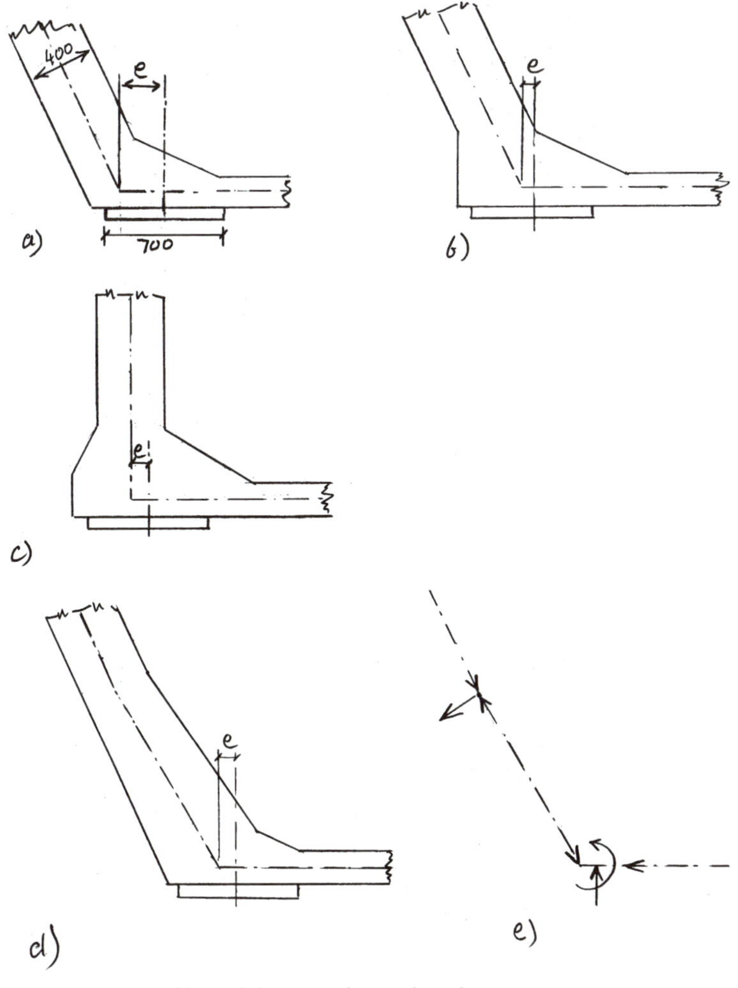

Figure 15.38 Eccentricity of launch bearings beneath web

As box section bridges are torsionally very stiff, it requires theoretically a misalignment of only 1 or 2 millimetres between the two slide bearings on a pier for the entire deck reaction to be carried by one bearing only. In reality this rigidity is mitigated by the use of laminated rubber (*15.8.10*) sliding pads, which have a small degree of vertical flexibility, and sometimes by the transverse flexibility of the pier. However, it should be assumed that each slide bearing may carry 75 per cent of the deck reaction, unless it can be demonstrated that the flexibility of the pier makes this impossible. The web should be checked at the ULS under the enhanced launch bearing reaction with a load factor of 1.33. Also, as it is undesirable that the web should crack significantly in bending during launching, the tensile stresses in the bending reinforcement of the web should not exceed 220 MPa under nominal launch bearing reactions. It is also necessary to check the shear in webs during launching, under the enhanced launch bearing reaction.

15.8.7 Casting area

The casting area is normally set back some 30 m behind the abutment, so that it is not affected by the deflections of the deck as it spans from the abutment to the first pier. If it is impossible to extend the construction site by this amount, one or two temporary falsework towers may be placed within the first span to reduce the deck deflections.

The most critical components of the casting area are the beams that carry the formwork for the strips of the soffit that engage with the launch bearings. These must be aligned to a tolerance of the order of a millimetre as they define the shape of the bridge deck in both the vertical and the horizontal planes. They usually consist of substantial steel or reinforced concrete beams, and must be either very solidly founded, or mounted on levelling jacks.

Usually a deck segment is one-third or one-half of a complete span. For a deck with regular spans, this simplifies construction, as special features of the deck such as web thickenings and pier diaphragms occur at the same place in the formwork for each successive span. The box deck is usually cast in two phases, bottom slab plus webs, followed by the top slab. Some contractors prefer to build each phase in consecutive casting areas as the labour force is less crowded and thus more productive, and the bottom slab concrete that is engaged by the launching system, is two weeks old when it is first put under stress, Figure 15.39. However this requires a longer casting area which may not be available.

The formwork is steel, and the internal shutter must be designed to cope with thickenings of the web and bottom flange, anchor pockets, tendon deviators, diaphragms and other items. This can become complicated when the bridge spans are not regular. It is worthwhile using high strength concrete for such bridges to allow webs and bottom flanges to remain of constant thickness, as this not only simplifies the mould but also reduces the interruptions to the programme.

In order to achieve the weekly cycle, some contractors prefer to prefabricate all or part of the reinforcing cage. The design for tender of an 800 m long viaduct for the rear access road to Sizewell B power station, prepared by Benaim for the contractor (Fairclough), was based on the construction of 30 m long complete spans on a weekly cycle, with prefabrication of the entire cage.

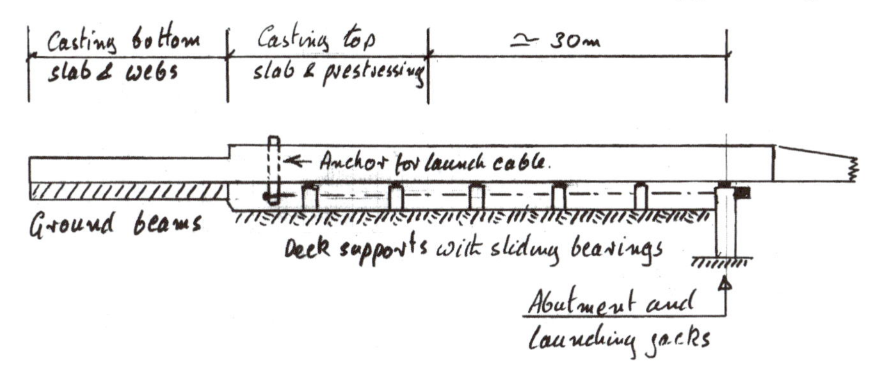

Figure 15.39 Typical arrangement of casting area

15.8.8 First stage prestress

Once a box segment is complete, it is attached to the previous segment by straight prestress tendons. These are situated in the top and bottom flanges, and the centroid of their force should correspond approximately with the centroid of the cross section, so the deck is centrally prestressed.

The central spans being effectively built-in, straight prestress will always be re-centred; that is the prestress parasitic moment *Mp* will equal the prestress primary moment *Pe*, but with the opposite sign (6.6). Thus this first stage prestress provides a uniform compression on the section even if its centroid does not lie precisely on the neutral axis. However, within the leading span of the launched deck, *Pe* will not be balanced by *Mp*, and this may be used to advantage by the designer in defining the additional prestress required.

As the self-weight hogging bending moments are twice the sagging moments, the ideal deck section properties would give a bottom fibre modulus that is half the top fibre modulus. The tensile bending stresses on the two extreme fibres would be identical and the launching prestress would then be minimised.

It is good practice for the tendons to be two or three segments long. This shortens the programme by reducing the number of tendons to be stressed at each cast, reduces the bursting stresses on the concrete, reduces the congestion of anchorages, and improves the economy of the project as the longer the tendons, the lower their unit cost. The tendons are stressed from live anchors at the rear of the new segment, and are dead anchored in pockets in the previous segment, Figure 15.40.

Bars are sometimes used for the first stage prestress. They are simple to detail and to couple, and avoid the necessity for internal anchorage blisters, simplifying the internal shutter. However they are much more expensive per ton of prestress force, and it is generally necessary to stress all bars at each construction stage, further increasing the cost. Their principal advantage would appear to be to the designer rather than to the contractor or the owner.

The uniform compression on the cross section due to the first stage prestress is in general beneficial for the deck in its permanent state, reducing the amount of second stage prestress required. Consequently, it is generally left in the deck, forming part of the permanent works. However, if the compressive stresses in service on the bottom

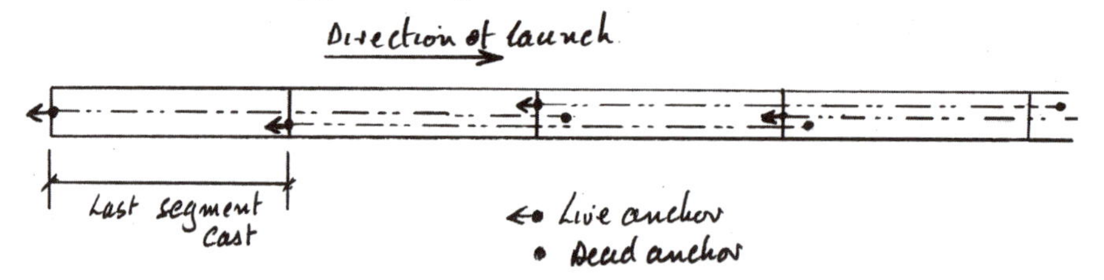

Figure 15.40 First stage prestress two spans long

fibre at the pier are close to the maximum permissible, some or all of the bottom fibre tendons located over the piers may be de-stressed after completion of launching.

With the exception of tendons due to be de-stressed, the first stage prestress should preferably be grouted as soon as possible after installation to provide greater security against unexpected events. However, it is perfectly feasible to launch a bridge using external, unbonded prestressing tendons, as long as the designer considers the issue of protection of the strand and the behaviour of the deck in the event of an accident.

As part of the total prestress of a launched deck is central, the total consumption of prestress steel is higher than for a deck built by other methods. Designers have frequently exercised their ingenuity to find ways of decreasing this disadvantage. For instance, decks have been launched with their permanent 'sinusoidal' prestress together with the temporary tendons installed on an opposing sinusoidal profile. The addition of the two sets of tendons gives the central prestress required for launching. At the end of launching, the temporary prestress is removed, and either re-used for a subsequent deck, or re-installed in the deck to complete the permanent prestress.

Another possibility frequently mooted, but never, to the author's knowledge actually adopted, is to use as temporary prestress external tendons installed along the neutral axis of the box, anchored in pier diaphragms every two or three spans. At the end of launching, these temporary cables would be deviated by hydraulic jacks, up at piers and down in the span, to form the profile of the permanent prestress.

A further option was adopted on the Liu To Bridge, and is described in *15.8.18*.

15.8.9 The launching nose

The launching nose usually consists of two tapering stiffened plate girders, typically 65 per cent of the length of the launching span, wind braced and cross braced together, Figure 15.37. The girders are usually 1.5 m deep at the front, and the same height as the deck at the rear. Due to the torsional stiffness of the deck, the nose should be designed so that one girder may take the entire reaction without permanent damage. When the nose of the launching girder approaches a pier, it is likely to have a deflection of at least 100 mm and it is common practice to equip it with landing jacks that lift it to the level of the slide bearings.

The launching nose is usually attached to the deck by short prestress bars. The connection to the deck must resist principally sagging bending moments and the shear force that arises just before the concrete span lands on the pier. Although in theory the launching prestress may be used to attach the nose, this is not usually very practical, as it is necessary to dismantle the nose, and hence de-stress these cables before the second stage prestress is installed. However, this does not mean that the designer cannot find an economical compromise, perhaps mixing short bars with some of the tendons of the launching prestress.

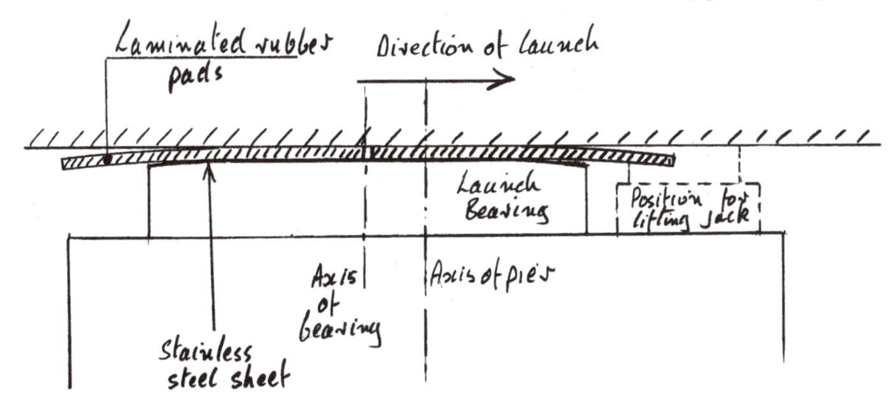

Figure 15.41 Launch bearings and pier head

15.8.10 The sliding bearings

The sliding bearings consist of concrete blocks covered with a polished stainless steel sheet. They have lead-in and exit tapers at each end. To allow the deck to move over them, pads consisting of two layers of rubber laminated with a central steel sheet, bonded to a PTFE lower surface are fed between the deck and the rear end of the bearings, Figure 15.41. The pads are collected as they are expelled from the front of the bearing and re-fed in at the rear. The bearings and rubber pads should be designed to work at 12 MPa under the nominal deck reaction, giving a stress of 18 MPa under the enhanced reaction described in *15.8.6*. The bearings should be as short as possible, as small deck rotations have to be accommodated, particularly at the front and rear of the launch. Making the bearings too wide, however, may increase the bending moments on the webs.

15.8.11 The launching jacks

The deck is usually moved forwards by one of two systems. In the first, two hollow ram jacks reacting against the abutment pull on cables that pass beneath the casting area and are anchored to the newly cast deck. Two jacks are used so that when one retracts, the other may hold the deck. Alternatively, a lifting jack located on the abutment reacts against the bridge deck through a sledge device that is equipped with the pushing jack. Again two devices working in tandem are normally used. Benaim designed the launching methods for the approach spans either side of the 450 m span cable-stayed Kap Shui Mun Bridge in Hong Kong, with a different contractor for each side. Although the two structures were identical, the two contractors adopted rival launching systems. There cannot be much to choose between them! The concrete approach spans used the first sections of the steel main span as a launching nose, Figure 15.42.

The jacks must have sufficient force to overcome the sliding friction added to any gradient effects. If the deck is launched downhill, it is essential to have hold-back systems. The friction experienced during launching is very variable. Generally it can be expected to be of the order of 2.5 per cent on the slide bearings and 10 per cent or more in the casting area. However the jacks should be designed for an average friction of 5 per cent. Friction coefficients of less than 0.5 per cent have been measured and holding-back systems are best designed on the assumption that friction is zero.

15.8.12 Effect of launching on the piers

As the bridge is launched, it applies a friction force to the top of the pier head, creating an overturning moment on the pier and its foundation. The friction force should be assumed to be 5 per cent of the deck reaction which is greatest on the leading pier. If the deck is on a longitudinal gradient, the reaction between the deck and the pier will be inclined to the vertical, giving a horizontal reaction on the pier head, Figure 15.43. This will be in addition to the friction.

In order to reduce the overturning moment on a pier during the launch, the launch bearings may be set behind the pier axis, providing a moment in the opposite direction to the friction, Figure 15.41.

Occasionally the sliding bearings may jam, principally due to a rubber pad being introduced upside down. If this is not spotted quickly, the pier could be severely damaged by the power of the launching jacks. It is essential that there is a system that stops the launch if the horizontal force at the head of any pier exceeds some pre-set limit. This may be achieved by surveying the deflection of each pier head during the launch, or by measuring the distance between each pier head, so any anomalous deflection of a single pier is quickly identified.

If the permanent deck bearings are sliding, the pier must be designed for a horizontal load in service of approximately 5 per cent of the reaction due to self weight + finishes,

Figure 15.42 Kap Shui Mun Bridge, Hong Kong: launching of approach span (Photo: Benaim)

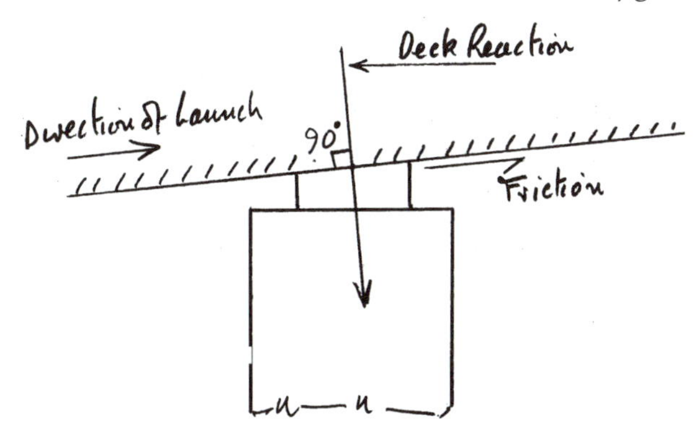

Figure 15.43 Forces on pier head

which is likely to be similar to the launching load, in the absence of gradient effects. However for flexible piers with pinned bearings, or piers with laminated rubber bearings, the launching load is likely to be the higher than the service load.

15.8.13 The pier head

The pier head must be designed to accommodate the jacks necessary to lift the deck in the event of a jammed rubber pad. As they may be required at any stage of the launch, these jacks must be situated close to the axis of the web. If the launch bearings have been set off centre on the pier as described above, it is often possible to accommodate a single jack per web in front of and in line with the launch bearings, Figure 15.41. Alternatively, two smaller jacks may be used, fore and aft of the launch bearings. Accommodating these jacks often leads the pier head to be elongated, which may look rather inelegant. When it is not desirable to increase the size of the pier head in this way, the jacks may be carried by falsework, either resting on the pier foundation, or stressed to the pier head. However, it is not economical to provide falsework for every pier for an eventuality that may not occur. Consequently, this falsework needs to be designed so that it may be erected quickly in the event of need, minimising delays to the construction programme.

The pier head must also be designed to accommodate the jacks required to transfer from the launch bearings to the permanent bearings. As the deck generally includes diaphragms at pier positions, these latter jacks may be situated in-board of the permanent bearings. These jacking positions may subsequently be used for maintenance of the permanent bearings. When the pier consists of two columns, the transfer jacks may be accommodated on falsework.

Every third pier is equipped with guides to avoid the possibility of the deck drifting sideways during the launch.

15.8.14 Launching procedure

As each segment is completed, it is attached to the previous segment by the first stage prestress and the whole deck is launched forwards, freeing the casting area.

Each sliding bearing must be attended during the launch to insert the laminated rubber pads at the back of the bearing and to collect them as they are ejected at the front of the bearing. If the bridge is long this can require a considerable labour force, and traditionally contractors mobilise their office staff for this duty.

The hydraulic pressures of the launching jacks and thus the launching force must be monitored continuously and compared with a predicted force. Any anomaly in the force must be investigated and explained. For instance, if a rubber pad jammed on a pier, it may be possible to register the increased force of the jacks before damage has been done. However, if there are many piers, the increased force in the launching jacks due to one jammed bearing may not be noticed before the pier has been damaged. Consequently, this is only a back-up to the security system described in *15.8.12*.

If the deck is launched over a vertical curve, the launching force may rise to a maximum and then reduce as the weight of the deck on the downhill side increases. Clearly this will only happen if the downhill gradient is in excess of the friction coefficient. However, as has been noted above, the design of the launching procedure must take into account the possibility that the friction coefficient may be only marginally above zero.

The launching nose is usually designed so that it may be dismantled in sections. This avoids having to accommodate its full length in an excavation behind the abutment at the end of the launch.

Once the deck is complete and in its final longitudinal position, the second stage prestress must be installed, and it must be jacked up pier by pier to install the permanent bearings. Generally these operations may be carried out simultaneously.

15.8.15 Alternative launching method

An alternative launching method, patented by Bouygues, may be used when launching up a steep gradient, where the design launching jack force using the traditional method would be $W \times 5 / 100 + W \times x/100$, where W is the weight of the deck and x the gradient in percent.

In this method, horizontal roller bearings are placed on the abutment and on each pier. Lifting jacks on each pier raise the deck by an amount that corresponds to x per cent of the horizontal stroke. The rollers are pulled back to the beginning of their travel, and placed into contact with the deck. The lifting jacks are released, transferring the weight of the deck to the rollers and the launching jacks at the abutment push the deck horizontally by one stroke. The lifting jacks then lift the deck off the rollers and the cycle is repeated. The friction of roller bearings is more predictable than PTFE, and is reported to be 0.5 per cent.

15.8.16 Second stage prestress

The second stage prestress may be internal or external. In any event, it is economical to plan for long tendons. If the bridge is less than 150 m long, tendons from end to end should be adopted. For longer bridges with internal prestress, tendons two or three spans long are usually used. Such tendons are often anchored at pier diaphragms, overlapping to provide continuity of effect. External tendons in excess of 400 m long have been used successfully by Benaim on the Broadmeadow and Blackwater bridges (*15.8.5*).

Some of the launching prestress, such as bottom tendons in the support zone and top tendons at mid-span, are in the wrong place for the long term. As they are outside the kern, they actually increase the amount of second stage prestress required, and could be designed to be removed or de-stressed. This is in fact rarely done, and it needs demonstrating that the saving in second stage prestress more than outweighs the cost of de-stressing.

A good solution, which is rarely adopted due principally to the additional work it creates for the designer, is to remove unwanted first stage tendons and to re-install them as second stage tendons. This can be done most elegantly by sliding them forwards from, say, the bottom fibre support position to the adjacent bottom fibre mid-span location. It may be necessary to cut off the ends of the tendons that have been marked by the first stage stressing jacks. It goes without saying that such a strategy depends for its success on very careful planning of the prestressing scheme. It was used on the Liu To Bridge designed by Benaim and is described in *15.8.18* below.

15.8.17 Construction programme

In a conventional launched bridge deck, where 15–25 m segments are being built, the following programme should be achieved. The first segment to be built is non-typical, as it includes the arrangements for attaching the launching nose. Building this segment is likely to take about 6 weeks. The first 'typical' segment may be expected to take 4 weeks, the second 3 weeks, the third and fourth 2 weeks each and from the fifth segment onwards the weekly cycle should be achieved.

A typical weekly cycle consists of:

Monday	Stress tendons, strip formwork, launch, clean formwork
Tuesday	Level formwork, fix reinforcement and tendons for bottom slab and webs
Wednesday	Place internal web formwork, cast concrete for bottom slab and webs
Thursday	Remove web formwork, place slab formwork, fix top slab reinforcement
Friday	Cast top slab
Saturday	Concrete curing
Sunday	Concrete curing

However, there are many alternative programmes that may be adopted. For instance, at the Upper Forth Crossing at Kincardine in Scotland, designed by Benaim, the contractor Morgan-Vinci has opted to build complete 45 m spans on a two-week cycle. Furthermore, there would appear to be no good reason why the techniques used for the prefabrication of complete spans at the Poggio Iberna Viaduct, described in *15.9.1*, could not be transferred to incremental launching. This would allow two complete spans to be launched per week.

Once the launch is complete, installing the second stage prestress should not take more than a week. The deck must also be transferred onto its permanent bearings, which, if well planned, should not take more than one or two days per pier. If the bridge is long, this operation may well be critical in allowing the completion of the contract, and several teams may be mobilised.

Other operations may well determine the contract duration. The launching nose and the casting area must be removed, and the two abutments completed. The deck finishes, including parapets, balustrades, verges, waterproof membrane and asphalt must be installed. When the programme is critical, the concrete parapet may be installed during the launch. This is best carried out forward of the deck casting area, typically over the first span of the deck. As launched decks are relatively stocky, the deflections imposed on the parapet by the second stage prestress will be imperceptible. If the parapet is cast in-situ it is necessary to create joints so that it is not subject to damaging bending stresses during the launch. The first stage launching prestress will need to be increased significantly to carry this extra weight, and the cost of this extra prestress must be judged against the shorter programme. Once the parapet is in place, the metal balustrade and the waterproof membrane may also be installed during the launch. Verges are likely to be too heavy to be built economically during the launch, while the asphalt is best laid in one operation once the launch is complete.

15.8.18 Variations on the theme

As with everything in this book, the author is describing how and why things are done, in the hope that it will stimulate the inventiveness of engineers to take matters further. Launching a concrete bridge into place is such a sensible solution, and there are many situations that differ from the standard where it can be adapted.

The Liu To Bridge in Hong Kong crossed a deep valley with spans of 50 m, 70 m and 50 m and had been designed to be built in cast-in-situ balanced cantilever. The bridge was straight in plan and elevation. Peter Baum, Technical Director of the French contractor Dragages et Travaux Publics, offered a tender with an innovative launched alternative, and once they had won, commissioned Benaim to carry out the detailed design.

The deck consisted of two adjacent box girders joined by a top slab. The launching formwork was sized to build 5 m lengths of one box at a time, on a 2–3 day cycle. A temporary tower divided the end span into two, with launching spans of 20 m and 30 m, and no launching nose was necessary. When the 50 m side span and half the 70 m main span had been built and launched forwards, the box was slid sideways across the pier and abutment, and the second box was built and launched. The top slabs of the two boxes were then stitched together. The falsework was moved across the valley and the operation was repeated. The deck was then joined at mid-span by a cast-in-situ stitch, Figure 15.44.

The launching prestress was designed so that a significant tonnage could be de-stressed and re-used as permanent prestress. The ends of the tendons that had been marked by the grips of the anchorages were discarded.

Each complete half bridge was statically determinate, and could be re-levelled by jacks on the abutment and pier. Thus the difficulty of creating an accurate longitudinal profile with a shutter only 5 m long was not relevant.

Figure 15.44 Liu To Bridge, Hong Kong (Photo: Dragages et Travaux Publics, Hong Kong)

15.9 Prefabrication of complete spans

15.9.1 The Poggio Iberna Viaduct

The ultimate form of prefabrication for the construction of bridge decks is when a complete span is built in a casting yard and then transported to the construction head and launched into place. The construction of box section decks by this method has been perfected by the Italian contractor Ferrocemento [4], on several major road and rail viaducts. One of these was the Poggio Iberna Viaduct on the Autostrada Livorno–Civitavecchia near Cecina in Italy, where the deck consisted of 42 m span units weighing some 900 tons.

On this site, Ferrocemento were building a pre-tensioned deck in a mould set in line with the bridge at an average rate of 12 spans per month, or a span every two working days. The reinforcing cage for the bottom slab and webs, including the prestressing strands, was assembled in the mould, while the cage for the top slab was prepared in a jig behind the mould. Once complete it was carried by a pair of gantry cranes and married up to the lower 'U'. The prestressing strands were then tensioned and the concrete poured and steam cured. Once the concrete was adequately cured, the core shutter was collapsed hydraulically and withdrawn in one piece, the pre-tensioning strands were cut, and a rail transporter rolled beneath the unit to lift it out of the mould and deliver it along the completed deck to the construction head. A very highly mechanised and automated gantry weighing some 740 tons, designed and fabricated by Paolo de Nicola [5] then picked up the deck, placed it on its bearings, and subsequently launched itself forwards to await the next span. The entire cycle of launching the deck and then moving the gantry forwards to the next span took no longer than 4.5 hours. This exceptional site demonstrated the possibilities of this form of construction where long viaducts need to be built quickly, Figure 15.45 and Figure 15.46.

It may be considered that the launcher was too sophisticated for its task, as it spent most of its time idle, waiting for the next span. It would have been better suited to a site equipped with multiple moulds, capable of producing several deck units per day.

The deck was designed to be continuous, the continuity being created by casting a narrow concrete stitch over the piers and stressing capping cables. However, as decks were delivered along the completed viaduct every two or three days there was no time to cast and cure the joint, and to install, stress and grout the capping tendons.

Figure 15.45 Poggio Iberna Viaduct, Italy: launching gantry (Photo: David Benaim)

Figure 15.46 Poggio Iberna Viaduct, Italy: deck unit on its transporter (Photo: David Benaim)

Consequently the entire operation of creating continuity had to be deferred until all the decks on one carriageway had been placed, significantly lengthening the construction programme. It would appear that continuity cannot be an economical option for such a bridge, as each deck carries, in its statically determinate state, the weight of a complete prefabricated deck plus the transporter, which is at least as great as its service load. However, there were clearly other reasons why continuity was adopted in this case.

As the method requires a substantial investment in temporary works, it is probably relevant to viaducts with at least 80 regular spans, and a deck area greater than 40,000 m^2, when it will become more economical than the precast segmental technique. If a contractor already had the falsework and the expertise from a previous contract, the method would be viable for considerably smaller projects.

15.9.2 The Guangzhou–Shenzhen–Zuhai Superhighway Project: the Pearl River Delta Viaduct

The author used the lessons learned in observing the Poggio Iberna site in responding to a commission from Tileman (S.E.) Ltd, a subsidiary of Slipform Engineering, itself a subsidiary of Hopewell Holdings of Hong Kong, to design a 17 km twin deck viaduct to carry the Superhighway across the Pearl River delta in Southern China [6]. The design was carried out with the assistance of Stuart Elliott of Tileman. The road consisted of dual three-lane carriageways, and the viaduct included three major river crossings and three interchanges. The non-typical interchange structures were designed by Tileman. Programme was of the essence, and the period allocated to the erection of some 28 km of deck, with a plan area of about 430,000 m^2, was initially of the order of seven months. Some 860 decks had to be built in 210 days, an average of over four spans per day.

Full span precasting was adopted as the only technique that could meet the programme. Placing gantries were designed that had a cycle time of eight hours. Thus two spans per day or three spans with 24-hour working, for each of two machines, could be achieved.

The deck consisted of a statically determinate single-cell box girder 32.5 m long, 15.75 m wide and 2.3 m deep, weighing 630 tons, Figure 15.47. As the core mould had to be withdrawn through the end of the deck, it was necessary to reduce to a minimum the end diaphragms. The diaphragm for the typical decks consisted of a 150 mm external thickening of the webs and bottom slab over a length of 1,300 mm, and an internal thickening of webs and top slab over a 1,500 mm length, creating a stiff frame. This geometry allowed the bearings to be placed close to the axes of the webs, and provided the strength necessary to resist the sway forces. The side cantilevers were stiffened at the deck ends by short brackets which also provided strong points for lifting the unit.

The casting yard was designed to produce one complete deck per day per mould. The precasting cell consisted of a soffit mould beneath the webs, raised to allow a transporter running on rails to penetrate beneath the segment, fixed steel side shutters and a hydraulically operated, collapsing core mould. The trapezoidal shape of the box would allow the deck to be lifted directly out of its side shutters. There were also suspended moulds for the inside face of the New Jersey parapets that were to be cast together with the deck. It was planned to prefabricate the bottom slab of the box, to shorten and simplify the casting of the concrete.

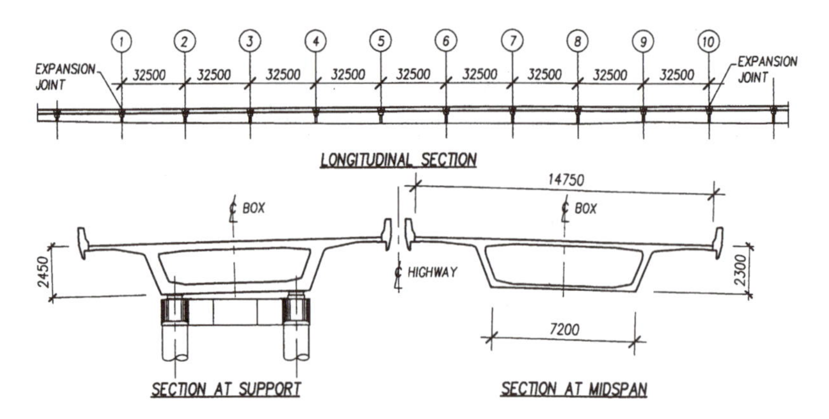

Figure 15.47 GSZ Pearl River Delta Viaduct, PRC: general arrangement of box deck option

In order to allow a daily turn round, each mould would be serviced by a production line, Figure 15.48. This consisted of a mould for the prefabrication of the bottom slab of the box and three jigs for the preparation of reinforcing cages for the webs and top slab of the box. Each production line would be serviced by a pair of gantry cranes that handled the reinforcing cages, bottom slab and the core mould. It was planned to install immediately four moulds together with their production lines, with the provision for four more if it was necessary to accelerate the works.

The prefabricated bottom slab would be placed in the mould followed by the prefabricated reinforcing cage which included the ducted post-tensioning tendons, then concrete for the deck and the New Jersey parapet would be cast and steam cured overnight. In the morning, when the concrete had reached a strength of at least 20 MPa, a first stage of prestress would be applied by partially stressing some tendons equipped with oversized anchor plates. The deck would be jacked up out of its mould, rolled forwards on rails and parked. This would free the mould for the next cycle. 24 hours later, the prestress would be completed, and the deck moved to storage.

With several units a day moving over the viaduct, it was considered impractical to carry out any sustained work on the decks once they had been erected. Consequently all cast concrete was excluded from the deck erection cycle. As the viaduct alignment included horizontal and vertical curves, the ends of some of the decks had to be angled slightly. Other specialisations included decks adjacent to expansion joints, occasional shorter spans and special parapet details. It was considered that if the units were to be prefabricated in the order in which they were to be used, there would be too much risk

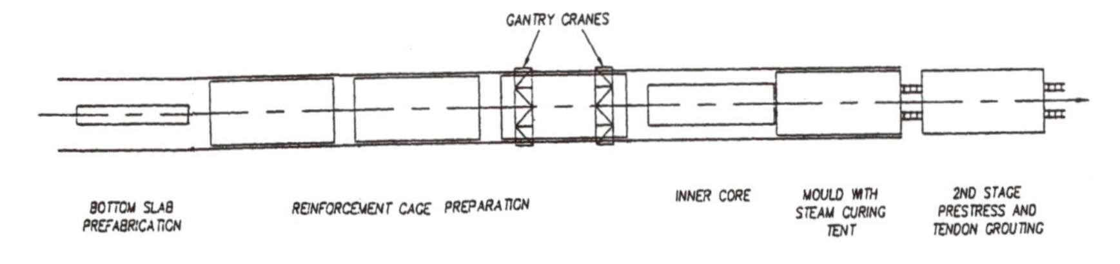

Figure 15.48 GSZ Pearl River Delta Viaduct, PRC: production line for prefabrication of one box per day

of delay to the programme. Consequently, it was necessary to invest in a storage area, serviced by a multi-wheeled pneumatic-tyred transporter, which would create a buffer between prefabrication and erection, and which would also give the opportunity to carry out any repairs to the decks.

Groups of up to nine statically determinate spans were linked together to create expansion lengths of some 300 m. They were joined by stainless steel Macalloy bars anchored on the end diaphragms. These bars were sized to resist the bearing drag, braking and seismic forces. They were to be stressed against a joint filler poured between the top slabs of the spans. This filler had to be strong enough to resist the prestress of the bars, flexible enough to accept the rotation of the decks ends in service, and had to set in a few hours. The filler used was an appropriate grade of Sika Rail currently marketed under the name ICOSIT [7], providing the required combination of flexibility and strength.

The gantry designed by Benaim for this project consisted of a 6 m deep steel truss 75 m long and weighing 300 tons, with legs at each end plus a third set of mobile legs, Figure 15.49. When waiting for a new deck unit, the gantry was carried by its front leg resting on the forward pier, and on its central leg resting on the rear pier. The rear leg was lifted to allow the new span to roll beneath the gantry on its rail transporter, Figure 15.49 (a). The rear leg was then lowered, and the centre leg disengaged and nested with the front leg. The gantry was now spanning 67 m. Two mobile crabs running on its top flange then lifted the new deck unit, Figure 15.49 (b). The crabs rolled forwards carrying the unit until it was over its bearings, and then lowered it into place. Every action involved in adjusting the position of the new span and placing it on the laminated rubber bearings had to be analysed and refined to fit in with the 8-hour cycle. The gantry launched forwards resting on its rear leg that rolled on short rails placed on the deck, and on the central leg that was equipped with rollers at its head, Figure 15.49 (c). All the operations comprising the placing and launching sequence were analysed in detail, and it was concluded that the cycle time was at most 8 hours, most probably reducing with practice.

Full working drawings of the decks, foundations and placing gantries were completed by Benaim while the design of the casting yard was completed to an advanced stage. Two of the moulds for the decks, Figure 15.50, were completed by Ninive Casseforme [8] of Italy and shipped to Hong Kong, and some of the heavy plant for the casting yard had been purchased. When many kilometres of the foundations for the viaducts had been completed, the project for the deck was modified to allow it to be built by local contractors rather than by a highly specialised company. It was finally built to a different time scale with a deck consisting of precast T beams, also designed by Benaim (*10.3.8*).

It is hoped that this brief description of the project makes clear that achieving a rate of deck erection of several spans per day per placing gantry is possible, but requires careful integration of the design of the deck and substructure with all stages of the construction process.

Benaim have carried out preliminary schemes that involve launching decks weighing up to 1,500 tons, such as that proposed to launch into place prefabricated two-level decks for Hopewell Holding's Bangkok Elevated Rapid Transit System, Figure 15.51.

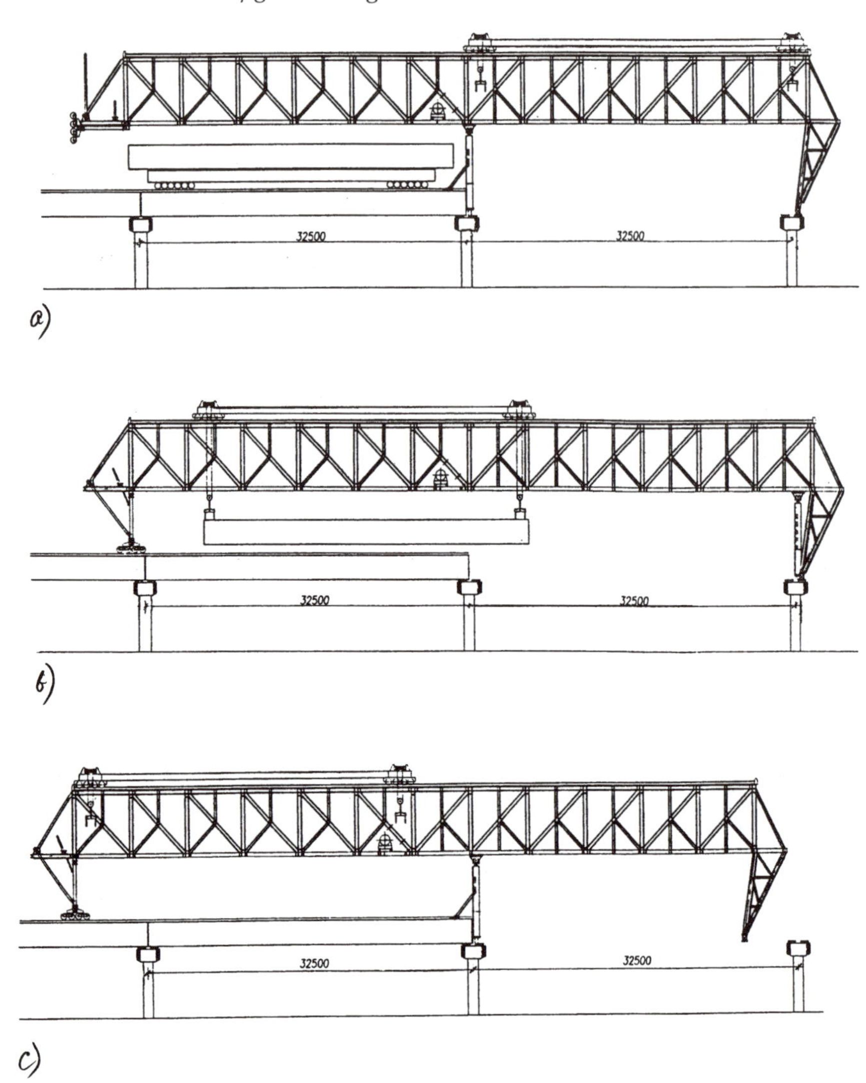

Figure 15.49 GSZ Pearl River Delta Viaduct, PRC: operation of launching gantry

Figure 15.50 GSZ Pearl River Delta Viaduct, PRC: mechanised mould for complete span (Photo: Ninive Casseforme srl)

Figure 15.51 Bangkok Elevated Rapid Transit System: proposed gantry for erection of 1,500 t decks (Image: Benaim)

15.9.3 Comparison between precast segmental and the whole-span methods

This comparison is extrapolated from work carried out in 1996 by Benaim for a project in South East Asia. The costs of moulds for the two systems were based on firm quotations at the time, and have not been updated. All the other equipment costs have been estimated. The number of men required for segmental construction is based on experience, while that for the whole-span site is estimated.

The project required the construction of two parallel viaducts 6 km long, each consisting of 170 spans of 35 m. The comparison was based on a requirement to build the deck in 8 months, a rate of 42.5 spans per month.

For the segmental option, 12 segments per span gave a total number of 4,080. It was assumed that segments could be produced at the average rate of 20 segments/mould/month. Thus for 510 segments per month, 25 moulds would be required. Each casting cell cost £130,000, and required a team of 15 men. Additional equipment, including foundations for the mould, jigs for reinforcement, roofs over the casting cells, overhead cranes for handling the reinforcing cages and delivering the concrete, survey equipment, vibrators etc were estimated at £60,000 per mould. Two straddle carriers were assumed, for handling the segments in storage at a cost of £100,000 each.

The segments were to be placed by overhead gantries weighing 150 tons that were assumed capable of placing two spans per week on average. Thus five gantries were required, each operated by an erection team of 10 men. Each gantry was assumed to cost £450,000 and was supplied with segments by purpose-made transporters costing £100,000 each.

For the whole-span method, it was assumed that each mould, supported by a production line, would be capable of producing one deck per day, or about 20 decks per month, giving a requirement of two moulds for this comparison. Each mould cost £350,000, while the supporting production line, including foundations for the mould, jigs for reinforcement, two 50 t capacity gantry cranes to handle the precast bottom slab and the reinforcing cages and the steam curing and concrete delivery equipment may be estimated to cost £500,000 per production line. It is estimated that 50 men would be required for each production line.

With only two moulds operating, it may be assumed that there is no need for a storage and sorting area; each precast unit moves directly onto the deck. Thus the only deck-handling equipment consists of the single rail mounted transporter weighing some 120 tons, and costing some £500,000. The transporter requires 6,000 m of rail, which may be second hand and sold on afterwards. As the rails weigh 50 kg/m each and four are required, there is a total of 1,200 tons at £200/ton, costing £240,000. Only one gantry capable of placing two decks per day of the type described in *15.9.2* would be necessary, weighing 300 tons, costing about £900,000 and operated by a team of 10 men.

The final balance sheet is as follows:

Segmental erection, equipment:	
25 casting cells and supporting equipment	£4.75 m
2 straddle carriers	£0.20 m
5 gantries	£2.25 m
5 segment transporters	£0.50 m
Contingency 15 per cent	£1.16 m
TOTAL	£8.86 m
Segmental erection, labour:	
Casting yard	375 men
Erection gantries	50 men
TOTAL	425 men
Whole-span erection, equipment:	
2 moulds and supporting production lines	£1.70 m
1 deck transporter	£0.50 m
1 deck placing gantry	£0.90 m
Rails	£0.24 m
Contingency 15 per cent	£0.50 m
TOTAL	£3.84 m
Whole-span erection, labour:	
Casting yard	100 men
Placing gantry	10 men
TOTAL	110 men

However approximate this comparison may be, it demonstrates the efficiency and economy of the whole-span erection system for large projects. In addition to the savings shown, the concentration of the activities of precasting and erection in very few locations limits the number of supervisors required, and makes more feasible 24-hour operations where necessary. In particular, the reduction in the amount of labour required is very striking. In fact, the Poggio Iberna site was remarkable for the small number of people visible, and for the impression of calm, orderly efficiency.

16

The effect of scale on the method of construction

16.1 General

In the previous chapters, the various methods of designing and building girder type bridge decks, both cast in-situ and precast have been described. There are many factors that determine which kind of bridge deck and which type of construction method are the most economical or most the appropriate. Among such factors are terrain, climate, appearance, local traditions, as well as the experience and prejudices of the client, designer and contractor. However, one of the most important factors is the scale of the project. In this chapter, a 22 m wide highway bridge deck will be considered, and the effect of the scale of the project on the design and construction decisions will be explored. Thus this chapter serves as a synthesis of the previous chapters describing the various types of girder bridge deck.

The width of 22 m has been chosen as it corresponds to a normal dual two-lane carriageway and may be built as one wide deck or as two narrower decks, either linked together or kept separate.

16.2 A bridge length of 130 m on four spans

16.2.1 Cast-in-situ

This bridge is too short to justify much investment in plant; the contractor needs to use what he has in the yard, or what he can hire readily.

The most likely form of construction for this bridge is a continuous structure cast in-situ on falsework. The creation of repetition in the construction of concrete bridges is in almost all cases highly desirable. As a labour force repeats a cycle of operations, it becomes more productive. In the author's experience, a complex sequence of operations repeated five times or more will at least double in productivity, that is the fifth operation will require half as many man-hours than the first; in many cases the productivity can increase by a factor of three or more. It is worth designing this bridge to generate such repetition.

Thus the most rational choice would be to opt for two adjacent decks, and to build them span-by-span on simple falsework, giving a sequence of operations repeated eight times. Building span-by-span means that after each phase, the bridge deck is rendered self-supporting, allowing the falsework to be moved forwards. The ideal ratio of side to main span for a continuous prestressed concrete deck of constant depth is 0.9, leading to spans of 30 m, 2 × 35 m, 30 m.

The temptation of many designers is to opt for a box section deck, as it is known to be the most efficient, and can, if well proportioned, yield the lowest quantities of concrete and prestressing (but not of reinforcement). However, as many experienced bridge designers know, the construction of box sections in-situ is not an easy process (*11.7.1*, *13.2*).

The most economical deck for this bridge would be a twin rib type, as described in Chapter 12 and Figure 12.1. For a span of 35 m, the deck should have a depth of about 2 m. One would expect the rate of construction on this low-tech site to be about four weeks per span on average, giving a total construction time for this very simple deck of about 32 weeks, to be followed by finishing works.

If it is necessary to reduce the depth of the deck, the ribs may be made wider and shallower, Figure 12.9. When it is essential to use the absolute minimum structural depth, original solutions may be found, such as the River Nene Viaduct in Northampton, described in *11.6* and Figures 11.16 and 11.17.

Most probably the falsework would consist of traditional birdcage scaffolding, when the mobilisation of two spans' worth would avoid the delay in dismantling and re-erecting it for each cycle. Clear passages 10 m or so wide may easily be made in the scaffolding, allowing it to span obstacles. If there are many obstacles to be crossed or if the columns are very tall, a more highly engineered falsework may be designed, but this will increase its cost, Figure 12.16.

Alternatively, the deck could be a single, super twin rib as shown in Figure 12.17. As there is less repetition, it is likely to require more labour to build and will need more falsework, although the overall construction programme would remain at about 32 weeks. However, it would benefit from much less fussy substructure having only two columns per pier; it would be the clear choice for an urban location.

16.2.2 Precast beams

If it is inappropriate to use falsework, statically determinate precast beams may be adopted. The 130 m length would be divided into four equal spans of 32.5 m, which would be linked at their top slabs, with expansion joints only at the bridge ends. As there is inadequate repetition, custom-designed Tee beams (Chapter 11) would not be economical, and the beams would be standard 'M' or 'Y' type. If sufficient depth was available standard beams could be spaced 2 m apart and 11 such beams would be required per span for the full width. If placed in the traditional manner side by side, 22 beams would be required per span, which is unlikely to be very economical.

16.3 A bridge length of 130 m on three spans

16.3.1 Cast-in-situ

If the bridge were crossing a deep gorge or river, it would be likely to have only three spans. This is ideal for balanced cantilever construction (*15.4* and Figure 15.15) when the spans would be arranged if possible as 35 m, 60 m, 35 m, with deck depths of 4 m at the supports (although 3 m would be possible) and 1.7 m at mid-span (for UK loading). Each balanced cantilever would consist of a 7 m hammerhead, seven pairs of 3.7 m long segments with a 1.2 m stitch in the centre of the main span, and a short in-situ section at each abutment which could be built by the side span traveller or on independent falsework.

The choice of one box or two is here more difficult to make. If two boxes were chosen, the form travellers would be lighter and more economical. However, two sets of substructure would be required, and critically, the construction time for the deck would be nearly doubled.

Construction of two narrow boxes, using only one pair of travellers, would typically follow the following programme:

- 6 weeks for the first hammerhead;
- 14 weeks for the segments of the first double cantilever (including an allowance for a learning curve *15.4.3*);
- 1 week for repositioning the travellers on the hammerhead for the next double cantilever (the second and subsequent hammerheads may be built concurrently with balanced cantilevering, and hence off the critical path);
- 7 weeks for the second balanced cantilever;
- 2 weeks for the stitch;
- 8 weeks for the third balanced cantilever (including the transfer of the travellers);
- 8 weeks for the fourth balanced cantilever;
- 2 weeks for the second mid-span stitch;
- total time approximately 48 weeks.

The in-situ ends of the side spans are assumed to be built on separate falsework concurrently with the other activities. The deck construction will be followed by the finishing works, including the construction of parapets and verges, road surfacing and expansion joints. Time could be saved by mobilising more than one set of travellers, although this would substantially increase the cost.

However, the designer should always be open to unconventional methods of construction. The Liu To Bridge in Hong Kong, had spans of 50 m, 70 m, 50 m, was just over 20 m wide and went to tender designed as a twin box balanced cantilever, but was built as a variation on the technique of incremental launching (*15.8.18* and Figure 15.44).

For a single box, a simple rectangular shape designed to carry British loading would suffer from redundant weight due to a thick top slab and to an over-wide bottom slab, Figure 16.1 (a), while a variable-depth, steeply trapezoidal deck would have the complications described in *13.4.4*. The form travellers would be considerably heavier, and the construction of such a wide deck would be difficult to enter into a weekly cycle for a pair of segments. In order to lighten and simplify the form travellers, it would be possible, for instance, to build only the box itself in balanced cantilever, adding the side cantilevers in a second phase of construction. This would lead one to a cross section that consisted of a relatively narrow box, rectangular or with slightly sloping webs, with transverse beams to support the slab and the side cantilevers, Figure 16.1 (b).

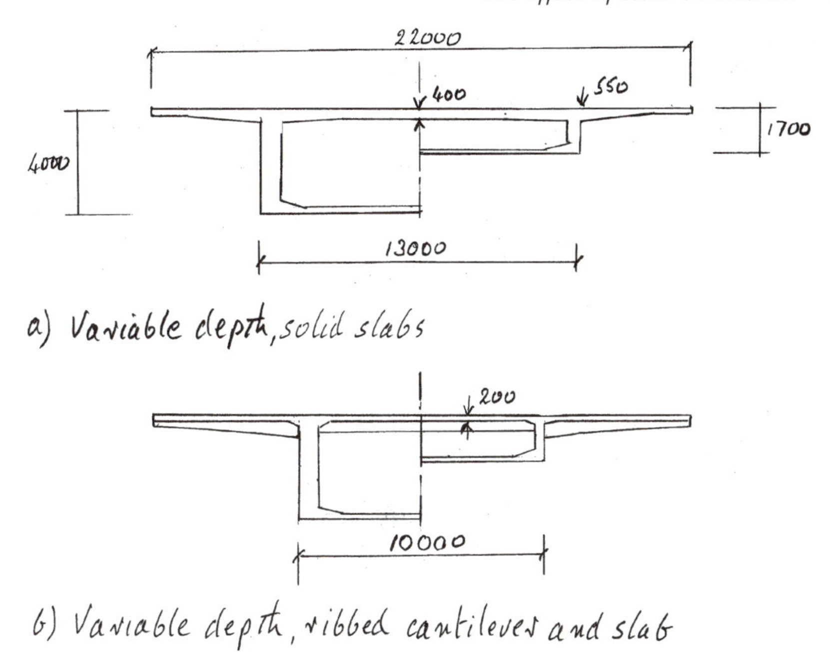

Figure 16.1 Typical alternatives for a variable-depth wide deck subject to UK loading

16.2.2 Precast

As the size of the bridge provides inadequate repetition for precast segmental construction, and the spans are too long for precast beams, precast construction is not appropriate for this bridge.

16.4 The bridge is 500 m long

16.4.1 Cast-in-situ

With a twin deck arrangement, there is 1,000 m of deck to build. If the spans can be less than 40 m, the twin rib, twin deck arrangement is likely to be the most economical. However, with at least 25 spans to build, it is worth investing in plant to speed up the method of construction and save on labour costs. The birdcage scaffold should be replaced with a semi-mechanised falsework rig that can build a span every 2 weeks (*15.2* and Figure 12.2).

The complete deck erection, allowing for a learning curve and the time to relocate the falsework for the second deck, would be about 65 weeks, or 2.6 weeks per span on average. If the rate of substructure construction may be designed to match the rate of deck erection, the substructure only needs a short start over the deck. The falsework and deck formwork are relatively quick to design and fabricate, so a relatively short mobilisation period helps compensate for the slow construction.

A disadvantage of the twin deck arrangement is the need for four columns per pier, or two columns with crossheads, or pier diaphragms in the deck. This may be overcome by adopting a single deck in a super twin rib layout, Figure 12.17. This type of deck, which may span typically up to 45 m, may also be built on a semi-mechanised rig, although there will be fewer re-uses. Assuming that each span of the wider bridge takes on average four weeks to build, the twelve spans will be built in about 50 weeks.

If the bridge was over difficult terrain or over water, cast-in-situ balanced cantilevering of box girders could be adopted. The spans could be 70 m, 3 × 120 m, 70 m, for instance. Due to the slow rate of progress of this method, it would be well worthwhile considering building the deck as one wide box or investing in at least two sets of travellers.

16.4.2 Precast Tee beam

A very economical method of construction applicable to bridges of this length is the precast post-tensioned custom-built Tee beam (*10.3*). As they weigh generally between 60 tons and 140 tons, they are normally built in a purpose-built casting yard on site, and are launched into place with the minimum of handling. Generally a minimum of 50 beams is required to amortise the precasting facility and the launcher.

Six beams at approximately 4 m centres would be appropriate for the full width. With 14 spans of 35.5 m, there would be 84 beams weighing some 80 t each. On such a relatively small site, the beams would be placed at the rate they are precast, avoiding the need for storage and double handling. Thus with say six precasting beds and one steel mould moved from bed to bed, it should be possible to cast the six beams in 12 working days, although 18 days would be a more relaxed target. The beams would then all be launched in 3 days. The following 18 days would be used to build the top slab for this span, and to precast the next set of six beams, giving a construction rhythm of a span in 4 weeks, or about 62 weeks for the entire deck including an additional 6 weeks as a learning curve.

16.4.3 Precast segmental

For a two-deck scheme, some 280 box section segments 3.5 m long, weighing up to 50 tons each would be required, which is at the lower limit of viability for this method (Chapter 14). A single casting cell working 5 days per week and producing 4.7 segments per week on average would cast these segments in about 60 working weeks, or 50 weeks for a 6-day week. Erection could be either span-by-span, or balanced cantilever.

Span-by-span construction of segmental decks is best suited to statically determinate spans, with dry joints if this is allowed (*15.3.3*). Erection would be expected to proceed at a rate of between one and two 40 m spans per week on average on this relatively small site, giving a deck erection period of about 20 weeks. Continuous bridges may also be built by this method, which was employed for the construction of the 400 m long twin box glued segmental deck of the Stanstead Abbotts Bypass (*15.3.4* and Figure 10.4) at a rate of approximately one span per week. However, even for a small site span-by-span construction needs a relatively sophisticated gantry that requires a design and construction time of probably 9–12 months.

The decks may be erected in balanced cantilever by crane or by shear legs. Crane erection can be very fast, erecting up to six segments per day, or three segments per day on average, taking into account the time taken to place the pier segment, build the mid-span stitch, hold-ups and maintenance, giving an erection period of some 20 weeks. However it can only be used on sites suitable for crane access. Shear legs may be used universally, as for the Byker Viaduct, Figure 15.31. However, this method is relatively slow, erecting generally not more than one pair of segments per day giving an average rate of about four pairs of segments per week, and a deck erection period of 35 weeks. Balanced cantilever construction requires only simple falsework that does not control the mobilisation period.

It would not be economically viable to adopt a single wide box, as 140 segments is well below the economical threshold for segmental construction, particularly as the segments would be 22 m wide and, at a length of 3.5 m would weigh over 100 tons.

16.4.4 *Incremental launching*

If the viaduct alignment and spans were suitable, this method of construction would be very economical, as described in *15.8*. A length of 500 m is well within the capabilities of the method, which has been used for viaducts over 1,000 m long. The deck could be either two adjacent boxes, or a single box.

Two small boxes would limit the costs of the mould and jacking equipment, although this option would lengthen the construction programme. 20 m long segments built at a rate of one per week, would give a construction period for the first deck of about 35 weeks, including a learning curve. The mould would then have to be moved sideways, and the second deck built, giving a total deck construction programme of about 65 weeks, exclusive of finishing works.

A single box, Figure 16.2, would economise on substructure and on time, although the mould and the launching nose would be more expensive. Adopting a 15 m segment on a weekly cycle for the much larger, more complex section would yield a construction programme of approximately 45 weeks. The single box option is likely to be the most economical for this length of bridge, due principally to the shorter programme.

Alternatively, the wide deck could be built in two phases, the launched box first, followed by the addition of the strutted side cantilevers. This would take considerably longer, but would need a simpler, more economical launching site, and a smaller labour force.

When necessary, this construction method may be speeded up as described in *15.8.7* for the Sizewell project where 30 m long complete spans were planned to be built at the rate of one per week, for an 800 m long deck, adopting complete prefabrication of the reinforcing cage.

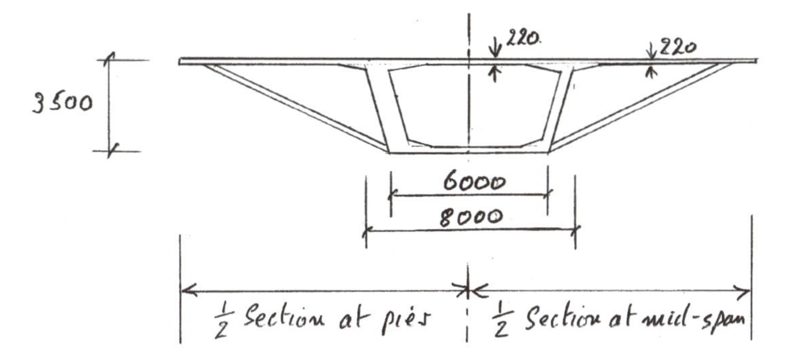

Figure 16.2 Constant depth with strutted cantilevers

16.5 A series of short bridges totalling typically 1,000 m

This scale of construction is typical of motorway contracts, where several over- and under-bridges are required, perhaps including a grade separated interchange. A large proportion of the cost of such contracts is in the highway construction, and the bridge element is often given insufficient attention. However the construction of many small bridges should be seen as an exercise in productivity where considerable savings may be made by adopting precast segmental techniques. Such contracts may be very large; the AREA contract in the French Alps in the 1970s included some 300 km of motorway, with the great majority of the over- and under-bridges standardised and prefabricated in a central casting yard.

As a single wide box is inappropriate for such a site, there would be some 2,000 m of box construction with about 570 segments 3.5 m long, which is more than adequate to justify setting up a facility for precast segmental construction. Alternatively, it may be worthwhile adopting shorter segments; 800 segments 2.5 m long and weighing typically about 35 tons could be transported on public roads without escort. The segments may then be fabricated remotely from the site at an already established precasting yard. The disadvantage of the shorter segments would be a longer casting and erection programme, although this may be immaterial for a site where the timescale is controlled by the construction of the highways.

An erection gantry would not be appropriate for a site consisting of several small bridges, due to the cost of moving it, and the segments would be erected by crane or by shear legs. The decks could be erected in balanced cantilever from each pier, or span-by-span on falsework. The author's preference is to adopt balanced cantilevering, as the falsework required is minimal, and there is great flexibility in building decks in any sequence. If there are several short structures such as motorway over-bridges, they may be erected on a full-length falsework and prestressed in one length. This would allow very rapid erection, with each deck taking typically four or five days. Such a method of deck erection would be particularly efficient if external prestressing could be used with dry joints.

If the contract involves many bridges of a similar family, the pile caps, piers and abutments may also be prefabricated. The piers could be made of match-cast segments, and erected with internal prestress, using a modular design allowing their height to be varied. This total prefabrication, coupled with the assembly of the segments on full-length falsework would allow typical over-bridges to be built at a rate approaching one per week, with the exception of piling or foundation excavations.

The prefabricated box section may readily be adapted to bridge decks of different width or of varying width. This may be achieved by cutting short or extending the side cantilevers of the box. The casting cell would be designed to make provision for the side cantilevers of the maximum width, and then stop-ends would be installed to provide the required width. Thus the same box could be adapted for use on rail bridges or narrow footbridges, by completely eliminating the side cantilevers, or for wider decks, initially by extending the cantilevers to the maximum compatible with the thickness of the cantilever root, and then by thickening the top slab upwards, allowing longer side cantilevers. A still wider deck may be provided by assembling two or three boxes in the cross section (*13.6*).

It is not uncommon for motorway interchanges to include some very wide decks, often of variable width and with crossfall crowns moving across the deck.

Multiple precast segmental boxes can solve such problems very elegantly (*11.7.3* and Figure 11.19).

An important decision in such contracts is the width of the basic box, as it is not generally economical to modify the core of the casting cell. On large contracts consisting of many small to medium sized bridges, it may be economical to adopt two basic widths of box, providing greater flexibility in adapting to the different deck widths. There should be at least 200 segments of each size of box to justify the cost of a dedicated casting cell. In some marginal cases, it is also possible to have only one size of casting cell, and to cast all the segments of that size. Then the casting cell may undergo the fundamental modification required to change the width of the box, to cast a further series.

16.6 The bridge is 1,000 m long

16.6.1 Cast-in-situ

Adjacent twin rib decks with a span length of 35 m built on a mechanised falsework rig, Figure 12.3, will provide an economical option. At an average rate of one week per span, using one rig, the construction of the two decks would take approximately 75 weeks including learning curve and transfer of the rig. Clearly, the programme could be approximately halved by investing in a rig for each deck.

As the bridge becomes longer, the attractions of the single deck, super twin rib become greater. The larger ribs may be made deeper, allowing the span to attain 45 m, further reducing the construction time. The construction process would probably proceed at a rate of two spans every three weeks, taking some 45 weeks in all.

16.6.2 Precast Tee beam

This method remains highly competitive for this length of viaduct, particularly for contractors or for regions of the world where a low-technology site is attractive. With a total of 168 beams, it may become appropriate to mechanise the site, with a single mould equipped with steam curing, producing one beam per day.

16.6.3 Precast segmental

Precast segmental construction is clearly viable for a bridge of this length. The area of deck being over 20,000 m^2, erection by self-launching gantry is likely to be competitive with erection by crane.

The use of two boxes in the cross section leads to a total of some 570 segments 3.5 m long. In order to cast the deck in a reasonable time, it is likely that at least two casting cells would be employed. Erection could be either span-by-span, or balanced cantilever. The latter method would be expected to proceed at an average rate of 3–4 segments per day, erecting the deck in about 35 weeks.

A mechanised gantry takes about 12–15 months to design, build and commission. During this time, of course, the substructure can be progressed. In fact, the programme for the substructure must be such that it will not be overtaken by the deck once the gantry is in full flow.

If only one box were used there would still be 300 segments 3.3 m long, weighing typically between 100 and 120 tons. This size of casting run is just about enough to amortise the large precasting facility necessary. A more substantial and expensive gantry would be required to handle such segments, but there is no reason that the rate of erection should be any slower. The single box scheme is here definitely in contention, although not a clear winner.

16.6.4 Incremental launching

1,000 m remains within the capabilities of a launched deck 11 m wide. It would be worthwhile investigating procedures to speed up the construction cycle. Techniques become frozen in an accepted routine, and sometimes it needs a designer or contractor with vision to break out of the conventional constraints. The author would not have believed it possible to cast on site complete 42 m long decks in two days until he had seen it being done by Ferrocemento on the Poggio Iberna Viaduct (*15.9.1*). There is no reason that a similar technology cannot be transferred to an incremental launching site, allowing the construction of two or three spans per week.

A deck consisting of a single box 22 m wide would weigh over 30,000 tons. Such a deck may be launched although it is probably approaching the economic limits, particularly if it is on an uphill gradient. An alternative launching technique, described in *15.8.15*, was used to build the 650 m long approach viaduct to the Pont de Normandie which weighed 26,000 tons and was launched up a 6 per cent gradient.

16.7 The bridge is 2,000 m long

16.7.1 Cast-in-situ

For a deck of this length, the rate of deck erection becomes very important in the economy of the construction. Cast-in-situ construction of two adjacent twin rib decks on 35 m spans would still be viable. As described in *15.2.5*, it is quite feasible to build twin rib decks in-situ at a rate of two spans per week, rivalling the rate of erection achievable by precast segmental methods. The 116 spans would be built in about 70 weeks using a single falsework rig, allowing for a learning curve. The time taken to reposition the rig for the second deck could be substantial. It may be worth designing a gantry that can slide sideways to build the two decks in parallel.

If it were necessary to link the two decks together with a cast-in-situ slab, this operation could be carried out either progressively as the second carriageway is erected, or once the two decks had been completed. The former option would require some periods of two days when the slab concrete cures. If these were co-ordinated with the casting and curing of the deck concrete and with the gantry launching, the effect on the programme may be minimised. If the latter option is adopted, the deck construction programme would be extended by several months, and it would be more economical to keep the two decks separate. For problems due to creep when two decks are linked, see *15.3.4*.

A super twin rib deck could also be built by a highly mechanised overhead rig. The construction process is likely to be slower, as the weight of reinforcement and prestress to be handled is greater. However, one span per week should be achievable, giving a similar overall time of deck construction to the twin deck option above. However, there is no repositioning of the rig, and no in-situ stitch to cast. On the debit side, the falsework would be heavier and more expensive.

Comments in *16.6.3* concerning the mobilisation period for a gantry apply also to the mechanised falsework, which if anything is even more complex.

16.7.2 Precast Tee beam

See the comments in *16.8.3* on using precast Tee beams for very long viaducts.

16.7.3 Precast segmental

With a deck area of over 40,000 m^2, this is prime precast segmental territory. A two-box scheme would require over 1,100 segments 3.5 m long. On the assumption that a gantry would place its first segment in month 15 of the contract (*15.5.10*), and would then proceed to place 75 segments per month, the final segment would be erected in month 30. As the last segment must be cast in month 29, and the first in say month 5, the required rate of segment casting must be 1,100/24 = 46 per month. At a casting rate of 20 segments per cell per month, 2.3 cells are required, meaning in reality three cells with one commissioned later in the programme, or a later start to the casting programme.

As the casting starts 10 months before erection, it would be necessary to store approximately 460 segments, a major expense. To reduce this storage requirement it is likely that it would be economical to start casting later using all three moulds. The casting rate would be 60 segments per month, giving a total casting period of 19 months. Casting must then start in month 29 – 19 = month 10. The storage requirement would then be reduced to only five months casting, or 300 segments. This requirement would be further reduced if 6-day working were adopted for the casting yard. If four moulds were used, the rate of segment casting would equal the rate of erection, and storage would be reduced to a nominal buffer.

Clearly these figures are somewhat theoretical, but demonstrate the mechanics of deciding on the number of moulds necessary and on the storage facilities required. As an example, the 2,700 m long first stage of the STAR Viaduct (*15.5.11*) required approximately 1,000 segments and adopted six moulds which were commissioned progressively. The 1,700 m long Dagenham Dock Viaduct (*15.5.12*), which had two or three boxes in the cross section and a total of 1,030 segments, was cast with four moulds.

The size of this viaduct makes a single-box scheme more viable. The extra cost of the handling equipment to manage the heavier segments could well be more than offset by the savings in manpower in the casting yard and on the erection front, which include eliminating the connection between the decks. Furthermore, potential savings in the cost of the piers and foundations would most probably make this the favoured solution. A single-box scheme would have 570 segments 3.5 m long and the deck could be erected in 8 months with a single gantry. Using the same mathematics as above, erection would be from month 15 to month 23, and casting with two moulds from month 7 to month 22, giving a requirement to store 8 months' production, that is 320 very large segments. It would probably be necessary to adopt three moulds to reduce the storage required. The overall deck construction time would be controlled by the time to commission the gantry and by the erection rate it could achieve, and may be estimated as 15 + 8 = 23 months using the above assumptions.

Erection coul be span-by span if the spans were sufficiently short and regular, or in balanced cantilever which can cope with virtually any span range.

16.7.4 Incremental launching

If the alignment were suitable, the decks could be incrementally launched, with a launching bay behind each abutment, and with the decks meeting in the centre. It would probably be economical to build one wide deck rather than two decks, principally due to the shortened programme.

16.7.5 Prefabrication of complete spans

At this length of viaduct, the prefabrication of complete box section spans becomes viable (*15.9*). A 40 m span, 11 m wide box section deck would weigh about 600 tons. A relatively simple mould, equipped with steam curing, can build such units at a rate of a span every two days. Thus, a half-width bridge could be built at a rate of one span every other day, say 12 spans per month. This was the rate of construction achieved at the Poggio Iberna Viaduct by Ferrocemento.

The bridge deck could be completed in a matter of 11 months from start of casting, allowing a 12 week addition for a learning curve. As the time for design and construction of the gantry and for the segment transporter would be similar to that for precast segmental placing gantries, the overall length of construction of the deck from commissioning the moulds and gantry would be about 26 months.

The spans would preferably be statically determinate, because creating continuity slows down the erection rate, and loses some of the advantage offered by the method. Also, it would be preferable to divide the two carriageways by a longitudinal joint. If a linking slab was required, it would most probably be cast once the two decks had been completed. The slab could be built at the same time as the parapet, if this had not been cast together with the spans.

In this technique, the deck units are built at the same rate they are placed, giving the possibility of a better use of labour, with elements of the same team being responsible for precasting and placing the units.

This method of construction requires major capital expenditure, including the mould and casting area, the transporter, rails and the placing gantry. However, the labour and supervision required to build the decks is less than for any other construction method, except possibly incremental launching.

16.8 The bridge is 10,000 m long

16.8.1 General

For a viaduct of this length, the rate of construction of both the foundations and the decks becomes the dominating design criterion. Although decks may be erected at a rate of several spans per week for each placing gantry, the construction of the foundations will be much slower. A mechanised system of bridge deck erection generally takes at least 15 months to mobilise, giving a good start to the foundation construction. It then remains to calculate how many foundation sites must be mobilised in order to have the piers finished in time to receive the deck.

In a highway deck of this length, there may be slip roads joining the main carriageway, and railway viaducts may even have stations or sidings, locally increasing the width of the elevated structure. If the deck is not of constant width, the method of construction and the type of deck may need to be chosen for their ability to adapt to such variations.

16.8.2 Cast-in-situ

The mechanised methods of deck erection for both the twin decks and the single decks described above are still applicable for this length of viaduct. However, the construction heads become remote from the site establishment, making it more difficult to supply construction materials and supervise the work force. Also, the rate of construction of two spans per week is most probably a maximum that cannot be accelerated. If twin decks with 40 m spans were adopted and one falsework rig was used to erect the 500 spans, the deck erection time would be some 70 months, including the learning curve plus the time to reposition the rig. Thus it would probably be necessary to mobilise at least four rigs, reducing the overall erection time to some 17 months. Multiple rigs would be best served by a site situated at the mid-point of the viaduct, so that they can all work away from their base, optimising their supply lines.

If the deck is of constant width, the logical choice would be to opt for a single wide deck, as a super twin rib. Two falsework rigs would build the deck in 27 months, assuming the spans were 45 m, and that the learning curve involves a 12 week addition to the programme for one rig.

16.8.3 Precast Tee beams

Precast beams remain a viable option for very long viaducts. They can cope well with increases of width of the deck at interchanges, as both the spacing and the number of beams can be changed from one span to another, even if the result is not very elegant.

With six beams across the cross section and with spans of 35 m, there would be 286 spans and 1,716 beams. As the rate of erection of precast beams is not very fast, it will be necessary to use several beam launchers. To avoid excessive transport of the beams that weigh upwards of 80 tons, it is likely to be cost effective to mobilise several casting yards along the length of the site.

For the GSZ project (*10.3.8*), which included 28 km of 15 m wide decks, there was one main casting yard with 24 steel moulds equipped for steam curing located near the centre of the viaduct, and several secondary yards at strategic positions along the route. This method of building a long viaduct is relatively labour intensive, and can achieve speed only by mobilising very large resources. This is likely to be cost effective in developing countries where labour is cheap compared with the cost of plant.

16.8.4 Precast segmental

If twin boxes are adopted, 5,700 segments 3.5 m long are required. At a rate of erection of 75 segments per month on average, 76 gantry/months are needed. Two erection gantries per deck would complete the deck erection in 19 months.

As it would not be feasible to store thousands of segments, the number of moulds needs to be calculated to match the rate of erection. Four gantries place 300 segments per month, requiring a minimum of 15 moulds working at 20 segments per month.

Allowing for some special moulds for pier segments and a reasonable contingency, probably at least 20 moulds would in fact be required. This compares with the 42 moulds actually mobilised to cast the 7,000 segments for the 22 km PUTRA LRT in Kuala Lumpur. This number is not strictly comparable, as there were two basic types of deck for single or twin track railways, with separate moulds for pier segments. However, it is an example of how a casting yard must be equipped to cast at the rate of erection for a long viaduct.

For a viaduct of this length, the single box scheme is likely to be the most economical if there are not too many changes in width due to slip roads or other special events. 2,850 segments would be required, needing two placing gantries for 19 months and at least 10 moulds.

Erection could be span-by-span or in balanced cantilever.

16.8.5 Incremental launching

In order to use incremental launching for a deck of this length, it would be necessary to set up five intermediate casting stations at 2,000 m intervals, equipped to launch in each direction. In reality, even if the deck alignment was suitable, it is unlikely that this would be an appropriate method of construction due to the cost of the elevated casting areas and the large workforce required. However, it needs a relatively low level of technology compared with precast segmental techniques or with the precasting of complete spans and it may be competitive where labour is cheap.

16.8.6 Precasting complete spans

This method of building is highly relevant for such a long viaduct. Whereas the precast segmental method cannot reasonably be accelerated beyond two spans per week per gantry on average, complete span placing gantries may work at up to three spans per 24-hour day, if the supply of decks can keep pace.

The Poggio Iberna site had a very high performance gantry, but its productivity was limited by a deck-casting rate of 12 per month. The first step in accelerating the construction rate for a bridge with twin decks would be to have two moulds feeding the single placing gantry, achieving 24 decks per month, erecting the 500 decks in 21 months. In order to avoid the need for the gantry to have to make the return journey along the completed first deck, it would be possible to adapt it with transverse sliding capability so that it may erect both decks in parallel.

In order to further accelerate the deck construction, it would be possible to add a third mould, leading to a theoretical erection rate of 36 decks per month, or about 1.6 decks per working day, and an erection period of 14 months. At a rate of casting of two days per span with three moulds operating, it is still feasible to introduce the minor specialisations required, and to produce the decks in the order required for erection, avoiding the need for expensive storage and sorting of decks. The three moulds would be arranged behind an abutment, with a system of railway tracks and points leading to the rails on the deck itself. Clearly, a very substantial and well-organised casting yard would be required to service these moulds.

An alternative strategy would be to adopt moulds that were capable of producing one deck per day, say 24 decks per month, as described in *15.9.2*. Two such moulds would produce 48 decks per month, completing the erection in less than a year. The single gantry would still be quite comfortable in placing the two decks per day. These moulds and their attendant production lines would be more expensive than the simpler moulds, and an economic comparison would need to compare the manpower and capital requirements (*15.9.3*).

It would be possible to design a single wide deck to be built in this way. The deck would weigh some 1,200 tons, a significant step up from the 850 tons of the Poggio Iberna Viaduct. However, from the studies carried out by Benaim on the prefabrication and launching of decks weighing up to 1,500 tons, Figure 15.51, the weight is not a barrier to feasibility. However, prefabrication of the deck will be slower; although the author has not carried out any detailed assessment, a rate of six decks a month would appear reasonable. The mould and its production line, the transporter and the placing gantry will all be approximately twice the cost of the smaller units. Two moulds would complete the deck in 16 months, marginally quicker than the three moulds for the smaller deck. Intuitively, it does not appear likely that the wider deck will offer any savings in construction technology, although as it could be carried by a single central column, there may be savings in the cost of substructure, and project advantages in adopting such a compact footprint.

17
The design and construction of arches

17.1 General

Loads may be carried to the foundations of a bridge by compression, tension or bending/shear, Figure 17.1, or by combinations of these three actions. Arches which work principally in compression, are among the most beautiful of structures, effortlessly spanning great distances. They may also be among the most economical, as they may be made from cheap and readily available materials such as stone or concrete, while tension and bending both require expensive materials such as steel or fibre composites.

However, arches can be labour intensive to build, and consequently had somewhat fallen out of favour. More recently, the growing awareness of the concept of sustainable development has renewed interest in this form of construction that minimises the materials employed, and has stimulated the search for more economical methods of construction.

A special feature of arch bridges is that they do not need any back span; if a large span is flanked by terrain that is suitable for much shorter spans, arches have a significant economic advantage. Most other types of bridge require side spans either side of a main span, which extends the costly long span technology over additional length.

Large-span stone and concrete arches have been used very extensively in China, where it is reported that some 70 per cent of all highway bridges are of this form of construction [1]. Chinese engineers have developed a variety of innovative methods of construction.

17.2 Line of thrust

A useful concept for understanding arches is the line of thrust. Any loads applied to an arch will create a line of thrust. If the arch is made exactly the same shape as the line of thrust, it can carry the loads without bending moments. The shape of the line of thrust may be visualised by applying the loads to a weightless suspended cable, Figure 17.2 (a). The shape of this loaded cable is known as a funicular diagram. Clearly, the sag of the loaded cable, and its tension, will depend on how taut it was before application of the loads. If the cable tension is varied, an infinite number of similar funicular diagrams may be drawn. This is an exact analogy to the line of thrust, Figure 17.2 (b), as the compression in the arch is inversely proportional to its rise.

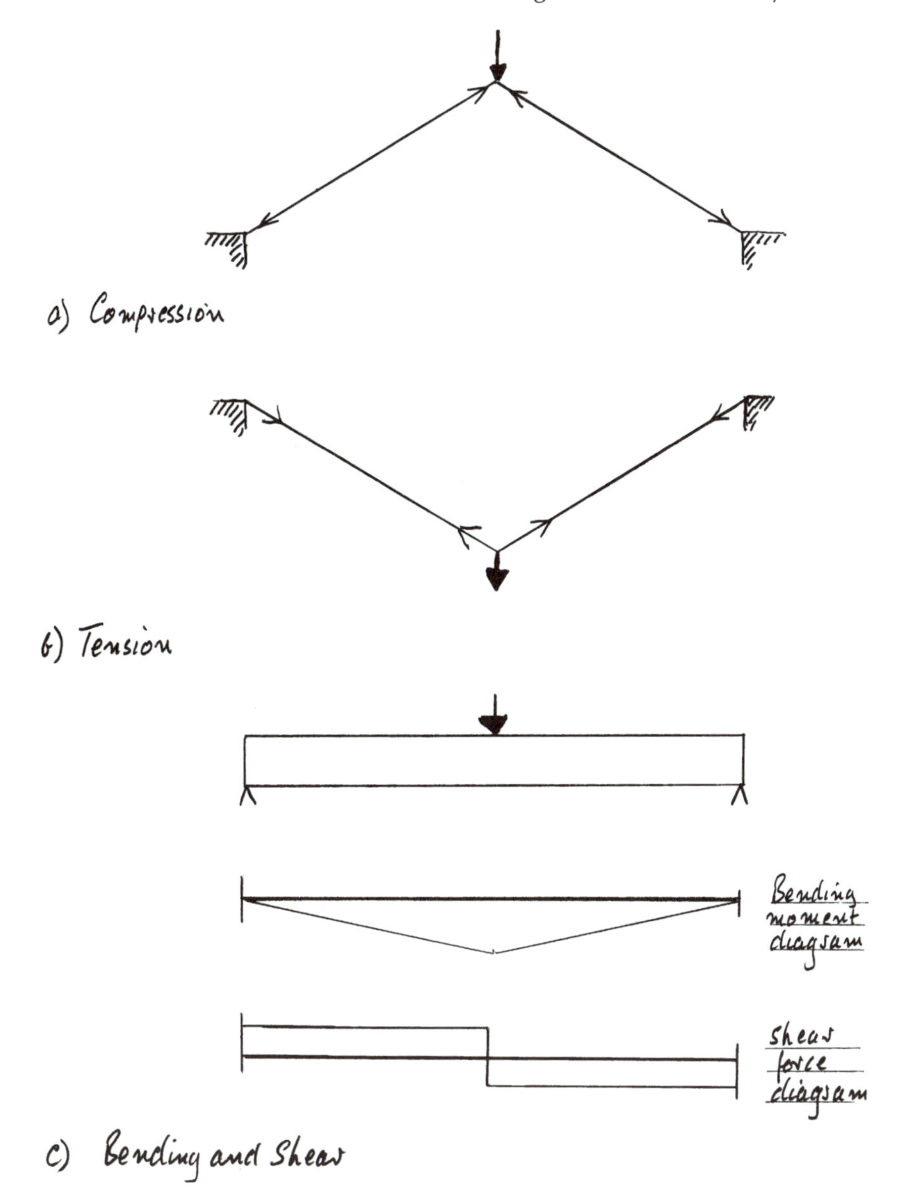

Figure 17.1 Compression, tension and bending

If the arch shape does not correspond exactly with the line of thrust, a local bending moment will be generated; M = N×d, where *N* is the force in the arch at that point, and *d* is the distance between the neutral axis of the arch and the line of thrust, Figure 17.2 (c).

If the arch has hinges in its length, the bending moment must be zero at the hinge positions, which means that the line of thrust must pass through those points. If the arch has three hinges, typically at the springings and at the crown, it is statically determinate. In this case, the line of thrust is also determinate, and can be drawn, superimposed on the arch. The line of thrust shown in Figure 17.2 (b) may be scaled

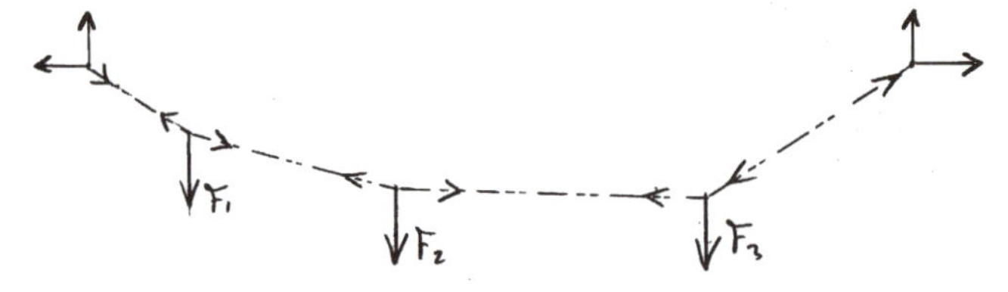

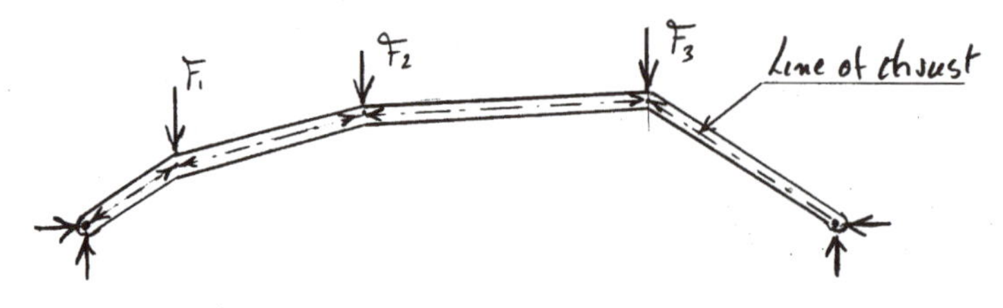

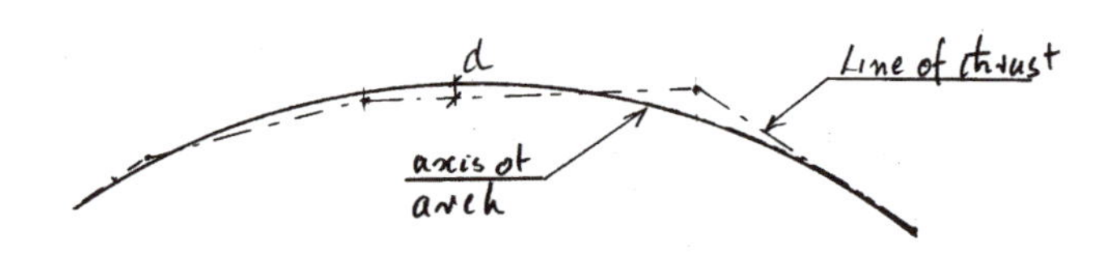

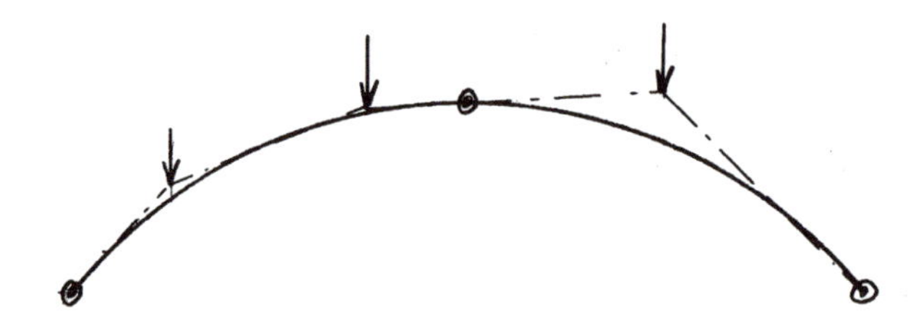

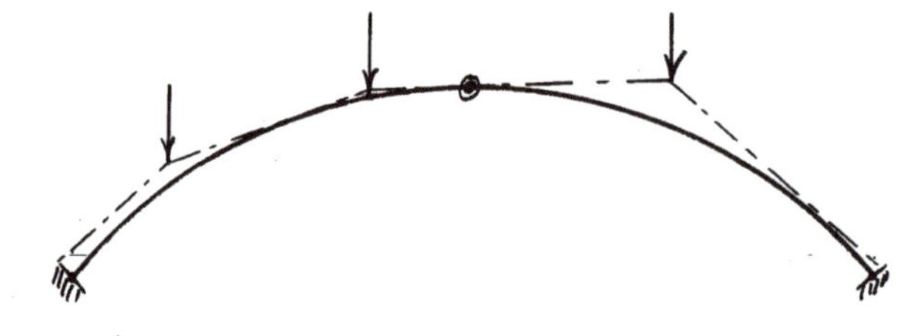

Figure 17.2 Concept of line of thrust

so that it passes through the three hinges, Figure 17.2 (d). The thrust at the arch springings and the compression at any point along the arch may then be calculated by simple statics. The bending moment at any point of the arch may also be calculated by taking moments about that point, or by measuring the distance between the line of thrust and the neutral axis of the arch.

If the arch has less than three hinges, and is consequently indeterminate, the line of thrust may not be drawn precisely without elastic analysis. Thus, for a fixed-ended arch, the line of thrust will not pass through the springings, where there will be a moment in the arch, Figure 17.2 (e). However, the line of thrust will always be a 'best fit' to the shape of the arch neutral axis, and so can be drawn approximately, allowing an early rough estimation of the bending moments in the arch, and also allowing a first approximation to the best shape for the arch.

The correct shape for an arch carrying a uniformly distributed vertical load is a parabola. For an arch of constant thickness carrying only its self weight, the correct shape is a catenary (which is very close to a parabola). For an arch carrying its self weight and external loads, the best shape is a compromise between a catenary and the funicular diagram of the loads.

17.3 Unreinforced concrete and masonry arches

The stability of a masonry arch is given principally by its shape and its thickness. When the shape of the line of thrust differs from the shape of the arch, the arch ring needs a finite thickness to contain it, Figure 17.3. Where the line of thrust is within the middle

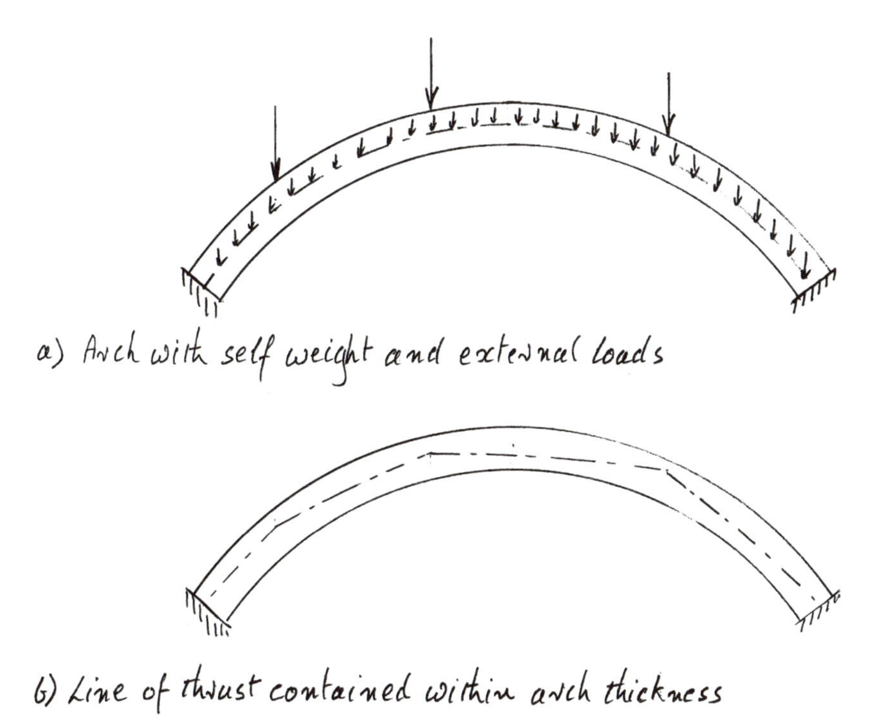

a) Arch with self weight and external loads
b) Line of thrust contained within arch thickness

Figure 17.3 Masonry arch with external loads

third of the arch ring, the complete cross section of the ring will be everywhere in compression. If the line of thrust is locally outside the middle third, one face of the arch ring at that section will be in tension, and the arch will crack. In stone or brick arches these cracks usually occur in the joints and may not be noticed. If the line of thrust touches or crosses the intrados or extrados of the arch, a hinge will form at those points. When four hinges form, the arch will become a mechanism and fail. However, as a certain thickness of material is necessary to resist the thrust, a hinge will in fact form when the line of thrust is at a finite distance within the arch structure. For masonry arches the cause of failure is almost always the mechanical instability described, not the inadequate strength of the masonry.

An unreinforced concrete arch will behave very much as a masonry arch. However, it should be remembered that if the line of thrust lies significantly outside the middle third, visible cracks are likely to occur. The arch is not threatened, but the cracks may be unsightly.

Masonry arch bridges have been built with spans up to 120 m, such as the Wuchao River Bridge in Hunan Province in China.

17.4 Flat arches

An arch does not have to be 'arched' in shape. Consider a flat 'arch' of constant depth spanning between rigid abutments, Figure 17.4. The line of thrust for a uniform load, for instance, will be parabolic. If it is possible to draw such a line of thrust within the rectangular profile of the arch with a rise/span of at least 1/10, the loads will be resisted in arch action. Clearly the line of thrust will be outside the middle third over much of the length, and the structure will crack. In fact, there is an 'incorporated' arch which is a finite thickness of material drawn around the line of thrust. The material above and below this incorporated arch is redundant, and could be removed, without affecting the performance of the arch.

If the arch was made of concrete, and reinforced as for a beam in bending and in shear, the shortening of the arch ring under the compression, and the consequent deflection of the arch, would induce bending action. The loads on the 'arch' would then be shared between arch and bending action, see *3.12* and *9.3.5*.

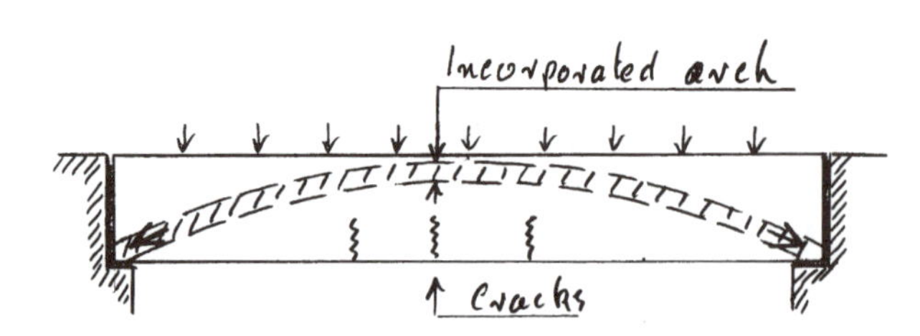

Figure 17.4 Flat arch

17.5 Reinforced concrete arches

17.5.1 General

Reinforced concrete arches differ from masonry arches principally in that they can resist bending tension, and consequently can remain stable with a line of thrust that lies outside the arch. They do not depend on their thickness and shape to give them

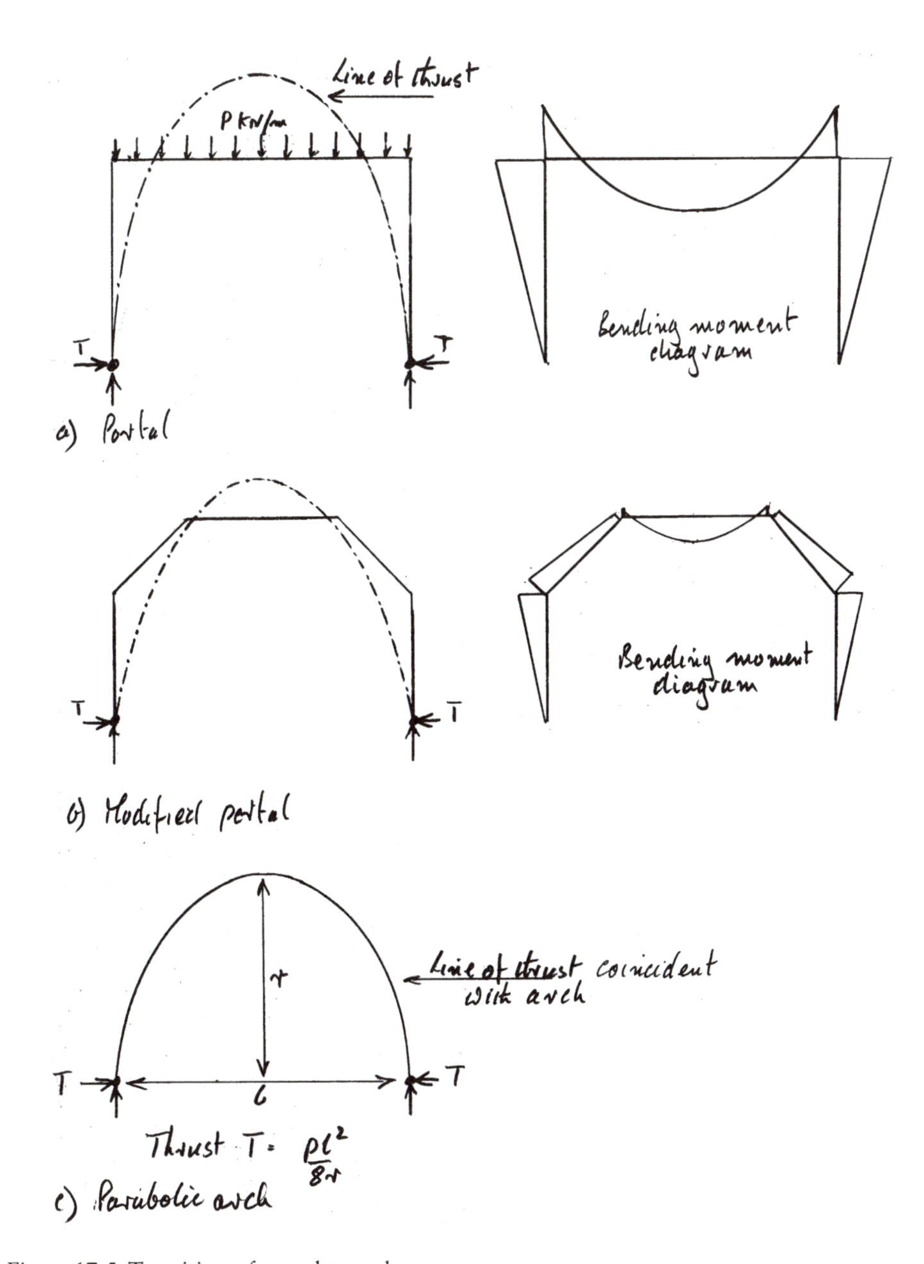

Figure 17.5 Transition of portal to arch

stability, but on their resistance to combined compression and bending. However, their economy depends on their shape being close to that of the line of thrust.

A two-pinned portal may be seen as a crude approximation to an arch. When subjected to a uniform load, the superimposition of the parabolic line of thrust on the portal illustrates graphically the bending moments that are proportional to the distance between its neutral axis and the line of thrust, Figure 17.5 (a). The loads are carried by a combination of arching action that are represented by the thrust at the arch feet, and bending. If the portal is refined by deflecting the cross-beam, Figure 17.5 (b), the distance between the line of thrust and the centroid of the portal frame is reduced, and the bending moments drop in consequence, with more of the load carried in arching action. Finally, if the portal is further modified to become a parabolic arch, the line of thrust and the neutral axis of the frame become coincident and the moments disappear, Figure 17.5 (c). At the same time the thrust T will increase as the rise of the line of thrust decreases, and as all the loads are carried in arching action.

17.6 Short-span reinforced concrete arches with earth fill

A thin reinforced concrete arch covered in earth fill is a very economical structure for carrying highway loads. The presence of the fill distributes point loads, reducing the bending moments on the arch.

If solid abutments are available, the arch may be simply an economical replacement for a slab roof. A good example is the 600 m long Byker Tunnel, part of the Tyne & Wear Metro in Newcastle-upon-Tyne, Figure 17.6, designed by the author when at Arup [2]. The 200 mm thick arch spanned 9.4 m, with a rise of 2.3 m, and the road above was subjected to full UK highway loading (45 units of HB + HA). The arch is three-centred, that is it had a larger radius over the central section than over the

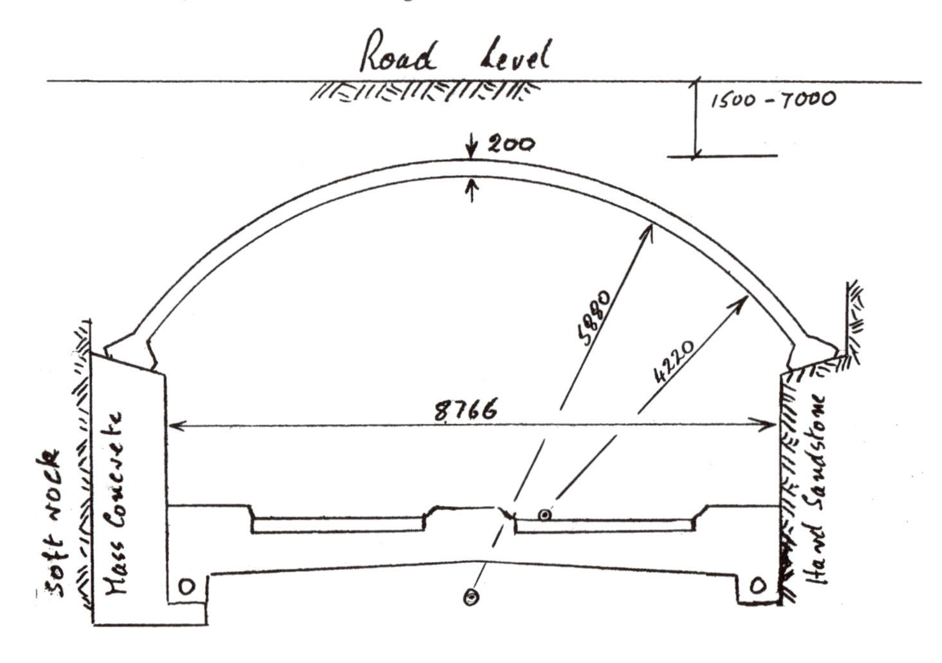

Figure 17.6 Byker Tunnel

Figure 17.7 Byker Tunnel under construction (Photo: Robert Benaim)

haunches, where there is a greater weight of earth. Over most of its length it rested on abutments cut into massive sandstone. As the hard rock fell away, it was replaced by unreinforced concrete walls, either dressing the soft rocks of the coal measures, or for short lengths acting as cantilever retaining walls.

The arch was cast in-situ on a set of steel travelling shutters, Figure 17.7. As the slope of the arch at the springings exceeded 18°, a partial top shutter was necessary. Such an arch needs only light reinforcement, and the bars should if possible be of small diameter such that they will lie to the arch radius without pre-bending, preferably under their own weight.

Two alternative designs carried out by Benaim demonstrate the economy of reinforced concrete arches. Both carried the M74 motorway in Scotland on very skew crossings of twin-track railways, Figure 17.8. The New Cowdens Bridge was 190 m long with an arch span of 21 m and crossed the electrified main line from England to Scotland, while the Maryville Bridge had a span of 16 m and crossed a lesser line. The arch thickness for New Cowdens was 450 mm, and 300 mm for Maryville. The greater thickness of the former was adopted to allay concerns about the impact resistance of the structure in the event of a derailment. These alternatives consumed approximately one-third of the materials of the conforming designs they replaced, Figure 17.9. The precast arches were erected in a matter of days, and once in place they protected the railway from the remaining construction activities. This minimised possession times and reduced the risk to the railways. An additional advantage of such arch bridges is that they require less maintenance; in particular they do not need roadway expansion joints.

The arch profiles consisted of two radii; a tighter radius is required over the lower part of the arch to accommodate the increasing weight of earth and lateral pressure.

Figure 17.8 New Cowdens arch (Photo: Alister Lynn/Balfour Beatty Civil Engineering Ltd)

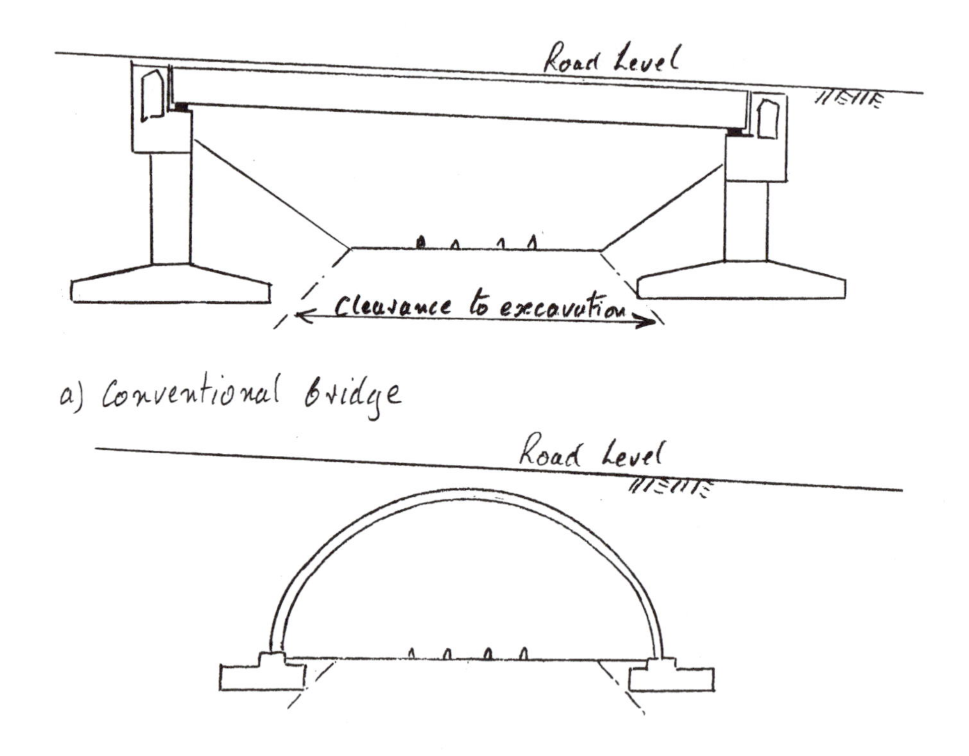

Figure 17.9 Comparison of an arch with a conventional bridge

However lateral earth pressure is not a unique value, varying over a range between the extreme limits of active and passive pressure. It is advisable to design for pressure at rest (K_o), and to check the sensitivity of the design to reasonable variations in the pressure. Heavy vehicles over the haunches will increase the lateral earth pressure on the arch as well as the vertical pressure. Such eccentric loading will cause the arch to sway sideways, away from the load, creating a countervailing earth pressure on the unloaded side. This may be simulated in analysis by modelling the soil as a series of springs. The arches also had to be designed for the situation near the portals where the embankment ramped down on one side.

The structures were precast in half arches 3.2 m long, and erected in pairs without falsework, Figure 17.10. They were placed in sockets on the foundations, on carefully levelled packs, and then grouted in. The precast units were then stitched at the crown to complete the structure. At the request of the client, on New Cowdens they were also stitched all the way down the arches to create a monolithic roof. Both arches were entirely covered with a waterproof membrane.

At the portals it is necessary to retain the earth above the arch. This may be done by building a conventional spandrel wall. However, this causes some difficulties, as the arch ring is likely to be too slender to provide moment fixity for a cantilevered retaining wall. One solution is to allow the spandrel wall to span horizontally between counter forts over the arch springings. Alternatively, the spandrel wall may be made of reinforced earth. There is also a need for wing walls to contain the fill in the approach to the arch.

Figure 17.10 Maryville arch under construction (Photo: Benaim)

Figure 17.11 Bridge on A16, France (Photo: Robert Benaim)

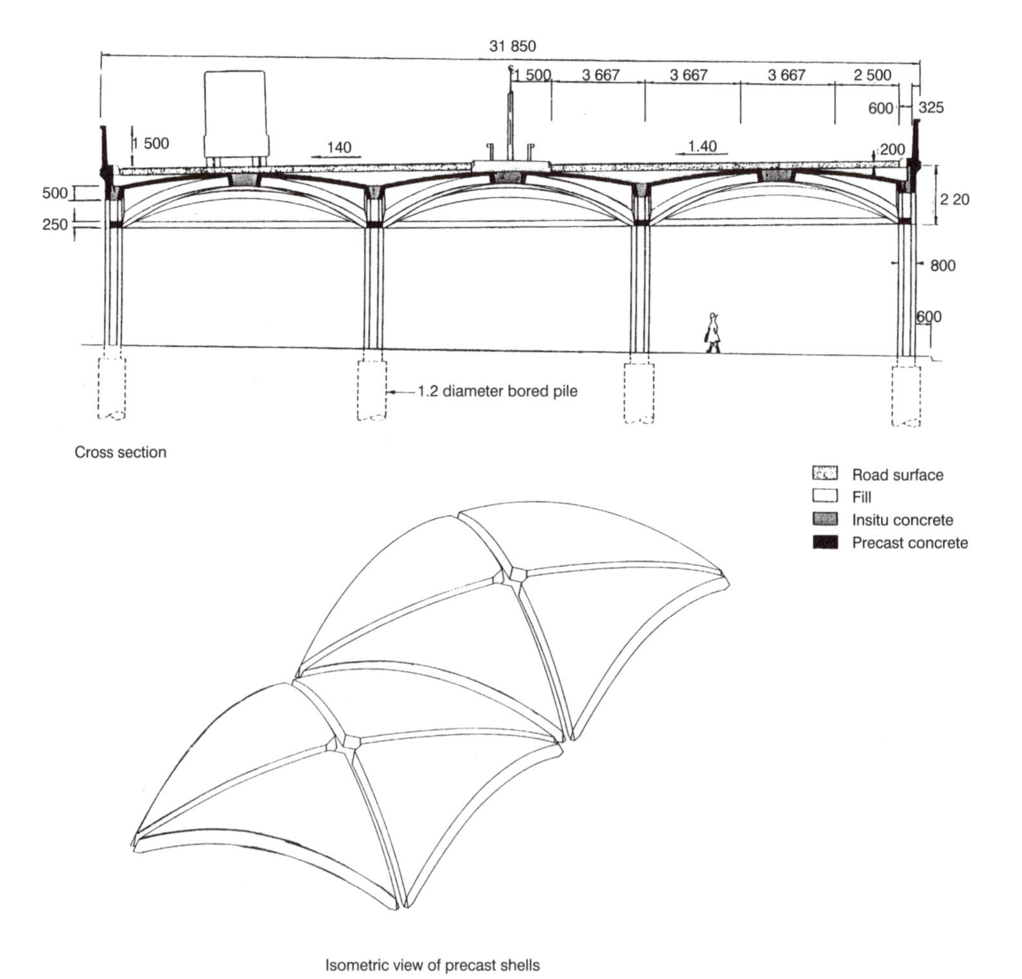

Figure 17.12 Project for Station Viaduct Middlesbrough

A spandrel wall can be avoided by extending the arch and allowing the earth to spill forward at its angle of repose. This solution was adopted for the bridges over the A16 motorway in France, Figure 17.11. An elegant way of detailing this last solution is to terminate the arch with an inclined plane at the angle of repose of the ground.

An original use of thin reinforced concrete vaults for a bridge deck was the project for the Station Viaduct in the heart of Middlesbrough, carrying a dual three-lane motorway between the railway station and a shopping centre, Figure 17.12, designed by the author when at Arup. The deck consisted of precast ¼ vaults, made of a 150 mm thick reinforced concrete shell. Each ¼ vault was trimmed with downstand permanent shuttering for arched stiffening beams. When assembled, the precast components created a series of 10 m square vaults with stiffening beams on the diagonals and around the perimeter, with the heads of the columns linked by a grid of prestressed concrete ties. The structure was completed by casting the core of the stiffening beams in situ. There were three vaults across the 30 m width of the motorway. PFA was placed above the vaults to provide the road platform. The use of precast arches led to a very economical structure, which would have competed favourably with more conventional bridge deck technology, while providing a useful and attractive space beneath the viaduct. Unfortunately this project was never built in this form.

17.7 Longer span reinforced concrete arches supporting bridge decks

17.7.1 Length changes of arch bridges

Arch bridges where the roadway is built on fill do not need expansion joints, as the length changes in the arch due to temperature variations are taken up by changes in the geometry of the arch; its crown rises or falls with the temperature. Longer span reinforced concrete arches generally carry a separate deck which requires provision to accommodate its length changes. Either the deck may be a series of statically determinate spans with joints every span, or it may be made continuous with expansion joints at the abutments. In the latter case, the columns linking the deck and the arch must either be sufficiently flexible to follow the deck movement, or must be pinned top and bottom, or must carry sliding or elastomeric bearings.

17.7.2 Relative stiffness of arch and deck

Highway and railway loads include both distributed and concentrated loads. Whereas distributed loads applied over the full length of the arch cause principally compressions in the arch, distributed loads applied over only part of the length and concentrated loads cause significant bending moments. As the arch ring and the deck are linked together, they will inevitably share these bending moments, in proportion to their stiffness.

The designer has a choice of designing the bridge such that the greater proportion of the bending moment is carried by the arch or by the bridge deck. For instance, if the deck is made deep and hence stiff in bending, while the arch is made as thin as possible, the greater part of the bending moments will be carried by the deck, and vice versa, Figure 17.13. Benaim's short-listed entry to the Poole Harbour competition, designed with the architectural assistance of Wil Allsop, Figure 17.14, was provided with a stiff deck, allowing the arch to be made as thin and unobtrusive as possible. On the other

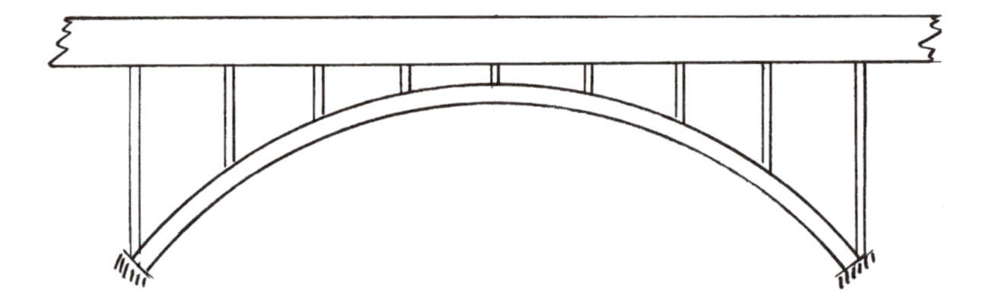

a) Stiff deck and flexible arch

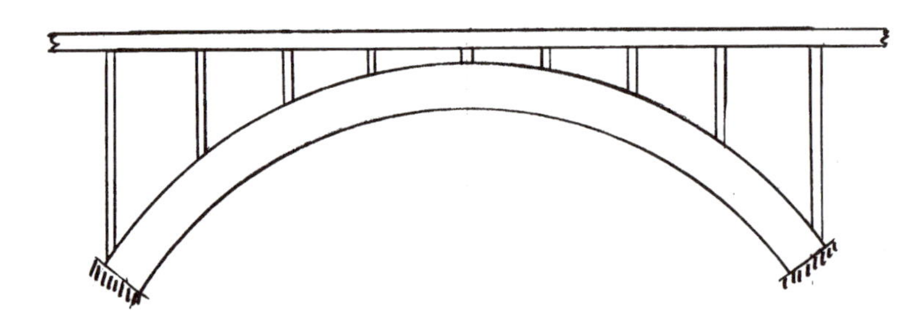

b) Stiff arch and flexible deck

Figure 17.13 Relative stiffness of arch and deck

Figure 17.14 Benaim entry to the Poole Harbour Crossing competition (Image: David Benaim)

hand, Benaim's Sungai Dinding Bridge in Malaysia, Figure 17.16, which had a slender steel composite deck, and Maillart's Salginatobel Bridge, Figure 1.18, are examples of bridges where the arch ring has been designed to carry most of the bending moments. It is also possible to design the bridge such that the bending moments are shared more or less equally.

For long-span bridges, such as the Krk Bridge, Figure 17.19, the arch needs to be deep to resist buckling and it naturally carries the bending moments. When the deck consists of a series of statically determinate spans, it clearly cannot relieve the arch of any bending moment.

17.7.3 Effect of shortening of the arch and spreading of the abutments

Reinforced concrete arches must be designed to resist the effects of shortening of the arch ring due to elastic and creep deformations under compression, concrete shrinkage and temperature drop. These effects cause the crown to drop and induce sagging bending moments at mid-span and hogging moments at the springings of a monolithic arch. Spreading of the abutments as a response to the thrust causes bending moments similar to those due to arch shortening. The thicker and stiffer the arch, the greater these moments will be. Consequently, it is important to keep the arch as thin as possible, compatible with stability and with the strength necessary to carry the bending moments due to point loads.

Reinforced concrete arches generally have a span/rise ratio of between 4 and 8. However arches with a rise of span/10 or even flatter can be designed. The flatter the arch, the greater the effects of shortening will be. This may be easily understood by the fact that the length of an arch with a rise of 1/5 is about 10 per cent greater than the span, while for a rise of 1/10, it is only 2.6 per cent greater.

The arch shortening/spreading of foundations are imposed deformations that can cause cracking, but in most cases cannot cause collapse. However, these actions change the geometry of the arch, and so induce second order effects. For arches with a span/rise ratio of about 8 or less, these second order effects are not significant. However, for very flat arches, the changes in geometry may be significant, increasing the bending moments in the arch and even threatening its survival.

17.7.4 Instability of arches

As arches are in compression, they must be checked against buckling instability. Generally, when the bridge has been designed so that the bending moments are carried by the arch, a thickness in the plane of the arch of span/60 at the crown is likely to provide a satisfactory factor of safety as the arch shape inhibits in-plane buckling. When the bending moments are carried by the deck, the arch thickness is governed by criteria other than general buckling, such as compressive stress, bending moments, local buckling between points of liaison with the deck or stability during construction.

Buckling perpendicular to the arch plane is similar to conventional strut behaviour, and the arch rib should be checked and reinforced as a strut as a first approximation. However, due to the curvature of the arch, its torsional strength is mobilised in any lateral deflection. A slenderness of 1/30 for an arch fixed in the plane of buckling, and 1/20 for a pinned structure are likely to be satisfactory, although more slender

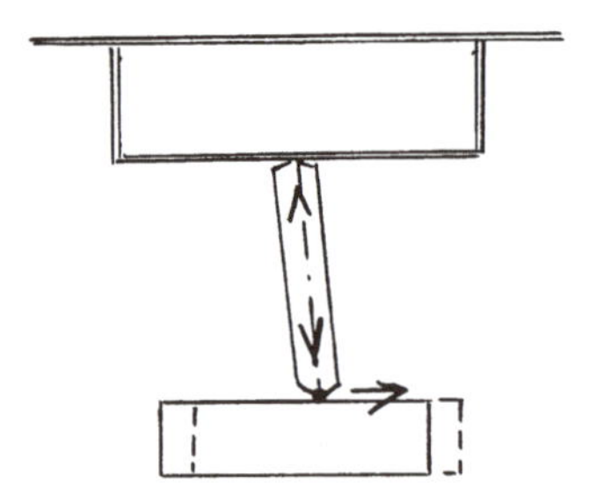

Figure 17.15 Lateral instability of arches

structures are possible. Lateral wind loading is also likely to be relevant to arch stability. A detailed analysis of stability in both planes is always required. For an arch consisting of several ribs, the critical condition for stability may well arise during construction before the links between ribs are in place.

Out of plane buckling of an arch cannot be considered in isolation from the bridge deck carried. If the deck is above or below the arch and consists of statically determinate spans without overall transverse stiffness, it will follow any lateral displacement of the arch and reduce its stability, in the same way as a weight on top of a cantilevered column will reduce its stability. A continuous deck has transverse stiffness and the effect on the arch will depend on the nature of the link between the deck and the arch. If it rests on the arch through columns effectively pinned at each end, the lateral deflection of the arch will incline the columns, imposing a horizontal disturbing force on the arch, decreasing its stability, Figure 17.15. However, if suspended from the arch, transverse movement of the arch will incline the hangers, applying a restoring force, Figure 17.21 (b), (*17.10*).

The stability of a concrete arch is affected fundamentally by creep under sustained loads. Preliminary elastic calculations of stability may be carried out using a Young's modulus of one-third of the short-term value, and a load factor of 3 against the critical load. Energy methods are very appropriate for a calculation of the stability of arches as are non-linear finite element programs, as long as the designer has thoroughly understood the statics of the problem, and is capable of checking the accuracy of the results. There are also charts available [3].

17.8 Construction of arches

Traditionally, arches were built on timber centring which were often major engineering structures in their own right. Removing this centring and allowing the arch to carry its own weight is a delicate operation. As the concrete is stressed in compression, the arch must shorten as it takes up its self weight, leading to a drop in the level of the crown. Ideally, the centring should be lowered incrementally over the full span simultaneously, so that potentially ruinous bending moments due to the self weight being applied over only part of the span are not introduced into the arch. This clearly is a difficult task for a timber centring consisting of a large number of members or for a large span. At the Plougastel Bridge with three 180 m spans, designed and built by Freyssinet in 1930, the problem was solved by the novel technique of introducing flat jacks into the cross section of the arch which, when expanded, lifted it off the centring.

The same method was used to strike the 305 m span Gladesville arch in Sydney, in 1964. The arch, which has a rise of 41 m, consists of four adjacent box section ribs, each 6 m wide, 4.25 m deep at the crown and 7 m at the springings. Each rib was made up of a series of precast voussoirs 3 m long jointed with cast-in-situ concrete, and was built on centring. It was struck by inflating flat jacks at the quarter points. The centring was then slid sideways for the next rib. The arch acts as an unreinforced concrete structure for its overall actions, and when built was the longest span concrete arch in the world. If the same arch were to be built today, it would be considered more economical to counter-cast the segments, avoiding the need for a jointing operation.

The cost and the time involved in building the centring, striking the arch and then removing the centring are one of the reasons that large-span arches have not been used more extensively. However new methods of construction avoid the need for centring.

For instance, Benaim's Sungai Dinding Bridge, Figure 17.16, had 13 arches with spans that varied from 45 m at the river banks to 90 m at the centre, plus approach viaducts. The arch ring was a box section, 8.8 m wide, with a depth that varied from 1.6 m for the smaller arches to 2.25 m for the largest. The arch spans were built in cast-in-situ balanced cantilever from each pier. A pair of half arch rings was temporarily hinged at the springings and suspended by stays from temporary falsework towers, Figure 17.17. When completed, the trailing half arch was stitched at the crown to the completed portion of the deck, and one of the two falsework towers could be leapfrogged forwards to the next pier. As each arch was completed, the temporary hinges were locked by cast-in-situ concrete. In this way the arch was built span by span, Figure 17.18. The deck was of continuous steel composite construction.

The 390 m span Krk Bridge had a rise of 60 m, and the arch was a constant 6.5 m deep and 13 m wide while the deck consisted of precast 'T' beams. It was built by cantilevering from the two abutments, Figure 17.19. A central box section was built first, from precast flange and web members stitched in-situ. The concrete and precast members were delivered by cable crane. The falsework was in the form of a truss, with the arch constituting the compression flange and the tension and shear members in

Figure 17.16 Bridge over Sungai Dinding (Photo: Benaim)

Figure 17.17 Bridge over Sungai Dinding: balanced cantilever construction of arch (Photo: Benaim)

Figure 17.18 Bridge over Sungai Dinding: span-by-span erection of arches (Photo: Benaim)

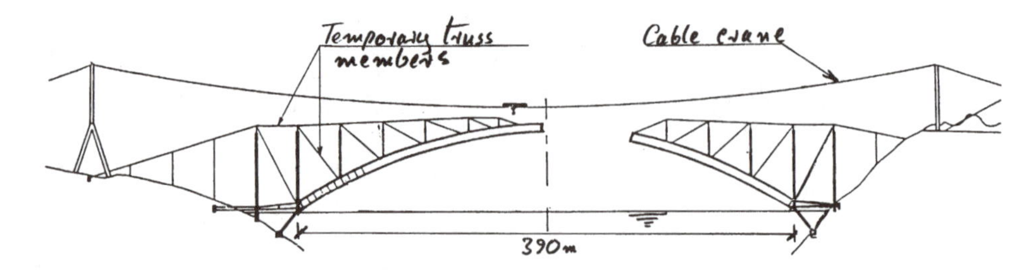

Figure 17.19 Construction of Krk Bridge

Figure 17.20 Construction of Yajisha Bridge (Photo: Engineer Leo K.K. Leung, Executive Director of Hopewell Highway Infrastructure Ltd (Hong Kong Stock Code: 737))

steel. Once this pre-arch had been closed, the steel falsework could be removed and the box was extended by adding two side cells.

Arches may also be prefabricated in sections weighing hundreds of tons, and assembled in cantilever using cable cranes. The Chinese have significant experience of this type of erection. Another method of erection for large-span arches used extensively in China is to erect a pre-arch consisting of steel tubes braced together which are subsequently filled with concrete that acts compositely with the tubes. The tubes are erected in cantilever with tie backs, or by cable crane. Clearly the falsework is much lighter than that required for the erection of the finished arch.

A recent example of this method of construction is the 360 m span Yajisha (an alternative rendition of Ah Kai Sha) Bridge over the Pearl River close to Guangzhou, Figure 17.20. Half arches consisting of steel tubes were built on falsework on either bank, and then rotated about a vertical axis into their final position. During the rotation, the steel half arches were counter-balanced by short concrete arches supporting the approach span. Once in position, the steel arches were filled with concrete [4]. This single-deck dual-carriageway bridge was built instead of the Ah Kai Sha twin-deck four-carriageway cable-stayed bridge designed by Benaim (*18.4.11*).

Yet another method of building large arches is to build each half arch vertically over the springings, and then to rotate them about a horizontal axis, lowering them into position. This requires restraining falsework of similar capacity to that required by cantilever erection, but for a much shorter time. The construction of each half arch may be carried out by slip-forming or jump-forming techniques that may be more economical than building in cantilever.

When an arch is built in cantilever, the temporary masts and cables have to carry the full weight of the arch before it is joined. These costly temporary works may be significantly reduced by building a structure that is hybrid between the arch and the truss. The total restraining bending moment needed to ensure stability is of course dependent solely on the weight of the composite structure, but economy is derived from the fact that much of the temporary works are incorporated into the permanent works. It should always be an ambition of the engineer to use his ingenuity to minimise the purely temporary components in a construction procedure. In addition to reducing the temporary works, this method of construction also builds the deck and the arch at the same time, giving the opportunity to reduce the overall construction programme. Many such bridges have been built in China [1].

It is hoped that this brief summary of the various methods of building large concrete arches will stimulate the imagination of the designer, as clearly other possibilities exist

17.9 Progressive collapse of multi-span arch bridges

The Romans designed each span of their multiple arch bridges to be stable in the event of the absence of an adjacent arch, either due to a span-by-span construction sequence, or to the destruction of a single span in war. Consequently, everyone else did likewise until the late eighteenth century when Perronet discovered that he could greatly lighten the piers of his bridges by building all spans at once. Thus he could design the piers only for the difference in thrust from adjacent spans. Now, modern engineering practice has returned to Roman concepts as it is considered prudent to design to limit the risks of progressive collapse, or of disproportionate damage in the event of a local accident. Thus a multi-arch bridge should be capable of carrying its self weight plus normal working live load without collapse in the event of the removal of one span.

In the Sungai Dinding Bridge described above, the progressive construction method imposed pier foundations that were designed to resist the unbalanced thrust of the dead weight of an arch. In fact, as the piers were designed for ship impact, the foundations did not require significant strengthening.

17.10 Tied arches

Conventional large-span arches require foundations to resist their thrust. If the ground is poor, such foundations are likely to be very expensive, and may rule out a conventional arch option. This problem can be overcome by providing a tie between the arch springings, leaving only vertical reactions on the foundations, Figure 17.21 (a). The tie is usually incorporated into the bridge deck that is suspended from the arch by hangers. The bridge must rest on bearings that allow for its length changes.

As the self weight of the arch is a significant proportion of the total load, it is likely to be economical to use very strong concrete working at high stress to reduce its size and weight. This is possible as the buckling of the arch may be restrained by the deck, as long as this has adequate bending strength in both planes. Any lateral deflection of the arch due to incipient buckling is resisted by the deviation of the hangers, which provide a restoring force, Figure 17.21 (b). However, although greatly improving the stability of the arch, this arrangement does not totally preclude buckling. Before the hangers can exert any restoring force the arch must deflect incrementally sideways. It needs to be proven by analysis that there is a stable outcome and that the deck can resist the small lateral loads that this action will impose on it, or that it will not deflect

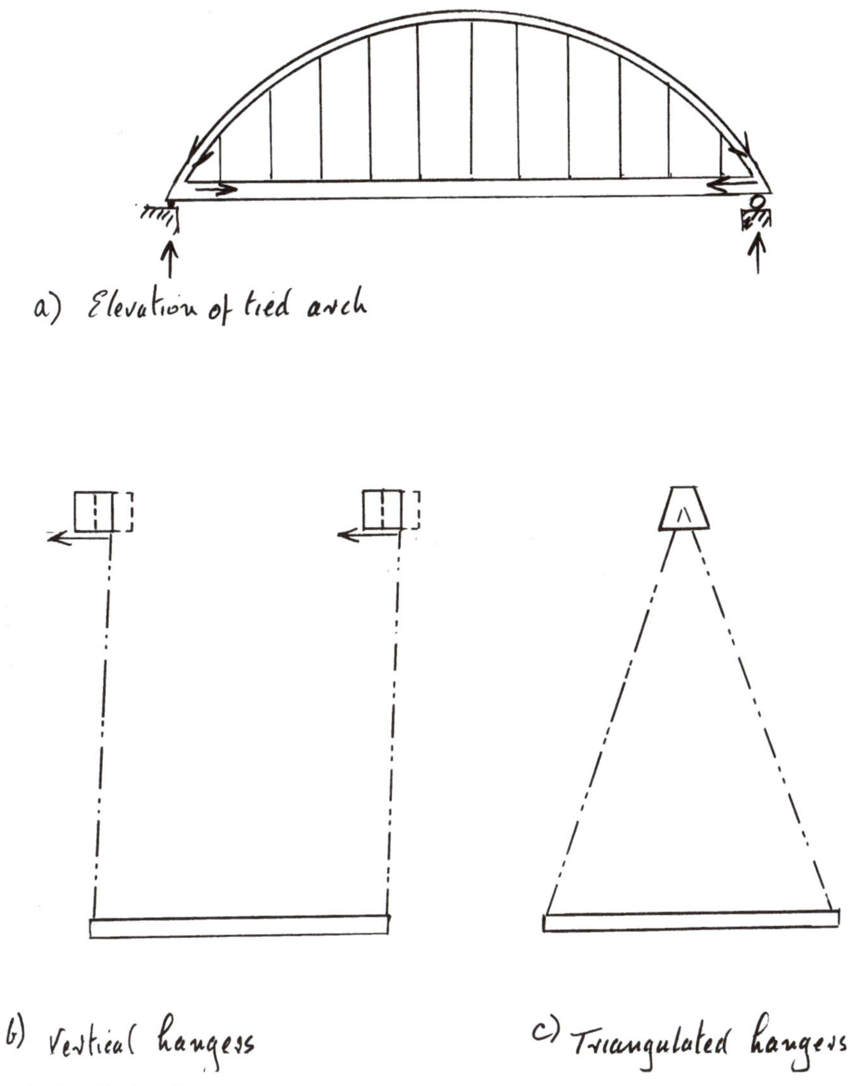

Figure 17.21 Tied arch

laterally so much that its stabilising effect is reduced. Any normal continuous highway deck is likely to fulfil these conditions, but light footbridges may not. If the hangers are effectively triangulated transversally, meeting at an apex at the arch, and if the deck has suitable torsional stiffness, the arch may be completely stabilised against lateral buckling, Figure 17.21 (c), allowing the most slender of structures. Even a narrow angle of triangulation will be effective in providing this stability, as the first incremental sideways movement of the arch immediately encounters a horizontal component of hanger force.

The bridge must of course be capable of carrying concentrated live loads which, if applied to the arch, would generate bending moments in it and significantly increase its required depth and weight. The logic of the tied arch is to provide a stiff deck that carries the moments, allowing the arch to be as light as possible.

The construction of a tied arch is dominated by the problem that it cannot be erected unless the tie is in place; but as the tie usually consists of the bridge deck, it needs the arch to carry it. Various solutions, other than ground-based falsework, are available.

For instance, the arch foundations may be designed to carry a small amount of horizontal thrust by the bending of vertical piles, augmented if necessary by temporary inclined ground anchors. A pre-arch made of steel or thin concrete shells may then be erected whose thrust is within the capacity of the foundations. The pre-arch may be erected in cantilever supported by temporary stays, or may be prefabricated in large sections and erected by crane. Once the pre-arch is complete, a tie would then be installed. One option for the tie consists of temporary cables in tension. However, as such cables would be much more flexible than the foundations any increase in thrust due to additional weight added to the arch would extend them, transferring most of the additional horizontal thrust to the foundations. This may be avoided by progressively increasing the tension of the cables to keep the horizontal force in the foundations between predetermined limits.

This is a fussy operation, and may be avoided if ties are installed that can also carry some compression. These may be slender steel or concrete tubes that cannot buckle, due to the internal prestressing (*4.2.8*), or the ties could be a first stage of the construction of the suspended deck. These ties would initially be compressed by the cables. The increasing thrust from the arch as it is completed will then de-compress them, with little elongation, hence adding only small additional horizontal thrusts to the foundations. However, ties capable of taking compression will be heavier than cables, and the thrust of the pre-arch due to this additional weight has to be taken by the foundations.

An alternative method of construction would be to erect the deck on temporary supports, either by launching or span-by-span, and then to use this as a platform for erecting the arch. If such temporary supports were acceptable and economical, it would of course beg the question as to whether the arch was indeed necessary.

18
Cable-supported decks

18.1 General

This chapter discusses so-called 'extradosed' concrete decks, cable-stayed bridges, under-trussed decks and stressed ribbons. All these bridge deck types fall outside the logic of sizing deck members given in Chapters 8 and 9.

18.2 Extradosed bridge decks

This title covers a family of bridge decks whose one common feature is that the prestressing cables at the pier section are raised above the extrados or top surface of the deck, hence their rather clumsy name. When these cables are encased in a concrete wall, the deck is typically called a 'fin back'.

Extradosed bridges are transitional between girders and cable-stayed bridges. Whereas cable-stayed decks have little stiffness or strength compared with the stays which carry most of the loads, extradosed decks are stiffer compared with their cable systems and the decks carry a significant proportion of the loads. The decks are designed as for any other prestressed girder bridge, with the difference that the eccentricity of the prestress over the supports is not limited to the deck envelope.

Most such bridges are erected in balanced cantilever, and may be cast-in-situ or precast segmental. When the deck erection is complete, there exists an array of extradosed cables. If the cables are not encased in concrete, the fluctuation of their stress under the effect of live loads may affect their fatigue life and should be checked. The Rambler Channel Bridge, designed by Benaim for tender, was intended to carry four tracks of the Hong Kong Mass Transit Railway on two levels. It spanned 205 m, and was 7 m deep, a span/depth ratio of about 1/30, Figure 18.1 and Figure 18.2. Although the live load was very heavy, the deck stiffness was such that the fluctuation in cable stress due to live loads was little greater than that experienced in conventional internal grouted prestressing tendons, and in this case did not constitute a fatigue risk. For more slender decks it is likely that the stress variation in the tendons will be greater.

It is also possible that the cables may suffer from vibrations due to wind, or to excitation by the deck, which in turn could give rise to fatigue problems. On the Rambler Channel, the cables rose 11 m above the deck and the longest were unsupported for 70 m. However, as the cables constitute a closely spaced array, with a wide range of lengths, the longer cables may be stabilised by attaching them to the shorter ones or

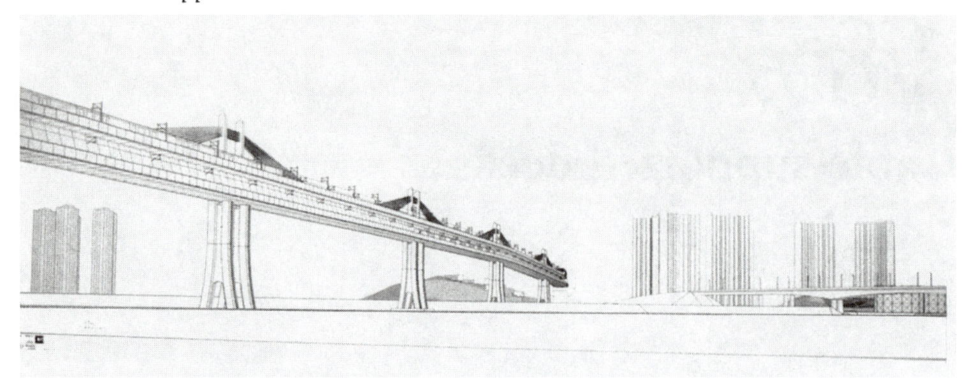

Figure 18.1 Project for Rambler Channel, Hong Kong (Image: Benaim)

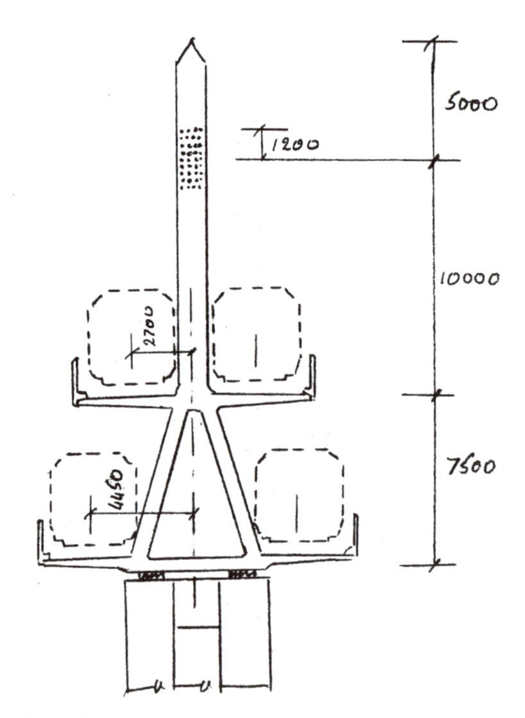

Figure 18.2 Rambler Channel, cross section

to the deck itself, making it most unlikely that any significant resonant vibration could occur.

Exposed tendons may be more vulnerable to accidental damage, when the performance of the deck with a number of tendons removed should be ascertained.

Cable stays typically cost between 2.5 and 3 times more than prestress tendons. This high cost is due to the use of fatigue-resistant anchorages, the difficulty of erecting the stays onto very high towers, the need for several layers of corrosion protection, the operations of grouting the very long stays with special products, the need to adjust the stay tension after installation and other factors. Furthermore, to control the risks of fatigue failure, stays are designed to operate at no more than 45 per cent of

Figure 18.3 Ganter Bridge (Photo: Christian Menn)

their breaking load under dead plus live loads. The viability of the Rambler Channel design depended on keeping the costs of the cables close to those of typical external prestressing. It was planned to use standard external prestress anchorages, and the tendons were designed to work at about 55–60 per cent of their breaking load, as for a normal prestressing cable. They were to be housed in grouted HDPE ducts that passed through the masts to anchor at both ends in the deck. Other than the need for a scaffold on the deck for their erection, there was no reason that these tendons should cost more than conventional external prestressing.

If the extradosed tendons are enclosed in a concrete wall it will behave as an integral part of the concrete deck, and will be subjected to tensile bending stresses under the effect of superimposed dead and live loads. Furthermore, as it is cast after the deck, it will have built-in tensile stresses due to heat of hydration shortening. Additional prestressing tendons will need to be stressed to prevent it from cracking. Also, its self weight increases significantly the dead load bending moments of the deck, requiring a further increase in prestress. The wall consumes considerable additional permanent materials and temporary works, and it must be cast in a separate operation which extends the construction period of the deck. Although they may give peace of mind, such walls are expensive, and it is well worth solving the problems of the exposed tendons.

The appearance of extradosed bridges needs careful treatment, as it is easy to make them very ugly. As the Ganter Bridge, Figure 18.3, by Christian Menn demonstrates, they can also be beautiful.

18.3 Undertrussed bridges

Undertrussed decks are virtually a mirror image of extradosed decks. The prestressing cables are not confined to the soffit of the deck, but extend below it and transfer their

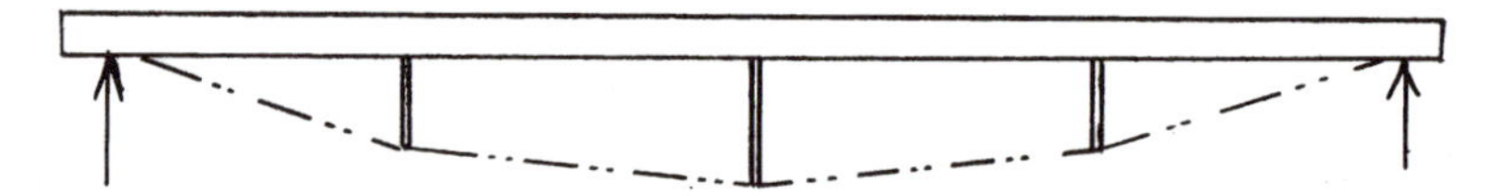

Figure 18.4 Undertrussed deck

reactions to the deck through downstand posts, Figure 18.4. Just as extradosed decks are essentially continuous as the additional strength is provided over the supports, undertrussed decks are essentially statically determinate.

As the author has no experience of designing such bridges, no detailed description will be given. However, it does appear that this is a rather under-used genre, and there is certainly scope for exploiting it for economy and for visual effect.

18.4 Cable-stayed bridges

18.4.1 General considerations

When compared with a girder bridge of equivalent span, a cable-stayed bridge trades off its cheaper deck against the additional cost of the expensive stays and towers. The tonnage of stay required is directly proportional to the total weight carried, of which a large proportion is the self weight of the deck. Thus in order to minimise the cost of stays, some decades ago, the deck of a cable-stayed bridge spanning, say, 400 m would have been a steel truss or girder, with a steel orthotropic deck.

More recently, the excessive first cost of such steel decks led designers to adopt composite construction for long-span cable-stayed bridges, with steel main girders and a concrete top slab. However, this is not a very rational design concept for a cable-stayed deck. The inclined stays impart large compressive forces to the deck. As the concrete slab is initially in tension due to restrained heat of hydration shortening, and is subject to creep when in compression, such decks are designed so that the steel girders carry all the compression, and as is well known, steel is not economical in compression.

The stays carry virtually all the overall shear force and bending moment, leaving the deck to carry only local actions. Consequently only two longitudinal girders are generally needed even for the widest of decks, and replacing the steel webs and bottom flanges of the girders by concrete need not represent a major increase in weight. The longitudinal compression in the deck then becomes beneficial, reducing the amount of prestress or reinforcement that is necessary to carry the local live load bending moments. By this reasoning, concrete decks are gradually being used for ever longer spans, currently up to about 500 m.

The towers are also sensitive to the weight of the deck, but less than the stays, as they have a very considerable self weight of their own and their design is heavily influenced by environmental loads.

18.4.2 Span arrangements and tower heights

The consensus is that the length of the side span should be 43 per cent of the main span while the height of the tower to the uppermost stay anchorage should be one-fifth of the main span. At this span ratio, the sagging bending moments due to live loads are about the same at the centre of the main span, and close to the end of the side span. However, as the side span is shorter than half the main span, the two or three uppermost stays coalesce to form a concentrated back stay, requiring a substantial counterweight or tension pile capacity to resist the uplift. If the side span is further shortened, this holding down force will increase. If the side span is lengthened, the live load bending moment may exceed that in the main span.

However, the aim of this book is to give engineers the tools necessary to question conventional wisdom. Consequently it is suggested that these figures are considered to be a starting point, but that they should not be taken as gospel. There exists in fact a continuum between extradosed bridges with low towers and stiff decks, and conventionally proportioned cable-stayed bridges.

Clearly, the ratio of main to side span is only valid if the side spans are fulfilling a useful function in extending the length of the bridge. At the extreme, when only a single long span is required, an arch bridge may be a better option. However, there are intermediate situations when it may be appropriate to shorten the side spans and provide the necessary tie downs, as in the Barrios de Luna Bridge, Figure 18.5, which had a 440 m main span and 102 m side spans.

Some cable-stayed bridges have intermediate supports in their side spans. These provide no benefit to the bridge in its completed state (except possibly when the cables are in the harp arrangement, *18.4.4*); on the contrary, they are expensive particularly as the deck has to be tied down to them. They can be useful during the construction of the bridge, particularly if the side spans are launched into place, but their cost must then be justified as temporary works.

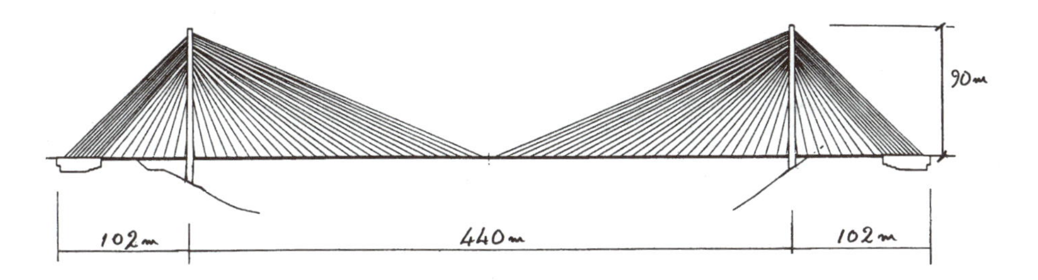

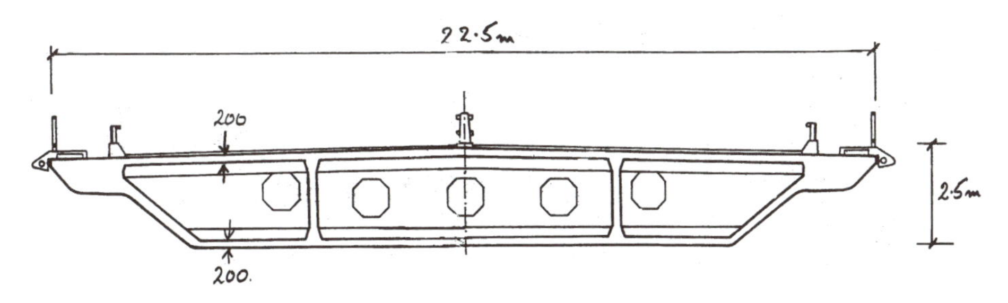

Figure 18.5 Barrios de Luna Bridge (Adapted from image in Walther, R., Houriet, B., Isler, M. and Moïa, P., *Cable Stayed Bridges*, p. 47)

The optimum tower height is a compromise between the cost of the tower and that of the stays. As the tower height is reduced the inclination of the stays becomes flatter and their force and thus their steel area must increase, while the deck becomes more flexible. If the tower is taller it will cost more and be more visually dominant, but the stays will be more economical and the deck stiffer.

Cable-stayed bridges derive much of their stiffness from the fact that the head of the tower is anchored by back stays to a fixed abutment. Where it is necessary to design a multi-span cable-stayed deck, means must be found to cope with the lack of this fixed point [1]. When one span is subjected to live load, the additional force in the stays must either be absorbed by bending in the tower, or transferred to the adjacent spans. Proposals have been made for rigid towers consisting of longitudinal 'A' frames, capable of resisting the overturning forces, Figure 18.6 (a). However, rather than this stiff solution, which will require a major increase in the cost of the piers and foundations, it is better to adopt a flexible solution, whereby a proportion of the live load tension in the stays is transmitted through a flexible tower to the adjacent spans. The live load bending moment is thus shared between tower and deck, requiring both to be strengthened and stiffened, Figure 18.6 (b). For a deck to have adequate stiffness it most probably has to be a box section. As box section decks have torsional strength, it would be logical to adopt a single plane of stay cables (*18.4.3*). In some bridges, notably the Ting Kau Bridge in Hong Kong, the towers have been stiffened by linking their summits to the base of adjacent towers with stays, Figure 18.6 (c), and it has

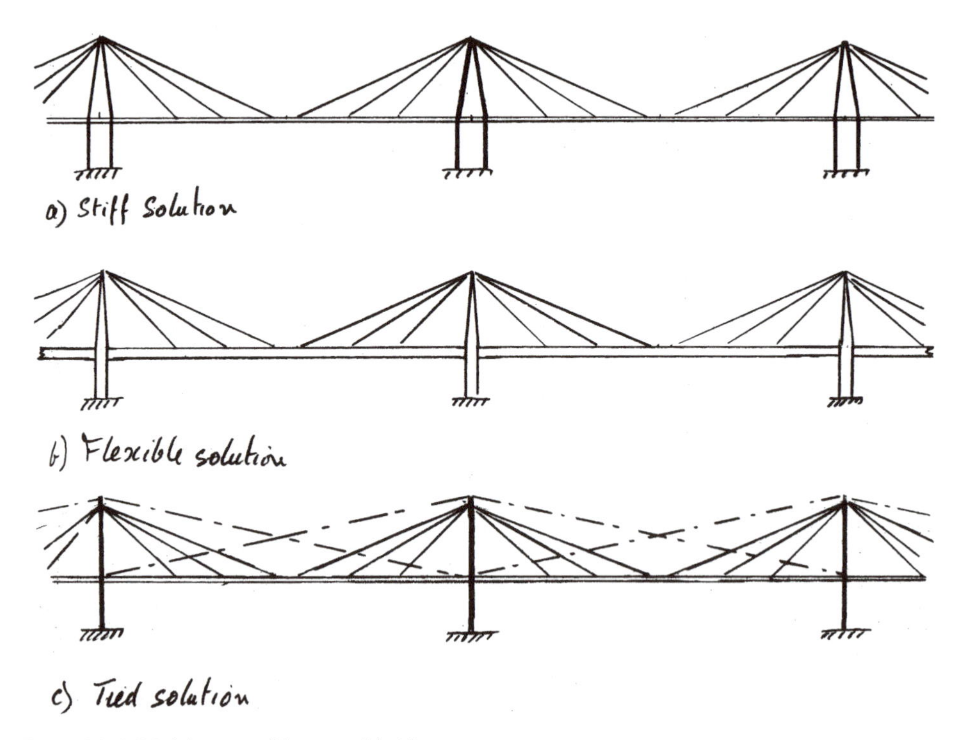

Figure 18.6 Multi-span cable-stayed bridges

been proposed on other projects to link all the tower heads together and thence to the abutments.

18.4.3 *Number of planes of cable stays*

a) Single cable plane

If one plane of stays is adopted, the deck must have torsional stiffness, both to resist eccentric live loads and to provide aerodynamic stability. Consequently the deck requires a box cross section. As the deck is supported only on its centre line, it cantilevers sideways for the half its width. This generally leads to the deck consisting of a trapezoidal box girder with long side cantilevers. The box usually has a trussed system of internal diaphragms, carrying the transverse shear from the side webs to the central points of attachment of the stays. The thickness of the side cantilevers and of the top slab and the sizing of the internal truss depend critically on the loading code to which the deck is being designed.

The tower consists generally of a single column on the deck centre line. This makes it necessary to widen the road, which adds deck area and weight. Consequently, great design effort is applied to making the tower as slender as possible, which requires a thorough understanding of the subject of elastic stability (*18.4.8*).

The tower may either pass through the deck, or be integral with it. An example of the latter arrangement is the 400 m span Pont sur l'Elorn in France designed by Michel Placidi of the contractor Razel, where the side spans of the bridge, including a short section of the main span incorporating the base section of the tower, were launched into place. This bridge had a main span of 400 m. The deck consisted of a single concrete box girder with propped side cantilevers, 23.1 m wide, Figure 18.7.

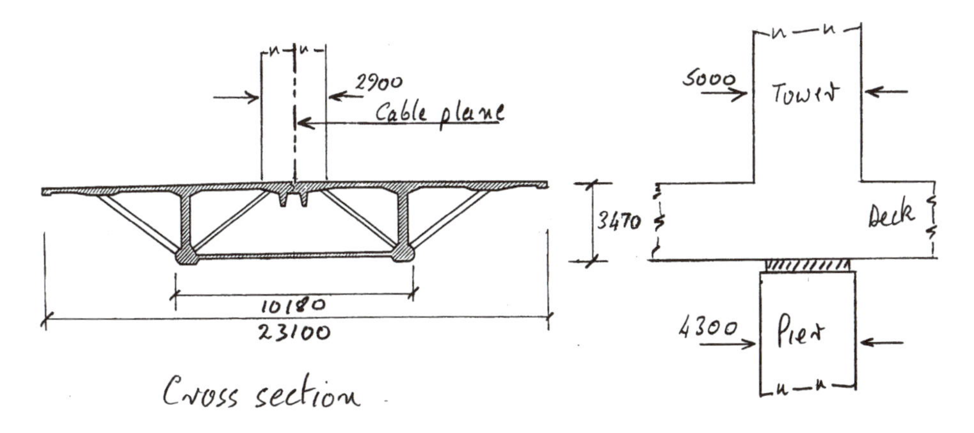

Figure 18.7 Pont sur l'Elorn

It is also possible to adopt 'A' frame towers with columns situated outside the deck and inclined to meet over the deck axis, with the stays attached close to the top in a fan arrangement.

The deck has to have significant depth to create an adequate torsion box, which attracts greater live load bending moments, as described in *18.4.5*. A torsion box generally has greater dead load than alternative decks, increasing the weight of stays required to carry the self weight. This disadvantage is more pronounced when one is designing to the lighter national loading codes, where very slender decks would otherwise be possible. However, the weight of stays necessary to carry live loads is minimised, as they are only carried once, and the greater stiffness of the deck distributes concentrated loads onto more stays. The deck is inherently more susceptible to wind induced torsional oscillations. Although it is clearly possible to design safe bridges with a single plane of stays, aerodynamic stability needs to be comprehensively proven each time.

Finally, a single plane of stays has some aesthetic advantages, as there is no visual interference between cable planes, and the tower may be very compact.

b) Two planes of stays

When a tower consists of two columns, they are usually placed outside the deck, while the stays are anchored on the deck edge. As a result, all the stays pull slightly inwards, creating a significant resultant force on the tower. This needs to be resisted either in transverse bending of the columns or, more economically, by a cross-beam between the columns located in the zone of stay anchorages, Figure 1.9. Alternatively, the columns

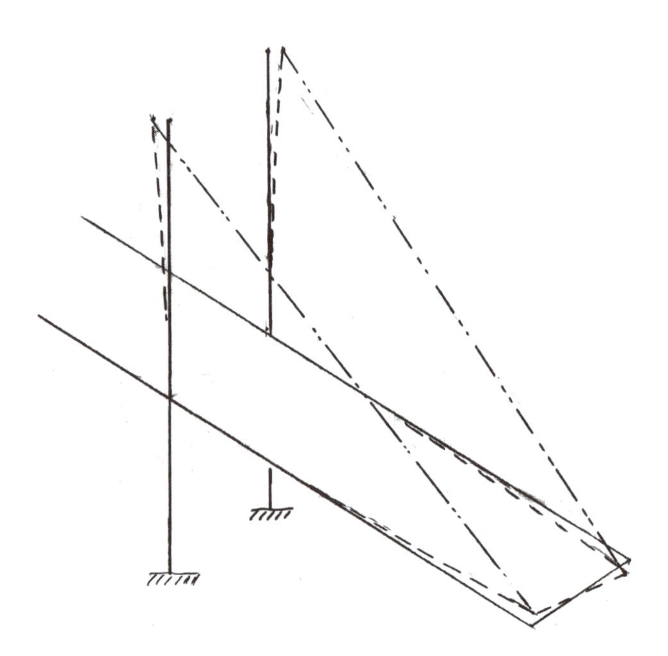

Figure 18.8 Linkage between tower and deck oscillation

may be cranked inwards above the deck, so the zone of stay anchorages is co-axial with the stays, Figure 1.10.

When a deck is carried by two planes of stays, it has a high degree of inherent aerodynamic stability. This does not mean that it cannot suffer wind-induced torsional oscillations. In particular, there is a possibility that the columns may oscillate out of phase along the line of the bridge, permitting a torsional deck oscillation, Figure 18.8.

The tower may also be an 'A' frame, with the anchors either concentrated at the top in a fan arrangement or distributed down the two legs in semi-harp or harp arrangements. This type of tower eliminates the possibility of the columns oscillating with the deck, providing the best resistance to wind-induced torsional oscillations. There must be a safe clearance, usually 1.5 m, between the stays and the highway clearance diagram which makes 'A' frame towers unsuitable for wide decks on short spans.

c) Three planes of stays

For very wide bridges, the transverse bending of the deck as it spans between two planes of stays becomes the main source of expense for the deck. Some designers have considered it viable to reduce the cost of this transverse bending by providing a third plane of cable stays. It is difficult to be dogmatic about the merits of this choice. It is certain that the more planes of stays adopted, the more times a concentrated live load is carried. If the tower for a three-plane arrangement consists of three columns, this will inevitably be more expensive than two columns, and in addition the deck will have to be widened. However, savings will be made in the self weight and the transverse bending of the deck; it is up to the designer to make his own assessment. The author's experience with the Ah Kai Sha Bridge (*18.4.11*) is that despite its 42 m width, it was more economical to maintain two cable planes and to find an efficient way of spanning the deck transversally. However, this was a particular case, and one should not generalise from it.

18.4.4 Arrangement and spacing of stays

Stays may be arranged in the harp, fan or semi-harp patterns.

(a) In the harp arrangement, all the stays are parallel, Figure 18.9 (a). The force in each stay is such that its vertical component balances the dead load of the section of deck it is supporting. Thus if the tower height is span/5, the longest stay has an inclination of 1/2.5, and the force in the stay is 2.7 times the weight it is carrying. As all the cables are parallel and equally spaced, they all have the same force.

    The application of live loads on one span will increase the force in the stays supporting that span and impose unbalanced loading on the tower which will generate in it substantial bending moments. The tower will also be subjected to bending moments if any stay is removed, either through damage or for maintenance. These bending moments are likely to govern its dimensions, and will certainly greatly increase the weight and congestion of reinforcing steel required. The towers of some cable-stayed bridges have required more than 600 kg of reinforcement per m^3 of concrete, which is close to the limit of the constructible. It is necessary for the designer to consider the implications of stay removal early

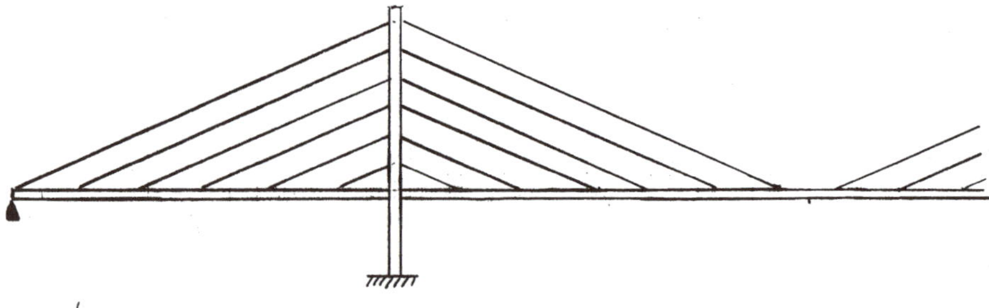

a) Harp

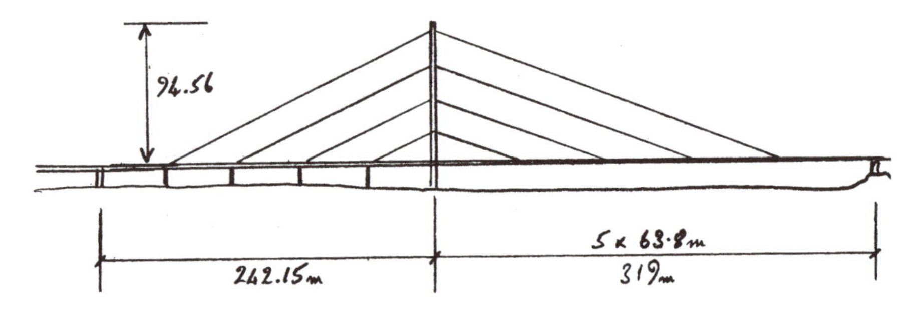

b) Knie Bridge; Dusseldorf (Adapted from Walther, R., Houriet, B., Isler, M. and Moïa, P., Cable stayed bridges, Thomas Telford 1999 London

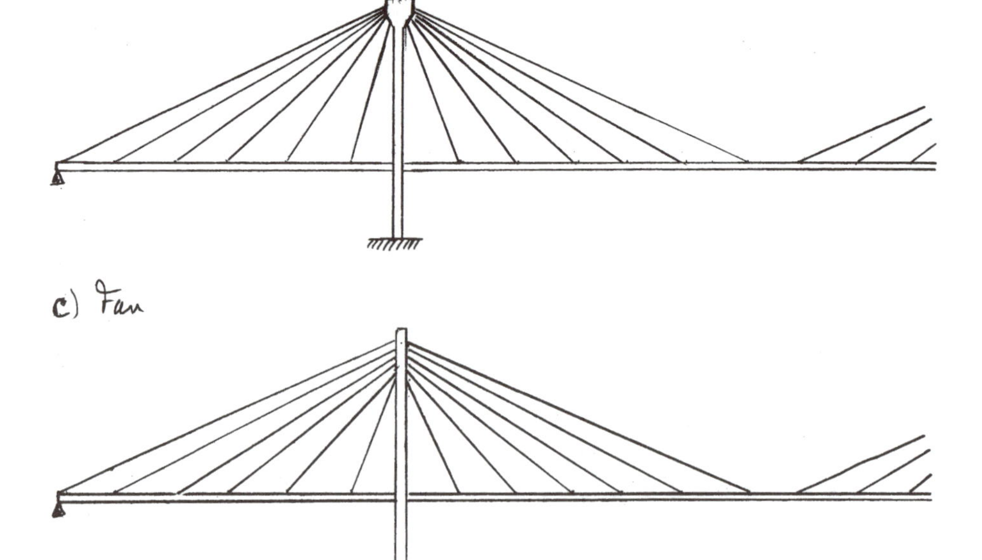

d) Semi-harp

Figure 18.9 Arrangement of stays

in the design process. One option suitable for a single plane of cables would be to use twin stays at each position.

If the site is suitable, the side spans may be provided with intermediate supports that anchor the cables and virtually eliminate bending in the tower, Figure 18.9 (b). Although allowing a slender tower, these supports add to the additional cost of the harp arrangement and rather defeat the logic of a cable-stayed deck. When the side spans do not require expensive long-span technology, an arch, tied or otherwise, with short-span approach viaducts may be a more economical structural type.

The weight of stay steel and the compression in the deck are both greatest in the harp arrangement. Despite its lack of economy, the harp is adopted for its harmonious appearance.

(b) In the fan arrangement all the stays radiate from the top of the tower, Figure 18.9 (c). Although the longest stay has the same inclination as in the harp, the stays become progressively more vertical and the force in them necessary to balance the load drops until the factor becomes unity for a vertical stay. When the main span is loaded, the additional force in the stays is transferred directly through the tower head to the backstays that are anchored to the end pier. Consequently, the application of live loads on the deck or the removal of a stay does not impose significant bending moments on the tower.

The principal disadvantage of the fan arrangement is the difficulty in housing all the stay anchors in a very compact space. Such bridges often require a pier head swelling that is not only odd in appearance, but is complex to design and build.

(c) Most modern bridges adopt the semi-harp arrangement, where the stay anchors are concentrated on the upper part of the tower, Figure 18.9 (d). The stays are spaced as closely as is possible, without resorting to expensive expedients such as combined base plates. This generally means that the upper stays that are more perpendicular to the tower will be spaced at about 1 m vertically, while the lower more acutely inclined stays will be progressively further apart. The weight of stay steel is marginally greater than in the fan arrangement. The bending of the tower is also greater, but experience shows that the compression in the tower is such that this bending moment does not cause difficulty in design or require dense reinforcement.

The spacing of the stay cables along the deck is a critical design decision. Early cable-stayed bridges tended to use few very powerful stays, requiring a deck that had the capacity to span 60 m or more between stays, Figure 18.9 (b). Such bridges are relatively complex to build, as the deck needs temporary support as it cantilevers towards the next stay, or it has to be erected in very long lengths that span from stay to stay. The stays are very powerful, and generally adopt expensive locked coil technology, while the transmission of their force to the deck also requires expensive details.

The trend for modern bridges has been to use smaller stays spaced at between 5 m and 10 m. This arrangement of stays reduces the self-weight bending moments in the deck to insignificant proportions, and opens the door to the very slender simple decks described in *18.4.6*. The stays are based on prestressing technology (*18.4.9*), and the smaller forces make the connection details simpler. Generally, the stay spacing is related to the construction module, such that as each new section of deck is cast or lifted into place, it is carried by a new set of stays.

18.4.5 Longitudinal bending in the deck

The deck may be entirely suspended, in which case the stays carry all the dead and live loads. Under the self weight of the deck the vertical component of the stay forces match exactly its weight; the only bending moments in the deck are as it spans between stays. However, anchoring steeply inclined stays on the tower is fussy.

Alternatively, the deck may be supported on the main towers, either by being built in or through bearings. When the superimposed dead load due to road surfacing and finishes is applied the stays will extend, creating some local bending moments adjacent to rigid supports. It is possible to adjust the force in the stays after application of the superimposed dead loads to cancel out these bending moments. However, the cost of this operation may well be more than any material savings.

The deck needs longitudinal stiffness in order to distribute a concentrated live load onto several stays. Under the effect of the live load, it behaves like a beam on elastic foundations, where the spring stiffness reduces as the stays become flatter. Consequently the largest local live load bending moments occur close to mid-span, and towards the end of the side spans. In addition, as the stays extend under the effect of live loads the deck will be subject to additional bending moments adjacent to rigid supports.

If the deck has a box cross section, it becomes economical to allow it to cantilever either side of the tower before reaching the first stay, Figure 18.10. In this way some savings in the tonnage of stays may be made and the awkward near-vertical stays may be omitted, at the expense of additional local bending moments in the deck.

The project illustrated in Figure 18.10 had spans of 140 m, 460 m and 230 m, and the 32 m wide deck cross section consisted of two 3.5 m deep boxes, increasing locally to 7.5 m adjacent to the towers, on which they were supported. The 60 m either side of the towers were not stayed. The deck was designed to carry British HB + HA live loading. The influence lines for bending moment at the centre of the main span and for a point 39 m to the left of mid-span, as well as the live load moment envelope for the main span are shown in Figure 18.11. The sharp peak of the influence lines shows how sensitive is such a design to a code of practice that adopts a very heavy single load. The large hogging moments close to the towers are due to the support of the deck on the towers and the local omission of the stays. However, the shape of the moment envelope on the central 80 per cent of the main span is typical of fully stayed bridges with a relatively stiff deck, subjected to British live loading. The dip in the bending moments at mid-span is due to the restraint of the tower top by the backstays. The large negative area in the side span for the influence line adjacent to mid-span shown in Figure 18.11 (b) translates into the negative bulge of the moment envelope to the left of mid-span. The large variation of moment ΔM in the deck defines the minimum section properties required for the deck, and hence its weight. The magnitude of these bending moments is dependent on the deck stiffness, the stiffer the deck the larger the moments. Thus one of the principal aims of the design is to keep the deck as slender and flexible as possible.

18.4.6 Deck design

The economy of a cable-stayed bridge with a concrete deck depends principally on the ingenuity of the designer in minimising its weight and in keeping the construction simple, to allow a rapid rate of erection. For shorter spans, simplicity of construction

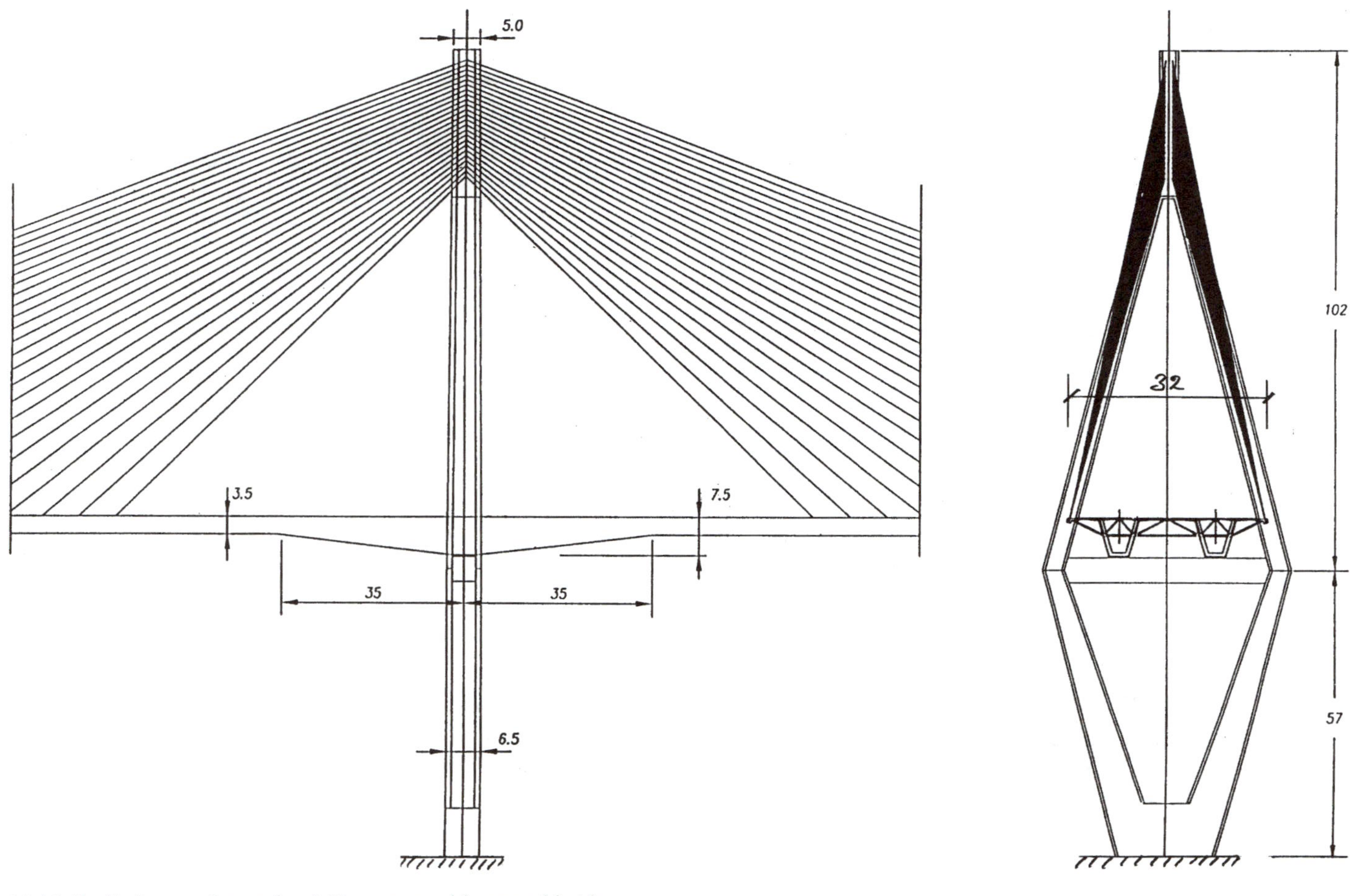

Figure 18.10 Preliminary scheme for 460 m span cable-stayed bridge

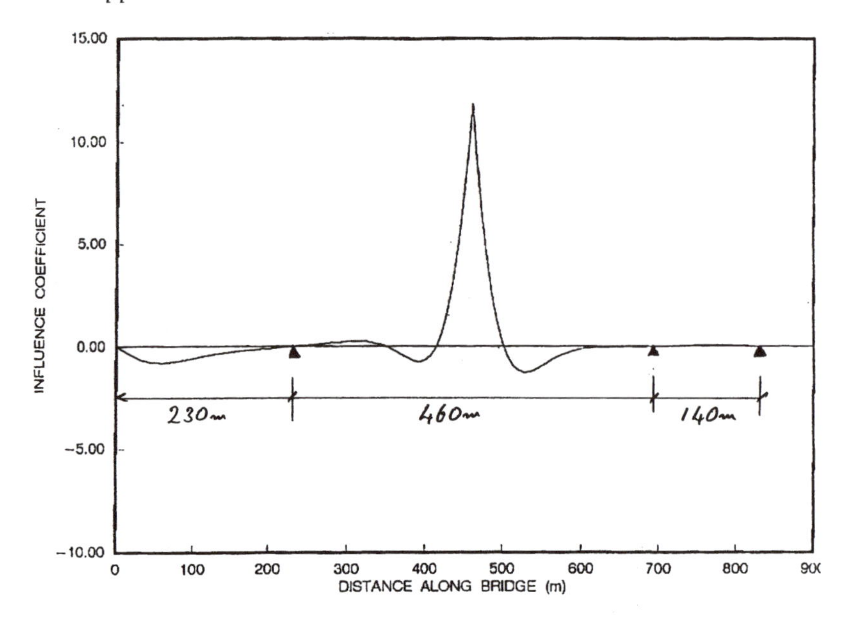

a) Influence line for centre of main span

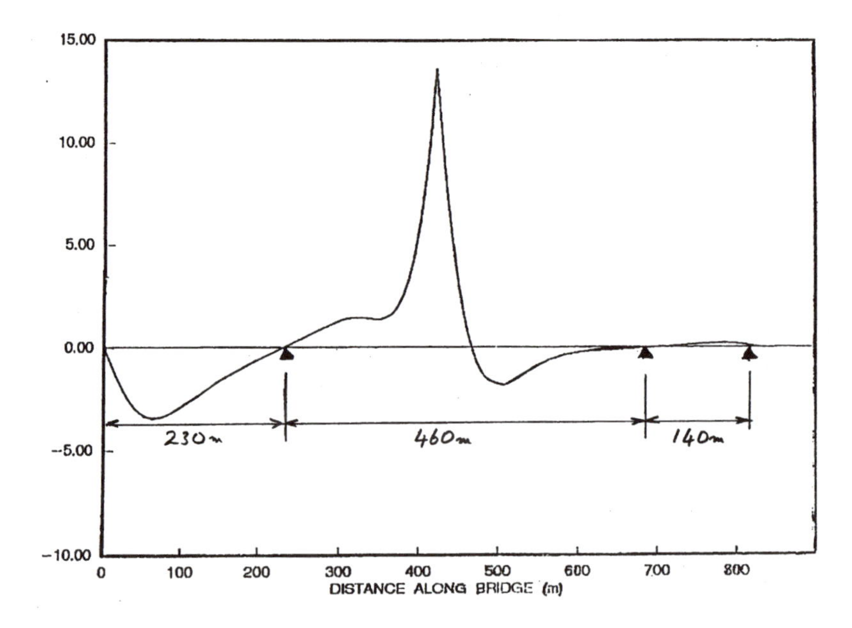

b) Influence line for point 39m to left of mid-span

Figure 18.11 Typical influence lines and bending moment envelope for bridge shown in Figure 18.10

Figure 18.11 continued...

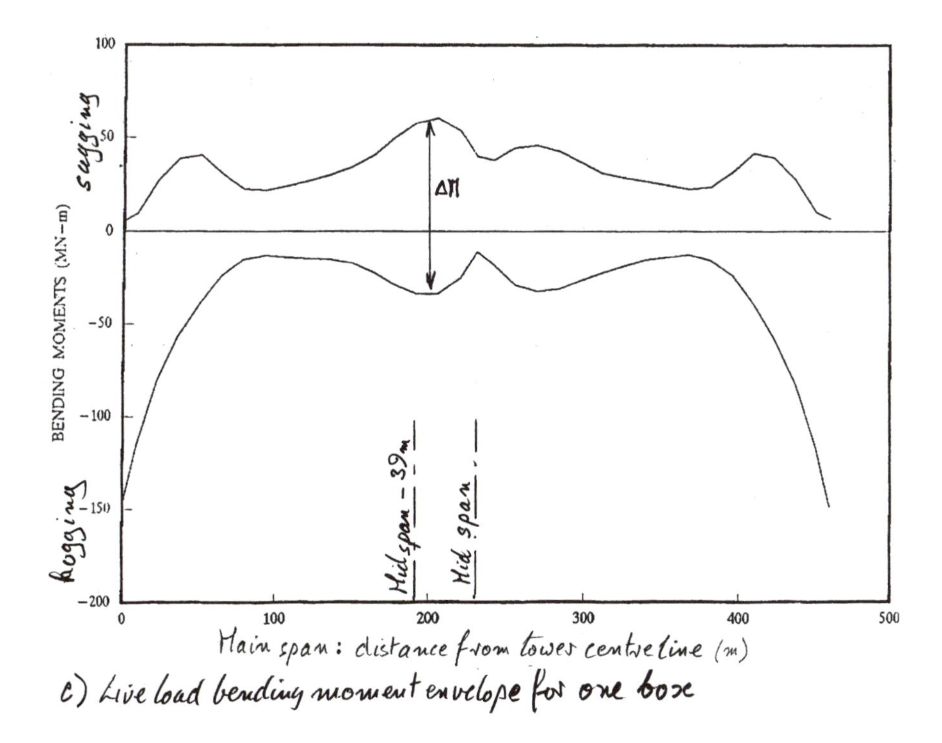

Figure 18.11 continued

may take precedence over weight. However, in a long-span cable-stayed bridge the cost of the stays, the towers and the foundations are all closely related to self weight and the pendulum swings back towards minimising the weight of the deck. It is very surprising to see the number of projects where this has not been understood, with long-span cable-stayed bridges equipped with crude, simplistic and heavy concrete decks.

The deck design is very dependent on the live loading code. For the example illustrated in Figures 18.10 and 18.11, subjected to the British HB + HA loading, the critical parameter is the variation of moment in the main span, ΔM. The variation of stress on the bottom fibre of the deck is $\Delta M/z_b$, where z_b is the bottom fibre section modulus (*5.3*). In a deck that is fully prestressed to Class 1 or Class 2, it is this variation of stress on the bottom fibre that will define the type of deck.

The characteristic of acceptable decks is that the z_b must be high compared with the moment of inertia in order to control the bottom fibre stresses. This requires adequate bottom fibre material, which means that economical beam and slab type decks are ruled out and one is led to box sections.

This conclusion would be different for live loads that are lighter than the British loadings, and which are not dominated by a single heavy vehicle, as such loading gives rise to a smaller ΔM. For such decks an acceptable z_b may be achieved with a wider variety of deck types. If partial prestressing is acceptable, the variation of bottom fibre stress is no longer critical, and the design reasoning will no longer be dominated by z_b.

a) Beam and slab decks

When there are two planes of stays providing torsional stability, a commonly used deck type consists of beams located close to the deck edges, joined by transverse beams and a top slab. The main beams may be solid concrete ribs, concrete 'I' beams or boxes. These edge beams or boxes may also be profiled to provide good aerodynamic performance. When sizing the longitudinal edge beams, the designer should remember that the deck also has to span horizontally from pier to pier under the effect of wind and seismic loads.

Decks carried by two planes of cables need a structure to span transversally between cable planes. Generally the spacing of transverse beams is the same as the spacing of the stay cables which is usually between 6 m and 10 m, or half that spacing. These beams may be either concrete or steel, the former being most suitable for narrower bridges of modest span, while the lower weight of the latter make them most appropriate for wide decks and long spans. The top slab, which should ideally be no more than 200 mm thick, will, under this arrangement, span longitudinally between transverse beams. Consequently, the local bending stresses in the slab will combine with the direct compression derived from the stays and with the overall bending of the deck. This could well make it necessary to limit its span to 3.5 – 4 m depending on the loading code, or to increase its thickness and weight, with consequences on the economy, Figure 18.12 (a).

Alternatively the concrete deck slab may be carried on longitudinally spanning stringers, Figure 18.12 (b). The advantage of this design is that it reduces the frequency, and hence the overall weight of the transverse beams, and allows the deck slab to span transversally, when it is not affected by the longitudinal compression in the deck, and can consequently span further. Generally, the stringers may be spaced at up to 5 m for a 200 mm slab, depending on the loading code. For a long span or wide deck, the transverse beams may be steel plate girders or trusses, and the stringers may be rolled steel sections or fabricated plate girders. Such beam and slab decks are generally very economical for most codes of practice.

However, as has been described above, solid ribs are not generally suitable for fully prestressed decks subjected to the heavy UK live loading, as the bottom fibre modulus is too low compared with the moment of inertia. As the width of the ribs is increased in an attempt to control the bending stresses, the moment of inertia of the deck increases and more live load moment is attracted in a vicious circle, while the deck becomes uneconomically heavy. On one such design for a 450 m long span, a deck with solid edge ribs and transverse steel beams was proposed. The application of HB loading led to the widening of the ribs to such an extent that the equivalent concrete thickness of the deck became over 700 mm ; that is definitely too heavy and uneconomical.

b) Box section decks

In order to achieve a fully prestressed light deck that is suitable for HB loading, it is difficult to avoid adopting a box section, with its more complicated construction

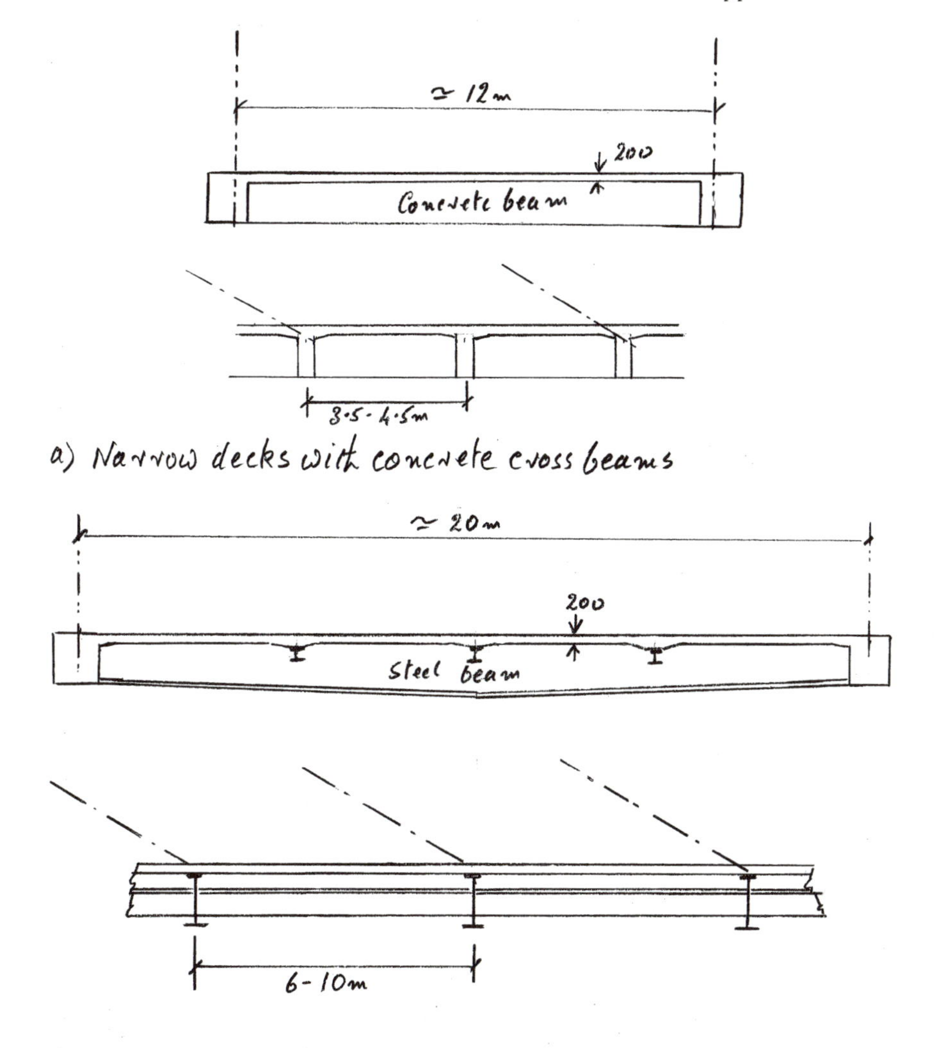

Figure 18.12 Transverse structure of deck

sequence. Box section decks are inherently torsionally strong. Consequently, where a deck is to carry British style heavy live loading, it would be logical to opt for a single central box carried by a single plane of cables. Such a deck may be very light, with an equivalent thickness close to 500 mm, and the complication of building the box section may be mitigated by adopting precast segmental technology. Alternatively the edge ribs may be replaced by boxes, or for a narrow deck, the box may be widened to the complete width of the deck, as at Barrios de Luna, Figure 18.5.

If the cable-stayed deck is part of a longer viaduct which is to be built in precast segmental box sections, the standard boxes may be adapted for use in the suspended deck. This was the case for the project shown in Figure 18.10, although it would have been more rational to have adopted a single plane of stays rather than the two planes shown.

c) Solid slab decks

Another way forwards for concrete decks has been given by R. Walther who has built a cable-stayed deck that consists of a solid concrete slab. The Diepoldsau Bridge has a 97 m main span, and an overall width of 14.5 m [2]. The two planes of stays are anchored close to the edges of the deck that is 550 mm thick at the centre, thinning to 360 mm at the edges, with an average thickness of 490 mm , which is similar to the average thickness of more conventional beam and slab or box section decks, but without the complications.

This is a very economical concept, and appears to offer clues to the best way forwards for narrow decks with spans up to at least 300 m. However, it needs proving that such thin decks are not at risk of buckling under the compression of the stays. Intuitively, it appears likely that such a deck will have a great resistance to buckling, as vertical displacement is inhibited by the stays which act as vertical springs. Clearly, the stays must be close enough to inhibit local buckling between their points of attachment.

A scale model was tested at the Swiss Federal Institute of Technology in Lausanne, representing at 1/20 scale a 200 m main span bridge, with a 14 m wide, 500 mm thick slab deck, with stays spaced at 6.4 m along the deck [3]. The deck thickness was thus 1/400 of the span. This test concluded that yielding of the stays would occur before the deck became unstable.

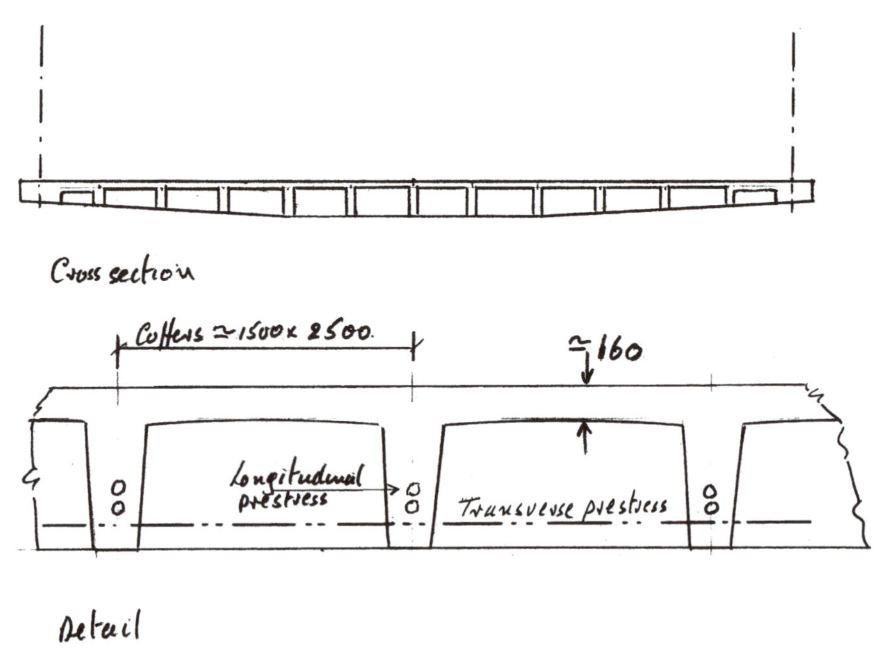

Figure 18.13 Proposed coffered deck

d) Proposed coffered slab decks

For longer spans or wider decks, a longitudinally prestressed coffered slab could be adopted, Figure 18.13. The author has carried out preliminary studies on such decks, which appear promising. Under UK loading, such a deck would probably need to be designed as partially or externally prestressed, due to the inherently low bottom fibre section modulus.

The size and aspect ratio of the coffers and the thickness of the top slab and ribs may be adapted to the loading code being used, to the width and span of the deck, and to the location in the deck. In highly stressed areas the roof and wall thicknesses could be increased, or the coffers omitted. For wide decks, the depth of the coffered slab may be varied, saving weight and improving economy. The transverse ribs would be internally prestressed. Such a coffered slab is likely to have an equivalent concrete thickness of less than 400 mm for a deck width of 20 m, which is lighter than most alternative systems.

18.4.7 Methods of construction of concrete decks

The decks of cable-stayed bridges are most frequently built by cast-in-situ balanced cantilevering. For instance, the Pont sur l'Elorn was built in this manner using an underslung traveller which cast 6.78 m long segments on a planned weekly cycle, Figure 18.14. As each segment was completed, a new stay cable was installed.

It is also possible to build decks by the precast segmental technique. This is most often adopted when the cable-stayed deck is part of a larger contract which justifies setting up a casting facility. For a stand-alone bridge to merit setting up a casting yard,

Figure 18.14 Pont sur l'Elorn: balanced cantilever construction (Photo: Robert Benaim)

it needs to involve at least 250 segments, which implies a total length of some 850 m for a single box cross section. A ratio of side to main span of 0.43 defines a main span of 450 m. If there are two or more boxes in the cross section, the viable length is clearly less.

Decks may also be built by partially precasting the deck, and assembling the precast elements by casting concrete in situ. For instance, for a deck consisting of longitudinal concrete beams and transverse steel beams carrying a concrete slab, the steel beams could be erected first, supported on the shutters for the longitudinal beams. The rear end of the shutters would be attached to the completed deck, while the forward end would be supported by a cable stay. This could be a temporary stay, or the permanent stay, subject to the ingenuity of the designer in finding a way to transfer the stay from the falsework to the permanent works. The slab would then be erected either as precast sections, or as pre-slabs. The deck would then be completed by casting the remainder of the concrete slab and the main beams.

18.4.8 Tower design

Towers for cable-stayed bridges are normally made of concrete. There seems so little justification for making a member that is principally in compression in steel that the author has no hesitation in dismissing the idea out of hand, except for small, architecturally driven bridges. However, the towers for the Tatara Bridge in Japan, which had, when it was built, the world's longest cable-stayed span of 890 m, are in steel, and several major suspension bridges also have steel towers. Consequently, there is clearly an aspect of tower design and construction that is outside the author's experience.

The transverse forces on a tower consist of wind on the tower itself, on the stays, on the deck and on the traffic. They also may include seismic forces, principally related to the self weight of the tower and deck. Longitudinally the forces include length changes of the deck, wind and seismic action and braking and acceleration of traffic.

The design of the towers depends critically on the articulation of the bridge deck. To illustrate one extreme, the Pont sur l'Elorn has one plane of stays, and the towers are an integral part of the deck, which is carried on bearings on traditional piers that provide the longitudinal fixity and transverse stability, Figure 18.7. This is unusual, and for most bridges the towers carry the loads of the deck to the foundations, and stabilise the deck both longitudinally and transversally.

Transversally the tower will tend to buckle as a free cantilever, while longitudinally its head will be held in place by the stays. Consequently, transverse stability is the most critical. If the stays are precisely aligned with the column axis, any lateral deflection of the column head will be resisted by an incremental horizontal component of the stay force, providing a system with considerable inherent stability. If the stays are pulling off the column axis, clearly they constitute a disturbing force. If a tower consists of two columns connected by a prop as at Ah Kai Sha (*18.4.11*), the geometric stability provided by the stays is re-established, as long as the two planes of stays are equally loaded. Clearly a curved cable-stayed deck will require careful consideration in this respect. An 'A' frame tower which is inherently stable transversally is well designed to resist transverse forces with minimum bending moments and materials.

In the author's experience, the towers of cable-stayed bridges are almost universally under-designed initially, and gradually increase in size as the designer meets and resolves the various problems. One of these is how to anchor the stays in the tower. The force must be transferred from the forward to the back stays. The conceptually simplest method of achieving this is for the stays to pass over saddles in the tower, anchoring in the deck at each end. In fact, this is only achievable on bridges with few cables. On a typical 350 m span highway bridge, there are likely to be at least 20 cables on each half span, and the practical problems of organising superimposed saddles are virtually insurmountable. Furthermore the fatigue performance of cables bent over the saddles and subjected to radial compression may be critical, and the cables would be more troublesome to replace.

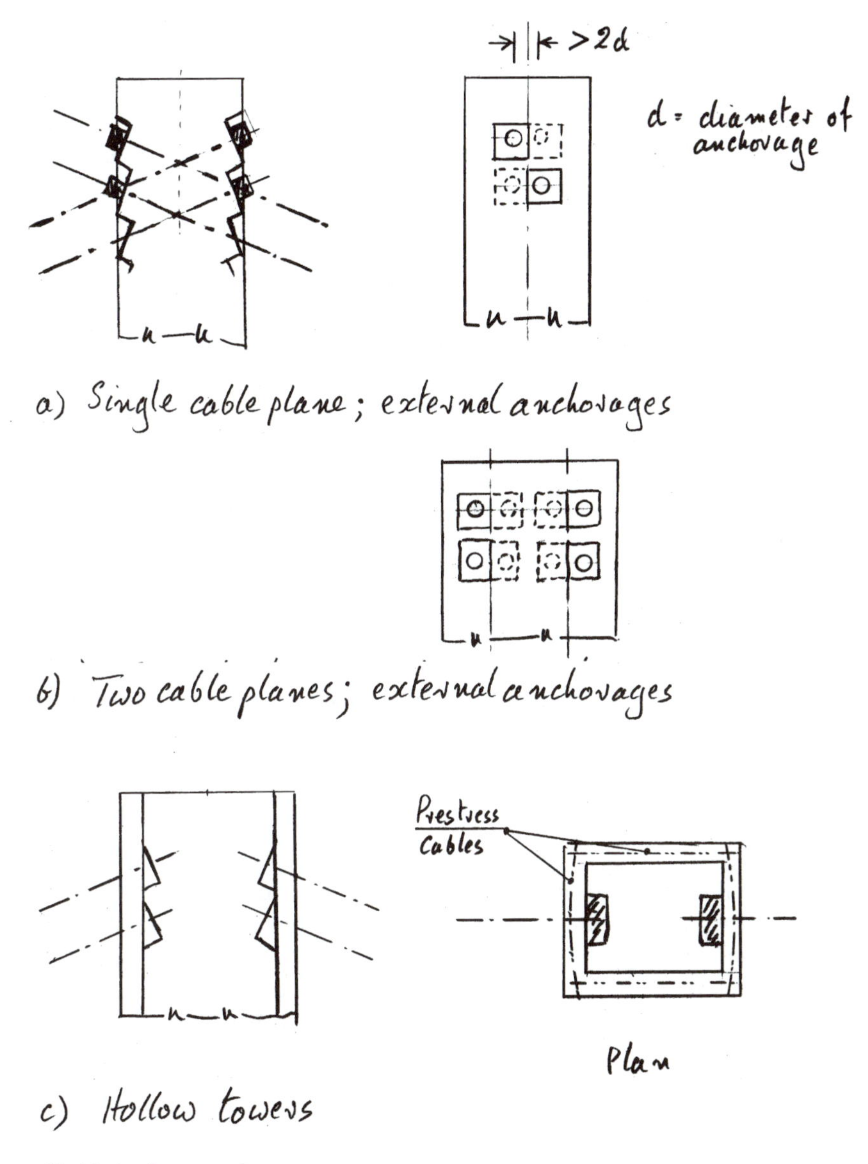

Figure 18.15 Anchorage of stays on towers

The next simplest method of transferring the force is to cross the stays and anchor them on the far faces of the tower, Figure 18.15 (a). This allows the use of a solid tower which may be very compact. Clearly the forward and back stays have to be offset to avoid each other. In order to avoid introducing a substantial torque into the towers, the planes of the stays should be alternated, such that the resultant passes through the tower axis. This alternation tends to be visually untidy. In a deck with two planes of cables and an inverted 'Y' frame tower with the stays anchored on the vertical leg, this problem does not arise, Figure 18.15 (b).

Placing the anchors on the outside faces of the tower gives rise to problems with respect to their weather protection, and complicates access for their initial installation, later inspection and eventual replacement. These problems are not insurmountable, and their solution depends on the expectations of the client, as well as on the size of the bridge. For a small footbridge, for instance, temporarily closing the bridge to allow the erection of a staging may be acceptable, while it may not on a higher tower or on a heavily used highway or railway bridge.

Alternatively, the stays may be anchored on the near faces of a hollow tower. This gives the best protection for the stay anchorages, and also offers the best facilities for inspection and maintenance. Clearly, the tower will not be as compact as a solid tower carrying the same loads. However, if the bridge is reasonably large, say with a main span of 250 m or more, a box section tower appears in proportion to the overall scale.

If the stays are to be stressed from this anchorage, there has to be enough room within the void to install and operate the jack. Even if the tower houses only dead anchorages, it must be possible to pull the stay through its ducting and to handle the heavy anchorage components. It is also necessary to accommodate either access stairs or, on large bridges, both stairs and a lift.

In order to transfer the stay force from the front to the back face of the tower, both the Pont de Normandie and the Pont sur l'Elorn used steel fabrications, where a pair of forward and back stays was connected by a steel box. These fabrications were bolted onto each other as the tower was erected. Concrete was then cast around them to complete the vertical strength of the tower.

The Ah Kai Sha Bridge and many others have been designed with all four faces of the box section tower prestressed to carry the stay forces to the side walls and then from front to back of the tower, Figure 18.15 (c).

It is necessary to position the stay anchorages very accurately so that they are truly co-axial with the stays. Whereas this is easily done with a steel fabrication, in a concrete solution it is more difficult. The solution adopted for the Ah Kai Sha Bridge is described in *18.4.11*.

A development of the conventional concept is the use of 'V' shaped towers. These lower the height of the tower which may offer environmental benefits in some cases and save a significant weight of stay cable. This option was adopted by Benaim as a competition entry for the Sundsvall Bridge in Sweden, Figure 18.16. Whether the savings in stays outweigh the extra cost of the towers needs to be proven by a fully-fledged preliminary design.

Figure 18.16 Sundsvall Bridge project (Image: David Benaim)

18.4.9 Design of stays

The design of stays must address the issues of control of their operating stresses, their method of erection, protection against corrosion, vibration and fatigue.

Stays for modern bridges may be composed of parallel strand, parallel wires or locked coil cables. The first two types are based on prestressing tendon technology, while the latter is based on suspension bridge hanger technology. As the author has no experience of using locked coil cables, the reader is referred to references [4] and [5].

This section offers only a brief overview of the design of stays. The brochures and technical documents prepared by the suppliers should be consulted to obtain the most up-to-date information in an evolving market.

a) Parallel wire stays

Parallel wire stays have anchorages which have evolved from the BBR prestressing system, adapted to improve their fatigue resistance. Stays may use up to 421 wires of 7 mm diameter which are usually galvanised, giving an ultimate load of approximately 27 MN and a working load of some 12 MN. This system is prefabricated, with the complete, sheathed stay being brought ready made to site. The prefabrication of the stays allows very high quality protection. A tightly fitting polyethylene sheath is extruded onto the bundle of wires. The extrusion process may give rise to variations in sheath thickness, and it is important to specify this thickness correctly and to check the results. Sheaths of different colours may be specified. The sheath is then pressure grouted with a petroleum wax to fill all voids. As the stay is horizontal and in a factory environment, several points of entry for the wax may be used, and the sheath may be re-injected to fill voids caused by the cooling of the wax. The stays are generally erected by a crane located at the top of the tower.

The anchors consist of a nut screwed onto the threaded outer casing and bearing on an anchor plate. The threaded length of the live anchor has to take into account the extension of the stay during stressing, including taking up its sag, as well as the extension due to any re-stressing that may be required during construction, and any tolerance on length. Clearly with this system, the cable lengths must be defined within close tolerances.

BBR also offer a stay system based on carbon fibre wires which have a tensile strength of 3,300 MPa, a Young's modulus of 165,000 MPa and a density of 1.56 t/m^3. They need no protection against corrosion, although they need to be housed in a duct to protect them from UV radiation. Their cost is likely to be high as long as such technology is in the prototype phase. It is to be hoped that they will become competitive, as corrosion-free stays is a goal worth working for.

b) Strand stays

Strand systems are offered by a variety of suppliers. The strand most commonly used is 15.2 mm or 15.7 mm diameter, and is usually galvanised. A stay may typically consist of up to 91 strands, giving an ultimate load of the order of 25 MN and a working load of approximately 11 MN. Then either the strand bundle is housed within a grouted duct, or each strand is individually protected in a grease-filled sheath. In the latter case, the bundle of sheathed strand is further encased in a plastic duct, which acts principally as a windshield. The strands are usually anchored by wedges.

When bare strand is used, the high density polyethylene (HDPE) duct is erected first using an external guide strand. The strands are then installed one by one or in pairs, with light equipment, each strand being stressed individually, Figure 18.17. It is necessary to adopt a programme of stressing that ensures that at the end of the process all strands have the same stress [6]. The whole stay may then be re-stressed by a single multi-strand jack. Once the stay is complete, the duct is grouted usually with a petroleum wax or other flexible material. Grouting such a long inclined stay is a difficult process, with considerable pressures being applied to the duct. Consideration must be given to the possibility of shrinkage of the grout, particularly if injected hot,

Figure 18.17 Pont sur l'Elorn: stay cable erection (Photo: Robert Benaim)

and of settlement of the grout down the inclined stay, both eventualities leaving voids and risking future corrosion.

Sheathed strands are generally installed as for bare strands. Clearly the sheaths must be removed adjacent to the anchorage where the strands are secured by wedges. This may be the weak point in the corrosion protection and must be carefully designed and built. The windshield/duct consists of half shells that have to be erected using a climbing rig or crane and is installed after the stay is complete.

There is absolutely no consensus on the best approach to the design of stays. This is underlined by two of the longest cable-stayed spans. The Normandie Bridge in France, with a main span of 856 m, uses individually sheathed galvanised 15.7 mm strand, assembled on site and then protected by half shell windshields. The Tatara Bridge in Japan with a main span of 890 m uses galvanised 7 mm parallel wires in an extruded sheath, with the stay prefabricated off site.

It is generally accepted that stays should work under full live loading at a direct tensile stress corresponding to 45 per cent of their breaking load. Although used as the basis for the design of most bridges, this does not appear to be a completely rational limit. One would expect the acceptable working stress to be dependent on a variety of factors including:

- the redundancy of the structural system;
- the fatigue performance of the stays and of their anchors;
- the spectrum of live loads.

The tension force in the stays under dead loads alone should not be too low; long lightly stressed cables will sag appreciably, and their effective Young's modulus will fall significantly below the theoretical value [7]. For instance, for a cable that is 200 m long in the horizontal projection, with a stress of 0.3 of its breaking load, the effective E will be about 97 per cent of the theoretical value, while for a 400 m long cable the figure would be about 88 per cent. At 0.2 of the breaking load, the two figures would become approximately 90 per cent and 70 per cent. This effect is most significant for long backstays, as their stress will be reduced below the dead load value by live loads applied to the side span, which pull the tower head backwards.

The protection of stays from corrosion is a continually evolving science. Not only are they very exposed to wind and rain, but the solar heating and subsequent cooling of the ducts causes a pumping action, tending to suck in damp air through any defects.

The angle at which the anchors are installed must be truly co-axial with the stay. This must take into account the sag of the stay under dead loads, as well as its precise inclination in both planes. The design of the anchorage housing must make it possible to achieve these very close tolerances, either by adjustment on site or by precise prefabrication.

The stay sag will vary as live loads are applied, and the stay will deflect under the effect of wind. To avoid these deflections applying bending moments to the stay anchors, which would compromise their fatigue resistance, the anchor is prolonged by a rigid tube providing a point of support to the stay about a metre from its forward end. Usually, the stay is separated from the tube by a neoprene ring that protects it from crippling, and also provides some damping, reducing the risk of harmonic vibrations. The tube end may also incorporate a deviator that allows the individual wires or strand to splay to their spacing at the anchor head. The length of this supporting tube may be calculated as a function of the bending stiffness of the component wires or strand [8]. In some systems, the stay anchorage is pre-grouted with resin, so that the full variation of stress in the stay is not transmitted to the mechanical anchorage.

The fatigue and ultimate load performance of stay anchors is very dependent on the fine detail of their design. On one major recent project, a proprietary anchor was tested by the contractor responsible for the detailed design and construction of the bridge and found to have an inadequate fatigue life. In a field where techniques are evolving rapidly, the engineer responsible for a major structure should not take for granted the claims of proprietary manufacturers, but should look at the testing evidence and exercise his own judgement, carrying out independent testing if necessary.

Stays may suffer from harmonic vibration, induced by wind or by vibrations transmitted from the deck. In exceptional circumstances, these vibrations may be so violent that adjacent stays have been known to clash with each other. It appears difficult to predict when a stay is likely to be vulnerable. The neoprene ring dampers incorporated in the anchorages may not be adequate on their own to control such vibrations. It is advisable for any cable-stayed bridge to make provision for retro-fitting further dampers if necessary. A variety of proprietary dampers is available. It is also possible to fit secondary cables that run approximately perpendicular to the stays, linking them together.

A particular form of instability that affects stay cables is so-called wind/rain vibration. Rain water running down the stay casing forms rivulets which disturb the air-flow. The Pont de Normandie has small spiral ribs attached to the stay windshields that break up these rivulets, while the Tatara Bridge has adopted a dimpled surface to the stay sheath that fulfils the same function. If stay cables are very close to each other in the horizontal plane, the wake of one may cause vibrations in the other. It is possible under these circumstances to link the pair of stays.

18.4.10 Progressive collapse

In the design of cable-stayed bridges it is essential to consider the risks of progressive collapse. The stays and their anchorages are inherently vulnerable to accidental damage or to hidden corrosion or fatigue failure. The failure of one stay could then put additional load on adjacent stays that causes them to fail in turn. The effect of the failure of a stay will depend on the stay and deck arrangement.

For instance, if a deck is supported by very few powerful stays either they must be designed, protected and inspected like suspension bridge cables, such that their unexpected failure by corrosion, fatigue or collision is inconceivable, or the structural system including the tower and the deck must be designed with adequate redundancy such that the failure of a stay does not result in the bridge becoming unserviceable.

Decks carried by multiple stays are inherently more able to survive the removal of a stay. Often, separate load cases are considered for accident and maintenance. For instance, the bridge may be designed such that it will survive, with a reduced factor of safety, the accidental removal of a stay when subjected to full live load. When a stay is removed for maintenance, the live load may be reduced, for instance by coning off critical highway lanes and the normal factors of safety maintained.

Whatever the design criteria, it is essential to have an effective regime of inspection of the stays and of their anchors to avoid the possibility that a series of stays may be weakened by undetected corrosion or fatigue damage.

Consideration of progressive collapse may also apply to other deck components, such as cross-beams or stringers. However, one cannot be specific about the design strategies to be taken to prevent disproportionate damage in the event of an accident; the risks need to be assessed case by case.

18.4.11 The Ah Kai Sha Bridge

The Ah Kai Sha Bridge constitutes a useful case history that illustrates many of the design features described above. Benaim were commissioned to design this bridge to carry the dual four lanes of the East-South-West Guangzhou Ring Road and the dual three lanes of a local road across the Pearl River in Guangdong Province in Southern China, Figure 18.18. The bridge was fully designed in detail and the construction of the main foundation piles for the towers had commenced before the project was abandoned as the highway project was modified. The design was carried out with the assistance of Leo Leung of the client organisation, Hopec Engineering Design Ltd, a subsidiary of Hopewell Holdings [9, 10].

The main span was 360 m and the side spans 170 m, and as the bridge provided 34 m navigation clearance over the river, it required approach viaducts. The towers rose 78 m above the deck, Figure 18.19. The bridge was situated in a zone subject to both seismic activity and typhoons.

The Ring Road was equipped with wide hard shoulders giving rise to 10 loaded lanes. Together with the local road there were thus 16 traffic lanes to carry. Various combinations of single and twin bridges with one or two decks were considered before it was concluded that the cheapest scheme would be to carry all 16 lanes on one

Figure 18.18 Project for Ah Kai Sha Bridge (Image: David Benaim)

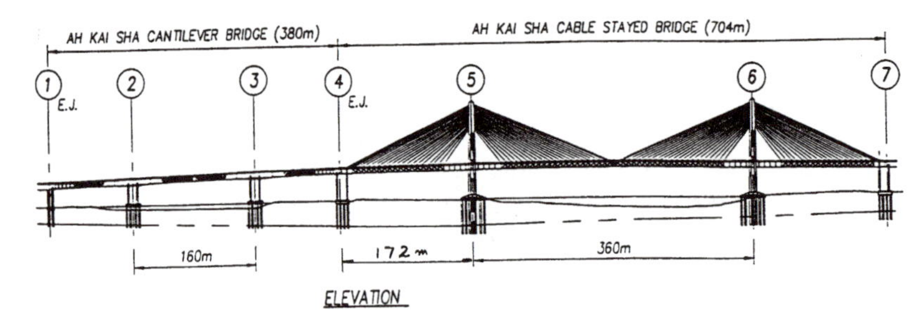

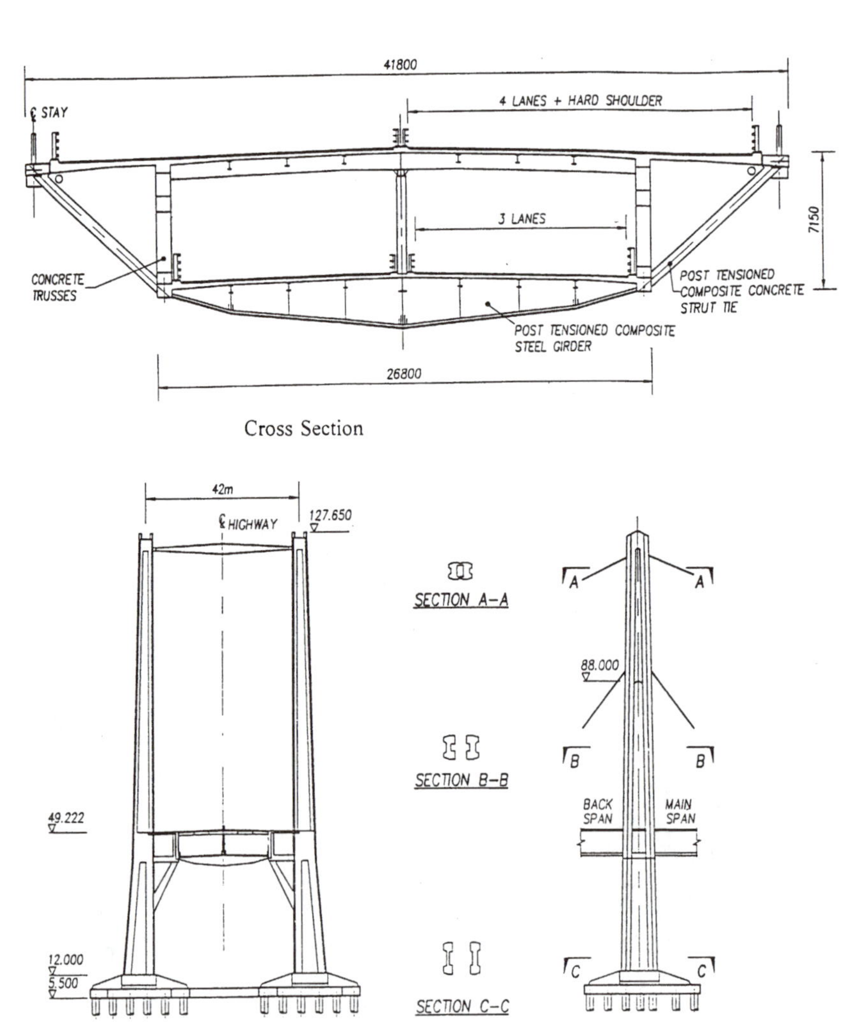

Cross Section

Elevations and Sections of Towers

Figure 18.19 Ah Kai Sha Bridge: general arrangement

double-deck cable-stayed bridge. Ten lanes were to be carried on the 42 m wide top flange and six lanes on a 26 m wide lower deck. The webs of the deck consist of prestressed concrete trusses. A truss arrangement was chosen both for economy and to provide natural ventilation to the lower traffic level. Clearance was left to install jet fans to ventilate the lower deck if this proved necessary. The total deck area of 48,000 m^2 carried by this bridge would have made it among the very largest cable-stayed structures in the world.

a) Articulation

The two 120 m high reinforced concrete towers are split longitudinally into two leaves. This arrangement gives the flexibility which allows the deck to be built into the towers, and provides a stable base to resist wind forces during the cantilever erection of the deck. The remote ends of the side spans are carried on flexible reinforced concrete piers built into the deck. These piers also carry the holding down tensions. Thus the bridge required no bearings either at the main towers or at the abutments.

b) Cable-stayed deck

The deck structure consists of a reinforced concrete box 7.15 m deep and 26.8 m wide. The box has 7.4 m wide propped side cantilevers extending to the cable planes. The top flange of the box is a 200 mm thick reinforced concrete slab thickening to 300 mm at the webs. Outboard of the webs the cantilevers remain at a thickness of 300 mm until they further thicken to 500 mm adjacent to the stay anchors, to control the shear stresses due to the transfer of the stay forces into the deck. The bottom slab of the box is also a 200 mm reinforced concrete slab thickening to 300 mm adjacent to the webs.

The top and bottom slabs are supported on steel stringer beams at 5 m centres that run longitudinally. As the slab spans transversally, it retains all its strength to resist the longitudinal compressions. The stringers are carried by transverse steel beams at 7 m centres. The upper transverse steel beam is 850 mm deep and is supported by a central steel stanchion, resting on the lower fish belly beam. This lower beam spans 25 m between the box webs and is 2.5 m deep at mid-span. Its end reaction, together with the end reaction of the upper steel beam transmitted down the webs, is picked up by inclined prestressed concrete ties which transfer the load to the cable stays. The prestress of the inclined ties is continued across the bridge, along the bottom flange of the lower transverse steel beam, effectively prestressing this beam.

The deck is prestressed longitudinally by internal grouted tendons housed in the haunches of the upper and lower slabs and anchored in blisters beneath the slabs.

The webs of the box consist of concrete 'N' trusses, Figure 18.20. The thickness of the truss members is 800 mm , which is adequate to resist accidental vehicle impact. The pitch of the truss is 7 m, while the height between the points of conjunction is 6.45 m. Both the diagonal and vertical members are 700 mm wide. The vertical tension members are prestressed, and the cables are given transverse eccentricities to carry the bending moments in the webs derived from the continuity with the upper and lower cross-beams. In areas near mid-span where the intensity of the 'reversed' shear increases, the diagonals are also prestressed. For the first 37.5 m of the deck

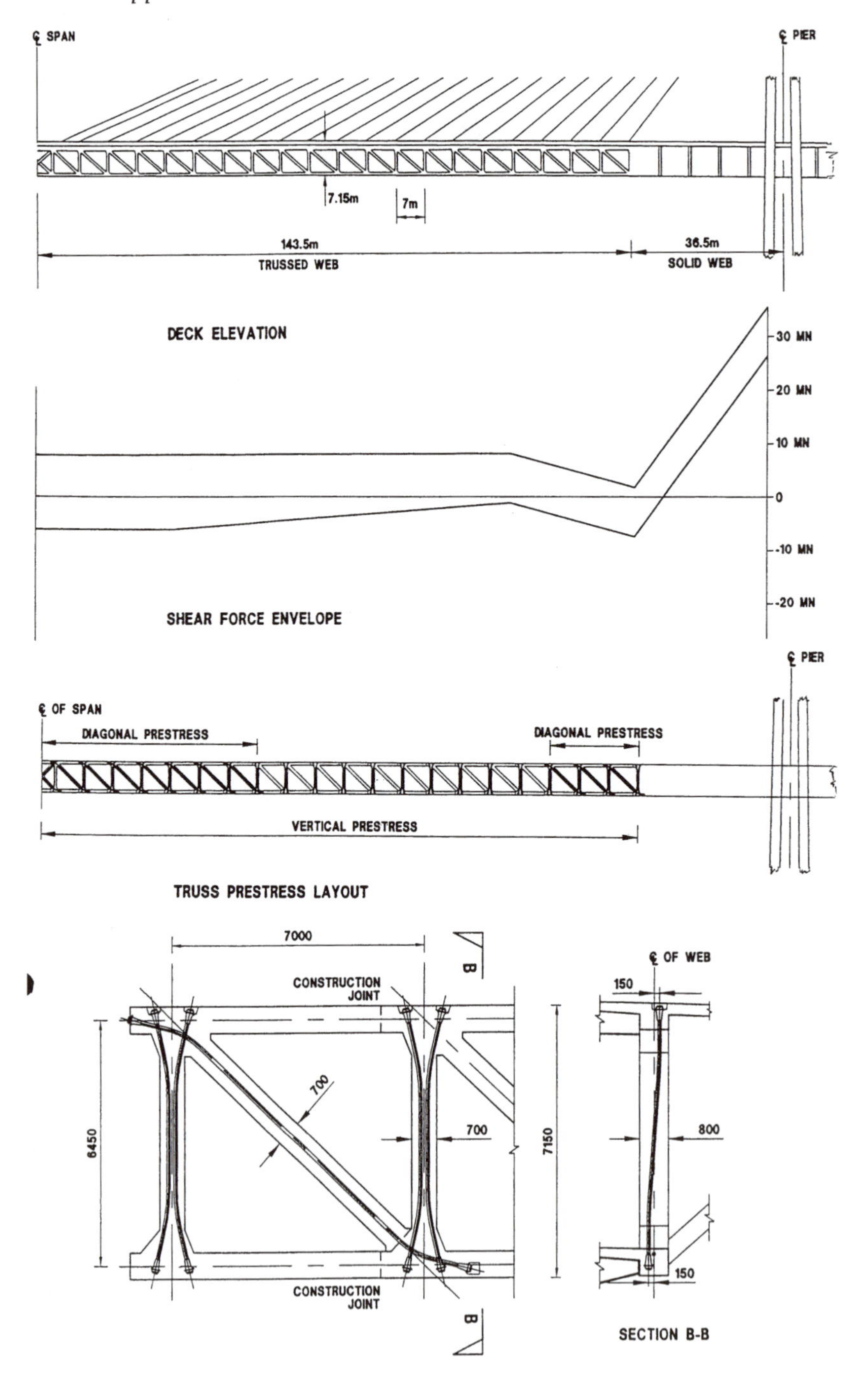

Figure 18.20 Ah Kai Sha Bridge: details of trussed webs

either side of the towers the 800 mm thick webs are solid. This zone of the deck is not stayed, and consequently is subjected to higher shear stresses. Also the connection of the deck to the towers is a highly stressed area and requires the solid sections. The deck was planned to be cast in-situ in 7 m long sections. The average deck thickness was 0.62 m; exceptionally thin for a twin deck structure.

It should be noted that an 'N' truss, when compressed as a column, deflects laterally as the loaded verticals shorten but not the unloaded diagonals. The compression caused by the stays would have caused a weightless deck to curl upwards. As the deck is in fact maintained straight by its weight and by the geometry of the stays, it results in a hogging moment in the deck that must be added to all the other effects.

c) Cable stays and transverse ties

There are 21 stays in each array, arranged in a semi-harp layout, attached to the deck at 7 m centres. Each stay consists of between 200 and 325 parallel 7 mm galvanised wires with button heads, and is designed to a working stress of 45 per cent of its breaking load. The stays are prefabricated under factory conditions and delivered complete to the site. This system was chosen as the durability of stays is critically dependent on the care with which the protective treatments are applied, and factory work was considered more reliable than site work in this instance. The disadvantage is that stays weighing up to 20 tons have to be erected on site.

The stays have dead anchors in the tower, with all the stressing anchors at deck level. This allows a compact array in the tower, with all the stays concentrated in a 27 m height, minimising bending moments when stays are removed for maintenance.

The tower anchor plates are welded in a workshop to a prefabricated steel frame. This is made in sections which are bolted together as the tower construction progresses, guaranteeing close tolerances for the orientation of the anchors. The base plates and guide tubes for the deck-level stay anchors are cast into precast anchor blocks that are then incorporated into the deck. The anchor block for each stay is different, and also incorporates the anchorages for the two tendons in the inclined prestressed concrete ties.

The horizontal component of the stay force is dispersed into the concrete deck slab, while the vertical component is taken down to the bottom of the web by a 700 mm wide by 500 mm thick prestressed concrete tie. This tie is prestressed by two internal strand tendons using double ducts and external prestress technology to allow replacement if necessary. It is sized so that it can resist the prestress force in the event of a stay cable being de-stressed. It is also reinforced so that one of its tendons may be removed while the bridge remains in service.

d) Main towers and foundations

Each tower consists of two twin-leaf columns in reinforced concrete. The columns are free-standing cantilevers, built into their pile caps. They are propped apart by the deck, and by a pin-ended concrete strut at their summit that carries the inwards component of the stay cable tensions. Each leaf has a dumb-bell cross section, 7 m wide at the base tapering to 4 m wide at the top. The leaves are 2.5 m thick, and their spacing reduces from 7.5 m at the base to 5 m at a point just below the stay-cable anchorage zone. The leaves are joined into a box section at the intersection with the deck and over the stay-cable anchorage zone. The load from the deck is transferred to the towers through inclined concrete struts. The thrust of the struts is carried in compression by the bottom slab of the deck and by transverse bending in the columns. Thus there are no portal beams linking the columns.

The twin-leaf layout allows the pylons to adapt to the deck shortening caused by concrete creep and shrinkage, and by temperature changes. The half decks were planned to be jacked apart before mid-span closure to preset the columns and reduce their bending moments. The foundations consist of twenty-six 2 m diameter hand-dug caissons beneath each column, capped by a 6.5 m deep, 32 m × 28 m cruciform reinforced concrete pile cap.

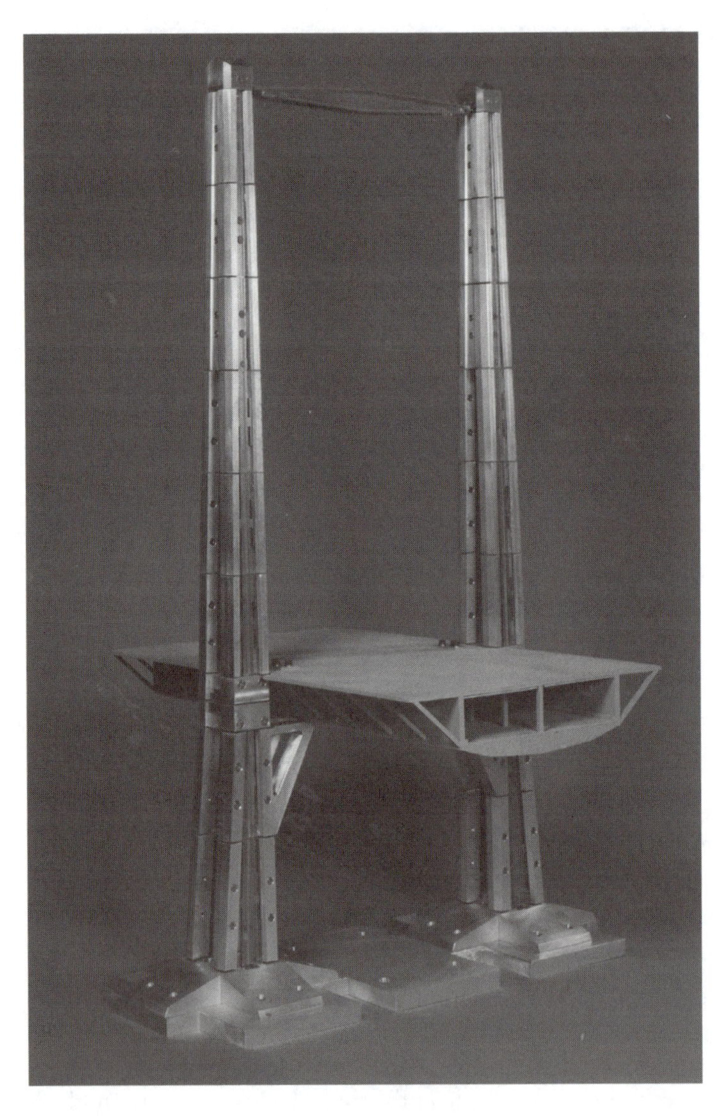

Figure 18.21 Ah Kai Sha Bridge: wind tunnel model of towers (Photo: BMT Fluid Mechanics Ltd)

e) Wind tunnel testing

The bridge is in a zone subject to typhoons. The deck and towers were tested by BMT Fluid Mechanics at their wind tunnel in Teddington, UK. The deck was tested both with and without traffic on both levels. It was shown to be stable in winds in excess of 120 m/sec, Figure 18.21. The towers were tested at all stages of construction, including the presence of tower cranes. They showed minor oscillations at some wind speeds during construction, but these were considered acceptable.

18.4.12 Hybrid structures

As cable-stayed decks surpass a span of 1,000 m, they occupy the same ground as suspension bridges. One of the major advantages of cable-stayed technology is that the stays are anchored on the deck itself; they do not need the massive and expensive cable anchorage structures of a suspension bridge. Cable-stayed decks also have a construction programme advantage, as once the towers have been completed, the deck erection in balanced cantilever may be started, while the deck erection of a suspension bridge must await the completion of the suspension cables. However, one of the principal disadvantages of cable-stayed bridges is the great height and hence cost of their towers; suspension bridge towers are typically half the height of cable-stayed towers for the same span.

Hybrid solutions, adopting a combination of the two technologies, appear to offer a promising way forwards for very long spans, Figure 18.22. The deck would be divided into cable-stayed zones, either side of the towers, and suspended zones at mid-span and at the end of the side spans. If the height of the towers is one-tenth of the main span, they are adequate to support approximately one-quarter of the span either side of each tower in cable-stayed technology. Suspension cables would then carry the central half of the main span and the end half of the side spans. The geometry of the stays may need to be arranged so that the deck has approximately the same vertical stiffness either side of the point of transition from stay to suspension.

As soon as the tower has been completed, erection of the cable-stayed deck may commence, as can the spinning of the suspension cables. These cables require a temporary external anchor, with sufficient strength to carry the tension forces caused by the self weight of the suspended half of the deck. This would require strengthening of the end piers or the abutments with temporary inclined ground anchors. However, this would be much less expensive than the permanent cable anchors required for a fully suspended solution.

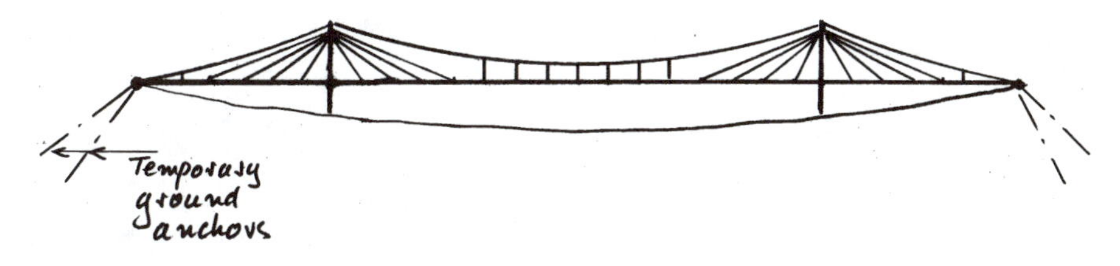

Figure 18.22 Self-anchored suspension/cable stayed hybrid

Once the suspension cables have been installed, the remaining sections of the deck may be erected. Once the deck is complete, the anchorage of the suspension cables is transferred to the deck, which must be designed with adequate buckling strength. In order to limit the force applied to the temporary anchors, the suspended portion of the deck could be built in two phases. For instance, a central core could be built first with adequate strength in compression to allow the cable anchors to be transferred to the deck. In the second phase side cantilevers would be added.

18.5 Stressed ribbon bridges

18.5.1 General

Stressed ribbon bridges consist of a thin flexible structural member that carries its own dead load and live loads principally in catenary action; that is in direct tension. The member may be made of steel or concrete. There are two basic types of concrete stressed ribbon bridges. For instance, the stressed ribbon of the Rio Colorado Bridge in Cost Rica, Figure 18.23, designed by T.Y. Lin has a deep sag, and the vehicle deck is carried by columns resting on the ribbon. The bridge is a mirror image of an arch. In such a bridge, the road alignment is completely divorced from the geometry of the ribbon.

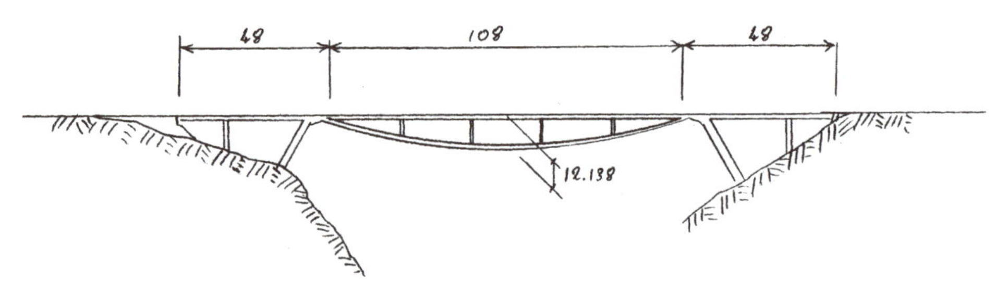

Figure 18.23 Rio Colorado Bridge, Costa Rica

Figure 18.24 Redding Bridge (Photo: Jiri Strasky)

On the other hand, the Redding Bridge, crossing the Sacramento River in California, designed by Jiri Strasky, Figure 18.24, [11, 12], has a much shallower sag and the stressed ribbon itself is used as the bridge deck. As the author only has experience of the latter type, the description will be limited to such bridges.

A correctly designed concrete stressed ribbon bridge has the same characteristics of durability as conventional reinforced or prestressed concrete decks. Although the torsional and bending strength of the slab give it good inherent stability, the aerodynamic behaviour and the risk of excitation due to pedestrian footfalls of such flexible structures should always be investigated.

18.5.2 Design of the deck

The deck consists of a thin concrete slab hanging in a catenary: a concrete cable. Loads are carried in direct tension, not in shear and bending as in a beam. As concrete cannot carry tension safely, the slab is prestressed with cables. The force in the cables must equal the tension in the catenary, plus an additional force which compresses the concrete. If more load is applied to the deck, the tension in the catenary increases. This will decrease the compression in the concrete; more of the force of the prestressing cables is used for resisting the tension, and less for compressing the concrete. If the load is further increased, the compression in the concrete will become zero, and the catenary tension will equal the force in the cables. At the ULS, when the loads are increased still further, the concrete will crack, and all the additional tension will be carried by the cables.

Theoretically, a stressed ribbon could be built in reinforced concrete. However, the concrete would be cracked through by the tension, exposing the reinforcement to corrosion and creating the risk of fatigue as every change in load or geometry would change the stress in the reinforcement. For a prestressed deck, variations in tension in the ribbon alter the compressive stress in the concrete, but do not change substantially the stress in the cables, and at working loads the concrete is not cracked through. Thus the prestressing cables are not exposed to the risk of fatigue or corrosion.

The tension in the catenary and its sag are related by the equation

$$T = pc^2/8s \qquad \text{(Equation 1)}$$

where:

T = tension in the band

p = the load per metre on the band

c = the chord length or span

s = the sag.

Variations in tension in the ribbon change its length and hence its sag. Temperature changes, and creep and shrinkage of the concrete also affect the length and the sag. For instance, if the temperature falls, the ribbon shortens and its sag s will decrease. If the applied load p remains constant, the tension T must increase.

From this brief description, it should be clear that the structure is completely non-linear; the principal of superposition does not apply as each new load or change of temperature changes the geometry of the structure. For instance, for the North Woolwich Road Footbridge described in *18.5.5*, the sag of the 27 m long main span varied from 542 mm to 766 mm from the cold, no-load to the hot, maximum load conditions.

Although the behaviour of the ribbon is non-linear, it is not difficult to carry out quite precise hand calculations for sizing the structure. These are based on the geometry of a parabola rather than a catenary, but at the small sags adopted for these bridges, the differences are insignificant. The length of the parabola L is given by:

$$L = c\left[1 + (8/3)(s/c)^2\right] \qquad \text{(Equation 2)}$$

This formula is approximate, but gives results that are within 0.1 per cent of the precise figure for sags of 1/10 and shallower.

Thus if the self weight and the sag of the ribbon are defined, the force in it may be calculated from Equation 1. If an additional distributed load is added, both the sag and the force will be increased, making it impossible to calculate them directly by hand. However, a new sag may be guessed allowing a new, increased force to be calculated. The increased force will decrease the compression stress in the concrete ribbon, causing it to extend. The change in stress in the ribbon $\Delta\sigma = \Delta F/A$ where ΔF is the change in force and A is the cross-section area of the ribbon. Then the extension of the ribbon δ, is given by:

$$\delta = \Delta\delta c / E \qquad \text{(Equation 3)}$$

where:

E is the appropriate Young's modulus for concrete

$\Delta\sigma$ is the change in stress in the ribbon

c is as before, the span of the ribbon, which is a close approximation to its length.

The new length of the ribbon is $L + \delta$.

By using the geometric Equation 2, a new length of the catenary corresponding to the guessed sag may be calculated and compared with that calculated from the force. If they differ, the guessed sag may be corrected, and the two calculations for length repeated until they converge on the correct value.

For instance, consider a prestressed concrete stressed ribbon deck 4 m wide and 0.3 m thick, with a chord of 100 m, a sag of 2 m and a self weight of 0.03 MN/m.

The length of the catenary from Equation 2 is 100.10667 m, and the tension force from Equation 1 is

$T = (0.03 \times 100^2)/8 \times 2 = 18.75$ MN.

If we add a live load of 0.012 MN/m, we have to find the new sag and tension. Guess a sag of 2.5 m. The new tension T will be

$T = (0.042 \times 100^2)/8 \times 2.5 = 21$ MN.

The change in stress in the deck is $(21 - 18.75)/(4\times0.3) = 1.875$ MPa.
From Equation 3, with $E = 30{,}000$ MPa, the extension of the catenary δ is

$\delta = (1.875 \times 100)/30{,}000 = 0.00625$ m.

The new length of the catenary calculated from the force is thus

$100.10667 + 0.00625 = 100.11292$ m.

However, using Equation 2, the length of a catenary with a sag of 2.5 m should be 100.16667 m. As this is higher than the figure calculated from the stress, the guessed sag is too great.

Try sag = 2.2 m.

$T = (0.042 \times 100^2)/8 \times 2.2 = 23.86$ MN,

the change in stress is $(23.86 - 18.75)/1.2 = 4.26$ MPa.

The extension of the catenary is 0.01420 m and its new length is thus

$100.10667 + 0.01420 = 100.12087$ m.

Calculating the length with a sag of 2.2 m gives 100.12907 m. The guessed sag is still slightly too great.

Try sag = 2.14 m.

$T = 24.53$ MN;

change in stress = $(24.53 - 18.75)/1.2 = 4.82$ MPa; extension = 0.01605 m.

New length of catenary calculated from force is

$100.10667 + 0.01605 = 100.12272$ m.

The length calculated from geometry is 100.12212 m. The guessed sag is now slightly too small, but the results for the sag and the force in the ribbon are sufficiently accurate.

The effect of creep, shrinkage and temperature change can be found by trial and error in the same way.

It is important that the sag of the catenary at the end of construction is not too small; a value of span/40 to span/45 is typical. As the deck shortens due to temperature drop, shrinkage and creep, the shortening is taken up by a reduction in the sag, by the extension of the deck under the increased tension, and in some cases by an elastic response of the abutments to the changes in tension. If a smaller initial sag is chosen, a greater proportion of the shortening must translate into an increased tension force. At the limit, for a straight slab with no sag and rigid abutments, all the shortening would translate into tension following the expression:

strain = stress/E.

The difference between the length of the ribbon and the chord length is the main driver of the behaviour of these bridges. The sensitivity of a concrete stressed ribbon to sag and to changes in length may be illustrated by the following figures. A 100 m long chord with a sag of 2 m (1/50) has a ribbon length of 100.10667 m, so the ribbon is 107 mm longer than the chord, while for a sag of 1.43 m (1/70) the length becomes 100.05453 m; the ribbon length would now be only 55 mm longer than the chord. The shortening of the ribbon due to temperature drop, shrinkage and creep would typically be of the order of 50 mm for a 100 m length, using mainly precast concrete elements for the deck, which minimises shrinkage and creep. It should be intuitively clear that this shortening will have a much greater effect on the flatter catenary. Thus the tension in a catenary that is too flat would rise uncontrollably when it shortens, leading to cracking of the concrete or the opening of construction joints.

The slope at the ends of a parabola may be expressed as 4 s/c. For a sag of 1/45, this gives an end slope of 1/11.25 at mean temperature and under dead loads. This gradient may conflict with regulations for use of the bridge by people in wheelchairs, which may be defined as a gradient not exceeding 1/20. At North Woolwich (*18.5.5*) this was the specified limiting gradient, and it was achieved by locally surcharging the deck. However, this solution is only applicable to short spans, where the thickness of surcharge remains only a few centimetres. A parabola with an end slope of 1/20 would have a sag of only 1/80, which is not compatible with a concrete stressed ribbon that would not be substantially cracked in cold weather.

Unlike a steel cable, the concrete catenary has a significant bending stiffness. At abutments and over piers, where the variations in sag cause changes to the curvature of the ribbon, local bending moments are created that are too large to be carried by the thin slab, making it necessary to thicken the deck locally with haunches. However, the haunches increase the local stiffness of the slab, increasing the bending moments. Consequently, they must be as shallow and as short as possible, and should have a thickness that varies parabolically rather than linearly, to match closely the shape of the bending moment diagram and to minimise their stiffening effect, Figure 18.27. It is not desirable to design these haunches to the normal rules of fully prestressed concrete, as this leads to excessive compressions in the deck, increasing the creep shortening. A partially prestressed solution should be adopted. Under normal service loads and at normal temperatures the deck should be uncracked, while under the effect of extreme temperatures the deck should not be decompressed by the direct tension, while the bending moments are carried in reinforced concrete action.

Concentrated live loads on the deck also create local curvature as it tries to behave like a cable, adjusting its gradient either side of the load. Consequently local sagging bending moments are generated beneath the load. If the stressed ribbon is only carrying pedestrians, moments will be negligibly small. However if it carries vehicles, such local moments may become significant.

As all the bending moments on the deck are caused by imposed changes in curvature, it is very important to use in the calculations an elastic modulus for the concrete that is relevant to the duration of the load; dead loads, shrinkage and creep call for the lowest modulus, seasonal temperature changes a higher modulus, daily temperature changes higher yet and live loads the normal short-term value. The cracked inertia should be used in the calculations in areas of high bending moment. Lightweight concrete that has a lower modulus than dense concrete should have a role in stressed ribbon bridges.

18.5.3 Intermediate pier crossheads

The design of intermediate pier crossheads further illuminates the particular characteristics of concrete stressed ribbon bridges. Consider the crosshead shown in Figure 18.25 (a), where the deck is carried by a pier consisting of a single central column. The deck slab cantilevers either side of the column. If the deck had consisted

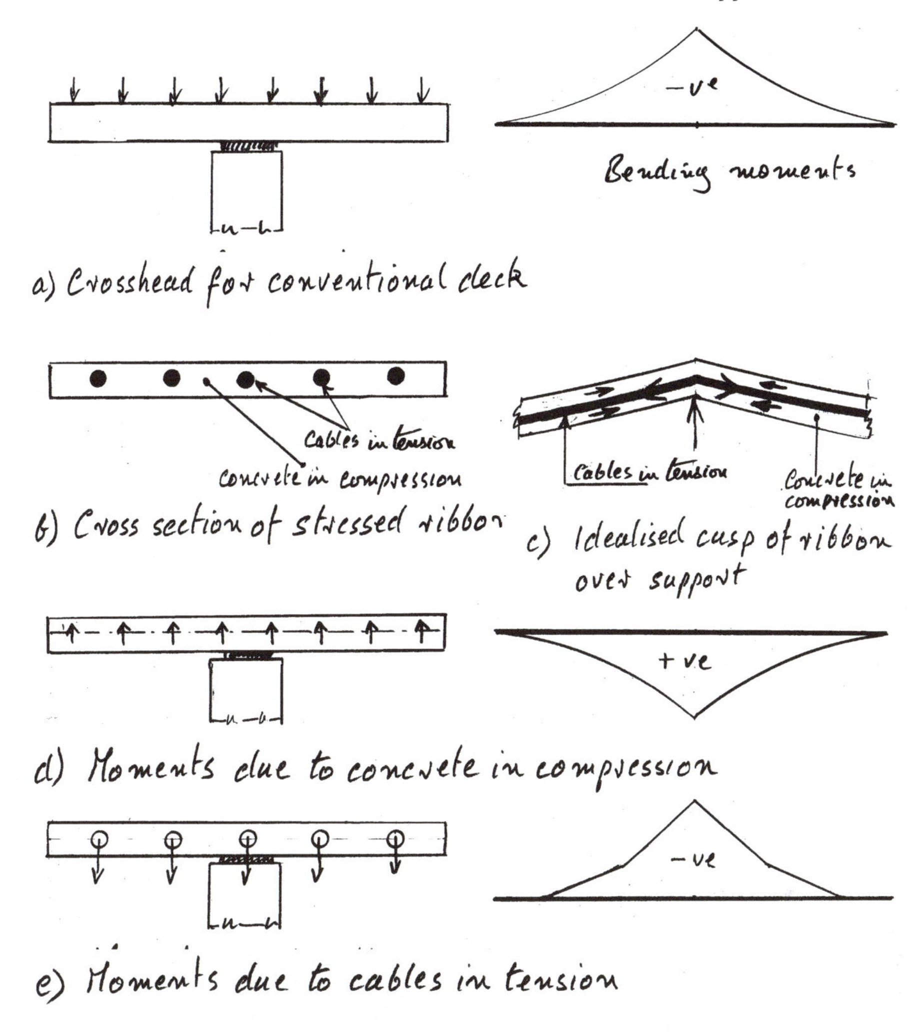

Figure 18.25 Forces on crosshead over pier

of a conventional slab bridge, carrying its loads in shear and bending, the crosshead would have hogging transverse bending moments.

However, the stressed ribbon carries loads in direct tension. The net tension of the ribbon is equal to the tension of the cables less the compression in the slab. The centre line of the slab of the two adjacent spans meets in a cusp over the pier. As the concrete of the slab is in compression, there is an upward resultant force across the crosshead, giving rise to transverse sagging bending moments, Figure 18.25 (d). The cables also meet in a cusp over the pier, and as they are in tension, they create a downward resultant force, giving rise to transverse hogging moments, Figure 18.25 (e). These two effects are in opposition, leaving a much reduced transverse moment in the crosshead. The spacing of the cables over the pier may be chosen by the designer, and need not be spread evenly across the deck at this location so that the net bending moment in the crosshead is optimised for the various load cases.

18.5.4 Abutments

The principal disadvantage of stressed ribbon bridges is the need to anchor the very considerable tension forces. In soft ground, this requires substantial arrays of inclined tension and compression piles, or the use of ground anchors, or bored piles in bending.

18.5.5 Case history: North Woolwich Road Footbridge

The North Woolwich Road Footbridge was designed to the stage of construction drawings, initially for the London Docklands Development Corporation, whose responsibilities in this respect were later transferred to English Partnerships. However the bridge was not constructed. It had a total length of 61.8 m, and carried a footpath across the North Woolwich Road to the Thames Barrier Park. After studying many options, Benaim, with the architectural assistance of Eva Jiricna, chose a prestressed concrete stressed ribbon design with a main span of 26.9 m and side spans of 17.45 m. The intermediate columns consisted of 700 mm diameter columns built into the deck, and the deck was also built into the two abutments, Figure 18.26.

The deck consisted of a 200 mm thick prestressed concrete slab 4.1 m wide, providing a 3.5 m wide walkway. The slab was thickened to 350 mm over a 2 m long haunch with a parabolic profile either side of the intermediate piers, Figure 18.27, and at the abutments. This constituted the very simplest possible bridge consisting of a solid concrete slab resting on cylindrical columns.

The deck was prestressed with six tendons of 7 No 15.7 mm strands. The main span had a sag of 625 mm under permanent loads, reducing to 542 mm under minimum temperature with live loads on the side spans, and increasing to 766 mm under maximum temperature and live loading on the main span. The deck was fully prestressed under normal working loads and at normal temperatures. At extreme

Figure 18.26 Project for North Woolwich Road Footbridge (Image: David Benaim)

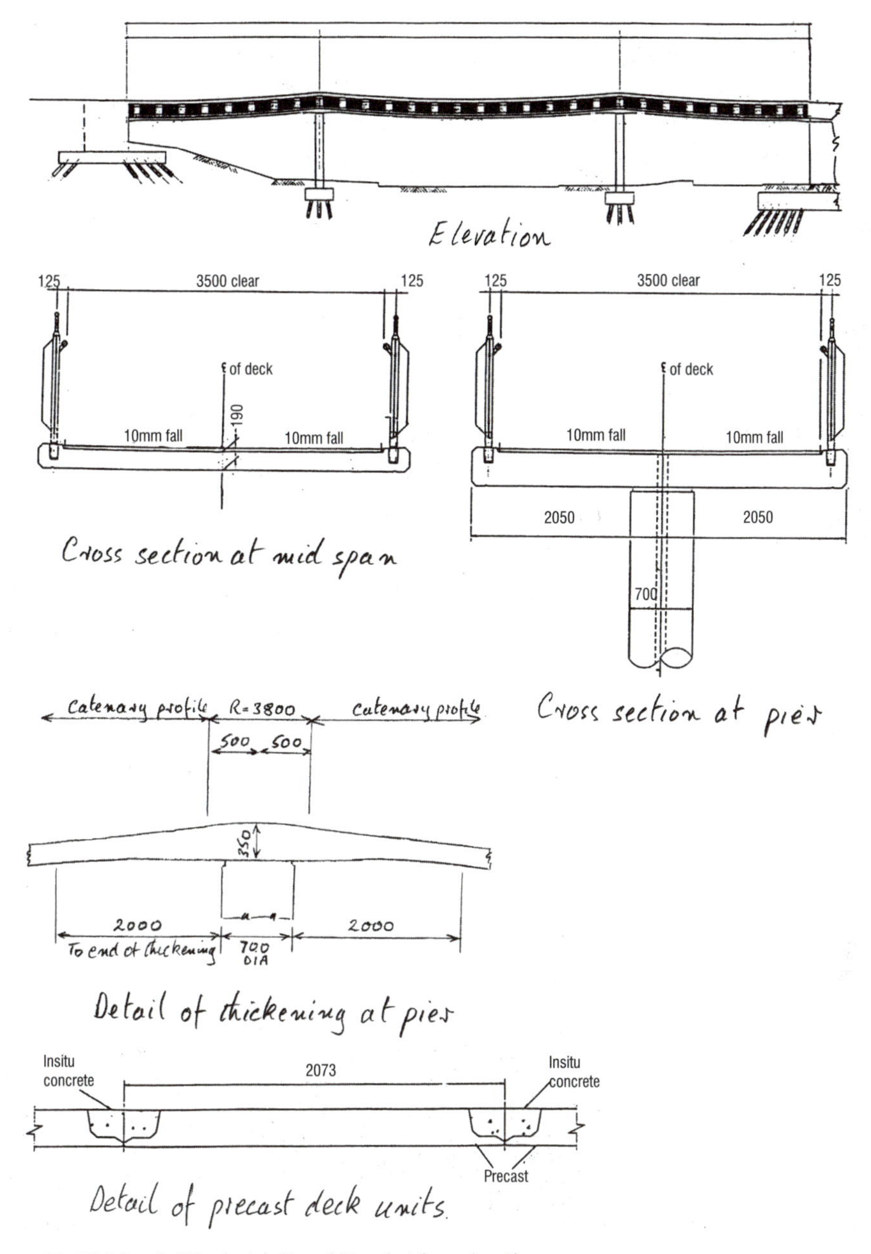

Figure 18.27 North Woolwich Road Footbridge: details

temperatures, the areas of the deck close to the piers and abutments were designed as partially prestressed. The deck was equipped with a very beautiful handrail that incorporated street lighting.

The reinforced concrete abutments were founded on an array of inclined micropiles resisting the tension in the catenary of approximately 6500 kN.

Construction consisted in erecting firstly 8,180 mm long precast sections over the two piers. The connection with the pier was then cast in-situ through a circular hole.

Two temporary erection cables, each consisting of 12 No 15.7 mm strands and located 150 mm above the extrados of the deck were draped in place, over saddles on the piers and attached to temporary anchors on the abutments. 2,073 mm long precast deck elements were then suspended from these erection cables, working from the piers outwards. When all the precast elements had been placed, the erection cables were re-stressed to correct the deck profile. The permanent prestress was then threaded into place and ducts were jointed in the 400 mm wide joint troughs at either end of the precast units, which were then filled with cast-in-situ concrete. After 24 hours, two tendons were stressed to avoid the risk of cracking of the jointing concrete. When this concrete had attained a strength of 35 MPa, the remainder of the permanent cables were stressed and the temporary strands removed.

18.6 Steel cable catenary bridges

It is possible to design a footbridge where the essential structure of the deck consists solely of highly tensioned steel cables anchored on abutments. The cables may be clad, for instance in precast concrete sections, which provide the walking surface, and anchorage for the handrail. Although superficially similar to stressed ribbons, as the concrete structure is not longitudinally compressed these bridges have different characteristics.

When additional load is applied to the catenary, the cables will extend under the greater tension and their sag will increase. The changes in sag under live loading will be greater than for a concrete stressed ribbon due to the higher stresses in the steel. When the temperature rises, the sag will increase and the tension in the cables will drop; when the temperature falls, the sag will decrease and the tension in the cables will rise. Thus the cables are subjected to continual variations in stress, and it is necessary to check their fatigue performance. The changes in sag will also cause angular changes where the cables pass over piers and at the face of abutments. If the cables are not protected from these angular changes, by for instance providing a stiffer duct locally that spreads the angular change over a length of the cable, their fatigue life is likely to be further reduced.

Such decks have no torsional or bending strength, other than that provided by the stiffness of the taut cables. Consequently, they are likely to be more susceptible to vibrations due to pedestrian excitation or to wind. Any such vibrations could be prejudicial to the fatigue life of the cables.

The transverse joints between precast sections are normally filled with a sealant, which must be sufficiently flexible to accept the changes in length and geometry of the cables. However, this sealant should not be relied upon to protect the cables from water seeping from the deck. The cables must be provided with corrosion protection similar to that of cable stays, and it must be possible to inspect and change cables if necessary.

This structural system is subject to the risks of progressive collapse; if one cable is out of service, its load will be shed onto the remaining cables. Consequently, there must be adequate reserves of strength.

In view of all these factors, it would be prudent to limit the maximum tension in the cables to 45 per cent of their breaking load, as for cable stays, and to make provision for cables to be inspected and replaced. However, the stress in the cables is likely to be still lower in order to provide sufficient steel area to control deflection.

The principal advantage of such decks is that as the cables are much more flexible in direct tension than the concrete band, they can be designed to a much flatter catenary and can respect a gradient of 1/20. Of course, the flatter catenary will increase the force on the abutments, with economic consequences.

The author has not designed such a bridge in detail, and consequently does not have the personal experience of overcoming the problems of durability and stability.

18.7 Flat suspension bridges

18.7.1 General arrangement

In a development of the steel cable catenary bridge described above, the alignment of the cables may be completely separated from that of the deck. Thus the cables may be given a greater sag, reducing their tension force and consequently the cost of the cables and abutments. Such bridges are in fact related to traditional suspension bridges, with very flat cables. As for the steel cable catenary, every variation of load or temperature causes changes to the stress and to the sag of the cables. As the sag varies or the cable vibrates, the cables will flex about their points of support, creating bending stresses which, as in a cable stay, may cause fatigue damage. Cable supports at intermediate piers and abutments should be designed to limit the bending moments applied to the cables.

The suspended deck may be given both torsional and bending strength if necessary to assist in controlling the tendency of the cables to oscillate under pedestrian or wind excitation and to carry lateral wind forces.

18.7.2 Suspension cables curved in the horizontal plane

The early version of the North Woolwich Road Footbridge shown in Figure 18.28 consisted of a shallow suspension bridge with cables that curved in both the vertical and the horizontal planes. This arrangement allows the cables to carry the horizontal

Figure 18.28 North Woolwich Road Footbridge: early version (Image: Benaim)

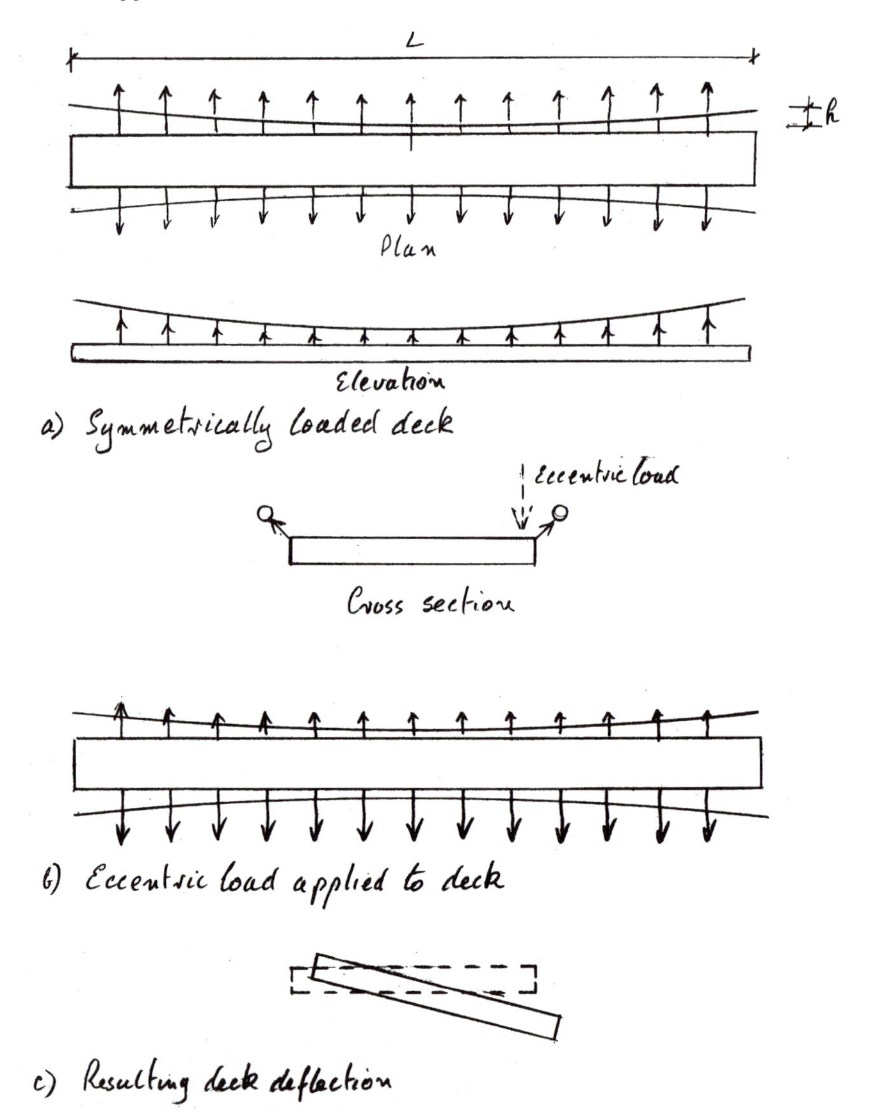

Figure 18.29 Effect of unsymmetrical loading on suspension bridge with cables curved in plan

wind loads. However it is still advisable for the deck to retain bending and torsional strength and stiffness to control the dynamic behaviour of the cables.

As the cables have a horizontal parabolic curvature, they also apply a distributed horizontal force to the deck equal to $8Fh/L^2$ units/m, where F is the force in the cables, h is the sag of the cables in the horizontal plane, and L is the span, Figure 18.29 (a).

When all vertical loads are equally shared between the cables, these horizontal forces are balanced. However, any eccentric vertical load on the deck will increase the tension in one set of cables with respect to the other, increasing the vertical sag of the more heavily loaded cable. As a result of the different tensions in each set of cables, the horizontal forces applied to the deck will be unequal, giving a net lateral force,

Figure 18.29 (b). If the deck is stiff it will be subjected to lateral bending moments and shearing forces. If the deck is infinitely flexible, it will deflect sideways until the horizontal sag in the cables has been adjusted so that once again the horizontal forces applied to the deck are in equilibrium. As this will increase the force in the less-loaded cable, it must reduce its vertical sag to maintain equilibrium, Figure 18.29 (c). The resulting overall movement of the deck is a combination of a vertical deflection, a lateral deflection and a rotation about its longitudinal axis. Although this behaviour is non-linear, the horizontal and vertical deflections of the cables may easily be calculated by hand using the techniques described in *18.5.2* above.

18.7.3 Dynamic performance

It should be noted that very shallow suspension bridges are extremely flexible, with deflections under full pedestrian loading that may attain span/150. It is likely that the sizing of the cables will be governed more by the need to control this deflection than to limit their tensile stresses. The dynamic behaviour of such flexible decks must clearly be investigated. It is well known that random pedestrian footfalls can cause vertical oscillations in a flexible deck. It follows that in bridges where eccentric vertical loads cause lateral deflections, eccentric vertical pedestrian footfalls are likely also to cause lateral oscillations. Examples of such decks are those carried by tall, flexible central columns or decks carried by a single suspension cable, Figure 18.30, as well as decks carried by cables curved in the horizontal plane as described above. Some studies have also suggested that the small horizontal component of footfalls may cause lateral oscillations in particularly susceptible bridges even when they are not predisposed to such movements by their geometry.

The bridge may also suffer from wind-induced oscillations. In a shallow suspension bridge where the stiffness of the cables may be large compared with that of the deck, or where the deck has no stiffness, it is unlikely that the classical theory of suspension bridge stability remains applicable. The wind tunnel testing of these bridges is essential.

In order to control oscillations due to pedestrian or wind excitation, it may be necessary to equip the bridge with dampers *ab initio*, or to make provision for retrofitting dampers in marginal cases.

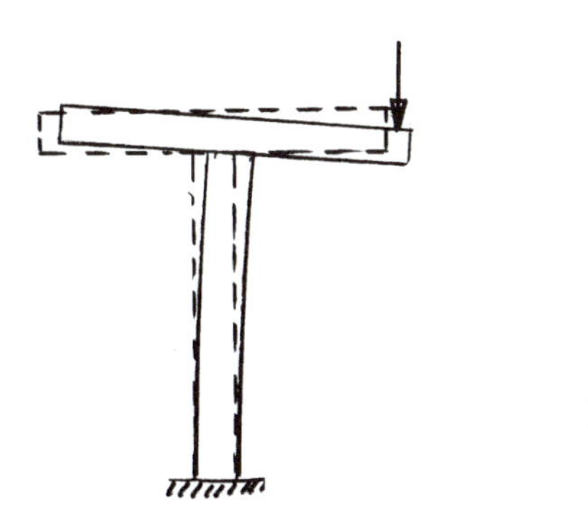

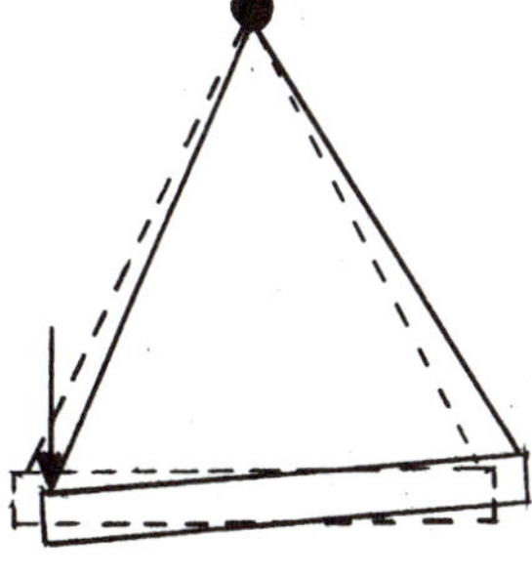

Figure 18.30 Examples of structures which deflect sideways under eccentric loads

Appendix

When carrying out preliminary designs it is frequently useful to be able to calculate the section properties of a bridge deck section quickly, by hand. The following simple tabular method is based on the *Formulaire du Beton Armé*, Volume 1, by R. Chambaud and P. Lebelle, published by La Société de Diffusion des Techniques du Batiment et des Travaux Publics in 1953. The method is exact if the section is described in sufficient detail, but is most useful as a first approximation, with the particular benefit that it is very quick to make small adjustments to a cross section in order to achieve desired section properties.

Figure A.1 (a) shows a typical bridge deck, which is the same as that used for the mid-span section of the sample deck in Chapter 6, Figure 6.7 and *6.10.1*, while Figure A.1 (b) shows the section reduced for calculation, which here has been drawn to scale for clarity, although this is not necessary in practice.

The calculation proceeds as shown in Table A.1.

Table A.1

1	2	3	4	5	6	7
Element no.	*Dimensions*	*Area*	l_a	*Product 3 × 4*	*k*	*Product 5 × 6*
1	2.20 × 0.70	1.540	1.1000	1.6940	1.4667	2.4846
2	11.30 × 0.20	2.260	0.1000	0.2260	0.1333	0.0301
3	5.30 × 0.20	1.060	2.1000	2.2260	2.1016	4.6782
4	3.60 × 0.20/2	0.360	0.2667	0.0960	0.2750	0.0264
5	2.0 × 0.20/2	0.200	0.2667	0.0533	0.2750	0.0147
6	1.6 × 0.30/2	0.240	1.9000	0.4560	1.9026	0.8783
Σ		5.660		4.7513		8.1123

Column 3 is the area of each element. Column 4 is the perpendicular distance of the centroid of each element from some convenient reference plane; in this case the top surface of the section. The reference plane should not cut the section. Column 5 is the first moment of area, and column 7 the second moment of area, of each element, both about the reference plane.

$$k = l_a + h^2/ql_a,$$

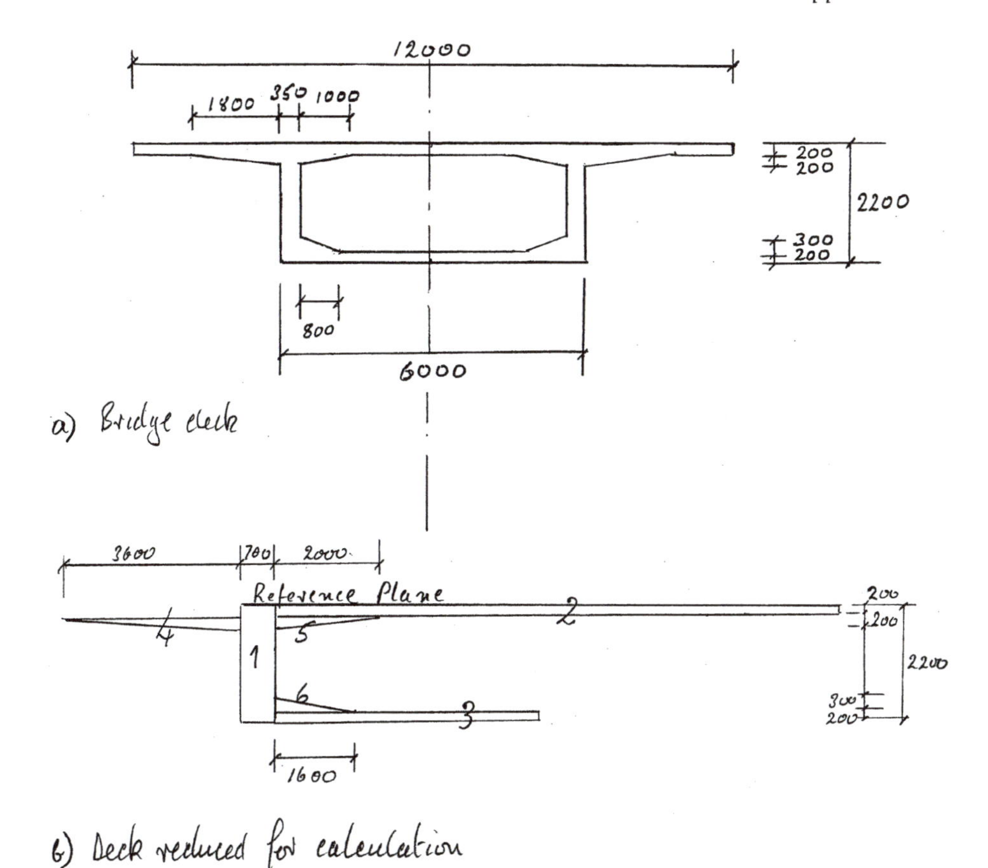

Figure A.1 Calculation of section properties

where q is a function of the shape of the element and h is the height of the element normal to the reference plane. The value of q for some frequently used element shapes is shown in Figure A.2. The value of q for a much greater selection of shapes may be found in the reference given above, or may be calculated from first principles. For rectangles which are based on the reference plane, such as Elements 1 and 2 in this example, $k = h/1.5$.

The calculation of the section properties proceeds as follows:

$A = \Sigma 3 \qquad = 5.660\ \text{m}^2$

$y_t = \Sigma 5/A \qquad = 0.8395\ \text{m}$

$y_b = 2.2 - y_t \qquad = 1.3605\ \text{m}$

$I = \Sigma 7 - \Sigma 5 \times y_t \qquad = 4.1236\ \text{m}^4$

Then:

$z_t = I/y_t \qquad = 4.91\ \text{m}^3$

$z_b = I/y_b \qquad = 3.03\ \text{m}^3$

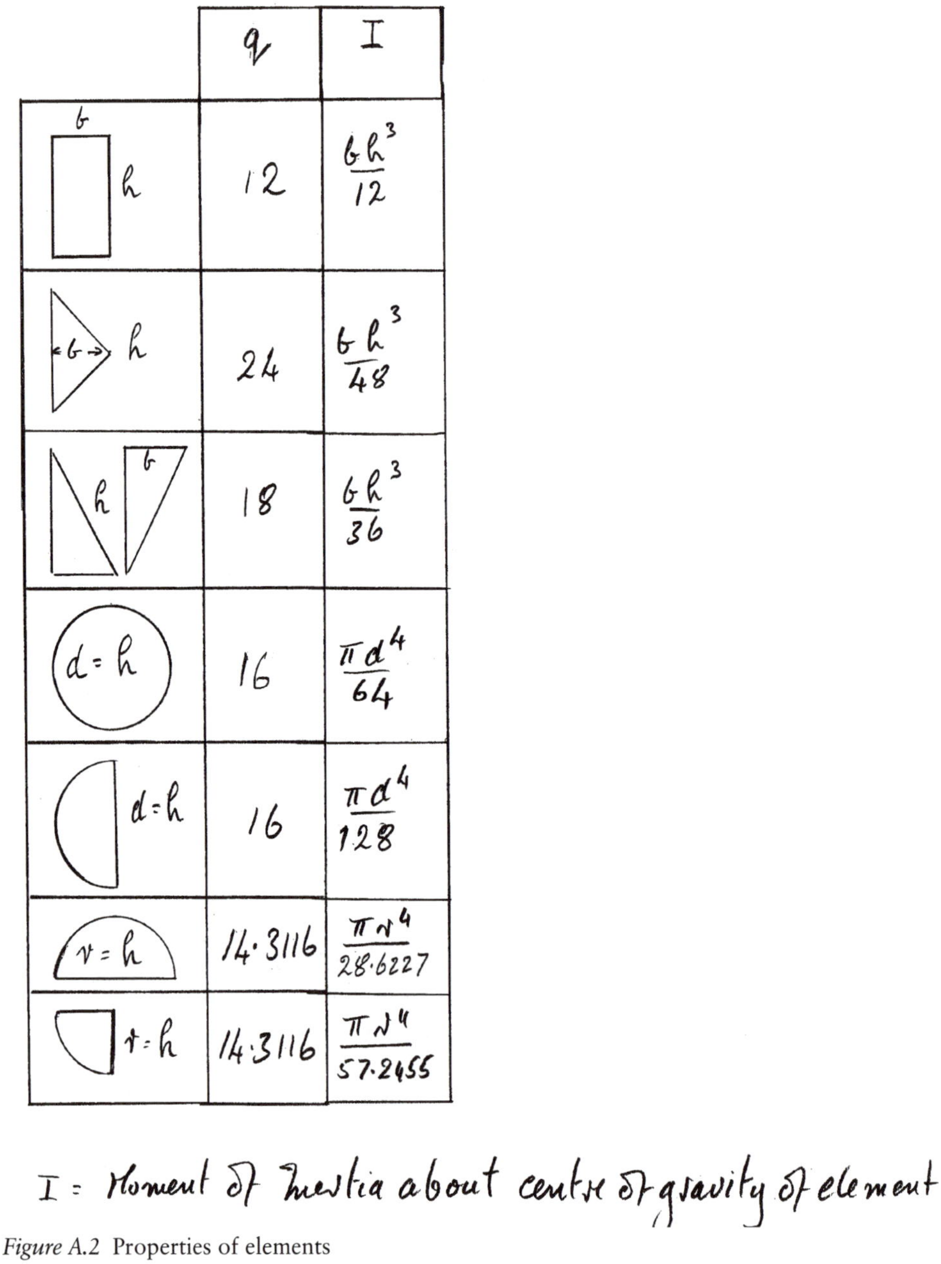

	q	I
Rectangle, b × h	12	$\frac{bh^3}{12}$
Triangle, base b, height h	24	$\frac{bh^3}{48}$
Right triangles, b, h	18	$\frac{bh^3}{36}$
Circle, $d = h$	16	$\frac{\pi d^4}{64}$
Semicircle, $d = h$	16	$\frac{\pi d^4}{128}$
Semicircle, $r = h$	14·3116	$\frac{\pi r^4}{28{\cdot}6227}$
Quarter circle, $r = h$	14·3116	$\frac{\pi r^4}{57{\cdot}2455}$

I = Moment of Inertia about centre of gravity of element

Figure A.2 Properties of elements

The efficiency of the section η should be calculated in order to provide a qualitative check of the result.

$\eta = I/Ay_t y_b = 0.638$, which is a credible result for a box section.

If it had been necessary to increase the width of the box to obtain a greater bottom modulus, it would only be necessary to change the calculation for Element 3, while changing the thickness of the bottom flange would entail recalculating only Elements 3 and 6.

The calculation is sensitive to rounding, and columns 4, 5, 6 and 7 should be calculated to four decimal places. In *6.10.1* the calculation of the deck inertia was carried out to three decimal places, while here four decimal places are used and it may be seen that the results differ by about 0.2 per cent.

Negative elements may be used, for instance in a deck with voids, or to facilitate the calculation for a box with sloping webs. Each of columns 3, 5 and 7 will then have negative components.

References

1 The nature of design

1 Cracknell, D.W., 'The Runnymede Bridge', *Proceedings of the Institution of Civil Engineers*, Vol. 25, July 1963, pp. 325–44.
2 Billington, D.P., *The Tower and the Bridge: The New Art of Structural Engineering*, Princeton University Press, Princeton, NJ, 1983.

2 Basic concepts

1 Hambly, E.C., *Bridge Deck Behaviour*, E&FN Spon, 2nd edition, 1991.

3 Reinforced concrete

1 Concrete through the ages, British Cement Association, 1999.
2 Neville, A.M., *Properties of Concrete*, 2nd edition, Pitman Publishing, 1978, pp. 475–6.
3 Neville, A.M., *Properties of Concrete*, 2nd edition, Pitman Publishing, 1978, pp. 483–485.
4 Neville, A.M., *Properties of Concrete*, 2nd edition, Pitman Publishing, 1978, pp. 309 *et seq*.
5 Neville, A.M., *Properties of Concrete*, 2nd edition, Pitman Publishing, 1978, pp. 318, 509.
6 Neville, A.M., *Properties of Concrete*, 2nd edition, Pitman Publishing, 1978, pp. 248, 471.
7 Neville, A.M., *Properties of Concrete*, 2nd edition, Pitman Publishing, 1978, p. 336.
8 Leonhardt, Dr.-Ing. Fritz, *Prestressed Concrete Design and Construction*, 2nd edition, Wilhelm Ernst & Sohn, 1964, pp. 439–42.
9 Menn, C., *Prestressed Concrete Bridges*, Birkhäuser Verlag, 1986.
10 Schlaich, J. and Schäfer, 'Design and detailing of structural concrete using strut-and-tie models', *The Structural Engineer*, Vol. 69, No. 6, 19 March 1991.

5 Prestressing for statically determinate beams

1 Detailing for post-tensioning, VSL International Ltd, Bern, Switzerland, 1991.
2 Leonhardt, Dr.-Ing. Fritz, *Prestressed Concrete Design and Construction*, 2nd edition, Wilhelm Ernst & Sohn, 1964, pp. 269 *et seq*.
3 *A Guide to the Design of Anchor Blocks for Post-tensioned Prestressed Concrete Members*, CIRIA Guide, 1 June 1976.
4 Guyon, Y., *Béton Précontraint: Etude théorique et expérimentale*, Editions Eyrolles, 1951, pp. 169 *et seq*.
5 Lin, T.Y., *Design of Prestressed Concrete Structures*, John Wiley & Sons; New York, Chapman & Hall, London, 1955.
6 Menn, C., *Prestressed Concrete Bridges*, Birkhäuser Verlag, 1986.

6 Prestressing for continuous beams

1 Neville, A.M., *Properties of Concrete,* 2nd edition, Pitman Publishing, 1978, p. 432.
2 Department of Transport, Highways and Traffic, Departmental Standard BD 37/88, Loads for Highway Bridges.
3 Emerson, M. *Temperature Difference in Bridges: Basis of Design Requirements*, TRRL Laboratory Report 765, Transport and Road Research Laboratory, Crowthorne, 1977.
4 Hambly, E.C., *Bridge Deck Behaviour*, 2nd edition, E&FN Spon, 1991, pp. 222–43.
5 Neville, A.M., *Properties of Concrete*, 2nd edition, Pitman Publishing, 1978, pp. 247, 471.
6 Maisel, B.I. and Roll, F., *Methods of Analysis and Design of concrete Box Beams with Side Cantilevers*, Cement and Concrete Association, Wexham Springs, 1974, Pub. No 42.494.
7 Hambly, E.C., *Bridge Deck Behaviour*, 2nd edition, E&FN Spon, 1991, p. 135 *et seq.*
8 Kermani, B. and Waldron, P., 'Behaviour of concrete box girder bridges of deformable cross section', *Proceedings of the Institution of Civil Engineers, Structures and Buildings*, 1993, **99**, May, pp. 109–22.
9 Guyon, Y., *Limit State Design of Prestressed Concrete*, Vol. 2, Applied Science Publishers, 1974, p. 36.
10 Kretsis, K., *Stress Distribution in Continuous Beams in the Neighbourhood of Internal Supports* MSc Thesis, University of London, 1961.
11 Low, A.McC., 'The preliminary design of prestressed concrete viaducts', *Proceedings of the Institution of Civil Engineers*, Part 2, 1982, **73**, June, pp. 351–64.
12 Burgoyne, C.J., 'Cable design for continuous prestressed concrete bridges', *Proceedings of the Institution of Civil Engineers*, Part 2, 1988, **85**, March, pp. 161–84.
13 Detailing for post-tensioning, VSL International Ltd, Bern, Switzerland, 1991.

7 Articulation of bridges and the design of substructures

1 Neville, A.M., *Properties of Concrete,* 2nd edition, Pitman Publishing, 1978.
2 Kauschke, W. and Baigent, M., *Improvements in the Long-term Durability of Bearings in Bridges*, 2nd World Congress of Joint Sealing and Bearing Systems in Concrete Structures, San Antonio, TX, 30 September 1986.
3 Fascicule Special No 81-26 bis, *Règles techniques de conception et de calcul des ouvrages et constructions en béton armé (CCBA 68); Conception, calcul et epreuves des ouvrages d'art*, Circulaire No 81-56, 19 June 1981.
4 The Highways Agency, *Technical Memorandum (Bridges), Rules for the Design and Use of Freyssinet Concrete Hinges in Highway Structures, BE 5/75.*
5 European Standard EN 1337-1:2000 Structural Bearings.
6 Hambly, E.C., 'Integral bridges', *Proceedings of the Institution of Civil Engineers*, Transp. 1997, 123, February, pp. 30–8.
7 Petursson, H. and Collin, P., *Composite Bridges with Integral Abutments Minimizing Lifetime Cost*, IABSE Symposium Melbourne, 2002.
8 England, G.L., Tsang, N.C.M. and Bush, D.I., *Integral Bridges: A Fundamental Approach to the Time-Temperature Loading Problem*, Thomas Telford, 2000.
9 Benaim, R., Watson, P.W. and Raiss, M.E., *Design and Construction of the City Centre Viaduct for the STAR Light Railway Transit System in Kuala Lumpur*, FIP Symposium, London, 1996.
10 Smyth, W.J.R., Benaim, R. and Hancock, C.J., 'Tyne & Wear Metro, Byker contract: planning, tunnels, stations and trackwork', *Proceedings of the Institution of Civil Engineers*, Part 1, Vol. 68, November 1980, pp. 689–700.

8 The general principles of concrete deck design

1 Menn, C., *Prestressed Concrete Bridges*, Birkhäuser Verlag, 1986, pp. 56, 57.
2 Podolny, W. and Muller J.M., *Construction and Design of Prestressed Concrete Segmental Bridges*, John Wiley & Sons, 1982, pp. 219–23.
3 Gee, A.F., 'Bridge winners and losers', *The Structural Engineer*, Vol. 65A, No. 4, April 1987.

9 The design of bridge deck components

1 Menn, C., *Prestressed Concrete Bridges*, Birkhäuser Verlag, 1986.
2 Podolny, W. and Muller, J.M., *Construction and Design of Prestressed Concrete Segmental Bridges*, John Wiley & Sons, 1982, pp. 202–3.
3 Podolny, W. and Muller J.M., *Construction and Design of Prestressed Concrete Segmental Bridges*, John Wiley & Sons, 1982, pp. 203–5.
4 British Standard BS5400: Part 4: 1990 *Code of Practice for Design of Concrete Bridges.*
5 Guyon, Y., *Calcul des hourdis de pont en béton précontraint*, September 1962. Documents Société Technique pour l'Utilisation de la Precontrainte (STUP).
6 Taylor, S.E., Rankin, G.I.B. and Cleland D.J., 'Arching action in high strength concrete slabs', *Proceedings of the Institution of Civil Engineers*, Structures and Buildings, 146, November 2001, Issue 4, pp. 353–62.
7 Benaim, R., Leung, K.K. and Brennan, G., *The Original Design of Ah Kai Sha and Dong Po Bridges: Guangzhou East–South–West Ring Road, Guangdong*, Structural Symposium, 2000 – Highway and railway structures, Hong Kong, 5 May 2000.
8 Benaim, R., Brennan, M.G., Collings, D. and Leung, L.K.K., *The Design of the Pearl River Bridges on the Guangzhou Ring Road*, FIP Symposium, London, 1996.

10 Precast beams

1 Kumar, A., *Composite Concrete Bridge Superstructures*, British Cement Association, 1988.
2 Connal, J., 'Developments for short to medium span bridges in Australia', *Structural Engineering International*, 1/2002.
3 Benaim, R., Brennan, M.G. and Raiss, M.E. *Design of a 17 km Viaduct in South China for Rapid Construction*, FIP Symposium on Post Tensioned Concrete Structures, London 1996.

12 Ribbed slabs

1 Hambly, E.C., *Bridge Deck Behaviour*, 2nd edition, E&FN Spon, 1991.

13 Box girders

1 Collings, D., Mizon, D. and Swift, P., 'Design and construction of the Bangladesh–UK friendship bridge', *Proceedings of the Institution of Civil Engineers*, Bridge Engineering 156, December 2003.

14 Counter-cast technology for box section decks

1 Ninive Casseforme s.r.l., www.ninivecasseforme.com.
2 FIP/9/2 Proposal for a standard for acceptance tests and verification of epoxy bonding agents for segmental construction, March 1978.

15 The construction of girder bridges

1 'Bridge across the Ahr Valley, Germany', *New Civil Engineer*, 24 April 1975.
2 Benaim, R., 'Design of the Byker Viaduct', *Trends in Big Bridge Engineering*, IABSE, Vienna, 1980.
3 DEAL, www.deal.it.
4 Ferrocemento – Costruzioni e Lavori Pubblici SpA, Via Feliciano Scarpellini, 20, 00197 Roma, Tel: 00 39 06 36 17 01.
5 Paolo de Nicola SpA, www.paolodenicola.com.
6 Benaim, R., Brennan, M.G. and Raiss, M.E., *Design of a 17km Viaduct in South China for Rapid Construction*, FIP Symposium on Post Tensioned Concrete Structures, London, 1996.
7 SIKA, www.sika.com.
8 Ninive Casseforme s.r.l., www.ninivecasseforme.com.

17 The design and construction of arches

1 *Bridges in China*, Tongji University Press, Shanghai, China, 1993.
2 Smyth, Benaim and Hancock, *Tyne and Wear Metro, Byker contract, planning, tunnels, stations and trackwork*, *Proceedings of the Institution of Civil Engineers*, Part 1, 1980, **68**, November, pp. 689–700.
3 O'Connor, C., *Design of Bridge Superstructures*, Wiley, Chichester, 1971.
4 Ding, D., 'Development of concrete filled tubular arch bridges, China', *Structural Engineering International*, 4/2001, pp. 265–7.

18 Cable-supported decks

1 Virlogeux, M., 'Bridges with multiple cable stayed spans', *Structural Engineering International*, 1/2001, pp. 61–82.
2 Walther, R., Houriet, B., Isler, M. and Moïa, P., *Cable Stayed Bridges*, Thomas Telford, London, 1999, pp. 49–51.
3 Walther, R., Houriet, B., Isler, M. and Moïa, P., *Cable Stayed Bridges*, Thomas Telford, London, 1999, pp. 191–6.
4 Walther, R., Houriet, B., Isler, M. and Moïa, P., *Cable Stayed Bridges*, Thomas Telford, London,1999, p. 84.
5 Podolny, W. and Scalzi, J.B., *Construction and Design of Cable Stayed Bridges,* 2nd edition, John Wiley and Sons, 1986.
6 Al-Qarra, H., 'Strand by strand installation of cable stays', *Structural Engineer*, 21 May 2002, pp. 31–4.
7 Walther, R., Houriet, B., Isler, M. and Moïa, P., *Cable Stayed Bridges*, Thomas Telford, London, 1999, p. 101.
8 Fürst, A., 'Bending of stay cables', *Structural Engineering International*, 1/2001, pp. 42–6.
9 Benaim, R., Leung, K.K. and Brennan, G., *The Original Design of Ah Kai Sha and Dong Po Bridges: Guangzhou East–South–West Ring Road, Guangdong*, Structural Symposium, 2000 – Highway and railway structures, Hong Kong, 5 May 2000.
10 Benaim, R., Brennan, M.G., Collings, D. and Leung, L.K.K., *The Design of the Pearl River Bridges on the Guangzhou Ring Road*, FIP Symposium London, 1996.
11 Redfield, C. and Strasky, J., 'Sacramento River Pedestrian Bridge', *Structural Engineering International*, 4/91, pp. 19–21.
12 Strasky J., *Stress Ribbon and Cable-Supported Pedestrian Bridges*, Thomas Telford, 2005.

Index